LECTURES ON THE PHYSICS OF HIGHLY CORRELATED ELECTRON SYSTEMS IV

Related Titles from AIP Conference Proceedings

483 High Temperature Superconductivity
Edited by Stewart E. Barnes, Joseph Ashkenazi, Joshua L. Cohn, and Fulin Zuo, August 1999, 1-56396-880-0

438 Lectures on the Physics of Highly Correlated Electron Systems
Edited by Ferdinando Mancini, July 1998, 1-56396-789-8

427 Lectures on Superconductivity in Networks and Mesoscopic Systems
Edited by Carlo Giovannella and Colin J. Lambert, April 1998, 1-56396-750-2

To learn more about these titles, or the AIP Conference Proceedings Series, please visit the webpage **http://www.aip.org/catalog/aboutconf.html**

LECTURES ON THE PHYSICS OF HIGHLY CORRELATED ELECTRON SYSTEMS IV

Fourth Training Course in the Physics of Correlated Electron Systems and High-Tc Superconductors

Salerno, Italy   11–22 October 1999

EDITOR
Ferdinando Mancini
Università degli Studi di Salerno
Salerno, Italy

Melville, New York, 2000
AIP CONFERENCE PROCEEDINGS ■ VOLUME 527

Editor:

Ferdinando Mancini
Dipartimento di Scienze Fisiche "E. R. Caianiello"
Unità INFM di Salerno
Università degli Studi di Salerno
Via S. Allende
84081 Baronissi (SA)
ITALY

E-mail: mancini@sa.infn.it

Authorization to photocopy items for internal or personal use, beyond the free copying permitted under the 1978 U.S. Copyright Law (see statement below), is granted by the American Institute of Physics for users registered with the Copyright Clearance Center (CCC) Transactional Reporting Service, provided that the base fee of $17.00 per copy is paid directly to CCC, 222 Rosewood Drive, Danvers, MA 01923. For those organizations that have been granted a photocopy license by CCC, a separate system of payment has been arranged. The fee code for users of the Transactional Reporting Service is: 1-56396-950-5/00/$17.00.

© 2000 American Institute of Physics

Individual readers of this volume and nonprofit libraries, acting for them, are permitted to make fair use of the material in it, such as copying an article for use in teaching or research. Permission is granted to quote from this volume in scientific work with the customary acknowledgment of the source. To reprint a figure, table, or other excerpt requires the consent of one of the original authors and notification to AIP. Republication or systematic or multiple reproduction of any material in this volume is permitted only under license from AIP. Address inquiries to Office of Rights and Permissions, Suite 1NO1, 2 Huntington Quadrangle, Melville, N.Y. 11747-4502; phone: 516-576-2268; fax: 516-576-2450; e-mail: rights@aip.org.

L.C. Catalog Card No. 00-104877
ISBN 1-56396-950-5
ISSN 0094-243X
Printed in the United States of America

Contents

Preface vii

Theory of the Spin-Polaron for 2D Antiferromagnets 1
A. F. Barabanov, L. A. Maksimov, and A. V. Mikheyenkov
I. Introduction 1
II. Local Spin-Polaron in Some Strongly Correlated Models 4
III. Antiferromagnetic $S=1/2$ Frustrated Heisenberg Model in the Spin-Rotation-Invariant Theory 33
IV. Complex Local Spin Polaron in a Spherically Symmetric Frustrated Spin-Background 52
V. Local Polaron Dressed by Antiferromagnetic Spin Waves 72
VI. Self-Consistent Born Approximation for a Local Spin-Polaron in the Effective Three-Band Model 92
VII. Summary 109
VIII. Acknowledgments 111

Ferromagnetism and Electronic Correlations 118
W. Nolting
I. Models of Magnetism 120
II. Itinerant Moment Systems: Hubbard Model, Basic Features 144
III. Ferromagnetism in Itinerant Moment Systems 170
IV. Ferromagnetism in Local Moment Systems 200
V. Conclusions 220
   Acknowledgments 223

Magnetic and Orbital Ordering in Cuprates and Manganites 226
A. M. Oleś, M. Cuoco, and N. B. Perkins
I. Correlated Transition-Metal Oxides with Orbital Degeneracy 227
II. Magnetic Interactions for Nondegenerate Orbitals 240
III. Spin-Orbital Model in Cuprates 247
IV. Spin Liquid due to Orbital Fluctuations 275
V. Spin-Orbital Model in Manganites 299
VI. Electronic Structure and Excitations in Doped Manganites 314
VII. Orbital Degrees of Freedom in a Ferromagnet 332
VIII. Phase Diagrams of Manganites—Open Problems 350
   Acknowledgments 373

Author Index 381

Preface

The ensuing volume contains the lectures delivered at the "Fourth Training Course in the Physics of Correlated Electron Systems and High-Tc Superconductors", held in Vietri sul Mare (Salerno) Italy, in October 1999.

Following the tradition of previous years, the meeting was devoted to promote the formation of young scientists by means of training through research. Different from usual workshops, the idea was to bring together for two weeks senior and young researchers in a close and informal atmosphere, paying special attention and effort to an active participation of the young researchers and to introduce them to some specific problems.

The course consisted of four lectures every morning, held by Professors A. F. Barabanov, W. Nolting, A.M. Oles and A. Ruckenstein, and afternoon activities aimed principally to increase discussions between the students and lecturers, the solving of particular exercises and oral exposition by the students. The outcome of this type of course was an active participation of the students and a large interchange of ideas due to the long afternoon time destined to discussions.

It is hoped that both the meeting, which brought together leaders in the field as well as bright and eager beginners, and the present volume may be useful as an up-to-date book for researchers interested in the field.

I wish to acknowledge and thank those whose support made the course and the proceedings possible. The main sponsor of the event is the European Commission, under the TMR Programme. For additional support I gratefully acknowledge the Consiglio Nazionale delle Ricerche and the University of Salerno. Finally, I wish to thank Professor Maria Marinaro, President of the International Institute of Advanced Scientific Studies, who hosted the event in the wonderful and warm venue of Vietri sul Mare.

Ferdinando Mancini
Salerno, 3 April 2000

Theory of the Spin-Polaron for 2D Antiferromagnets

A.F.Barabanov[*], L.A.Maksimov[**], A.V.Mikheyenkov[*]

Institute for High Pressure Physics, RAS,
142092, Troitsk, Moscow Region, Russia
**Russian Research Center "Kurchatov Institute",*
Kurchatov sq. 46, 123182, Moscow, Russia

Abstract. The aim of these lectures is to describe theoretical ideas concerning spin-polaron scenario for the charged excitations in a two-dimensional antiferromagnet in normal state. The distinctive feature of our approach consists in treating a local polaron (not a bare hole) as a zero approximation for a quasiparticle. Then we dress this excitation by antiferromagnetic spin waves. The realization of this concept in the framework of several models demonstrates that many unusual features may have a natural explanation through the spin-polaron picture. As the presented spin-polaron concept gives a rather simple description of the important lowest excitation band (even in mean-field approach) we strongly believe that in the near future this concept will be useful for theoretical investigations of such a phenomena as superconductivity and normal kinetic properties of HTSC and other strongly correlated systems.

I INTRODUCTION

During the last years a lot of experimental and theoretical work has been devoted to the investigation of strongly correlated electrons systems. To these systems are related the compounds with d and f electrons, such as heavy-fermion systems, mixed-valence systems. A great stimulation to such investigations was initiated by the discovery of high-temperature superconducting cuprates.

All these systems demonstrate a non-Fermi-liquid behavior and the theories which try to describe them are far from completion at present time. It is well known, that the main difficulty for the description of elementary excitations in these systems consists in the absence of Bose or Fermi calumniation relations for the operators of the elementary excitations. That is, for example, that the carrier's energetically low-lying excitations are far from the bare electron (hole) operator. As a result, it is rather difficult to construct a regular diagrammatical technique for such excitations.

One of the most explicit examples for such a situation is a doped two-dimensional antiferromagnet which reflects the main features of CuO_2 plane in high-temperature superconductors. This system has been the most experimentally investigated in the last years. In particular many surprising features were demonstrated by recent angular resolved photoemission spectroscopy (ARPES) experiments [192,201,75,1,44,109] and [130,2,3,47,121,166] which give in principle the single particle spectral function of the carriers [93].

We shall discuss a theory of two-dimensional systems with two subsystems: localized electrons, which form a spin subsystem, and band carriers (electrons or holes), whose movement strongly depends on the state of the spin subsystem. We shall try to illustrate that many features of ARPES experiments can be explained in the frames of a so-called spin polaron concept.

The main idea of this concept is well known, but its realization in the framework of particular models is not trivial, as it will be seen below. The idea is that a good elementary excitation for the discussed systems must be a bare particle (electron or hole) surrounded by some deformation of the spin dielectric background. The simplest realization of such quasiparticle (we shall name it **local spin polaron**) must be given by the solution of a cluster problem where the cluster contains, for example, four sites with one hole and three spins in the case of a square lattice. Taking into account the energetically low states of such a cluster one must secondly describe the movement of a local spin polaron on an antiferromagnetic background. The situation of strong correlations is explicit in the case when there are few cluster eigenstates which in energy spectra are separated enough from the other states.

Since the motion of a spin polaron strongly depends on the state of the antiferromagnetic background and the correlation functions of the spin subsystem, the problem of $S = 1/2$ Heisenberg antiferromagnet will be discussed separetely. Below we shall represent a spherically symmetrical approach for this problem both in the cases of $D = 2$ and $D = 3$ dimensions.

The main goal will be to study the frustration in the spin system which appears by introducing the competing spin-exchange interactions between the first and second nearest-neighbours. It is well known that the increase of doping in copper-oxygen plane leads to the increase of the frustration in the spin subsystem. Due to this circumstance we can qualitatively describe the doping dependence of the quasiparticle spectrum by means of frustration introduced by competing spin-exchange interactions in a magnetic subsytem.

Then studying the small radius polaron excitations one can reproduce the following interesting ARPES features of the quasiparticle spectrum: 1) A single hole doped in such an antiferromagnet has the bottom of the lowest band close to the point $(\pi/2, \pi/2)$ in **k**-space. The spectrum is isotropic in the $k-$ space near the band bottom. 2) At relatively small doping the Fermi surface center shifts to (π, π) point. 3) As the doping increases, a non-rigid band evolution is observed. 4) The existence of very flat bands close to the Fermi surface (the existence of such flat regions is often used to explain a large value of superconducting temperature).

The concept of the local polaron excitation will be presented in the framework of several commonly used models — the Hubbard model, the Kondo-lattice model, the effective three-band Hubbard model (spin-fermion model) and the generalized

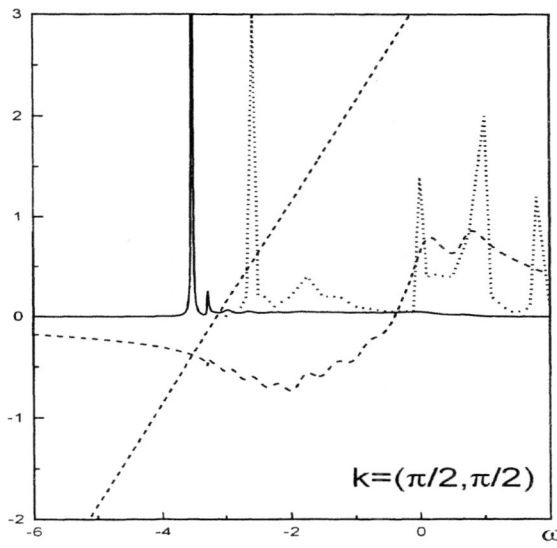

FIGURE 1. Local spin polaron spectral density $A_p(k,\omega)$ – thick line and real part of the self-energy $\Sigma(k,\omega)$ – dashed line, sloping straight lines represent the function $\omega - \Omega(k)$, where $\Omega(k)$ is a mean-field local polaron energy, Ref. [18]. The dotted line is a spectral density $A_h(k,\omega)$ for the calculations starting from a bare hole, Ref. [106].

$t - J$ model. One of these models — the effective three-band Hubbard model — will be studied more thoroughly since it is the most adequate for the copper-oxygen plane.

Intuitively it is clear that the movement of a small polaron must depend on the presence (or absence) of a long-range order in the spin subsystem. This means, that the second important step in the development of a spin polaron concept is to take into account some coupling of a local spin polaron with infinite spin waves with the quasimoment $\mathbf{Q} = (\pi, \pi)$. This demands the introduction of **complex spin polaron** operators — local spin polarons dressed by spin waves with quasimoment q close to $\mathbf{Q}$. Then it turns out that the lowest local polaron band is split, the effective Fermi surface reflects the strongest violation of the Luttinger theorem and the following interesting ARPES experimental results can be reproduced — sudden drop in the intensity of ARPES peaks as $\mathbf{k}$ goes from $(\pi/2, \pi/2)$ to (π, π) or $(0,0)$ and the possibility of the existence of a so-called "shadow band" effect and a so called pseudogap on the Fermi surface.

Let us mention, that methodically we shall mainly use the projection method for Green's functions which is in some way a mean field approach. But we think that even the results of this rather simple method can give a picture which is difficult to reproduce by numerical calculations in finite lattices. This concerns firstly the importance of the coupling of a spin **Q**-wave with the local spin polaron: it seems that the description of such coupling demands to perform calculations on a rather large lattices.

Let us also note that the treating of the strongly correlated systems is often based on the decoupling of a bare hole on a spinless charged fermion and a bose spin excitation. This treatment is in some sense alternative, but not equivalent to the approach which we discuss. It is important that the approach presented below takes into account the short range correlation between a hole and a spin subsystem from the very beginning by the introduction of a complex local excitations.

To test the validity of the results based on the mentioned coupling we shall reproduce the calculation of a bare hole spectral function $A_h(\mathbf{k}, \omega)$ in a self-consistent Born approximation for a local spin polaron (not a bare hole). We shall show that the so-called quasiparticle peak and its intensity is well given by the lowest band of our complex mean-field spin polaron. As to the bands which correspond to the excited states, they effectively correspond to the non-coherent part of $A_h(\mathbf{k}, \omega)$.

In order to illustrate that namely a local spin polaron is a good zero order approximation for the description of strongly correlated systems, we show in Fig. 1 two results for spectral function calculation in the framework of the same model (three-band effective Hubbard model). In one case the spectral function of elementary excitations $A(\mathbf{k}, \omega)$ is calculated for a Zang-Rice local polaron by the self-consistent Born approximation [18]. In the second case the same function is calculated in the same approximation for the same $\mathbf{k} = (\pi/2, \pi/2)$, starting from a bare hole [106]. It is explicitly seen, that the quasiparticle peak for a small polaron is much sharper and its incoherent part is much less than for a bare hole.

So, the aim of the lectures is to present a detailed picture for a spin polaron structure, starting from local spin polaron. The lectures cover a relatively narrow region of the theory for strongly correlated systems and is mainly based on the works of the authors. We don't review many important questions and interesting papers due to the focused nature of the present lectures. We refer to other review material, which covers other general problems of the spin polaron theory for strongly correlated systems and related questions [72,33,41,128,175,102,103].

II LOCAL SPIN-POLARON IN SOME STRONGLY CORRELATED MODELS

In this Section we shall clarify the main idea of a simplest local spin-polaron in the framework of an ordinary Hubbard model, the Kondo-lattice model and three-band Hubbard model on a two-dimensional square lattice. The last one is the most realistic model for CuO_2 plane in HTSC. All these models will be treated in the strongly correlated regime. In order to construct a simplest spin polaron one must consider the smallest cluster and restrict oneself to the energetically low-lying states. In the Hubbard model such a cluster consists of four sites and the states related to different clusters don't overlap. In the three-band Hubbard model the cluster consists of one copper site and four oxygen sites and due to the originality of the CuO_2 plane structure the states of the neighbouring clusters overlap. The most simple example for a local spin-polaron state will be given for the case of the Kondo-lattice model where the cluster corresponds to one site. This model most explicitly and rigorously demonstrates some important features of the spin-polaron concept such as strong deviation from the Fermi statistic for bare carriers. Some possible mechanisms for the formation of bound states of two polarons are also discussed.

A Local spin polaron in the Hubbard model–four-site cluster approach in strongly correlated limit

In this section we discuss the Hubbard model for the square planar lattice in the limit $t/U \ll 1$. The ground state is chosen to be the product of the Hubbard model's exact solutions for four-site blocks. One-particle excitations appear to be Fermi particles with spin 1/2. Their effective mass near the bottom of the band coincides with the effective mass of excitations in the paramagnetic state in the 'Hubbard-I' approximation. However, minima of the spectrum are situated at other points in the Brillouin zone. The possibility of binding two excitations into a singlet pair with charge $2e$ is demonstrated. Cohesive energy is calculated to be equal to $0.1t$. The superfluid state of the Bose gas of such pairs is assumed to be connected with high-T_c superconductivity.

1 Variational ground-state function

The Hubbard model [92] for the square planar lattice with resonating valence bonds (RVBs) [6] has been recently treated in connection with the high-T_c superconductivity problem. In particular, the possibility of hole pairing has been discussed. For this purpose it is important to construct the RVB ground state Ψ_{gr} as well as to determine the elementary excitation spectrum.

The RVB state Ψ_{gr} is usually constructed on the basis of the two-electron nearest-neighbour singlet Φ_0 [67,7,87,110]:

$$\Phi_0 = \frac{1}{2}\left[(a_1^+ b_2^+ - b_1^+ a_2^+) - 2(t/U)(a_1^+ b_1^+ - b_2^+ a_2^+)\right] \quad (1)$$

where t and U are standard model parameters, and a_i^+ and b_i^+ create an electron on the ith site with $+1/2$ and $-1/2$ spin projections, respectively. We shall construct Ψ_{gr} originating from the singlet ground state $|\,4^0\rangle$ of a square four-site block with four electrons. In comparison with Φ_0 (equation (1)) the singlet state $|\,4^0\rangle$ is more appropriate for the square symmetry of the problem and from the very beginning takes better account of short-range order. In the following, we determine the variational energy value for the ground RVB-like state on the two-dimensional square lattice with one electron per site: $\alpha = t/U \ll 1$. We also calculate the one-particle elementary excitation spectrum. In our block approach excitations are naturally described as small magnetic polarons – the lowest-energy states of the block with three electrons. We demonstrate that it may be energetically favourable for two such excitations to be bound into a localized pair, i.e. a bipolaron. It should be emphasized that our results have a variational character; the approximations involved are discussed at the end of each section. We shall also discuss the special case of the RVB state for a square ladder, which was considered in [66].

The Hubbard Hamiltonian has the form

$$\widehat{H} = t \sum_{<ij>} (a_i^+ a_j + b_i^+ b_j) + U \sum_i a_i^+ a_i b_i^+ b_i \qquad (2)$$

As usual, t and U are the hopping integral and the on-site Coulomb repulsion, and $<ij>$ denote thr nearest neighbours. To build the ground state let us now divide the plane into square four-site blocks and rewrite the Hamiltonian in block form:

$$\widehat{H} = \sum_n \widehat{h}_n + \sum_{<nm>} \widehat{t}_{nm} \qquad \widehat{h}_n = \sum_{S,i} \widehat{X}_n^{S^i S^i} \varepsilon_S^i$$
$$\widehat{t}_{nm} = \sum_{Sij,Pkl} \left(\widehat{t}_{nm,SP}^{ij,kl} \widehat{X}_n^{(S-1)^i S^j} \widehat{X}_m^{(P+1)^k P^l} + h.c. \right) \qquad (3)$$

where $\widehat{h}_n$ is the Hamiltonian of the block $\mathbf{n}$, $\widehat{t}_{nm}$ is the inter-block hopping operator, $<\mathbf{nm}>$ are the nearest-neighbour blocks, $\widehat{X}_n^{\lambda\mu}$ is Hubbard's operator [92] transferring the block $\mathbf{n}$ from state μ to state λ, $|S\rangle$ is the eigenstate of the four-site block with S electrons ($S = 0 \div 8$) and energy ε_S^i, i is the set of quantum numbers and $\widehat{t}_{nm,SP}^{ij,kl}$ are the corresponding hopping matrix elements. Transformation from (2) to (3) is exact because (3) is simply the representation of Hamiltonian 2 in the basis of the block operators $\widehat{X}_n^{\lambda\mu}$.

It is at first necessary to find eigenstates $|S\rangle$ of a block containing S electrons. It is appropriate to choose indexes i to be the eigenvalues q, v of the block's Hamiltonian symmetry operators: $\widehat{C}_4$ indicates rotation by $\pi/2$, and $\widehat{\sigma}_y$ – reflection by the Oy axis. The ground state $|4^0\rangle$ for the block with four electrons (one electron per site) appears to be a singlet. It has $q = -1, v = +1$ and energy $\varepsilon_4^0 = -12\alpha t [1+O(\alpha^2)]$. The expression for its wavefunction is presented in the last subsection. Let us note that the first excited state $|4^1\rangle$ has spin equal to unity and $\varepsilon_4^1 = -8\alpha t$.

We now introduce a variational wavefunction of the ground state of the system with $4N$ electrons (N is the number of blocks). It is constructed on the base of $|4^0\rangle$ and all states $|3^i\rangle$, $|5^j\rangle$ with three and five electrons in a block:

$$\Psi_{gr} = \sum_n \left(1 + \sum_{g,i,j} \beta_g^{ij} X_n^{3^i 4^0} X_{n+g}^{5^j 4^0} \right) \Psi_0 \qquad \Psi_0 = \prod_n |4_n^0\rangle \qquad (4)$$

where $\beta_{\mathbf{g}}^{ij}$ are variational parameters, and $\mathbf{g}$ is the nearest-neighbour vector for the block lattice. The zero approximation Ψ_0 is built from independent blocks with four electrons in the singlet state $\mid 4^0 \rangle$ with lowest energy ε_4^0. Straightforward calculations of $\langle \Psi \mid H \mid \Psi \rangle / \langle \Psi \mid \Psi \rangle$ with the trial function 4 involving all states $\mid 3^i \rangle, \mid 5^j \rangle$ lead to a variational value of energy per block, i.e. $\varepsilon^0 = \varepsilon_4^0 + \tilde{\varepsilon}_4$, $\tilde{\varepsilon}_4 = -16\alpha t + O(\alpha^3 t)$; this is the same value of energy per site, i.e. $e_0 = \varepsilon_0/4 = -4\alpha t = -4t^2/U$, as for the classical Neel state (e_N). However, the state (4) has the following typical RVB properties: magnetic sublattices are absent, the average spin projection on each site is zero, while the anti-ferromagnetic correlators are non-zero.

For a quasi-one-dimensional square ladder the same calculation yields $e^0 = -3.5\alpha t$ for the energy per site, in contrast with $e_N^0 = -3\alpha t$. For a one-dimensional chain (two-site block), $e^0 = -2.5\alpha t$ [25] and lies much closer to exact solution $e_{exact} = -2.77\alpha t$ [116] than to $e_N = -2\alpha t$.

The function Ψ_{gr} 4 accounts for correlation mainly inside the 2×2 block. If Ψ_{gr} has the form (4), the energy e^0 in the linear approximation in α is not lowered because of the complicated nature of state $\mid 4 \rangle$, i.e. by admixing other states $\mid 4^i \rangle$. In order to account for long-range correlation, one must enlarge the block size in (4), dividing the plane into blocks of 16 sites (compare [25]). It is difficult to diagonalise the Hamiltonian of such a block, but we may construct its states approximately. The states of the 4×4 block may be considered as non-factorized combinations which include products of the states of four 2×2 blocks. Taking into consideration the hops between them (states $\mid 15^i \rangle$ and $\mid 17^j \rangle$), we obtain, in the simplest case, $e = -4.06\alpha t$ for the energy per site. An analogous complication for the quasi-one-dimensional square ladder leads to an energy $e = -3.56\alpha t$.

Note that the variational estimate of energy $e = -3.73$ for a square ladder obtained in [66] is lower. The trial function in [66] is a superposition of different bonds (1) connecting pairs of nearest-neighbour sites and it accounts for correlation on the scale of several blocks better than function (4) does. Nevertheless, the trial function in [66] is essentially quasi-one-dimensional and is difficult to generalize for the plane case; this function should also not be used in considering one-particle excitation. The block structure of trial function (4) allows us to take better account of correlation, but at the same time the block structure is the drawback of function (4), because it leads to an artificial change in the lattice period (which doubles in the present case).

2 One-particle excitations

Let us consider one-particle hole-type excitations above ground state 4. One must use block states $\mid 3^i \rangle$ with three electrons in a block to construct wavefunctions of such excitation (we shall classify these states, as all the other block states, using the block's Hamiltonian symmetry group representation). Every state $\mid 3^i \rangle$ is characterized, besides the symmetry indexes q, v, by the value of energy ε_3^i and spin of the block in this state. The ground state $|\,3^0\rangle$ has a spin 3/2 (i.e. it is a quartet) and energy $\varepsilon_3^0 = -2t + O(\alpha t)$; its value of $q = v = -1$. The next four states $|\,3^{v,\sigma}\rangle$ are transformed by the two-dimensional representation. $\widehat{C}_4\,|\,3^{v,\sigma}\rangle = -v\,|\,3^{-v,\sigma}\rangle$, $v = \pm 1$.

The spin is 1/2 (two degenerate doublets), and the energy $\varepsilon_3^v = -\sqrt{3}t + O(\alpha t)$. The wavefunctions of these states are presented in the last subsection.

The state $|\,3^i\rangle$ is in fact an immovable magnetic polaron, restricted by the block sizes. Let us take into consideration a possible motion of polaron, which may lead to energy lowering. Then, in the framework of the ground state involved, the trial function of hole-type excitation is as follows:

$$\Psi_p^j = \sum_n \exp(i\mathbf{p}\cdot\mathbf{n})\widehat{X}_{\mathbf{n}}^{3^j 4^0}\Psi_{gr} \qquad (5)$$

where $\mathbf{n}$ is the block's vector, and the quasi-momentum $\mathbf{p}$ lies in the reduced Brillouin zone.

The spectrum of states (5) is defined by the matrix elements $\langle 4_{\mathbf{n+g}}^0 3_{\mathbf{n}}^i\,|\,\widehat{H}\,|\,4_{\mathbf{n}}^0 3_{\mathbf{n+g}}^j\rangle$. It is easy to show that these matrix elements containing states $|\,3^0\rangle$ are equal to zero, because the state $|\,3^0\rangle$ has spin, equal to 3/2. That is why the energy of the state ψ_p^0 is independent of the momentum and is equal to $-2t + 0(\alpha t)$. The spectrum corresponding to states built from $|\,3^{v,\sigma}\rangle$ is equal to

$$\begin{array}{l}\varepsilon_v(\mathbf{p}) = \varepsilon_3^v + \tilde{\varepsilon}_3 \qquad \tilde{\varepsilon}_3 = -2\gamma v\,[\cos(p_x 2a) - \cos(p_y 2a)] - \tilde{\varepsilon}_4 \\ \gamma = |\,\langle 4_{\mathbf{n+g}}^0 3_{\mathbf{n}}^{v,\sigma}\,|\,\widehat{H}\,|\,4_{\mathbf{n}}^0 3_{\mathbf{n+g}}^{v,\sigma}\rangle\,| = 0.125t\,[1 + 2\sqrt{6}\alpha + O(\alpha^2)]\end{array} \qquad (6)$$

where a is the lattice constant. Thus the hole in the state with the lowest self-energy $|\,3^0\rangle$ is immovable, and the energy of the state $|\,3^{v,\sigma}\rangle$ is lower than the energy value of $|\,3^0\rangle$, because of motion through the crystal. The hole in the state $|\,3^{v,\sigma}\rangle$ has the lowest energy $\varepsilon^v = -(\sqrt{3}t + 4\gamma) + O(\alpha t) = -2.23t$ and band width $8\gamma = t$. Note that, unlike the RVB case, motion of the holes in the Neel state is difficult [193]. The effective mass of the hole $|\,3^{v,\sigma}\rangle$ near the bottom of the band is $1/8\gamma a^2$. It coincides with the 'Hubbard-I' approximation. The hole band (6) is degenerate in v, because the representation of the $|\,3^{v,\sigma}\rangle$ transformation is two-dimensional. In the following this feature appears to be essential when considering the bound states of two holes.

The spectrum of electron excitations (five electrons in a block) can be found in a similar way. This calculation is simplified by the electron-hole symmetry of the problem and leads to the following expression for the gap in the one-particle excitation spectrum: $\Delta = U - 4.46t$. It should be noted that Hubbard's scheme for Green's function decoupling in the paramagnetic ground state yields $\Delta = U - 4t$, and for the Neel state $\Delta \simeq U - t^2/U$. The corresponding gap values in the one-dimensional case are $U - 3.27t$ (present approach), $U - 2t$ (decoupling) and approximately $U - t^2/U$ (the Neel state). Rigorous solution, which is known in one dimension only, gives $\Delta = U - 4t$ [116].

Note the following restrictions of the variational deduction of spectrum 6.

(i) The magnetic polaron size is restricted by the 2×2 block

(ii) The wavefunction (5) does not account for the hybridization of state $\mid 3^{v,\sigma}\rangle$ with higher states $\mid 3^i\rangle$. Such hybridization could take place because of the inter-block hopping part of the Hamiltonian.

(iii) Spectrum (6) is sensitive to the specific form of the ground state Ψ_{gr}. It can be seen from (6) that the kinetic energy γ directly contains block states from Ψ_{gr}.

Nevertheless, we consider, that the spectrum of excitations above the RVB-like ground state near the bottom of the zone is described sufficiently well by equation (6), and this expression is considerably different from the spectrum in the tight-binding or 'Hubbard-I' approximation.

3 Two-particle excitation spectrum

Now we shall consider the case of two holes in the system. We shall describe the holes localized in different blocks by the states $\mid 3^{v,\sigma}\rangle$ which were found earlier. One must also consider the situation of two holes in one block, involving block states $\mid 2\rangle$ with two electrons, i.e. with two holes.

State $\mid 2^0\rangle$ with the lowest energy $\varepsilon_2^0 = -2\sqrt{2}t$ is a singlet; it has $q = v = +1$ (detailed discussion is in the last subsection). In fact, state $\mid 2\rangle$ is a bipolaron, in the same sense as $\mid 3\rangle$ is a polaron. It is noteworthy that the first excited state $\mid 2^1\rangle$ has the energy $\varepsilon_2^1 = -2t$. The effective interaction of two neighboring holes is determined by their hopping into one block $\mid 3_n\rangle \mid 3_{n+g}\rangle \rightarrow \mid 4_n\rangle \mid 2_{n+g}\rangle$ and by the inverse process. The relation $2\varepsilon_3^v < \varepsilon_2^0 + \varepsilon_4^0$ of eigenenergy values takes place and, from the viewpoint of the block's self-energies, the interaction of holes has a repulsive character. However, it will be seen in the following that consideration of the dynamical processes, i.e. the matrix elements $\langle 2_{n+g}4_n \mid \widehat{H} \mid 3_n 3_{n+g}\rangle$ may lead to the attraction of two holes. The symmetry of states $\mid 2^0\rangle, \mid 4^0\rangle$ is such that this matrix element is non-zero only in the case when states $\mid 3_n\rangle$ and $\mid 3_{n+g}\rangle$ have opposite spin projections σ and equal parities v. Simple but cumbrous calculation yields $\tau = \langle 2^0_{n+g}4^0_n \mid \widehat{H} \mid 3^{v,\sigma}_n 3^{v,-\sigma}_{n+g}\rangle = 0.197t$.

We shall consider the problem of two holes in the system against a background of Ψ_{gr} (equation (4)). Describing holes, we shall restrict ourselves to the states with lowest energy, i.e. $\mid 2^0\rangle, \mid 3^j\rangle$ ($j = v, \sigma$). Then the Hamiltonian in block projection Hubbard operators is rewritten in the form

$$\widehat{H} = \sum_n (\varepsilon_2 \widehat{X}_n^{22} + \sum_j \varepsilon_3 \widehat{X}_n^{jj}) + \gamma \sum_{n,g,j} vS_g \widehat{X}_n^{j4} \widehat{X}_{n+g}^{4j} \\ + \tau \sum_{n,g,j} S_g \sigma(\widehat{X}_n^{2\bar{j}} \widehat{X}_{n+g}^{4j} + \widehat{X}_n^{j4} \widehat{X}_{n+g}^{2\bar{j}}) \quad (7)$$

$$j = (V, \sigma) \quad \bar{j} = (V, -\sigma) \quad v, \sigma = \pm 1 \quad S_g = (\mid g_x \mid - \mid g_y \mid)/\mid g \mid$$
$$\varepsilon_2 = -2\sqrt{2}t \quad \varepsilon_3 = -\sqrt{3}t$$

S_g accounts for the fact that matrix elements with hops τ along the Ox and Oy axes have different signs. The Hamiltonian 7 describes the self-energies of polarons and the bipolaron, free motion of polarons (partly by the matrix element γ from

(6)) and the formation of the bipolaron. The Hubbard operators on each block are under the constraint $\widehat{X}_n^{44} + \sum_j \widehat{X}_n^{jj} + \widehat{X}_n^{22} = 1$.

The spectrum of the system is determined by the poles of the retarded Green's functions

$$F_n = \langle\langle \widehat{X}_n^{42} | \widehat{X}_n^{24} \rangle\rangle \qquad G_{n,n+m}^v = \sigma \langle\langle \widehat{X}_n^{4j} \widehat{X}_{n+m}^{4j} | \widehat{X}_n^{24} \rangle\rangle \quad j = (v, \sigma) \qquad (8)$$

In the equations of motion for these Green's functions we shall omit commutators in the right-hand side, which are unimportant for spectrum determination:

$$\begin{aligned}
(z - \varepsilon_2) F_n &= \tau \sum_{g,v} (G_{n,n+g}^v + G_{n+g,n}^v) \\
(z - 2\varepsilon_3) G_{n,n+m}^v &= \tau \sum_g \delta_{m,g} S_g (F_n + F_{n+g}) \\
&+ (1 - \delta_{m,0}) \gamma \sum_{g,v} v S_g (G_{n,n+m+g}^v + G_{n+g,n+m}^v)
\end{aligned} \qquad (9)$$

where $z = \omega + i\delta$. These equations are deduced as a result of the simplest decoupling of 'Hubbard-I' type. We ignored states with more than two holes in the system and assumed that $\langle \widehat{X}_n^{44} \rangle = 1$. The first equation of (9) describes the break-up of the bipolaron into two polarons in neighbouring blocks. The second equation corresponds to the inverse process and to the free motion of polarons. System (9) is analogous to the system of equations appearing in the problem of the bound state of two spin waves. It can be solved exactly if a Fourier transformation is made:

$$\begin{aligned}
F^{\mathbf{q}} &= N^{-1} \sum_{\mathbf{n}} \exp(i\mathbf{q}\mathbf{n}) F_{\mathbf{n}} \\
G_{\mathbf{k}}^{\mathbf{q}v} &= N^{-2} \sum_{\mathbf{n},\mathbf{m}} \exp(i\mathbf{q}\mathbf{n} + i\mathbf{k}\mathbf{m}) G_{n,n+m}^v \\
(z - \varepsilon_2) F^{\mathbf{q}} &= \tau \sum_{\mathbf{k},v} P_{\mathbf{k}}^{\mathbf{q}} G_{\mathbf{k}}^{\mathbf{q}v} \\
(z - 2\varepsilon_3) G_{\mathbf{k}}^{\mathbf{q}v} &= \tau P_{\mathbf{k}}^{\mathbf{q}} F^{\mathbf{q}} + \gamma v G_{\mathbf{k}}^{\mathbf{q}v} - \gamma v \sum_{\mathbf{k}_1} P_{\mathbf{k}_1}^{\mathbf{q}} G_{\mathbf{k}_1}^{\mathbf{q}v} \\
P_{\mathbf{k}}^{\mathbf{q}} &= \sum_{\mathbf{g}} [1 + \exp(i\mathbf{q}\mathbf{g})] \exp(-i\mathbf{k}\mathbf{g})
\end{aligned} \qquad (10)$$

$\mathbf{q}$ is the total momentum of two polarons. The solution of system 10 leads to the following equation, which determines the spectrum $z = \omega(q)$:

$$\begin{aligned}
\Theta \widetilde{U} + (1 - \Theta)\omega &= w_{\mathbf{q}}^{-1} \qquad \Theta = \tfrac{1}{2}(\gamma/t^2) \qquad \widetilde{U} = \varepsilon_2 - 2\varepsilon_3 \\
w_{\mathbf{q}} &= \sum_{\mathbf{k}} [\omega + i0 - \varepsilon(\mathbf{q}/2 + \mathbf{k}) - \varepsilon(\mathbf{q}/2 - \mathbf{k})]^{-1} \\
\varepsilon_{\mathbf{k}} &= 2\gamma [\cos(k_x 2a) - \cos(k_y 2a)]
\end{aligned} \qquad (11)$$

The spectrum in (11) is measured from the center of the zone of the two holes of energy $2\varepsilon_3$.

One type of solution of equation (11) describes the free motion of two particles with total momentum $\mathbf{q}$ and the lowest energy -8γ. However, we are interested in the solution corresponding to the stable bound state of two holes ($\omega_{\mathbf{q}} < -8\gamma$). Analysis of equation (11) shows that such solutions appear when the following inequality between the energy parameters of the Hamiltonian is fulfilled:

$$8\gamma(\Theta^{-1} - 1) > \widetilde{U} \qquad (12)$$

(8γ in the left-hand side is replaced by 4γ in the case of a square ladder.)

This condition is true for the parameters $\gamma = 0.125t$, $\Theta = 0.26$ and $U = 0.64$ of the problem obtained earlier. The bottom of the zone of bound states is at $\mathbf{q} = 0$. Equation (11) with these parameters yields for the gap $\Delta = -8\gamma - \omega_{\mathbf{q}=0}$ the value $\Delta \simeq 0.1t$. When $\Delta \ll 8\tau$, the effective mass m is $2m_0$, where m_0 is the mass of one hole. An analogous consideration for one chain based on two-site blocks does not lead to the bound state because condition (12) is violated. Thus, in the framework of the approximations involved, we are led to the conclusion that two independent holes on a plane are energetically favoured to be bound into a singlet pair with charge $2e$. This pair is the state $\mid 2^0\rangle$ with a coherent admixture of two states $\mid 3^{v,\sigma}\rangle$.

The results of this section are obtained with the following approximations:

(i) the approximations noted earlier;

(ii) on the assumption that it is enough to consider only the ground state $\mid 2^0\rangle$ to describe the bipolaron;

(iii) by considering only two holes, which corresponds to the case of a low hole concentration.

To summarize, we have constructed a simple RVB-like state on the base of square blocks, found the one-particle elementary excitation spectrum and have shown the possibility of binding two holes in the framework of the planar Hubbard model. We should emphasize that our RVB approach contains a reformulation of the one-site Hamiltonian (2) to the block form 3. In such approach a hole is considered as a magnetic polaron of size equal to the block size. Two holes in the system may correspond either to two polarons (holes in different blocks) or to a bipolaron (holes in one block). In this case the block Hamiltonian is equivalent to the Hubbard Hamiltonian with on-site (where the new 'site' is the block) repulsion, i.e. $\tilde{U} = \varepsilon_2 - 2\varepsilon_3 > 0$, which is of the hopping t order. Another difference from the usual Hubbard Hamiltonian is that the matrix element τ of hole hopping to the block containing another hole appears to be larger than the matrix element γ of hopping to the block without holes. This dynamical mechanism may be responsible for pair formation.

Our method has a variational character. Only the lowest-energy states of polarons and the bipolaron were considered when calculating the excitation spectrum. So the conclusion that it is energetically favorable for two holes to be bound into a pair in the limit $t \ll U$ is obviously inconclusive.

At low temperatures, the Bose gas of such pairs undergoes superfluid transmission, which corresponds to the possibility of superconductivity in the model discussed.

Let us evaluate the lowest value of the parameter $\alpha = t/U$ when the formation of a 'ferromagnetic bag' [143] can be neglected. For this purpose we compare at fixed α the energy of a RVB state with one hole (the minimal value $-(\sqrt{3}+0.5)t$ was formerly obtained for this energy) and the energy $\varepsilon_F(N)$ of the hole localised in the ferromagnetic region containing N sites. The energies $\varepsilon_F(N)$ for $N = 1, 4, 8$ and ∞

are equal to $0, -2t, -2.62t$ and $-4t$, respectively. If we take into account that for each site in the ferromagnetic region one has energy deficit $4\alpha t$ relative to the RVB state, we easily conclude that the RVB state is stable at least when $\alpha > 0.025$. Thus, there is a sufficiently large α range where the involved approximation $\alpha \ll 1$ holds.

4 Explicit expressions for block states

Bellow are presented some states of the four-site block discussed in the text.
a_i^+ and b_i^+ create an electron on the ith site ($i = 1, 2, 3, 4$, clockwise from the higher left site) with $1/2$ and $-1/2$ spin projections. $\hat{C}_4$ is a rotation by $\pi/2$ (change in indexes $1 \to 2, 2 \to 3, 3 \to 4, 4 \to 1$. σ_y is a reflection by the Oy axis ($1 \leftrightarrow 2, 3 \leftrightarrow 4$). $|0\rangle$ is the vacuum state.

The lowest energy four-electron state $|4^0\rangle$ is

$$|4^0\rangle = w \sum_{k=1}^{4} \nu_k |X_k\rangle$$

$$|X_1\rangle = (1/\sqrt{2})(1 - \hat{C}_4)\, a_1^+ b_2^+ a_3^+ b_4^+ |0\rangle$$
$$|X_2\rangle = (1/2)\hat{\Theta}_- \, a_1^+ a_2^+ b_1^+ b_2^+ |0\rangle$$
$$|X_3\rangle = (1/2\sqrt{2})\hat{\Theta}_-(1 - \hat{C}_4^{-1}\hat{\sigma}_y)\, a_1^+ b_1^+ a_2^+ b_2^+ |0\rangle$$

$$\hat{\Theta}_q = \sum_{m=0}^{3} q^m \hat{C}_4^m \qquad q = \pm 1 \qquad w = \left(\sum_{i=1}^{4} \nu_i^2\right)^{-1/2}$$

$$\nu_1 = -\sqrt{2} \qquad \nu_2 = 1 \qquad \nu_3 = -6\alpha + O(\alpha^3) \qquad \nu_4 = 6\alpha^2 + O(\alpha^3)$$

The three-electron ground state $|3_0\rangle$ is a quartet. The next four states $|3^{v,\sigma}\rangle$ are two degenerate doublets. In the following, only states with spin projection $-1/2$ are given.

$$|3_0\rangle = w_\xi \left(\xi_1 |h_{--}^1\rangle + \xi_2 |h_-^2\rangle + \xi_3 |h_{--}^3\rangle + \xi_4 |h_-^4\rangle\right)$$
$$w_\xi = \left(\sum_{i=1}^{4} \xi_i^2\right)^{-1/2} \qquad \xi_1 = \sqrt{2} \qquad \xi_2 = -1 \qquad \xi_3, \xi_4 \sim \alpha$$

$$|3^{v,\sigma}\rangle = (1/\sqrt{2})(|K^-\rangle + v |L^-\rangle)$$
$$|K^-\rangle = w_\mu \left(-\mu_1 |h_{4-}^1\rangle + \mu_2 |h_{3-}^1\rangle + \mu_3 |h_3^2\rangle - \mu_4 |h_{4-}^3\rangle + \mu_5 |h_{3-}^3\rangle - \mu_6 |h_4^4\rangle\right)$$
$$|L^-\rangle = w_\mu \left(\mu_1 |h_{3+}^1\rangle + \mu_2 |h_{4+}^1\rangle + \mu_3 |h_4^2\rangle + \mu_4 |h_{3+}^3\rangle + \mu_5 |h_{4+}^3\rangle - \mu_6 |h_3^4\rangle\right)$$
$$w_\mu = \left(\sum_{i=1}^{6} \mu_i^2\right)^{-1/2} \qquad \mu_1 = \frac{1}{\sqrt{6}} \qquad \mu_2 = \frac{1}{\sqrt{2}} \qquad \mu_3 = \frac{1}{\sqrt{3}} \qquad \mu_4, \mu_5, \mu_6 \sim \alpha$$

$$|h_{qs}^{1(3)}\rangle = (1/2\sqrt{2})\hat{\Theta}_q \hat{T}_s |h^{1(3)}\rangle$$
$$|h_{3s}^{1(3)}\rangle = (1/4)\hat{\Theta}_3 \hat{T}_{-s} |h^{1(3)}\rangle \qquad |h_{4s}^{1(3)}\rangle = (1/4)\hat{\Theta}_4 \hat{T}_s |h^{1(3)}\rangle$$
$$|h_3^{2(4)}\rangle = (1/2\sqrt{2})\hat{\Theta}_3 |h^{2(4)}\rangle \qquad |h_4^{2(4)}\rangle = (1/2\sqrt{2})\hat{\Theta}_4 |h^{2(4)}\rangle$$

$$\hat{\Theta}_3 = 2(\hat{C}_4 - \hat{C}_4^3) \qquad \hat{\Theta}_4 = 2(1 - \hat{C}_4^2) \qquad \hat{T}_s = 1 + s\hat{C}_4^{-1}\hat{\sigma}_y \qquad q, s = \pm 1$$

$$|h^1\rangle = b_2^+ b_3^+ a_4^+ |0\rangle \qquad |h^2\rangle = b_1^+ a_2^+ b_3^+ |0\rangle$$
$$|h^3\rangle = a_2^+ b_2^+ b_3^+ |0\rangle \qquad |h^4\rangle = a_2^+ b_2^+ b_4^+ |0\rangle$$

The singlet two-electron ground state $|2^0\rangle$ is

$$|2^0\rangle = w\left(|d_+^1\rangle - |d_{++}^2\rangle + \eta |d_+^3\rangle\right)$$

$$|d_+^1\rangle = (1/2)\hat{\Theta}_+ a_2^+ b_4^+ |0\rangle \qquad |d_{++}^2\rangle = (1/2\sqrt{2})\hat{\Theta}_+ \hat{T}_+ a_2^+ b_3^+ |0\rangle$$
$$|d_+^3\rangle = (1/2)\hat{\Theta}_+ a_2^+ b_2^+ |0\rangle \qquad \eta \sim \alpha \qquad w = (2 + \eta^2) - 1/2$$

B Local spin polaron in Kondo-lattice–one-site cluster approach

If the interaction of bare carriers with surrounding magnetic background is strong, we must from the very beginning describe the elementary excitations by operators which take onto account the coupling of bare carriers with spins. Here we shall try to demonstrate this point for a simple case of a local polaron in the regular Kondo model.

According to the common knowledge the polaron effect in metals can be understood as a renormalization of the spectrum due to interaction with lattice vibrations (usual polarons) or with localized spins (spin polarons). Usually, one assumes that the polaron quasi-particles obey to the Fermi statistics. On the contrary, we recently observed [124] strong deviations from the Fermi distribution such that one additional electron leads to the filling of several states in momentum space. To investigate that effect in its clearest form, we study here the Kondo-lattice with vanishing exchange interaction between the localized spins. The kinetic energy t is assumed to be only a small perturbation with respect to the dominating Kondo-exchange $J \gg t$.

The example of such situation is the CuO_2 plane of cuprate superconductors. Here, the Kondo-exchange J corresponds to the energy gain due to the building of the Zhang-Rice singlet within one CuO_4 plaquette [206,85] being of the order of 2-3 eV. The kinetic energy t of the motion of the singlets is roughly 0.5 eV and considerably larger than the exchange between neighbouring spins of approximately 0.15 eV. Here the main aim is the discussion of a simple model showing the deviations from Fermi statistics most clearly. A similar effect is also known under the name of "spectral weight transfer" [64]. A band with a filling dependent decreased weight in comparison with a free fermion band is observed. This means that less electrons are needed to fill the whole band. The Kondo-lattice model is one of the most simple models showing this effect.

1 Bare carrier fillings of exact one-site polaron states

If we completely neglect the motion of electrons and the exchange interaction between neighbouring spins, the Hamiltonian of our model has a very simple form:

$$\hat{H} = \sum_r \hat{H}_r$$
$$\hat{H}_r = 2J\vec{S}_r \cdot \vec{s}_r = J\left(S_r^z(n_{r\uparrow} - n_{r\downarrow}) + S_r^+ a_{r\downarrow}^\dagger a_{r\uparrow} + S_r^- a_{r\uparrow}^\dagger a_{r\downarrow}\right) \quad (13)$$

where $\vec{S}_r$ denotes the operator of the localized spin $S = 1/2$ at the lattice site r, $a_{r\sigma}$ ($\sigma = \uparrow (+), \downarrow (-)$) is the annihilation operator of the electron with spin projection $s_r^z = \sigma/2$, $n_{r\sigma} = a_{r\sigma}^\dagger a_{r\sigma}$. Each local Hamiltonian has 8 eigenstates: 2 states without electrons corresponding to two directions of the localized spins and also 2 states with 2 electrons, all at zero energy $E_0 = E_2 = 0$. Besides these we have 4 states with one electron per site: one lower singlet state and 3 triplet states with the energies:

$$E_{10} = \varepsilon_0 = -\frac{3}{2}J, \qquad E_{11} = \varepsilon_1 = \frac{1}{2}J. \quad (14)$$

The thermodynamic potential of that model is given by

$$\Omega = -TN_L \ln Z, \qquad Z = 2 + e^{-\beta(\varepsilon_0 - \mu)} + 3e^{-\beta(\varepsilon_1 - \mu)} + 2e^{2\beta\mu}, \quad (15)$$

(N_L - number of lattice sites, μ - chemical potential, β - inverse temperature). From Ω we can derive the averaged number of particles (it is determined by the value of μ) and the averaged energy at a given site,

$$N = -\frac{1}{N_L}\frac{\partial\Omega}{\partial\mu} = \frac{1}{Z}\left(e^{-\beta(\varepsilon_0-\mu)} + 3e^{-\beta(\varepsilon_1-\mu)} + 4e^{2\beta\mu}\right)$$
$$\langle\hat{H}_r\rangle = \frac{1}{Z}\left(\varepsilon_0 e^{-\beta(\varepsilon_0-\mu)} + 3\varepsilon_1 e^{-\beta(\varepsilon_1-\mu)}\right) \quad (16)$$

At high temperature ($T \gg J$) the filling of all one electron states (14) is identical and we have

$$N = \frac{2}{e^{-\beta\mu} + 1} \quad (17)$$

At low temperatures ($T \ll J$) and $\mu < 0$ (the zero of energy scale is taken $E_1 = E_2 = 0$) the two electron states are not occupied. Furthermore, due to $e^{-\beta\varepsilon_0} \gg e^{-\beta\varepsilon_1}$, also the one-electron triplet states are empty and we find in that limit

$$N = \frac{1}{2e^{\beta(\varepsilon_0-\mu)} + 1}, \qquad \langle\hat{H}_r\rangle = \varepsilon_0 N \quad (18)$$

For $\mu > 0$ the two electron states become occupied. However, no special treatment of that case is needed due to the symmetry of the Hamiltonian (13) with respect to the particle-hole transformation ($a_\sigma = \sigma \tilde{a}^\dagger_{-\sigma}$). All formulas in hole representation are given by the replacement $\mu \to -\mu$, in particular

$$N_h = \frac{1}{2e^{\beta(\varepsilon_0+\mu)}+1}, \qquad \langle \hat{H}_r \rangle = \varepsilon_0 N_h, \qquad (19)$$
$$N_h = 2 - N$$

For comparison let us assume that we have free fermions at the level ε_0 with a twofold spin degeneracy. In that case we would have instead of (18)

$$N^{\text{free}} = \frac{2}{e^{\beta(\varepsilon_0-\mu)}+1} \qquad (20)$$

Therefore, from (18), for $\mu < \varepsilon_0 - T$, $\exp(\beta(\varepsilon_0 - \mu)) \gg 1$ we find

$$N \cong \frac{1}{2} e^{-\beta(\varepsilon_0-\mu)} = \frac{N^{\text{free}}}{4} \qquad (21)$$

but for $\mu > \varepsilon_0 - T$, $\exp(\beta(\varepsilon_0 - \mu)) \ll 1$ we get

$$N \cong 1 = \frac{N^{\text{free}}}{2} \qquad (22)$$

This suggests a different filling of the singlet states also in the case when they are broadened into a singlet band by the electron hopping.

In order to generalize the above result for the more realistic case including a weak kinetic energy we now solve the same problem (13) in terms of two-time Green's functions. Using the operators

$$\begin{aligned} a &= a_{r\uparrow}, & b &= S^z_r a_{r\uparrow} + S^-_r a_{r\downarrow}, \\ c &= n_{r\downarrow} a_{r\uparrow}, & d &= S^z_r n_{r\downarrow} a_{r\uparrow} + S^-_r n_{r\uparrow} a_{r\downarrow}, \end{aligned} \qquad (23)$$

a closed set of equations of motion can be written down

$$\begin{aligned} \tilde{\omega}\langle\langle a \mid a^\dagger \rangle\rangle &= 1 + J\langle\langle b \mid a^\dagger \rangle\rangle, \quad \tilde{\omega} = \omega + \mu \\ (\tilde{\omega} + J)\langle\langle b \mid a^\dagger \rangle\rangle &= \tfrac{3}{4} J\langle\langle a \mid a^\dagger \rangle\rangle + 2J\langle\langle d \mid a^\dagger \rangle\rangle, \\ \tilde{\omega}\langle\langle c \mid a^\dagger \rangle\rangle &= c_0 + J\langle\langle d \mid a^\dagger \rangle\rangle, \\ (\tilde{\omega} - J)\langle\langle d \mid a^\dagger \rangle\rangle &= d_0 + \tfrac{3}{4} J\langle\langle c \mid a^\dagger \rangle\rangle \end{aligned} \qquad (24)$$

Here, we introduced the following expectation values

$$\begin{aligned} c_0 &= \langle\{c, a^\dagger\}\rangle = \langle n_{r\downarrow} \rangle = \tfrac{N}{2}, \\ d_0 &= \langle\{d, a^\dagger\}\rangle = -\tfrac{1}{2J}\langle H_r \rangle = -\langle a^\dagger b \rangle. \end{aligned} \qquad (25)$$

The last correlator is easily calculated from Green's function $\langle\langle b \mid a^\dagger \rangle\rangle$. We restrict ourselves in the following to low temperatures $T \ll J$ and $\mu < 0$. In this limit we can guess the selfconsistent value of d_0 taking it from the preceding thermodynamic analysis which gives (see (18))

$$d_0 = \frac{3}{4}N. \tag{26}$$

The system of equation (24) results in the solution for the electron Green's function $\langle\langle a \mid a^\dagger \rangle\rangle$ and Green's function $\langle\langle b \mid a^\dagger \rangle\rangle$.

By using (25,26) we also get :

$$\langle\langle a \mid a^\dagger \rangle\rangle = \frac{1+N}{4}(\tilde{\omega} + \tfrac{3}{2}J)^{-1} \\ + \frac{3(1-N)}{4}(\tilde{\omega} - \tfrac{1}{2}J)^{-1} + \frac{N}{2}(\tilde{\omega} - \tfrac{3}{2}J)^{-1}. \tag{27}$$

$$\langle\langle b \mid a^\dagger \rangle\rangle = -\frac{3(1+N)}{8}(\tilde{\omega} + \tfrac{3}{2}J)^{-1} \\ + \frac{3(1-N)}{8}(\tilde{\omega} - \tfrac{1}{2}J)^{-1} + \frac{3N}{4}(\tilde{\omega} - \tfrac{3}{2}J)^{-1}. \tag{28}$$

These three poles correspond to the transitions $E_0 \to E_{10}$, $E_0 \to E_{11}$ and $E_{10} \to E_2$. At low temperatures $T \ll J$ and $\mu < 0$ there is no occupation of triplet states which explains why there is no corresponding pole in (27). We can find the average number of electrons at site r with spin $s_r^z = +1/2$ from the result (27) using the fluctuation-dissipation theorem:

$$\langle n_{r\uparrow} \rangle = \frac{1+N}{4}(e^{\beta(\varepsilon_0-\mu)} + 1)^{-1} \\ + \frac{3(1-N)}{4}(e^{\beta(\varepsilon_1-\mu)} + 1)^{-1} + \frac{N}{2}(e^{\beta(-\varepsilon_0-\mu)} + 1)^{-1}, \quad \mu < 0. \tag{29}$$

For $T \ll J$ and $\mu < 0$ we only have to keep the first term on the right hand side

$$\frac{N}{2} = \frac{1+N}{4} \frac{1}{e^{\beta(\varepsilon_0-\mu)} + 1}, \tag{30}$$

whose self-consistent solution gives (18). Let us note that by calculating the value of $\langle a^\dagger b \rangle$ from (28) can be seen that (26) is a self-consistent solution of the system (24).

So we have found that at $T \ll J$ and $\mu < 0$ only the lowest pole of the electron Green's function (27) is important. For the following, it is more convenient to define a singlet operator and the corresponding Green's function which has only that one pole. Choosing the combination

$$\alpha = \frac{1}{\sqrt{2}}\left(\frac{a}{2} - b - \frac{c}{2} + d\right) \tag{31}$$

we note that the equations of motion (24) (using also (25, 26)) lead to the simple relation

$$(\tilde{\omega} + \tfrac{3}{2}J)\langle\langle \alpha \mid a^\dagger \rangle\rangle = \langle\{\alpha, a^\dagger\}\rangle = \frac{1+N}{2}, \tag{32}$$

for the singlet Green's function. With (23), we can write the singlet operator (31) in a different form

$$\alpha = \frac{1}{\sqrt{2}} \left(Z_r^{\downarrow\downarrow} X_r^{0\uparrow} - Z_r^{\downarrow\uparrow} X_r^{0\downarrow} \right), \tag{33}$$

where $Z_r^{\sigma\sigma} = \frac{1}{2} + \sigma S_r^z$, $Z_r^{\downarrow\uparrow} = S_r^-$ and $X_r^{0\sigma} = a_{r\sigma}(1 - n_{r-\sigma})$, the so called Hubbard projection operators. Also the combined operator (33) is a projection operator between a state with no electron and the *exact local singlet polaron state*. Taking into account (30) we can see that the expectation value

$$\langle \alpha^\dagger \alpha \rangle = N \tag{34}$$

gives the average electron number.

2 The lowest polaron band

Let us add to the Hamiltonian (13) the operator of the kinetic energy

$$\hat{T} = -t \sum_{g\sigma} a^\dagger_{r+g\sigma} a_{r\sigma}, \tag{35}$$

where g denotes the nearest neighbours and z its number. In the limit

$$t \ll J, \qquad T \ll J, \qquad \mu < 0, \tag{36}$$

one can restrict the consideration to the broadening of the singlet band and replace (35) by

$$\hat{T} = -t \sum_{g\sigma} X_{r+g}^{\sigma 0} X_r^{0\sigma} \tag{37}$$

This means that we excluded the transitions to (from) two-electron states. Going in (32) to the momentum space and adding the contribution of the kinetic energy, we find the equation

$$(\tilde{\omega} + \frac{3}{2} J) \langle\langle \alpha_p | \alpha_p^\dagger \rangle\rangle = \frac{1+N}{2} + \langle\langle [\alpha_p, \hat{T}] | \alpha_p^\dagger \rangle\rangle \tag{38}$$

The last term of this equation will be considered within the projection technique. That means we project the result of the commutation of the singlet operator with the kinetic energy onto the singlet operator itself which leads to

$$[\alpha_p, \hat{T}] \cong E_p \alpha_p, \qquad E_p = \frac{\langle\{[\alpha_p, \hat{T}], \alpha_p^\dagger\}\rangle}{\langle\{\alpha_p, \alpha_p^\dagger\}\rangle} \tag{39}$$

The last expression is calculated in lowest order of t, i.e. we decouple correlation functions connecting different sites. In this case one can calculate all expectation values using the results of the previous Section:

$$\begin{aligned}\langle Z_r^{\uparrow\uparrow} X_r^{00}\rangle &= \langle Z_r^{\downarrow\downarrow} X_r^{00}\rangle = (1-N)/2\\ \langle Z_r^{\uparrow\uparrow} X_r^{\downarrow\downarrow}\rangle &= \langle Z_r^{\downarrow\downarrow} X_r^{\uparrow\uparrow}\rangle = N/2\\ \langle Z_r^{\uparrow\downarrow} X_r^{\downarrow\uparrow}\rangle &= \langle Z_r^{\downarrow\uparrow} X_r^{\uparrow\downarrow}\rangle = -N/2\\ \langle X_r^{\uparrow\uparrow}\rangle &= \tfrac{N}{2}\ ,\qquad \langle X_r^{00}\rangle = 1-N\end{aligned} \qquad (40)$$

So we arrive at

$$E_p = -(tz\gamma_p)\frac{1+N}{4}\ ,\qquad \gamma_p = \frac{1}{z}\sum_g e^{ipg}\ . \qquad (41)$$

An important point of this solution is the doping dependence of the bandwidth ranging from $1/4$ to $1/2$ of the free fermion bandwidth for electron numbers N between 0 and 1. A very similar singlet Green's function (38,39,41) was obtained in [85] for a more complex situation in the CuO_2 plane.

We should stress that we neglected spin correlations at neighbouring lattice sites ($\langle \vec{S}_r \vec{S}_{r+g}\rangle = 0$) in deriving (41). This means that we neglect the superexchange of neighbouring localized spins by means of the conduction electrons. This superexchange is of the order of t^2/J. In fact, there are also other possible magnetic ordering processes besides the superexchange, especially processes leading to a ferromagnetic coupling. The calculation of the competition between different magnetic ground states is beyond the aims of the present work. Therefore, defining a magnetic ordering temperature $T_m \sim t^2/J$ we can state that the result (41) is valid only at the temperatures

$$T_m \ll T \qquad (42)$$

For $t < T$ the width of the singlet band does not play any role and the problem is traced back to the previous Section. In the temperature region $T_m \ll T \ll t$ we have a filling of the singlet band similar to the Fermi distribution, but not identical. Let us call it quasi-Fermi distribution. From (38,39) we get

$$N_p = \langle a_p^\dagger a_p\rangle = \frac{1+N}{2}\frac{1}{e^{\beta(\varepsilon_p-\mu)}+1}\ ,\qquad \varepsilon_p = \varepsilon_0 + E_p \qquad (43)$$

At very low filling $N \ll 1$ we find from (43) a Fermi-like distribution with the weight $1/4$:

$$N_p = \frac{1}{2}\frac{1}{e^{\beta(\varepsilon_p-\mu)}+1} \qquad (44)$$

similar to the result (21) in the previous Section. This is correct if the chemical potential is low enough:

$$\mu < \min \varepsilon_p + O(tN/N_L)\ ,\qquad N \ll 1 \qquad (45)$$

For higher chemical potential we have to find the number of electrons $N(\mu)$ from (34) and (43). This leads to

$$N = N_L^{-1} \sum_p N_p = \frac{F}{2-F}$$
$$F = N_L^{-1} \sum_p (e^{\beta(\varepsilon_p - \mu)} + 1)^{-1} \tag{46}$$

This also defines a quasi-Fermi surface as the region where the quasi-Fermi distribution (43) drops from $(1 + N)/2$ to zero. Due to the weight factor between $1/4$ and $1/2$ (depending on the value of N) the quasi-Fermi surface is larger for a given number of electrons than the usual Fermi-surface. One finds from (46) that the number of electrons goes from zero to unity by increasing the chemical potential. In the region $\varepsilon_0 + tz < \mu < |\varepsilon_0| - tz$ we find a filling $N = 1$, similar to the situation in the Hubbard model if the chemical potential lies between the two Hubbard sub-bands.

Therefore, even for the very simple model considered there is a strong deviation from the Fermi-distribution function if we fill the singlet band with electrons. The spectral weight of the band appears to be only from one fourth up to one half of the free fermion case. Spectral weight and the bandwidth depend on the filling ("spectral weight transfer" [64]). This effect should be observable as a decreased spectral weight in photoemission experiments, for instance. One can also expect a reduced spectral weight below the magnetic ordering temperatures, but the dispersion relation (41) may change. A further consequence of the reduced spectral weight is a faster filling of the singlet band in comparison with free fermions.

C Local spin polaron in realistic model for CuO_2 plane–five-site cluster approach in the three-band Hubbard model

In the present Section the Hamiltonian of the generalized Hubbard model will be considered. We shall show that with some reasonable assumptions regarding the energy parameters of the model, the magnetic subsystem of localized copper atom spins can be described by an effective antiferromagnetic Hamiltonian. The free hole movement is possible due to the hybridization of the copper ion $d_{x^2-y^2}$ orbital and bonding orbitals of the closest oxygen ions. The quasiparticle (spin-polaron) spectrum is found for the Neel-type ordering of the magnetic subsystem. In a first approximation the bottom of the spectrum coincides with the boundary of the magnetic Brillouin zone. The possible mechanisms of formation of bound states of such quasiparticles are also analyzed.

1 The Hamiltonian of the three-band Hubbard model

Among the many experimental works on the physical properties of high temperature superconductors, several favour the idea that hole doping leads to the formation of the oxygen hole band in the vicinity of the Fermi level. This is equivalent to the generally accepted idea of the absence of Cu^{3+} ions in the CuO_2 planes.

According to this concept, in the absence of doping the fundamental state of the copper and oxygen ions is Cu^{2+} and O^{2-}; i.e., on each copper ion there is exactly one hole in a d-orbital, if we take the state with filled d^{10} and p^6 orbitals (Cu^+, O^{2-}) as a vacuum state. The crystal field of the tetragonal symmetry splits the copper d-orbital series, so that the energy of an electron on a $d_{x^2-y^2}$ orbital is maximal and this state is singly occupied. In turn, the p_x oxygen orbital, having a nearest-neighbour copper ion in the x-direction, has a lower energy in the crystal field than the nonbonding p_y and p_z orbitals, which may favour the appearance of free holes on these orbitals. However, a sufficiently strong hybridization of the $d_{x^2-y^2}$ and p_x states of copper and oxygen can form an electronic level lying higher than the nonbonding ones; then the holes will occupy this level first. We can convince ourselves of this by examining the simplest two-site Hamiltonian, taking into account the hybridization of $d_{x^2-y^2}$ and p_x states and intrasite repulsion:

$$H = \sum_\sigma [-t(d_\sigma^+ c_\sigma + c_\sigma^+ d_\sigma) + \varepsilon_d n_d^\sigma + \varepsilon_p n_p^\sigma] + U_d n_d^+ n_d^- + U_p n_p^+ n_p^- \qquad (47)$$

where $d, c(d^+, c^+)$ are annihilation (creation) operators for holes on copper and oxygen ions. The features of the hole state, described above, in the presence and absence of doping impose the following conditions on the energy parameters:

$$\varepsilon = \varepsilon_p - \varepsilon_d > 0, \quad U_d > \varepsilon \qquad (48)$$

According to the values of [199], $U_d \gg \varepsilon, t, U_p$. We will take $U_d = \infty$ for simplicity. In this case the minimum energy of two holes in the hybridized hh states is:

$$E_{hh}^{(2)} = \varepsilon_d + \varepsilon_d + (\varepsilon + U_p)/2 - [(\varepsilon + U_p)^2/4 + 2t^2]^{1/2} \qquad (49)$$

To answer the question of which orbitals form the hole band, we must compare this energy with the energy of two holes, one of which is in the hybridized state with energy $E_h^{(1)}$, described by the Hamiltonian (47), and the second is on a nonbonding orbital with energy ε_p', $\varepsilon_p' < \varepsilon_p$. In this case we have

$$\begin{aligned} E_{nh}^{(2)} &= \varepsilon_p' + E_h^{(1)}, \\ E_h^{(1)} &= \varepsilon_d + (\varepsilon + U_p')/2 - [(\varepsilon + U_p')^2/4 + 2t^2]^{1/2} \end{aligned} \qquad (50)$$

Here U_p' is the Coulomb repulsion of holes occupying different p-orbitals.

Comparison of expressions (49) and (50) shows that $E_{hh}^{(2)}$ can be less than $E_{nh}^{(2)}$. In particular, if $t \ll \varepsilon$ and $U_p = U_p' = 0$, this will happen if the inequality $t^2 > \varepsilon(\varepsilon_p - \varepsilon_p')$ is fulfilled. When the real environment of a copper atom, surrounded by four oxygen atoms, is taken into account, the energy of the hh-state should be further reduced compared to the energy of the nh-state.

We note that in the case $t/\varepsilon \ll 1$ (this is just the situation studied below) the degree of hybridization is small. Our remarks about sufficiently strong hybridization should not be taken literally. They only mean that the band of hybridized states is much wider than the band formed by the overlap of nonbonding orbitals.

Under the assumptions taken above, the Hamiltonian describing the hole states is a simple generalization of (47) and has the form

$$H = \varepsilon_d \sum_{\mathbf{R},\sigma} n_{\mathbf{R},\sigma} + \varepsilon_p \sum_{\mathbf{r}} n_{\mathbf{r},\sigma}^c + U_d \sum_{\mathbf{R}} n_{\mathbf{R},\sigma}^d n_{\mathbf{R},-\sigma}^d + U_p \sum_{\mathbf{r}} n_{\mathbf{r},\sigma}^c n_{\mathbf{r},-\sigma}^c \\ + \sum_{\mathbf{R},\mathbf{a},\sigma} t_{pd}(\mathbf{a})(d_{\mathbf{R},\sigma}^+ c_{\mathbf{R}+\mathbf{a},\sigma} + H.c.) + V \sum_{(r,R)} n_p(r) n_d(R) \quad (51)$$

Here $\mathbf{R}$ are the positions of copper sites, which form a square lattice with constant g (see Fig.2).

Each elementary cell $\mathbf{R}$ contains one Cu atom in the site $\mathbf{R}$ and two O atoms with positions $\mathbf{R} + \mathbf{a}_{x,y}$, $\mathbf{a}_x = g(1/2, 0)$, $\mathbf{a}_y = g(0, 1/2)$. $\mathbf{r}$ relates to oxygen sublattice sites and $\mathbf{R} + \mathbf{a}$ are positions of four oxygen atoms nearest to the copper site $\mathbf{R}$: $\mathbf{a} = (\pm \mathbf{a}_x, \pm \mathbf{a}_y)$. The operators d, c (d^+, c^+) are annihilation (creation) operators for holes on copper and oxygen ions relative to filled d^{10} and p^6 shells. They describe $3d_{x^2-y^2}$ and $2p_{x(y)}$ hole orbitals with energies ε_d and ε_p respectively. The orbital $2p_x$ relates to the operators $c_{\mathbf{R}\pm\mathbf{a}_x}$, and $2p_y$ – to the operators $c_{\mathbf{R}\pm\mathbf{a}_y}$. The orbital energies are separated by the charge-transfer energy $\varepsilon = \varepsilon_p - \varepsilon_d > 0$. U_p and U_d are Coulomb on-site repulsion on copper and oxygen sites. The Coulomb p-d repulsion V on neighbouring sites is also introduced.

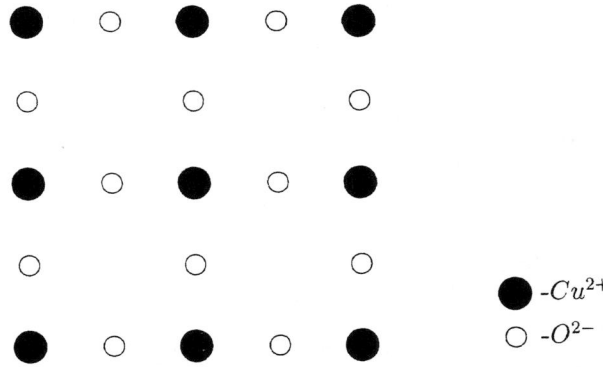

FIGURE 2. The structure of the CuO_2 plane.

A hole can hop from the oxygen $p-$ orbitals to the nearest copper $d-$ orbitals with a hopping amplitude $t_{pd}(\mathbf{a})$. This amplitude is negative for copper-oxygen jumps within a unit cell ($\mathbf{a} = \mathbf{a}_x, \mathbf{a}_y$) and positive for neighbouring cells ($\mathbf{a} = -\mathbf{a}_x, -\mathbf{a}_y$).

In order to make the signs of hopping integrals in (51) to be independent of the hopping direction, one must perform a unitary transformation changing the phases of the $Cu(d_{x^2-y^2})$ and $O(p)$ wave functions. [133,29,69]. This is done by multiplying local orbital wave functions by the factor $\exp(i\mathbf{Q} \cdot \mathbf{R})$, where $\mathbf{R}$ is the radius-vector of the corresponding elementary cell and $\mathbf{Q} = (\pi, \pi)$ is the antiferromagnetic vector (hereinafter the copper sublattice constant is taken $g = 1$). After this transformation we can take in (51) a hopping amplitude t which does not change sign.

In order to restore the initial translation symmetry of local wave functions, we replace $\varepsilon(\mathbf{k})$ by $\varepsilon(\mathbf{k+Q})$ and present our results in these variables everywhere below where we discuss the three-band Hubbard model.

As to the values of the energy parameter set ε, t, U_d, U_p in HTSC, it may be extracted from the band structure or from cluster calculations. Two typical sets are: $\varepsilon = \varepsilon_p - \varepsilon_d = 3.6\text{eV}$, $t = 1.3\text{eV}$, $U_d = 10.5\text{eV}$, $U_p = 4.0\text{eV}$ [95], or $\varepsilon = 2.0\text{eV}$, $t = 1.0\text{eV}$, $U_d = 8.0\text{eV}$, $U_p = 4.0\text{eV}$ [151,135].

In the usual Hubbard model for a half-filled band in the insulating phase ($U \gg t$ and $\bar{n}_d = 1$) the effective exchange (antiferromagnetic) interaction of particles with amplitude $\propto t^2/U$ arises in second-order perturbation theory. In the problem considered here the exchange H_m appears only in the fourth-order in the kinetic energy t:

$$H_m = \sum_{<\mathbf{R},\mathbf{R'}>} t^4 (\varepsilon + V)^{-2} [4(2\varepsilon + U_p)^{-1} + 2U_d^{-1}](2\mathbf{S_R S_{R'}} + 1/2) \qquad (52)$$

In Fig.3, two of the virtual processes leading to the expression (52) are shown.

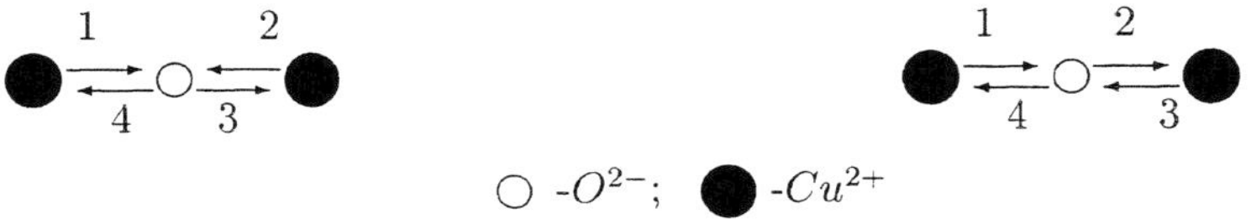

FIGURE 3. Two types of the virtual processes contributing to the exchange interaction (52); the numbers denote the order of the hole hops.

It is known that the ground state of a spin system described by the Hamiltonian (52) in the three-dimensional case is close to the AFM Neel state, and in one dimension is an RVB-type state.([7]) In two dimensions the type of spin system ground state is unknown. In the present section, we will assume that it is a two-sublattice state, but in future we will treat a more realistic RVB-type state.

It is more important for our purposes to obtain the effective Hamiltonian H_t describing hopping of free holes along the oxygen sublattice. It is obtained in second-order perturbation theory. The simplest processes which contribute to the effective Hamiltonian are following (see Fig.4): firstly a copper hole hops from the Cu-site $\mathbf{R}$ on the O-site $\mathbf{R} + \mathbf{a}_2$ forming a virtual state with two O-holes with energy $2\varepsilon_p$, secondly the O-hole hops from site $\mathbf{R} + \mathbf{a}_1$ to the copper site $\mathbf{R}$. This effective movement of the hole from site $\mathbf{R} + \mathbf{a}_1$ to site $\mathbf{R} + \mathbf{a}_2$ is described by the expression $t^2 \varepsilon^{-1} c^+_{\mathbf{R}+\mathbf{a}_2 \sigma_2} Z^{\sigma_1 \sigma_2}_{\mathbf{R}} c_{\mathbf{R}+\mathbf{a}_1 \sigma_1}$.

The virtual processes shown in Fig.4 lead to the following expression:

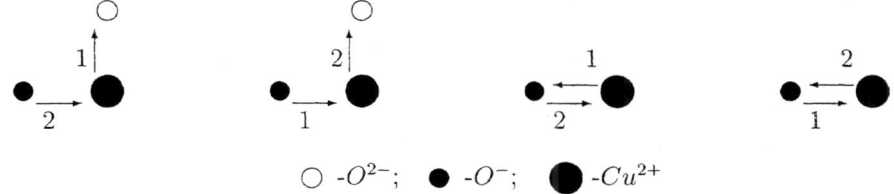

FIGURE 4. The contribution of virtual processes to the hopping effective Hamiltonian (53). Black circles–the sites occupied by one hole.

$$H_t = -\sum_{R,a_1,a_2,\sigma} t^2 (U_d - \varepsilon - 2V)^{-1} c^+_{R+a_2,\sigma} c_{R+a_1,\sigma}$$
$$+ \sum_{R,a_1,a_2,\sigma_1\sigma_2} t^2 \{\varepsilon^{-1} + (U_d - \varepsilon - 2V)^{-1} \qquad (53)$$
$$- \delta_{a_1,a_2} [\varepsilon^{-1} - (\varepsilon + U_p)^{-1}]\} c^+_{R+a_2,\sigma_2} c_{R+a_1,\sigma_1} Z_R^{\sigma_1\sigma_2}$$

Here the Hubbard operator $Z_R^{\sigma_1\sigma_2}$ produces a spin change $\sigma_2 \to \sigma_1$ at site $\mathbf{R}$ of the copper sublattice. It is important, that the hole movement can lead effectively to the Kondo-type processes, which destroy the initial magnetic order of Cu-spin subsystem.

Energies of the excitations described by the effective Hamiltonians (53) and (52) differ in their scale. Theoretically, in setting the parameter t/ε to zero, we arrive at the quantum-mechanical problem of particle (hole) motion superposed on the magnetic ordering created by the hole itself. A similar situation is well known in the Hubbard model: for $U_d = \infty$ a single hole tends to establish a ferromagnetic order in the whole space [143]. It is also known [142] that for $U_d \gg t$, but $U_d \neq \infty$. spins order ferromagnetically in the neighbourhood of a hole. The radius of such *polaron* is $\propto (U/t)^{1/4}$ in the two-dimensional case. Outside this neighbourhood the ordering is antiferromagnetic. It follows that one should consider that the dimension of the ferromagnetic neighbourhood of a hole for real values of the parameter U/t is not large ($R \sim 1 - 2$ for $U/t \sim 5$). An oxygen hole described by the Hamiltonian (53) creates a *polaronic* state in its surroundings also, but of a more complex type. In [73] a variational approach was used to determine the ground state of a system of an oxygen hole and a copper ion spin, in the limiting case $U_d = \infty$ and $U_p = 0$. A tendency was observed towards the formation of a nonmagnetic polaron, but it is impossible to form a polaron with unsaturated magnetization (compare with the results of [99]). It may be shown that a ferromagnetic polaron is also energetically favourable in the case $U_p \neq 0$ [23].

The effective three-band Hamiltonian (53), (52) can be obtained more consistently by canonical transformation. According to [171] it is defined by:

$$H' = e^{-S} H\, e^S = \qquad (54)$$
$$H_0 + \tfrac{1}{2}[H_1, S] + \tfrac{1}{3}[[H_1, S], S] + \tfrac{1}{8}[[[H_1, S], S], S] + \ldots$$

provided that the generator S of this Schrieffer–Wolf transformation satisfies the condition

$$H_1 + [H_0, S] = 0 \qquad (55)$$

The detailed form of the effective Hamiltonian within the second order by the oxygen hole density for the cases of hole and electron doping is given for example in [133]. Let us mention here the so-called p-d model which is close to the three-band Hubbard model which is also used to treat the properties of cuprate oxides. The only difference in this model Hamiltonian relative to (51) is in a structure of p-orbitals sublattice: it is taken that p-orbitals take the same sites as d-orbitals. For an arbitrary value of U_d this model was analyzed in [127].

Now our basic goal is to construct a realistic variational function which reflects the fact that the polaron radius (even nonmagnetic!) $R \propto (\varepsilon^2/t^2)^{1/4}$, and for reasonable values of ε/t is of the order of one-two lattice constants.

2 Variational wave function and single-particle excitation spectrum

The hopping Hamiltonian (53) describes the movement of one doped O-hole in the CuO_2 plane which has one localised d-hole per each Cu-ion $n_d(\mathbf{R}) = 1$. The state of magnetic ordering of these d-holes strongly influences the properties of the oxygen hole band.

Putting $U_d = \infty$ the Hamiltonian (53) takes the form

$$H_f = \sum_{R,a_1,a_2,\sigma_1,\sigma_2} \{[\tau_1 - \delta_{a_1,a_2}(\tau_1 - \tau_2)] c^+_{R+a_2,\sigma_2} c_{R+a_1,\sigma_1} Z^{\sigma_1,\sigma_2}_R\}$$
$$\tau_1 = t^2 \varepsilon^{-1}, \quad \tau_2 = t^2(\varepsilon + U_p)^{-1} \tag{56}$$

In [205] the case $U_d = 2\varepsilon$ was studied, and it was shown that the effective hole Hamiltonian reduces to the usual Hubbard model, with a spectrum of elementary excitations corresponding to the strong-coupling approximation. The correctness of such an approximation depends on the choice of the ground state. In the limit $U_d \to \infty$ examined below the Hamiltonian (56) clearly depends on the state of the spin system through the operator Z and, as it will be shown below, the spectrum of elementary excitations turns out to be different.

We note that it is energetically favourable to form a singlet state of the spin belonging to the copper ion and the spin of the hole "smeared out" over the oxygen surroundings (see below the section devoted to the Zhang-Rice polaron). If we consider the state of the copper spin subsystem to be antiferromagnetic, then particles (holes) having a given spin projection are localized near the sites of the copper sublattice with the opposite magnetization. Then the effective particle hopping is possible only between the sites of this sublattice.

To construct a Bloch eigenfunction of the strong-coupling Hamiltonian, it is common to use a set of Wannier functions (see, for example, [205]). We will build Bloch functions in the subspace of a set of site functions which are not necessarily orthogonal:

$$| \mathbf{R}\sigma > = \sum_a (f(\mathbf{R},\mathbf{a}) c^+_{R+a,\sigma} + g(\mathbf{R},\mathbf{a}) c^+_{R+a,-\sigma} Z^{\sigma,-\sigma}_R) | G > \tag{57}$$

where $|G>$ is the ground state wave function of the copper spin subsystem, and **a** is the vector connecting the nearest Cu and O ions. The function (57), centered near site **R**, having a particle spin projection σ and accounting for the possibility of copper ion spin flip, is the simplest realization of a magnetic polaron. The coefficients f and g introduced in (57) are variational. For a simple magnetic lattice, they must only depend on **a**. For different types of AFM states the set of coefficients is doubled according to the number of magnetic sublattices: f_+, g_+ and f_-, g_-.

On the Bloch wave function (**k** belongs to the magnetic Brillouin zone)

$$|\mathbf{k}\sigma> = A_\mathbf{k} \sum_\mathbf{R} e^{i\mathbf{kR}} |\mathbf{R}\sigma> \tag{58}$$

must satisfy orthonormality conditions, equivalent to the relationship

$$1 = A_\mathbf{k}^2 \sum_{\mathbf{R},\mathbf{R'}} e^{i\mathbf{k}(\mathbf{R}-\mathbf{R'})} <\sigma\mathbf{R'}|\mathbf{R}\sigma> \tag{59}$$

Having chosen $\sigma = -1/2$ to fix the projected particle spin, we will rewrite (59) in the form:

$$1 = 0,5 N_0 A_\mathbf{k}^2 \{4(f_+^2 + g_+^2 Z_+^{+,+} + f_-^2 + g_-^2 Z_-^{+,+}) \\ + 4\phi(\mathbf{k})[2f_+ f_- + g_+ g_- (<Z_\mathbf{R}^{+,-} Z_\mathbf{R'}^{-,+}> + <Z_\mathbf{R}^{-,+} Z_\mathbf{R'}^{+,-}>)]\} \tag{60}$$

where **R** and **R** are the nearest neighbours in the copper sublattice, $Z_+^{+,+} Z_-^{-,-} = 1 - Z_+^{-,-} = <Z_\mathbf{R}^{+,+}>$, **R** belongs to the + sublattice, and $\phi(\mathbf{R}) = 0,5(\cos(2k_x a) + \cos(2k_y a))$. The particle spectrum

$$\varepsilon(\mathbf{k}) = <\mathbf{k}|\mathbf{H}_t|\mathbf{k}> \tag{61}$$

will be found using the variational functions (57). On the right-hand side of formula (61), the following matrix elements differ from zero:

$$<\mathbf{R}|H_t|\mathbf{R}>, \quad <\mathbf{R}|H_t|\mathbf{R}+2\mathbf{a}>, \quad <\mathbf{R}|H_t|\mathbf{R}+2\mathbf{a}+2\mathbf{a'}> \tag{62}$$

Straightforward calculations of terms (62) lead to the following expressions depending on the spin correlation functions of the ground state of the antiferromagnet:

$$<\mathbf{R}|H_t|\mathbf{R}> = 12\tau_1(f_\mathbf{R}^2 <Z_\mathbf{R}^{--}> + 2 f_\mathbf{R} g_\mathbf{R} <Z_\mathbf{R}^{--}>) \\ + 4\tau_2[f_\mathbf{R}^2(<Z_\mathbf{R}^{--}> + <Z_\mathbf{R'}^{--}>) + f_\mathbf{R} g_\mathbf{R}(<Z_\mathbf{R}^{--}>] \\ + <Z_\mathbf{R}^{+-} Z_\mathbf{R'}^{-+}> + <Z_\mathbf{R}^{-+} Z_\mathbf{R'}^{+-}> + g_\mathbf{R}^2 <Z_\mathbf{R}^{+-} Z_\mathbf{R'}^{-+}> \tag{63}$$

(here, as in the following formulas. **R** and **R'** signify a pair of the nearest copper sites);

$$<\mathbf{R'}|H_t|\mathbf{R}> = 3(\tau_1 + \tau_2)[f_{\mathbf{R'}} f_\mathbf{R}(<Z_\mathbf{R}^{--}> + <Z_\mathbf{R'}^{--}>) \\ f_{\mathbf{R'}} g_\mathbf{R}(<Z_\mathbf{R}^{++}> + <Z_\mathbf{R'}^{+-} Z_\mathbf{R}^{-+}>) + g_{\mathbf{R'}} f_\mathbf{R}(<Z_\mathbf{R}^{++}> + <Z_\mathbf{R'}^{+-} Z_\mathbf{R}^{-+}>)] \tag{64}$$

$$< \mathbf{R}'' \mid H_t \mid \mathbf{R} > = \tau_1(f_{\mathbf{R}'}f_{\mathbf{R}}(< Z^{--}_{\mathbf{R}'}> + g_{\mathbf{R}'}f_{\mathbf{R}} < Z^{+-}_{\mathbf{R}''}Z^{-+}_{\mathbf{R}'} > \\ + f_{\mathbf{R}''}g_{\mathbf{R}} < Z^{+-}_{\mathbf{R}'}Z^{-+}_{\mathbf{R}} > + g_{\mathbf{R}''}g_{\mathbf{R}} < Z^{+-}_{\mathbf{R}''}Z^{++}_{\mathbf{R}'}Z^{-+}_{\mathbf{R}} >) \qquad (65)$$

In the last matrix element the three site spin correlation function appears due to the hole hopping from the site $\mathbf{R}$ to the site $\mathbf{R}''$ through the site $\mathbf{R}'$.

For an Ising AFM state the matrix elements (65), with vectors $\mathbf{R}''$ equal to $\mathbf{R} + 2\mathbf{a}_x + 2\mathbf{a}_y$ and $\mathbf{R} + 4\mathbf{a}_x$ are identical. In the case of a Heisenberg state their magnitudes differ slightly, which is the reason for the removal of degeneracy in the energy spectrum along the nesting line.

Below, we produce the values of the magnetic correlators that are needed to calculate the single-particle spectrum for a Heisenberg ground state of the copper sublattice. They have been calculated with the aid of a method applied in [94] to find the energy of a Heisenberg two-sublattice state. The method consists of using a variational wave function in combination with a Monte-Carlo method, which allows estimation of energy and calculation of the correlator with great accuracy for such essentially quantum systems as the Heisenberg antiferromagnet with spin 1/2. In the qualitative analyzis of the hole spectrum it is sufficient to use a single-parameter variational function. Then we find $< Z^{++}_{\mathbf{R}}>= 0,9181$ ($\mathbf{R}$ belongs to the " + " sublattice); $< Z^{+-}_{\mathbf{R}}Z^{-+}_{\mathbf{R}''} + Z^{-+}_{\mathbf{R}}Z^{+-}_{\mathbf{R}''}>= -0.2592$; $< Z^{++}_{\mathbf{R}}Z^{++}_{\mathbf{R}''}>= 0.0594$.

To calculate numerical values of the correlator

$$K = < Z^{++}_{\mathbf{R}'}(Z^{+-}_{\mathbf{R}}Z^{-+}_{\mathbf{R}''} + Z^{-+}_{\mathbf{R}}Z^{+-}_{\mathbf{R}''}) >$$

we must consider four cases:

a) $\mathbf{R}'$ belongs to the " + " sublattice, but $\mathbf{R}$ and $\mathbf{R}''$ are the next-nearest copper sites; for instance $\mathbf{R}'' = \mathbf{R} + 2\mathbf{a}_x + 2\mathbf{a}_y$; $K = 0,0499$;

b) $\mathbf{R}'$ belongs to the " − " sublattice, $\mathbf{R}$ and $\mathbf{R}''$ are the same as in (a); $K = 0,0371$;

c) $\mathbf{R}'$ belongs to the " + " sublattice, $\mathbf{R}$ and $\mathbf{R}''$ are the next-nearest sites, but horizontally or vertically, for example, $\mathbf{R}'' = \mathbf{R} + 4\mathbf{a}_x$; $K = 0,0206$;

d) $\mathbf{R}'$ belongs to the " − " sublattice, $\mathbf{R}$ and $\mathbf{R}''$ are the same as in (c); $K = 0,0371$.

The result of a numerical calculation shows that the degree of splitting of the energy level along the nesting line is very small and amounts to $\sim 0,5 \times 10^{-4}\tau_1$; the minima are distributed in the corners of the magnetic Brillouin zone.

At first we shall analize the dispersion $\varepsilon(\mathbf{k})$ for the simplest case of the AFM ordering taking it to be of Ising type in which $Z^{++}_+ = 1$. Then

$$\varepsilon(\mathbf{k}) = \{\tau_1[3f^2_- + 6f_+g_+ + 6\phi(\mathbf{k})(f_+ + g_+)f_- + (4\phi^2(\mathbf{k}) - 1)f^2_+] \\ + \tau_2[f^2_- + f^2_+ + 2f_+g_+ + 2\phi(\mathbf{k})(f_+ + g_+)f_-]\}/ \\ (f^2_+ + g^2_+ + f^2_- + 2f_+f_-\phi(\mathbf{k})) \qquad (66)$$

The essential feature of expression (66), which varies with f_+, g_+ and f_-, is that the bottom of the band corresponds to $\phi(\mathbf{k}) = 0$, that is, to a line in $\mathbf{k}$-space coinciding with the boundary of the magnetic Brillouin zone. In the vicinity of this line the spectrum has the form

$$\varepsilon(\mathbf{k}) = \varepsilon_0[1 - \beta\phi^2(\mathbf{k})] \qquad (67)$$

with $0 < \beta \sim 1$, and $\varepsilon_0 = \varepsilon_0(\tau_1, U_p)$ as a function of U_p is inside the limits

$$\varepsilon(\tau_1, 0) = -4\tau_1, \qquad \varepsilon(\tau_1, \infty) = -(1 + 37^{1/2})\tau_1/2 \qquad (68)$$

In the direction perpendicular to the line $\phi(\mathbf{k}) = 0$, the spectrum is quadratic in $\delta\mathbf{k}$. This means that in the limit of exactly zero particle concentration, the boundary mentioned above is a Fermi surface near which the density of states has a quasi-one-dimensional character with a square-root singularity. This singularity can cause the appearance of instabilities, including superconductivity. A system with a similar excitation spectrum, but of a different nature, was studied in [112], where it was shown that the Coulomb interaction does not inhibit the superconducting channel. We note that near the bottom of the band the wave functions (57) are close to a spin singlet ($f_+ = -g_+, f_- = g_- = 0$), in other words, the holes on the copper atom and in its oxygen surroundings almost form a singlet pair.

The more complex Heisenberg type of the magnetic interaction leads to the appearance of the quantum "zero-point" fluctuations in the spin subsystem and different values of spin correlators. As a result the expression (66) for the energy spectrum has, naturally, changed: the degeneracy along the whole line of minimal energy disappears, the bottom of the band becomes isolated equivalent points lying in the corners of the magnetic Brillouin zone. In this case the quasi-one-dimensional singularity in the density of states disappears, although the density of states itself remains large near the lower edge of the band on account of the small changes in excitation energy along the nesting line given by the boundary of the magnetic Brillouin zone.

The variational wave function (57) describes the copper spin flips only near the doped oxygen hole. To clear up the question whether the quasi-one-dimensional singularity in the density of states $\rho(E)$ is stable upon increased polaron size, we must add the trial functions which describe the spin reversals on copper sites other than those nearest to holes. To do this we simultaneously take into account the exchange interaction of spins determined by the Hamiltonian (52). Such analysis of spectral stability is difficult to carry out analytically. The simple numerical calculations can be done for the case of an Ising-type AFM ground state of the copper subsystem. We will also take $U_d = \infty, U_p = V = 0$ in (52); then the effective exchange interaction differs from the hopping parameter $\tau = t^2/\varepsilon$ by the factor t^2/ε^2, which must be less than unity.

The expansion of the basis of variational functions will be accomplished by successive action of the operator H_t of (53) on the function $\mid R, \sigma >$.

$$H_t^n \mid R, \sigma > \rightarrow \{\Psi\}_n \qquad (69)$$

Here $\{\Psi\}_n$ is a class of orthonormal site functions which is formed by n-fold action of H_t on $\mid R, \sigma >$: $\{\Psi\}_n$ allows $n+1$ reversed spins on copper ions relative to the ground state. For the cases $n = 1, 2, 3$ the number of site functions is 4, 15, and

46 , respectively. Having built Bloch states out of these functions, we can find the spectrum and the density of states $\rho(E)$ of the lower hole band that interests us.

Such calculations (with a value $t^2/\varepsilon^2 = 0,25$) show that near the bottom of the band the singularity in $\rho(E)$ is qualitatively preserved for increase in the polaron dimension.

3 Two-particle excitation spectrum

Here we shall examine the two-particle singlet states corresponding to a small polaron; we limit ourselves to the case $U_p = \infty$. This means that instead of the operators c_r we must use the Hubbard operators $X_r^{\sigma,0}$ ($X_r^{0,\sigma}$) which describe the creation (annihilation) for a hole with projected spin σ on the oxygen site $\mathbf{r}$. Unfortunately, an attempt to use the expanded basis of functions $\{\Psi\}_n$ encounters major computational difficulties. Therefore, the chief goal of this part is to demonstrate the dynamic mechanism of hole pairing at small distances. As a basis for describing particles spaced far apart, we shall use single-particle hole states corresponding to small polarons. The contribution of large distances ($|\mathbf{R}-\mathbf{R}'| \gg a$) to the asymptotic form of the wave function is

$$\Phi \mid G >= g(\mathbf{R}-\mathbf{R}')A_+(\mathbf{R})A_-(\mathbf{R}') \mid G > \qquad (70)$$

where the amplitude $g(\mathbf{R}-\mathbf{R}')$ must become exponentially small for $|\mathbf{R}-\mathbf{R}'| \gg a$ in the case of a bound state. The single-particle wave functions $A_\sigma(\mathbf{R})$ must coincide with $\mid \mathbf{R}, \sigma >$ [see (57)]. According to the comment at the end of the preceding section, these functions must be approximately of the form of a singlet state of two holes on a copper ion and its oxygen surroundings:

$$A_\sigma(\mathbf{R}) = 8^{-1/2} \sum_\mathbf{a}(X_{\mathbf{R}+\mathbf{a}}^{\sigma,0} - Z_\mathbf{R}^{-\sigma,-\sigma} - X_{\mathbf{R}+\mathbf{a}}^{-\sigma,0} - Z_\mathbf{R}^{\sigma,-\sigma}) \qquad (71)$$

where $\mathbf{R}$ belongs to the $-\sigma$ sublattice.

When $|\mathbf{R}-\mathbf{R}'|$ becomes of the order of a lattice constant, the two-particle wave function Φ changes. Thus, where both holes are localized near one copper ion, the function Φ takes the form

$$\Phi(\mathbf{R},\mathbf{R}) = \sum_{j=1}^{3} f_j(\mathbf{R})A_{+-}^{(j)}(\mathbf{R}) \qquad (72)$$

where the function $A_{+-}^{(j)}(\mathbf{R})$ describes the formation of a complex of three holes: one on the copper ion at $\mathbf{R}$ and two on the four oxygen ions at $\mathbf{R}+\mathbf{a}$. The index j numbers the three degenerate ground states of this complex with energy $\varepsilon_j = \varepsilon_f = 3\tau_1$ ($j = 1 \div 3$) and with spin projection $\pm 1/2$ (the sign depends on the "sign" of the magnetic sublattice $\mathbf{R}$). We give an expression for one of the representations of the degenerate set of functions $A_{+-}^{(j)}(\mathbf{R})$:

$$A^{(1)}_{+-}(\mathbf{R}) = -40^{-1/2} \sum_{l=1}^{3} (-1)^l X^{-0}_{\mathbf{R}+\mathbf{a}_{1+l}} [(X^{+0}_{\mathbf{R}+\mathbf{a}_{2+l}} + X^{+0}_{\mathbf{R}+\mathbf{a}_{4+l}} + 2X^{+0}_{\mathbf{R}+\mathbf{a}_{3+l}}) Z^{--}_{\mathbf{R}}$$
$$- 2X^{-0}_{\mathbf{R}+\mathbf{a}_{2+l}} Z^{+-}_{\mathbf{R}}]$$
(73)

where $\mathbf{R}$ belongs to the " $-$ " sublattice.

Configurations with two free holes localized near two neighboring copper ions are described by wave functions

$$A_{+-}(\mathbf{R}, \mathbf{R}+2\mathbf{a}_s) = 6^{-1/2} \sum_{t \neq s} (X^{+0}_{\mathbf{R}+\mathbf{a}_t} Z^{--}_{\mathbf{R}} - X^{-0}_{\mathbf{R}+\mathbf{a}_t} Z^{+-}_{\mathbf{R}})$$
$$\times \sum_{v \neq s} (X^{-0}_{\mathbf{R}+2\mathbf{a}_s - \mathbf{a}_v} Z^{++}_{\mathbf{R}+2\mathbf{a}_s} X^{+0}_{\mathbf{R}+2\mathbf{a}_s - \mathbf{a}_v} Z^{-+}_{\mathbf{R}+2\mathbf{a}_s})$$
(74)

where $\mathbf{R}$ belongs to the " $-$ " sublattice. The energy of such configuration is $\varepsilon_b = -4\tau$. The amplitude corresponding to this function will be designated as $p(\mathbf{R}, \mathbf{R}+2\mathbf{a}_s)$. We note that the energy of the asymptotic state (70) is equal to $2\varepsilon_g = -6\tau_1$. The increase of the two-particle states energy $\varepsilon_f > \varepsilon_b > 2\varepsilon_g$ with the decrease of the interparticle distances $\mathbf{R} - \mathbf{R}'$ can be treated as an effective repulsion between holes on the oxygen. Together with this, mechanisms exist for the formation of a bound state. One of these is the increased amplitude for particle hops when they get closer to each other. The matrix element for "decay" of the state $A_{+-}(\mathbf{R})$ and the state $A_{+-}(\mathbf{R}, \mathbf{R}+2\mathbf{a}_j)$ is equal to $(2/5)^{1/2}\tau_1$, while a free-particle hop has the amplitude $\tau_1/8$. A similar situation for the standard Hubbard model was discussed in [20].

Another important mechanism [23] of attraction results from the Coulomb interaction of a hole on oxygen with holes localized on copper sites. In our model [see the Hamiltonian (51)] this energy is denoted by V. The role of an interaction of this type in binding holes into pairs is indicated in [90]. If two holes on an oxygen are next to one copper ion, as shown in Fig.5a, then in second-order perturbation theory a term exists which lowers the energy ε_f. The hole hopping process, drawn in Fig.5a, has the amplitude $t^2/(\varepsilon - V) = \tau'_1$; as a result we get $\varepsilon_f = -3\tau'_1$. One more term from second-order perturbation theory is shown in Fig. 5b and is only connected with virtual hopping of a copper hole. For any configuration of the type shown in Fig.5b, it is equal to $-2\tau'_1 + 6\tau_1 - 4\tau''_1$, where $\tau''_1 = t^2/(\varepsilon + V)$, and becomes negative for $V > \varepsilon/3$. Too large values $V > \varepsilon/2$ lead to the formation of large hole clusters on oxygens and to demixing. The configurations shown in Fig.5c and 5d, corresponding to the latter situation, are in resonance with each other.

To find the two-particle spectrum $\varepsilon^{(2)}(\mathbf{q})$ we shall solve the variational problem with the Schroedinger equation with the wave function

$$\Phi_q = \sum_{|\mathbf{R}-\mathbf{R}'|>2a} g_q(\mathbf{R}, \mathbf{R}') A_+(\mathbf{R}) A_-(\mathbf{R}') + \sum_{|\mathbf{R}-\mathbf{R}'|=2a} b_q(\mathbf{R}, \mathbf{R}') A_{+-}(\mathbf{R}, \mathbf{R}')$$
$$+ \sum_{j,\mathbf{R}} f_q^{(j)} A^{(j)}(\mathbf{R}) \mid G >$$
(75)

Thus, the problem is reduced to the solution of a simultaneous system of equations for the coefficients g, b, and f. Due to the awkwardness of this system, it was solved for the special cases $\mathbf{q}(q, \pm q)$, $\mathbf{q} = (\pi/2a, 0)$ and equivalent points. The dispersion equation has the form

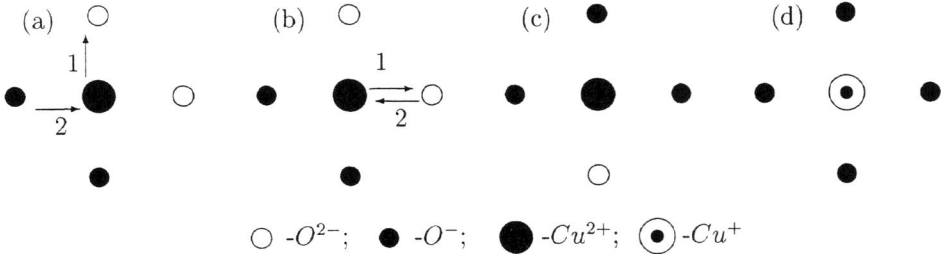

FIGURE 5. Copper ions in the environment of few oxygen holes; explanation in the text.

$$[\tilde{\varepsilon}_b - 2\varepsilon_g + 2\tau_1 + 2f_1(\mathbf{q})]^{-1} = N_0^{-1} \sum_{\mathbf{k}} f_2(\mathbf{k})/[\varepsilon_{\mathbf{q}}^{(2)} - \varepsilon(\mathbf{k}+\mathbf{q}/2) - \varepsilon(\mathbf{k}-\mathbf{q}/2)] \quad (76)$$

where

$$\varepsilon(\mathbf{k}) = \varepsilon_g + 4\tau_1[4\phi^2(\mathbf{k}) - 1], \quad \tilde{\varepsilon}_b = \varepsilon_b + 28\tau_1^2/[5(\varepsilon_{\mathbf{q}}^{(2)} - \varepsilon_f)], \quad (77)$$
$$f_1(q,q) = (4\tau_1 \cos^2(qa))/9, \quad f_1(\pi/2a, 0) = 0$$

$$f_2(\mathbf{k}) = \begin{cases} 4\sin^2[(k_x+k_y)a]\sin^2[(k_x-k_y)a] & \text{for } q_x = \pm q_y = q \\ 2\cos^2(2k_x a) & \text{for } q_x = \pi/2a, \ q_y = 0 \end{cases}$$

Here $\tilde{\varepsilon}_b$, describes the renormalization of the energy ε_b, due to hybridization with the states $A_{+-}(\mathbf{R})$. The energy denominator on the right side of (76) corresponds to the motion of two noninteracting particles with total momentum q. It is not difficult to see that (76) always has a solution corresponding to free motion of an unbound pair.

We are interested in the other type of solution, which describes bound states of particles. The right side of (76) has a logarithmic singularity in ε_q^2 at the point $\varepsilon_0^2 = 2\varepsilon_g - 8\tau_1$, corresponding to the bottom of the band for two free particles, which in our model do not depend on the total wave vector. For a decrease $\varepsilon^{(2)} < \varepsilon_0^{(2)}$ the right side of (76) tends to zero, remaining negative. From this, it is easy to see that the condition for the existence of a bound state is determined by the inequality

$$\varepsilon_f - \frac{25}{5}\tau_1^2[\tilde{\varepsilon}_b - 2\varepsilon_g + 2\tau_1 + f_1(\mathbf{q})]^{-1} < \varepsilon^{(2)} \quad (78)$$

In the case we examine we always have $f_1(q,q) \geq 0$. This means that when condition (78) is fulfilled, the bottom of the band for a bound state must lie at $q_x = \pi/2a, q_y = 0$ and equivalent points, and not in the center of the zone. This fact is connected with the form of the single-particle spectrum.

For the energy parameters of the model presented above (energies and matrix elements τ_1, τ_1') one can check that fulfillment of the inequality (78) is equivalent to imposing the condition $V \geq \varepsilon/3$.

We shall try to summarize some results obtained above. The Hamiltonian considered here for the description of CuO_2 planes is quite realistic, although it contains several parameters $(U_d, U_p, U'_p, \varepsilon, \varepsilon_p, \varepsilon'_p, t)$ affecting the electronic and magnetic properties of the system. It seems most realistic to assume that all quantities U and ε are larger than the "width" of the band t, and also, that U_d is the largest energy parameter. Reasonable quantum-chemical estimates show that the nonhybridized level ε'_p lies below the hybridized level ε_p. However, a careful analysis of two-hole hybridization demonstrates the reality of the reverse idea, according to which transport occurs through the hybridized oxygen band. In this case, the real dependence of the hole spectrum on the magnetic structure of the copper sublattice is of interest. Above we have limited ourselves to considering the Neel-type AFM state. Calculation is somewhat simpler for the limit $U_d \to \infty$: most of our results correspond to this situation, although the case of final U_d comparable to ε is still of interest.

The singularities in the single-particle spectrum found here, in particular its quasi-one-dimensional character, show the importance of analysis of the pairing mechanism because of the strong singularity in the density of states at the edge of the magnetic Brillouin zone. Other mechanisms are connected with formation of a pair of small radius from the quasiparticles-magnetic polarons.

There is another very important question on the form of the hole spectrum in the magnetic RVB state, which is attracting a great deal of interest at present due to the work of Anderson [6,7] The method used here to construct the variational wave function leads to a universal dependence of its characteristics on the correlators of the magnetic subsystem. This method will be developed below in the case of RVB states.

4 Zhang-Rice polaron

Here we shall clarify in more detail the main properties of local spin polaron in the effective three-band model. Considering the movement of an extra oxygen hole in the framework of the Hamiltonian (52) we are interested firstly in the lowest energy band. In order to describe the band correctly one must start from some set of local operators which corresponds to the most energetically favorable state of an extra hole.

Let us consider a fragment of CuO_2 plane which corresponds to a single CuO_4 plaquette with one hole on Cu site $\mathbf{R}$ and one extra O-hole, belonging to four oxygen sites $\mathbf{R} + \mathbf{a}_i$, $i = 1 \div 4$. Let the plaquette Hamiltonian contain only $\hat{T}$ term (53) with $U_d = \infty$, $U_p = 0$. It is immediately seen that 12 plaquette eigenstates correspond to one singlet ground state φ with the energy $\varepsilon = -4\tau$, three singlet states $\varphi_{1 \div 3}$ with the energy $\varepsilon_{1 \div 3} = 0$, nine triplet states ψ with the energy $\varepsilon = 0$ and three triplet states $\psi_{0;\pm 1}$ with the energy $\varepsilon_\psi = 4\tau$. The explicit form of the singlet is the following

$$| \varphi_{\mathbf{R}} \rangle = 2^{-1} \sum_{i=1}^{4} | \varphi_{\mathbf{R},\mathbf{a}_i} \rangle$$
$$| \varphi_{\mathbf{R}}^{(1,2)} \rangle = 2^{-1} (| \varphi_{\mathbf{R},\mathbf{a}_1} \rangle \pm | \varphi_{\mathbf{R},\mathbf{a}_2} \rangle - | \varphi_{\mathbf{R},\mathbf{a}_3} \rangle \mp | \varphi_{\mathbf{R},\mathbf{a}_4} \rangle) \quad (79)$$
$$| \varphi_{\mathbf{R}}^{3} \rangle = 2^{-1} (| \varphi_{\mathbf{R},\mathbf{a}_1} \rangle - | \varphi_{\mathbf{R},\mathbf{a}_2} \rangle + | \varphi_{\mathbf{R},\mathbf{a}_3} \rangle - | \varphi_{\mathbf{R},\mathbf{a}_4} \rangle)$$

where $| \varphi_{\mathbf{R},\mathbf{a}_i} \rangle$ is a singlet state of two holes on $\mathbf{R}$ Cu and $\mathbf{R} + \mathbf{a}_i$ O sites.

$$| \varphi_{\mathbf{R},\mathbf{a}_i} \rangle = 2^{-1/2} \left(c^+_{\mathbf{R}+\mathbf{a}_i,\sigma} Z_{\mathbf{R}}^{-\sigma,0} - c^+_{\mathbf{R}+\mathbf{a}_i,-\sigma} Z_{\mathbf{R}}^{\sigma,0} \right) | 0 \rangle_{\mathbf{R}} \quad (80)$$

$| 0 \rangle_{\mathbf{R}}$ denotes the vacuum of a plaquette $\mathbf{R}$.

The triplet states $\psi, \psi^{(1,2,3)}$ (with zero spin projection) have the same form as (79) but instead of $| \varphi_{\mathbf{R},\mathbf{a}_i} \rangle$ one must write $| \psi_{\mathbf{R},\mathbf{a}_i} \rangle$, where $| \psi_{\mathbf{R},\mathbf{a}_i} \rangle$ is

$$| \psi_{\mathbf{R},\mathbf{a}_i} \rangle = 2^{-1/2} \left(c^+_{\mathbf{R}+\mathbf{a}_i,\sigma} Z_{\mathbf{R}}^{-\sigma,0} + c^+_{\mathbf{R}+\mathbf{a}_i,-\sigma} Z_{\mathbf{R}}^{\sigma,0} \right) | 0 \rangle_{\mathbf{R}} \quad (81)$$

So the ground singlet state $| \varphi_{\mathbf{R}} \rangle$ is separated from the higher lying states by an energy difference 4τ. The importance of the singlet $| \varphi_{\mathbf{R}} \rangle$ formation for the low-lying excitations in the three-band Hubbard model was recognized by Zhang and Rice [205] and has been confirmed by exact diagonalization studies of Cu_nO_m clusters [91,65,186].

It is obvious that an extra hole motion and elementary excitations strongly depend on a Cu-spin background that doesn't appear explicitly in the expression for $| \varphi_{\mathbf{R}} \rangle$. In order to construct an operator of extra hole elementary excitation we must firstly eliminate a copper spin $\mathbf{R}$ (for example with spin projection $-\sigma$) from the background state $| G \rangle$ to form a vacuum $| 0 \rangle_{\mathbf{R}}$ in $\mathbf{R}$-plaquette, acting by an operator $Z_{\mathbf{R}}^{0,\sigma}$, and then create a state $| \varphi \rangle$ on the plaquette. This procedure must be described by the following operator $A_{\mathbf{R}}^+$ of a local spin-polaron

$$A_{\mathbf{R},\sigma}^{+} = | \varphi_{\mathbf{R}} \rangle Z_{\mathbf{R}}^{0,-\sigma} = 8^{-1/2} \sum_{i=1}^{4} \left(c^+_{\mathbf{R}+\mathbf{a}_i,\sigma} Z_{\mathbf{R}}^{-\sigma,0} - c^+_{\mathbf{R}+\mathbf{a}_i,-\sigma} Z_{\mathbf{R}}^{\sigma,0} \right) Z^{0,-\sigma} = \\ 8^{-1/2} \sum_{i,\sigma_1} \sigma_1 c^+_{\mathbf{R}+\mathbf{a}_i,\sigma_1} Z_{\mathbf{R}}^{-\sigma_1,-\sigma} \quad (82)$$

Let us note, that the action of the creation operator $A_{\mathbf{R},\sigma}^+$ on any background state $| G \rangle$ with zero spin projection of hole plane leads to the state $A_{\mathbf{R},\sigma}^+ | G \rangle$ with $\frac{1}{2}\sigma$ spin projection of the plane.

So the operator $A_{\mathbf{R},\sigma}^+$ is an analogue of a Zhang-Rice plaquette polaron when we treat an excitation on a spin background.

In Section IV B we discuss the generalization of the local set of cell operators $A_{\mathbf{R},\sigma}^+$ which represents the above mentioned 16 states of a plaquette. The set is given by ten cell operators such that six of them are given by one term from the right-hand part of (82) with fixed $\mathbf{a}_i$, and the bare hole operators $c^+_{a_i}$.

III ANTIFERROMAGNETIC $S = \frac{1}{2}$ FRUSTRATED HEISENBERG MODEL IN THE SPIN-ROTATION-INVARIANT THEORY

Let us keep in mind the reasons why it is important to have a decription of the aniferromagnetic $S = \frac{1}{2}$ Heisenberg model in the framework of the spin-rotation-invariant (or spherically symmetric) theory and why the frustration is important.

The movement of the carries on the spin background strongly depend on the features of the background. As it may be seen for the simplest treating of local spin polaron the conventional two-sublattice spin approach leads to spectrum periodicity relative to the magnetic Brillouin zone [156,161,173]. But the APRES experiments indicate the absence of such a periodicity in the 2-D case — the spectrum is periodic relative to a full Brillouin zone of CuO_2 plane. As it will be seen namely in the spin-rotation-invariant approach, the periodicity relative to a full Brillouin zone is preserved for the Green functions of spin excitations and as a result for the carrier spectrum.

The typical behaviour of the 2-D magnetic subsystem corresponds to undoped CuO_2 planes in La_2CuO_4. Such a system is well described by the two-dimensional (2D) antiferromagnetic $S = 1/2$ Heisenberg model [191,36] with interaction between the nearest spins on the square lattice.

The antiferromagnetic exchange interaction between the spins of the first-nearest-neighbour Cu^{2+} ions in a given CuO_2 plane is extremely large (in the order of $0.13 eV \cong 1500 K$ for La_2CuO_4 [178]) and considerably greater than the interplanar exchange. The interplanar exchange is primarily responsible for the onset of the long-range order that is observed in the dielectric phase of the CuO_2 planes (for La_2CuO_4 the characteristic Néel temperature is $T_N \sim 300$ K [179,32]). However, in the case of comparatively light doping of the system by holes the antiferromagnetic long-range order disappears over the whole range of temperatures. It is customary to assume that the doping leads to antiferromagnetic interaction between second-nearest neighbours in the Cu^{2+} subsystem in a given plane, i.e. to frustration [97]. Note, that the cluster calculation indicates a large enough value of frustration parameter $J_2/J_1 \sim 0.1$ even for the undoped La_2CuO_4 [8].

In view of this, intensive studies are currently being carried out on the frustrated Heisenberg model. We shall also discuss the effect of temperature in the spin subsystem.

A Two-dimensional antiferromagnet

1 Different descriptions of the phase transitions on frustration

For the aniferromagnetic $S = \frac{1}{2}$ frustrated Heisenberg model we take the Hamiltonian

$$H = \frac{J_1}{2} \sum_{i,g} S_i S_{i+g} + \frac{J_2}{2} \sum_{i,d} S_i S_{i+d} \qquad (83)$$

where $J_1, J_2 > 0$ are the antiferromagnetic interactions between the first-nearest neighbour ($\mathbf{g} = \pm \mathbf{g}_x, \pm \mathbf{g}_y$) and the second-nearest neighbour ($\mathbf{d} = \pm \mathbf{g}_x \pm \mathbf{g}_y$) on a square lattice (see Fig.6).

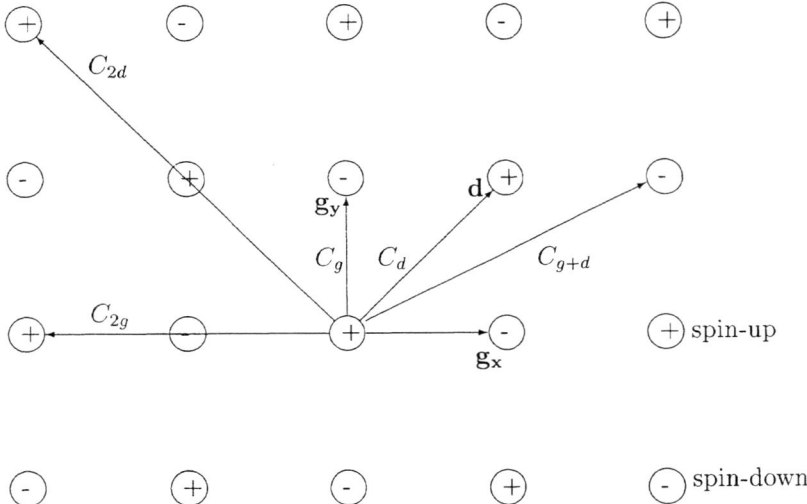

FIGURE 6. Classical state of the spin system — the Néel phase I (the corresponding correlation functions $C_\mathbf{r}$ are indicated).

For convenience, J_1 and J_2 can be expressed in terms of the frustration parameter p:

$$J_1 = (1-p)J, \quad J_2 = pJ, \quad 0 \le p \le 1, \quad J > 0 \qquad (84)$$

The frustration parameter p can be regarded as the analog of the number x of holes per copper atom. An estimate based on the single-band Hubbard model with realistic values $U/t \sim 5$ leads, e.g., to a value of $p \sim 0.26$ for $x = 0.1$. We note, that in the case of $La_{2-x}Sr_xCuO_4$ the spin system of the CuO_2 plane loses its long-range order for $x > 0.02$.

In the case of the model without frustration the Mermin-Wagner theorem [136] asserts the absence of spontaneous magnetization at finite temperatures (i.e. the absence of long-range order), but does not rule it out at $T = 0$ (see the review [128]).

In the classical approximation ($S \gg 1$) the frustrated model (83) has at $p = 1/3$ a first-order transition between two ordered phases. Phase I ($p < 1/3$) is characterized by the order vector $\mathbf{q}_I = (\pi, \pi)$ and corresponds to the Néel state

(Fig. 6). Phase II (Fig. 7) is realized for $p > 1/3$ and corresponds to the so-called "stripe" state (order vector $\mathbf{q}_{II} = (\pi, 0)$ or $(0, \pi)$).

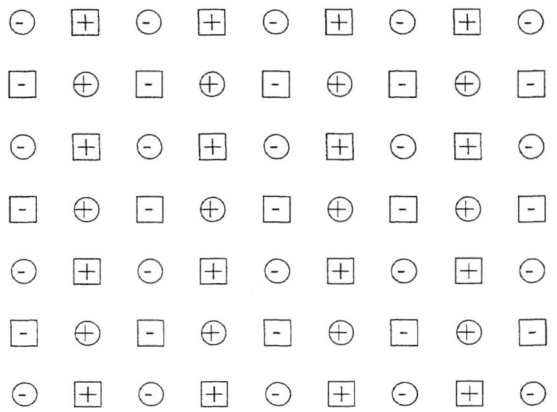

FIGURE 7. Classical state of the spin system — the "stripe" phase II, with two mutually penetrating square sublattices, represented by circles and squares.

The spin correlation functions in these phases have the form

$$C_{\mathbf{r}}^{I} \propto (-1)^{n_x + n_y}, \tag{85}$$

$$C_{\mathbf{r}}^{II} \propto (-1)^{n_x} \text{ or } (-1)^{n_y}, \tag{86}$$

$$C_{\mathbf{r}} = \sum C_{\mathbf{r}}^{\alpha}, \quad C_{\mathbf{r}}^{\alpha} = \langle S_{\mathbf{i}}^{\alpha} S_{\mathbf{i}+\mathbf{r}}^{\alpha} \rangle,$$
$$\alpha = x, y, z \quad \mathbf{r} = n_x \mathbf{g}_x + n_y \mathbf{g}_y,$$

where n_x and n_y are integers. We note that in the case $p = 1$ ($J_1 = 0$) the spin system takes the form of two superimposed noninteracting Néel lattices (Fig. 7), for each of which a situation analogous to the case of phase I at $p = 0$ is realized.

In the quantum limit $S = 1/2$ at $T = 0$ we may regard it as established that the system possesses long-range order of type I and type II in the limiting cases at $p = 0$ and $p = 1$ Ref. [128]. In this section we are interested in the properties of the system and the phase transitions between these states in the parameter p.

We shall discuss here the results given by different approaches for the frustrated model with the Hamiltonian (83).

In Refs. [37,149,140] this model was considered in terms of the theory of linear spin waves. In this approach for $p < 1/3$ the operators of spin-wave excitations are constructed relative to the Néel state, and for $p > 1/3$ they are constructed relative to the stripe state. From the theory of linear spin waves it follows that for $p < p_{1L} = 0.274$ the phase I with sublattice magnetization $m_{1L} \propto (-1)^{n_x + n_y}$ is realized, while for $p < p_{2L} = 1/3$ the phase II with $m_{2L} \propto (-1)^{n_x}$ takes place.

At points p_{1L} and p_{2L} the corresponding sublattice magnetizations m_{1L} and m_{2L} vanish. On this basis it was postulated in Ref. [37] that in the interval $p_{1L} < p < p_{2L}$ the system loses its long-range order and turns into a spin-liquid state. However, the theory of linear spin waves is unable to describe the state of the system at $T = 0$ for this range of values of p. In the framework of this theory, for any value of p, it is also not possible to consider the case of nonzero temperatures (because of the well-known divergence of m at $T \neq 0$ in the two-dimensional case). We note that in Ref. [149] correlation functions $C_\mathbf{r}$ for both phases were found with allowance for biquadratic terms in the Dyson-Maleev transformation [53,54,126]. Values of $C_\mathbf{r}$ over the whole range of p, obtained by numerical calculations on finite blocks of up to 20 spins, are also given in Ref. [149].

In Ref. [88] the frustrated model was studied on the phase-I side in the framework of the theory of linear spin waves with the additional condition that the sublattice magnetization is equal to zero (a detailed account of the sublattice-symmetric theory of linear spin waves can be found in Refs. [190,89]). In the case of $T = 0$ it was shown that for $p < p_{1L}$ the spin-wave spectrum $\omega(\mathbf{q})$ is gapless at $\mathbf{q} = 0$, i.e. there is a long-range order determined by Bose condensation of the spin waves with $\mathbf{q} = 0$. Hereinafter we consider Bose condensation to give the anomalous (as $T \to 0$) contribution to the expression for the spin correlators from a small region of wave-vector values in the neighbourhood of the points $\mathbf{q}_i$ at which $\omega(\mathbf{q}_i) = 0$. For $p > p_{1L}$ a gap appears in the spin wave spectrum. This implies that the system loses its long-range order and turns into a spin-liquid state, in which the spin excitations have nonzero mass. The sublattice-symmetric theory of linear spin waves can be generalized to the case of $T \neq 0$. However, for $p > 0.27$ the correlation functions $C_\mathbf{r}$ over short distances, found on the basis of this theory, differ strongly from the corresponding values of $C_\mathbf{r}$ from numerical calculations for finite blocks [88].

The theory of modified spin waves, [190,89,187] which takes into account terms proportional to zero power of spin value S in the Dyson-Maleev or Holstein-Primakoff transformations, leads to different results. In Refs. [202,146,26] it was shown that phase I possesses a long-range order in the interval $0 < p < p_{1M}$ ($p_{1M} = 0.38$), while phase II possesses a long-range order for $p_{2M} < p < 1$ ($p_{2M} = 0.35$). Here the average spin $\langle \mathbf{S_i} \rangle$ per site is zero, while long-range order corresponds to a nonzero effective spin m, defined in terms of the spin correlation functions at large distances by $m^2 = |C_{\mathbf{r} \to \infty}|$. In phases I and II $C_{\mathbf{r} \to \infty}$ has a form analogous to (85, 86). Thus, in the theory of modified spin waves the long-range order is preserved for all values of p, and in the region $p_{2M} < p < p_{1M}$ a transition should occur between two phases with a different long-range order. We emphasize that in the theory of modified spin waves, as in the theory of linear spin waves, the spin-excitation operators in the different phases are constructed relative to different classical states. As a consequence, in the framework of these theories the transition between phases I and II should be a first order transition.

The approaches listed above are not spherically symmetric, in the sense that there is no spherical symmetry of the spin correlation functions in them, i.e. $C_\mathbf{r}^z \neq$

$C_\mathbf{r}^x \ne C_\mathbf{r}^y$. In Ref. [139] the model (83) was studied in terms of a spherically symmetric theory based on a mean-field approximation for Schwinger bosons [9,10]. For $S = 1/2$ the results are the same as in the theory of modified spin waves. It is easy to see that in the approach of Ref. [9,10] the way the mean field is introduced for the two phases can only yield a first-order transition, analogously to the theory of modified spin waves. We also note that at nonzero temperatures both theories lead to a nonphysical first-order phase transition at values of $T \sim J$ (Refs. [188,138]).

The possibility of a first-order transition from a state with long-range order to a spin-liquid state at the comparatively small value $p \approx 0.2$ was pointed in [97] and [48]. In these a variational approach was used, and for the spherically symmetric ground state of the spin-liquid phase a state of the resonance-valence-bond type was constructed. A second-order transition from the state with long-range order to the spin-liquid state at $p \approx 0.2$ was obtained assuming spherically symmetric spin correlation functions in Ref. [21], in which the spin excitations were considered on the basis of four-spin blocks covering an infinite lattice. However, a detailed analysis turns out to be complicated in the region $p > 1/3$, when the inversion of the energy levels of a block sets in.

Finally, we also mention a number of theoretical approaches that predict the absence of long-range order in a certain interval near $p = 1/3$. There are numerous papers on exact diagonalization on finite systems [149,140,164], the renormalization-group method combined with the mean-field approximation for Schwinger bosons [59] (this approach gives a transition from the state with long-range order to the spin-liquid state at $p \approx 0.13$), a self-consistent spin wave approach on the basis of the dimer classical state [39], and, finally, an analysis based on the technique of the $1/N$-expansion [162,163]. As it can be seen, different analytical approaches give contradictory results for the frustrated model.

In this section the model is discussed following the spherically symmetric approach for two-time retarded Green's functions. This approach was proposed by Kondo and Yamaji for one-dimensional unfrustrated case [111]. The method was developed for two-dimensional case in [176,27] and for frustrated 2D and 3D cases in [13,14]. Unlike most of the analytical approaches mentioned above, this treatment makes it possible to describe both the two phases with long-range order and the intermediate spin-liquid phase over the whole range of p in a single approximation.

Below we present the equations of motion for Green's functions, which make it possible to express the spin-excitation spectrum $\omega(\mathbf{q})$ and the correlation functions. Closed expressions for Green's functions, and a self-consistent system of equations for the correlation functions $C_\mathbf{r}$, are found by a decoupling procedure with effective allowance for vertex corrections [176]. In the next Section we give the results for solving the self-consistent system of equations at $T = 0$ in the entire range of the frustration parameter p, namely, the value of the effective spin m, the energy of the ground state, the magnitude of the gap in the spin-excitation spectrum at points $\mathbf{q}_\mathrm{I} = (\pi, \pi)$ and $\mathbf{q}_\mathrm{II} = (\pi, 0)$, and the correlation functions $C_\mathbf{r}$ for the first five nearest neighbours. It is shown that as the frustration increases a continuous transition occurs in the system, from a phase with long-range order (the analog of phase I) to a spin-liquid phase, and then to a phase with long-range order (the analog of phase II). The temperature dependence of the static uniform susceptibility $\chi(T)$ is also presented.

2 Self-consistent spin excitations

To determine the spin correlation functions we shall consider the two-time retarded Green's functions (and their Fourier transforms with respect to time) for the spin operators [207]:

$$G_{ij}^{\sigma}(\omega) = \left\langle S_i^{\sigma} \mid S_j^{\bar\sigma} \right\rangle_{\omega} = -i \int_0^{\infty} dt\, e^{i\omega t} \left\langle [S_i^{\sigma}(t), S_j^{\bar\sigma}] \right\rangle \tag{87}$$

$\sigma = \pm 1$, $\bar\sigma = -\sigma$.

For the operators S^{σ} we introduce the site correlation functions C_r^{σ}, and also the spatial Fourier transforms of these functions and of Green's functions $G_{ij}^{\sigma}(\omega)$:

$$\begin{aligned} C_r^{\sigma} &= \langle S_i^{\sigma} S_{i+r}^{\bar\sigma} \rangle = N^{-1} \sum_{\mathbf{q}} e^{-i\mathbf{qr}} C_{\mathbf{q}}^{\sigma}; \\ G^{\sigma}(\mathbf{q},\omega) &= \sum_{\mathbf{r}} e^{i\mathbf{qr}} G_{i,i+r}^{\sigma}(\omega) \end{aligned} \tag{88}$$

The functions $C_{\mathbf{q}}^{\sigma}$ and $G^{\sigma}(\mathbf{q},\omega)$ are connected by the relation

$$C_{\mathbf{q}}^{\sigma} = -\frac{1}{\pi} \int_{-\infty}^{\infty} d\omega \, [1 + n(\omega)] \, Im\{G^{\sigma}(\mathbf{q}, \omega + i\varepsilon)\} \tag{89}$$

where $n(\omega) = [\exp(\omega/T) - 1]^{-1}$.

The first-order and second-order equations of motion for Green's functions $G_{ij}^{\sigma}(\omega)$ have the form

$$\omega G_{ij}^{\sigma}(\omega) = 2\sigma \delta_{ij} \langle S_i^z \rangle + \sigma \sum_{r1=g,d} J_{r1} \left\langle S_i^z S_{i-r1}^{\sigma} - S_{i+r1}^z S_i^{\sigma} \mid S_j^{\bar\sigma} \right\rangle_{\omega} \tag{90}$$

$$\begin{aligned}
\omega\sigma \left\langle S_i^z S_{i+r1}^{\sigma} - S_{i+r1}^z S_i^{\sigma} \mid S_j^{\bar\sigma} \right\rangle_{\omega} &= \left(2\langle S_i^z S_j^z \rangle + \langle S_i^{\sigma} S_j^{\bar\sigma} \rangle \right) \delta_{i+r1,j} \\
&\quad - \left(2\langle S_{i+r1}^z S_j^z \rangle + \langle S_{i+r1}^{\sigma} S_j^{\bar\sigma} \rangle \right) \delta_{i,j} + + \tfrac{1}{2} J_{r1} \left(G_{ij}^{\sigma}(\omega) - G_{i+r1,j}^{\sigma}(\omega) \right) \\
&\quad + \tfrac{1}{2} \sum_{r2=g,d\ r2\neq r1} J_{r2} \Big\{ \left\langle 2S_i^z \left(S_{i+r1}^z S_{i+r1-r2}^{\sigma} - S_{i+r1-r2}^z S_{i+r1}^{\sigma} \right) \mid S_j^{\bar\sigma} \right\rangle_{\omega} \\
&\quad - \left\langle 2S_{i+r1}^z \left(S_i^z S_{i+r2}^{\sigma} - S_{i+r2}^z S_i^{\sigma} \right) \mid S_j^{\bar\sigma} \right\rangle_{\omega} \\
&\quad + \sigma \sum_{\gamma=\pm 1} \gamma \left\langle S_i^{\gamma} S_{i+r2}^{\bar\gamma} S_{i+r1}^{\sigma} - S_{i+r1}^{\gamma} S_{i+r1-r2}^{\bar\gamma} S_i^{\sigma} \mid S_j^{\bar\sigma} \right\rangle_{\omega} \Big\}
\end{aligned} \tag{91}$$

where $J_g = J_1$ and $J_d = J_2$.

For Green's functions that arise in the right-hand side of (91) we use a decoupling procedure analogous to the one performed in [176]. This decoupling has the following features.

First, it preserves the local correlation on a single site. This is expressed in the fact that for Green's functions the decoupling is performed only when the three site spin operators in the left part of the bracket pertain to different sites.

Second, when identifying averages of the form $\langle S_i^z S_{i+r}^z \rangle$ and $\langle S_i^\sigma S_{i+r}^{\bar\sigma} \rangle$ ($\mathbf{r} \neq 0$) we introduce extra factors α_1, α_2 and α_3 that have the meaning of vertex corrections. The factor α_1 is introduced if the sites $\mathbf{i}$ and $\mathbf{i} + \mathbf{r}$ are nearest neighbours, i.e. $r = g$; α_2 is introduced if $\mathbf{r}$ corresponds to a third-, fourth-, or fifth-nearest neighbour ($r > d$); $\alpha_3 = (1 - p)\alpha_2 + p\alpha_1$ is used in the intermediate case when $r = d$.

The method of taking vertex corrections into account makes it possible to pass correctly from the limiting case $p = 0$ to the limiting case $p = 1$. As noted above, for $p = 1$ we have two noninteracting superimposed antiferromagnetic sublattices, each of which is physically equivalent to the complete lattice for $p = 0$. The decoupling scheme satisfies this equivalence. For the case $p = 0$ (i.e. in absence of frustration) we take the very simple scheme that was discussed in detail in Ref. [176]. If correlation functions with operators on nearest neighbours arise in the decoupling, we introduce α_1, while for more distant neighbours we introduce α_2. For $p = 0$ we have $\alpha_3 = \alpha_2$, and our scheme coincides with the one adopted above. For $p = 1$ we have $\alpha_3 = \alpha_1$, and it is easily verified that for each of the sublattices our scheme goes exactly over into the scheme of Ref. [176]. In particular, it leads to the result that the effective spins m found below for $p = 0$ and $p = 1$ coincide.

Furthermore, by virtue of the assumption that the correlation functions are spherically symmetric, we assume

$$C_{\mathbf{r}} = 3\langle S_i^z S_{i+r}^z \rangle = \frac{3}{2}\langle S_i^\sigma S_{i+r}^{\bar\sigma} \rangle = \langle \mathbf{S}_i \mathbf{S}_{i+r} \rangle \qquad (92)$$

and also $\langle S_i^z \rangle = 0$.

After the decoupling Eq.(91) takes the following form:

$$\sigma\omega\left[\langle S_i^z S_{i+r_1}^\sigma \mid S_j^{-\sigma}\rangle_\omega - \langle S_{i+r_1}^z S_i^\sigma \mid S_j^{-\sigma}\rangle_\omega\right] =$$
$$\frac{2}{3}\Big\{2C_{r_1}(\delta_{i+r_1,j} - \delta_{ij})$$
$$+[J_{r_1}(\alpha_2 C_{2r_1} + 2\alpha_{r_1}C_{r_{1x}+r_{1y}} + 1/2) \qquad (93)$$
$$+2\tilde{J}_{r_1}(\alpha_1 C_g + \alpha_2 C_f)](G_{ij}^\sigma(\omega) - G_{i+r_1,j}^\sigma(\omega))$$
$$+\tilde{\alpha}_{r_1}C_{r_1}\left(\sum_{r_2=g_2,d_2\ r_2\neq r_1}\left(G_{i+r_1-r_2,j}^\sigma(\omega) - G_{i+r_2,j}^\sigma(\omega)\right)\right)\Big\}$$

where

$$J_g = J_1, \quad J_d = J_2, \quad \tilde{J}_g = J_2, \quad \tilde{J}_d = J_1,$$
$$\alpha_g = \alpha_3, \quad \alpha_d = \alpha_2, \quad \tilde{\alpha}_g = \alpha_1, \quad \tilde{\alpha}_d = \alpha_2,$$

An expression for Green's function $G^\sigma(\mathbf{q},\omega)$ can be obtained by going over to Fourier components in Eqs.(90), (91) and (93). As a result, we have

$$G^\sigma(\mathbf{q},\omega) = -(16/3)[J_1(1-\gamma_g)C_g \\ +J_2(1-\gamma_d)C_d](\omega^2 - \omega^2(\mathbf{q}))^{-1} \tag{94}$$

where the excitation spectrum $\omega(\mathbf{q})$ has the form

$$\omega^2(\mathbf{q}) = \frac{8}{3}[(1-\gamma_g)A_1 + (1-\gamma_g^2)A_2 + (1-\gamma_d)A_3 + (1-\gamma_d^2)A_4 + \gamma_g(1-\gamma_d)A_5] \tag{95}$$

$$A_1 = J_1^2(\alpha_2 C_{2g} + 2\alpha_3 C_d + \frac{3}{4} + 3\alpha_1 C_g) + 2J_1J_2(\alpha_2 C_{g+d} - \alpha_1 C_g)$$

$$A_2 = -4J_1^2\alpha_1 C_g$$

$$A_3 = 2J_1J_2(3\alpha_1 C_g + \alpha_2 C_f) + J_2^2(\alpha_2(C_{2d} + 2C_{2g}) + \frac{3}{4} + 3\alpha_3 C_d)$$

$$A_4 = -4J_2^2\alpha_3 C_d$$

$$A_5 = -4J_1J_2(\alpha_1 C_g + \alpha_3 C_d)$$

$$\gamma_d = \cos(q_x g)\cos(q_y g); \quad \gamma_g = \frac{1}{2}[\cos(q_x g) + \cos(q_y g)]$$

Using the expression (94) for Green's functions, and also Eqs.(88) and (89), we can obtain a self-consistent system of equations for the five correlation functions C_g, C_d, C_{2g}, C_f and C_{2d} in terms of which the spectrum (95) is determined:

$$C_\mathbf{r} = N^{-1}\sum_\mathbf{q} F(\mathbf{q})\omega(\mathbf{q})^{-1}e^{-i\mathbf{qr}}[n(\omega(\mathbf{q})) + 1/2] \\ F(\mathbf{q}) = (-8)(J_1(1-\gamma_{gq})C_g + J_2(1-\gamma_{dq})C_d) \tag{96}$$

The sum rule $C_0 = \langle S_i S_i \rangle = 3/4$ gives an additional condition for the determination of α_1. Thus, putting α_2 aside, we have six self-consistent equations for the six unknowns $C_\mathbf{r}$ ($\mathbf{r} = \mathbf{g}, \mathbf{d}, 2\mathbf{g}, \mathbf{f}, 2\mathbf{d}$) and α_1. Here $\mathbf{f} = 2\mathbf{g}_x + \mathbf{g}_y$.

The last unknown decoupling parameter α_2 is found from the following condition. In absence of frustration at $T = 0$ it is well known that the effective spin has the value $m \approx 0.3$, and we require that the quantity m calculated for $p = 0$ and $T = 0$ coincides with this value. In the problem without frustration a phenomenological choice of this type for the parameter α_2 was used in Ref. [176]. Also in [176] it was noted that as T increases the vertex corrections tend to unity, and for $T \neq 0$ it was suggested to fix the value of the parameter

$$r_\alpha = (\alpha_1 - 1)/(\alpha_2 - 1) = \text{const} \tag{97}$$

by using the additional condition

$$m(r_\alpha, T = 0, p = 0) = 0.3 \tag{98}$$

Let us mention that there is another way to fix the value of parameter r_α that is by requiring that the average energy at $T = 0$, $p = 0$ is equal to the well-known value.

Analogously, for arbitrary frustrations and temperatures, we also determine α_2 from the condition (97). The parameter r_α remains equal to the value given by Eq.(98). This scheme for the determination of α_2 not only satisfies the requirement that the vertex corrections become equal to unity with the increase of T, but it also allows us to obtain at $T = 0$ the correct values of m in two limiting cases, $p = 0$ and $p = 1$.

3 Results

At $T = 0$, depending on the value of the frustration parameter p, for the self-consistent system of equations (96) two types of solutions are possible. The forher corresponds to the case when the spectrum has no gap at $\mathbf{q} \neq 0$. The latter corresponds to $\omega(\mathbf{q}) > 0$ for all $\mathbf{q} \neq 0$.

A solution of the first type occurs in two intervals of values of the frustration. For $p < p_1$ ($p_1 = 0.1$) the spectrum is found to be gapless at the point $\mathbf{q}_\mathrm{I} = (\pm\pi, \pm\pi)$, while for $p > p_2$ ($p_2 = 0.62$) it is found to be gapless at the point $\mathbf{q}_\mathrm{II} = (\pm\pi, 0)$ or $(0, \pm\pi)$. In these cases, in Eqs. (96) we can separate the condensate part $m_{\mathbf{q}_0}^2$ corresponding to Bose condensation on the corresponding antiferromagnetic vector $\mathbf{q}_0$, equal to $\mathbf{q}_\mathrm{I}$ or $\mathbf{q}_\mathrm{II}$:

$$C_\mathbf{r} = \frac{1}{2N} \sum_\mathbf{q} F(\mathbf{q}) \omega(\mathbf{q})^{-1} e^{-i\mathbf{q}\mathbf{r}} + e^{-i\mathbf{q}_0 \mathbf{r}} m_{\mathbf{q}_0}^2 \qquad (99)$$

The new unknown $m_{\mathbf{q}_0}^2$ is determined from the equation $\omega(\mathbf{q}_0) = 0$. From the expression (95) for $\omega^2(\mathbf{q})$ we can see that $\omega^2(\mathbf{q}_\mathrm{I})$ is proportional to A_1, and $\omega^2(\mathbf{q}_\mathrm{II})$ is proportional to the sum of the first three terms in the right-hand side of Eq.(95). A nonzero value of $m_{\mathbf{q}_0}$ is the effective spin, since

$$C^\mathrm{I}_{r \to \infty} = m_{\mathbf{q}_0}^2 (-1)^{n_x + n_y}, \qquad p < p_1 \qquad (100)$$

$$C^\mathrm{II}_{r \to \infty} = (1/2) m_{\mathbf{q}_0}^2 [(-1)^{n_x} + (-1)^{n_y}], \quad p > p_2 \qquad (101)$$

Thus, in two different intervals of the frustration parameter we describe two different states with long-range order, analogous to the Néel phase I and the stripe phase II. We draw attention to the fact that the long-range order (101) in phase II preserves the square symmetry, in contrast with the approaches mentioned above, which lead to a long-range order of the form (86). In addition, unlike the theories of modified spin waves and the mean-field approximation for Schwinger bosons, in our approach Bose condensation at $\mathbf{q} = 0$ is absent. The dependence of the effective spin on the frustration parameter is presented in Fig. 8.

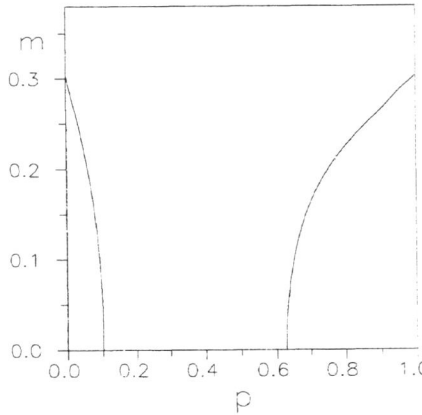

FIGURE 8. Dependence of the effective spin m on the frustration parameter p. Ref. [13]

Solutions of the second type, when $\omega(\mathbf{q}) > 0$ holds for all $\mathbf{q} \neq 0$, are found in the interval $p_1 < p < p_2$. In this region of p values the effective spin is equal to zero, the correlation functions fall exponentially at large distances, and the system is in the spin-liquid state.

We shall explain how the system of equations (96), (97) is solved for $T = 0$. In the cases $p = 0$ and $p = 1$ this system (with allowance for (99)) is solved analytically. The values obtained for the spin correlation functions, r_α and α_1 are used as the starting point for a numerical iterative method of solution of the system of equations for $p \neq 0$ and $p \neq 1$. After each step in p we first solve the truncated self-consistent system (96), (97) for α_1, the correlator $C_\mathbf{g}$, and also m or the gap in the spectrum (95) (in the case when a solution with the gap arises). Here we first use the values obtained for the long-range correlators $C_\mathbf{d}$, $C_{2\mathbf{g}}$, $C_\mathbf{f}$ and $C_{2\mathbf{d}}$ in the previous step in p. The resulting self-consistent values of α_1, $C_\mathbf{g}$, and m (or the gap) are used to find new values of the long-range order correlators, and with the latter we again implement self-consistency for the truncated system. This procedure leads to rapid convergence of the results.

In Fig. 9 we show the dependence of the gaps $\omega(\mathbf{q}_\mathrm{I})$ and $\omega(\mathbf{q}_\mathrm{II})$ on the frustration, which determine the properties of the system. At $p = p_1$ a gap opens up at the point $\mathbf{q}_\mathrm{I}$. With a further increase of p the gap $\omega(\mathbf{q}_\mathrm{I})$ remains nonzero, but behaves nonmonotonically. The gap $\omega(\mathbf{q}_\mathrm{II})$ decreases monotonically with an increase of p, and vanishes at point p_2. At points p_1 and p_2 a continuous transition is realizes between a phase with long-range order and the spin-liquid phase.

The energy per site and the first two correlation functions are presented as functions of p in Figs. 10 and 11.

Here, for comparison, we give results of calculations on finite blocks [149,164] with up to 20 sites. As it can be seen, our results qualitatively agree with the results of exact diagonalization on finite systems. It is difficult to speak of quantitative

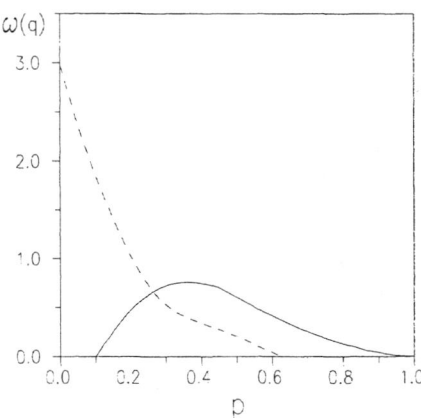

FIGURE 9. Spectrum $\omega(\mathbf{q})$ at the Bose-condensation points, as a function of the frustration parameter p. The solid curve shows $\omega[\mathbf{q} = (\pi, \pi)]$, while the dashed curve shows $\omega[\mathbf{q} = (0, \pi)]$. Ref. [13].

agreement, especially for $p > 1/3$. This is because, as p increases, the system approaches two noninteracting square sublattices. For example, a block of 16 sites decomposes into two noninteracting blocks of 8 sites. Consequently, in calculations on finite lattices the correlators are calculated in different approximations for small and large p. In particular this leads to the result that the relations of the form $C_{\mathbf{d}}(p = 1) = C_{\mathbf{g}}(p = 0)$ are not fulfilled. Our results, however, satisfy this condition.

In the case of finite temperatures ($T \neq 0$), for all values of the frustration parameter p, in the spectrum $\omega(\mathbf{q})$ there is a nonzero gap everywhere except for $\mathbf{q} = 0$. Solving the system of equations (96) makes it possible to determine the uniform static susceptibility $\chi(T, p)$. The latter is Green's function $G^{\sigma}(\mathbf{q}, \omega)$ at $\mathbf{q} = 0$ and $\omega = 0$.

Figure 12 shows the temperature dependences $\chi(T)$ for values of the frustration parameter p that correspond to the ordered phase I and the spin liquid. It can be seen that for small p the susceptibility $\chi(T)$ has a broad maximum at $T \sim J$. With increase of the frustration this maximum is shifted in the direction of lower T. We shall regard the frustration parameter p as the analog of the doping x for $\text{La}_{2-x}\text{Sr}_x\text{CuO}_4$. It can be seen that the dependence $\chi(p, T)$ in Fig. 12 qualitatively coincides with the experimental dependence $\chi(x, T)$ in Ref. [105].

In conclusion we shall formulate the principal results and note the essential difference between the method discussed and other treatments.

Using Green's functions and approximations of spherically symmetric spin correlations it is possible to describe, within the framework of a single theory, two phases with different long-range order and the spin-liquid phase separating them, and also the continuous transitions between them in the frustration parameter. For small values $p < 0.1$ the system is in a state with long-range order, analogous to the Néel

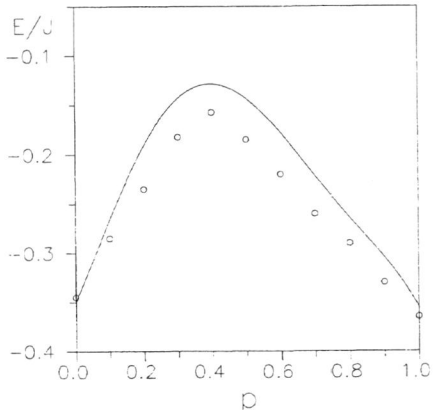

FIGURE 10. Energy per bond (E/J) as a function of the frustration parameter p Ref. [13], and results from a calculation in Ref. [149] on blocks of 20 spins.

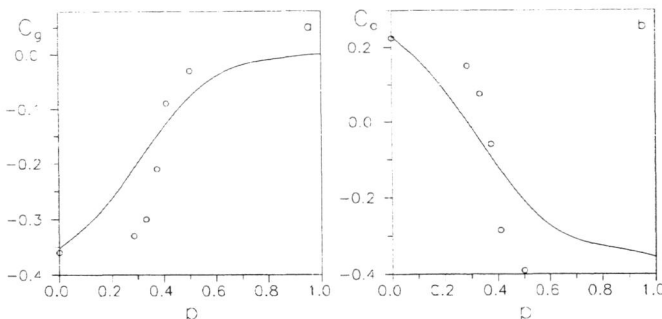

FIGURE 11. Correlation functions C_r as functions of the frustration parameter p Ref. [13]. The points are the results from a calculation in Ref. [164] on blocks of 20 spins: (a) C_g ; (b) C_d .

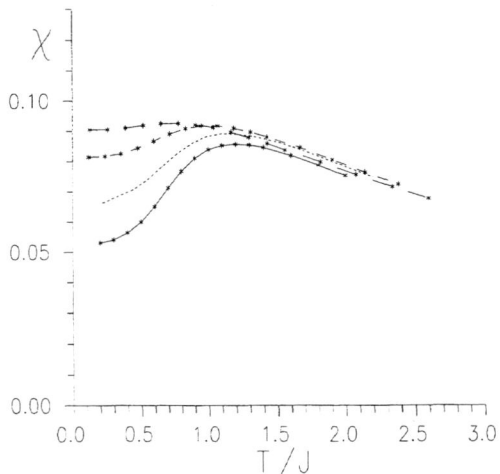

FIGURE 12. Temperature dependence of the uniform static susceptability $\chi(T)$ for different values of the frustration parameter: $p = 0$ (the solid curve), $p = 0.09$ (the dotted curve), $p = 0.16$ (the dashed curve) and $p = 0.23$ (the dashed-dotted curve). Ref. [13].

state, and the spin-excitation spectrum is gapless at point $\mathbf{q}_I = (\pm\pi, \pm\pi)$, at which Bose condensation of the spin waves occurs. At $p_1 = 0.1$ a gap opens up at point $\mathbf{q}_I$, the system loses its long-range order, and the ground state is the spin-liquid state. With an increase of p the gap at points $\mathbf{q}_{II} = (\pm\pi, 0), (0, \pm\pi)$ decreases, and vanishes at $p_2 = 0.62$. For $p_2 > 0.62$ Bose condensation occurs at point $\mathbf{q}_{II}$, and the system is in a state with long-range order, analogous to the stripe phase but without loss of the square symmetry. At points p_1 and p_2 second-order transitions occur in the system between the phases described above. We draw attention to the fact that the parameter value $p = 0.1$ for the first transition between a phase with long-range order and the spin-liquid phase differs greatly from the results obtained by the other approaches (0.274 in Refs. [37,88], 0.2 in Refs. [48,21], and 0.38 in Refs. [202,146,26]). A close value $p = 0.13$ was obtained in Ref. [59].

The temperature dependence obtained by this method for the static uniform susceptibility $\chi(p, T)$ for different frustrations p qualitatively agrees with $\chi(x, T)$ for the high-temperature superconductors (x is the parameter specifying the doping of the CuO_2 plane by holes), if we assume that the frustration is the analog of the doping. At the same time, the temperature behaviour found for $\chi(T)$ for the spin-liquid state ($p > 0.1$) differs fundamentally from the results of the theory based on the mean-field approximation for Schwinger bosons [139], in which, in the spin-liquid state, $\chi(T) \to 0$ as $T \to 0$. The reason for the discrepancy could be the fact that, in contrast to our approach, in this theory, the spin-excitation spectrum is found to be symmetric with respect to the boundaries of the magnetic Brillouin zone both in the Néel phase and in the stripe phase. Point $\mathbf{q}_0 = (0, 0)$, is identical to the corresponding Bose-condensation points $\mathbf{q}_I$ and $\mathbf{q}_{II}$ in these phases. Therefore, in the mean-field approximation for Schwinger bosons, when the long-range order is lost as $T \to 0$ a gap simultaneously opens up at points $\mathbf{q}_I$ and $\mathbf{q}_0$, and, as a consequence, $\chi(T) \propto (1/T)\exp(-\Delta/T) \to 0$ as $T \to 0$ $[\Delta = \omega(\mathbf{q}_0)]$.

Let us mention another interesting question in the problem discussed — calculation of the damping of the described magnetic excitations. This question is open even in traditional techniques [196]. In the discussed approach the problem of excitations damping was considered in [19].

B Quantum phase transitions in 3-D frustrated model

In this section we show that the spherically-symmetric Green's function approach can also describe the 3-D case when the system has a finite Neel temperature. We discuss the phase transitions for the $J_1 - J_2$ model on the cubic lattice. The usual Hamiltonian for $S = 1/2$ frustrated AFM Heisenberg model has the form:

$$H = \frac{J_1}{2}\sum_{i,g}\mathbf{S}_i\mathbf{S}_{i+g} + \frac{J_2}{2}\sum_{i,d}\mathbf{S}_i\mathbf{S}_{i+d}, \qquad (102)$$

where $J_1 = (1-p)J$, $J_2 = pJ$, $(0 \leq p \leq 1)$ are the AF exchange interactions between NN (vectors $\mathbf{g} = \pm \mathbf{g}_x, \pm \mathbf{g}_y, \pm \mathbf{g}_z$) and the next NN (vectors $\mathbf{d} = \pm \mathbf{g}_x \pm \mathbf{g}_y, \pm \mathbf{g}_y \pm \mathbf{g}_z, \pm \mathbf{g}_z \pm \mathbf{g}_x$).

To examine the spin correlation functions $C_\mathbf{r} = \langle \mathbf{S}_i \mathbf{S}_{i+r}\rangle$, $(\mathbf{r} = n_x\mathbf{g}_x + n_y\mathbf{g}_y + n_z\mathbf{g}_z)$ we introduce as before the two-time retarded Green's functions for spin site operators and their Fourier transforms:

$$\begin{aligned}G^\sigma_{i,j}(\omega) &= \langle\langle S^\sigma_i | S^{\bar\sigma}_j\rangle\rangle = -i\int_0^\infty dt e^{i\omega t}\langle [S^\sigma_i(t), S^{\bar\sigma}_j]\rangle, \\ G^\sigma(\mathbf{q},\omega) &= \sum_\mathbf{r} e^{i\mathbf{qr}}G^\sigma_{i,i+r}(\omega),\end{aligned} \qquad (103)$$

where $\langle ...\rangle$ means the thermal average, σ - spin index ($\sigma = \pm 1, \bar\sigma = -\sigma$).

Below we take the notations close to those that were used in a previous Subsection when we considered the 2D case. We don't give the detail equation which is close to that case.

For four site Green's functions in the equations of motion we use the same decoupling procedure as introduced in [177]. Firstly, we decouple only the spin operators on different sites in order to retain the local correlations. Then, within the framework of the spherically symmetric concept, we consider $C_\mathbf{r}$ independent on $\mathbf{r}$ direction, take $\langle \mathbf{S}_i\rangle = 0$ and as before use the identity of the decoupled correlation functions:

$$C_\mathbf{r} = 3\langle S^z_i S^z_{i+r}\rangle = \frac{3}{2}\langle S^\sigma_i S^{\bar\sigma}_{i+r}\rangle = \langle \mathbf{S}_i\mathbf{S}_{i+r}\rangle. \qquad (104)$$

Finally we take into account the ambiguity in the decoupling procedure by introducing the parameter α which may be regarded as the vertex correction [177]. For example:

$$\langle S_{i+r1}^z S_i^z S_{i+r2}^\sigma | S_j^{\bar{\sigma}} \rangle = \alpha \langle S_i^z S_{i+r1}^z \rangle \langle S_{i+r2}^\sigma | S_j^{\bar{\sigma}} \rangle =$$
$$= \tfrac{\alpha}{3} C_r \langle S_{i+r2}^\sigma | S_j^{\bar{\sigma}} \rangle, \quad r1 \neq r2. \tag{105}$$

As a result of the decoupling and Fourier transformation one can obtain an expression for Green's function and spin excitation spectrum:

$$G^\sigma(\mathbf{q}, \omega) = -2 \frac{J_g z_g K_g (1 - \gamma_g) + J_d z_d K_d (1 - \gamma_d)}{\omega^2 - \omega^2(\mathbf{q})}, \tag{106}$$

$$\begin{aligned}\omega^2(\mathbf{q}) = &\; \alpha J_g^2 z_g (1 - \gamma_g) \left(z_g K_g + \tfrac{1}{2\alpha} - K_g (1 + z_g \gamma_g) \right) + \\ &+ \alpha J_d^2 z_d (1 - \gamma_d) \left(z_d K_d + \tfrac{1}{2\alpha} - K_d (1 + z_d \gamma_d) \right) + \\ &+ \alpha J_g J_d z_g z_d \left((1 - \gamma_g)(R_{gd} - \gamma_d K_g) + (1 - \gamma_d)(R_{gd} - \gamma_g K_d) \right);\end{aligned} \tag{107}$$

$$\gamma_g = \frac{1}{z_g} \sum_g e^{i\mathbf{q}g}, \quad \gamma_d = \frac{1}{z_d} \sum_d e^{i\mathbf{q}d},$$

$$R_g = \frac{1}{z_g} \sum_{g1 \neq -g} K_{g+g1}, \quad R_d = \frac{1}{z_d} \sum_{d1 \neq -d} K_{d+d1}, \quad R_{gd} = \frac{1}{z_g z_d} \sum_{g,d} K_{g+d},$$

where $z_g = 6$ – the number of NN, $z_d = 12$ – number of next NN. $K_r = \langle S_i^\sigma S_{i+r}^{\bar{\sigma}} \rangle = 2C_r/3$; $J_g = J_1$, $J_d = J_2$

The same expression for $\omega^2(\mathbf{q})$ may be obtained if we consider Green's functions $\langle S_i^x | S_j^x \rangle$, $\langle S_i^y | S_j^y \rangle$ and $\langle S_i^z | S_j^z \rangle$ [27].

The explicit expression for $G^\sigma(\mathbf{q}, \omega)$ and the sum rule ($\mathbf{S}_i^2 = S(S+1) = \tfrac{3}{4}$) allow to obtain six self-consistent equations for α and for the correlation functions K_g, K_d, R_g, R_d and R_{gd} using the following relationships:

$$\begin{aligned}\langle S_j^{\bar{\sigma}} S_i^\sigma \rangle &= \tfrac{1}{N} \sum_\mathbf{q} e^{i\mathbf{q}(j-i)} \int d\omega I^\sigma(\mathbf{q}, \omega), \\ I^\sigma(\mathbf{q}, \omega)(e^{\omega/T} - 1) &= -\tfrac{1}{\pi} \mathrm{Im} G^\sigma(\mathbf{q}, \omega + i\varepsilon),\end{aligned} \tag{108}$$

$I^\sigma(\mathbf{q}, \omega)$ is a spectral density.

After integrating over ω in (108) we convert the sum over the Brillouine zone into integral and solve the system of equations numerically. It turns out that at a high temperature the system of self-consistent equations has the solution with the spectrum which is gapless in the whole Brillouine zone except $\mathbf{q} = (0, 0, 0)$. Let us mention that the absence of a gap at this point doesn't lead to any LRO because the numerator of $G^\sigma(\mathbf{q}, \omega)$ (106) is zero at $\mathbf{q} = (0, 0, 0)$. As the temperature decreases the spin wave spectrum behaves in two possible ways. For the values of frustration parameter $(0 \leq p \leq 0.146)$ and $(0.262 \leq p \leq 0.755)$ (see phase diagram on Fig.13) the spectrum becomes gapless at finite temperature $T_N(p)$ at AF vectors $\mathbf{q}_R = \pi/g(1,1,1)$ or $\mathbf{q}_M = \pi/g(1,1,0), \pi/g(0,1,1), \pi/g(1,0,1)$ respectively. This manifests the appearance of LRO with the AF structure determined by the AF

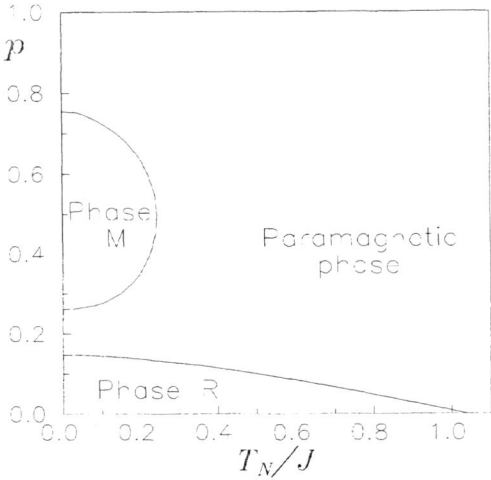

FIGURE 13. Phase diagram of frustrated $S = \frac{1}{2}$ antiferromagnet. LRO exists at temperatures $0 \leq T \leq T_N(p)$ for $0 \leq p \leq 0146$ (AF vector is $\mathbf{q}_R = (\pi/g)(1,1,1)$) and for $0.26 \leq p \leq 0.755$ (AF vectors $\mathbf{q}_M = (\pi/g)(1,1,0)$, $(\pi/g)(1,0,1)$ and $(\pi/g)(0,1,1)$). For other values of the frustration parameter antiferromagnet is in SL state even at $T = 0$. Ref. [14].

vector (phase R and phase M). Another situation takes place for $(0.146 \leq p \leq 0.262)$ and $(0.755 \leq p \leq 1)$ when the spectrum remains gapless at all temperatures (including $T = 0$). For this case the correlation functions exponentially decrease with distance and the system is in SL state even at zero temperature.

Below $T_N(p)$ there is no solution of self-consistent equations without Bose-like condensation of spin-wave excitations. To take into account the condensate we introduce the transformation which doesn't violate the relationship (108) between $I^\sigma(\mathbf{q},\omega)$ and $G^\sigma(\mathbf{q},\omega)$:

$$I^\sigma(\mathbf{q},\omega) \to I^\sigma(\mathbf{q},\omega) + N \sum_{\mathbf{q}_A} \Delta_{\mathbf{q},\mathbf{c}_A} Q_A \delta(\omega), \qquad (109)$$

where the sum only runs over such vectors $\mathbf{q}_A$ of Brillouine zone at which $\omega(\mathbf{q}_A) = 0$. As a result at $T < T_N$ the self-consistent system of equations has the solution with $\omega(\mathbf{q}_A) = 0$ and with a finite value of the condensate $Q_A = Q(\mathbf{q}_A)$ only at the AF vectors. Below T_N the system remains rotationally invariant.

The procedure of solving six coupled self-consistent equations is a non-trivial problem. We start from the case $T = 0$. For $p = 0$ the system may be solved analytically and it has a unique solution with $\omega(\mathbf{q}_R) = 0$ and finite value of the condensate Q_R. Formally Q_R is an additional variable which corresponds to an additional equation for α and the five correlators mentioned above. As a result we have seven equations for seven variables. Beginning from $p = 0$ we continuously increase the frustration parameter from p to $p + \delta p$. To solve the system of equations numerically at $p + \delta p$ we take it's solution at p as a starting point in the iteration procedure of the direction set (Powell's) method [157]. At $p = 0.146$ the solution leads to the zero value of condensate Q_R. For $p > 0.146$ the solution with $Q_R > 0$ disappears and the system turns to the system of six equations (with $Q_R = 0$ and $\omega(\mathbf{q}_R) = 0$). Further as p increases we examine at each step the behaviour of the spectrum $\omega(\mathbf{q})$. When $\omega(\mathbf{q})$ turns to zero at $\mathbf{q} = \mathbf{q}_M$ we introduce the condensate Q_M and again solve the system of seven equations. For $T > 0$ and fixed value of p we use the same numerical method [157] starting from the previously obtained solution at $T = 0$.

The finite value of Q_A leads to the appearance of LRO and the finite effective magnetization M:

$$\lim_{r \to \infty} \langle \mathbf{S}_i \mathbf{S}_{i+r} \rangle = (-1)^{n_x+n_y+n_z} \tfrac{3}{2} Q_R, \quad M = \sqrt{\tfrac{3}{2} Q_R}; \quad \text{phase } \mathbf{R}$$
$$\lim_{r \to \infty} \langle \mathbf{S}_i \mathbf{S}_{i+r} \rangle = ((-1)^{n_x+n_y} + (-1)^{n_y+n_z} + (-1)^{n_z+n_x}) \tfrac{3}{2} Q_M, \quad (110)$$
$$M = \sqrt{\tfrac{9}{2} Q_M}; \quad \text{phase } \mathbf{M}$$

The spin structure of the phase R reminds a classical Neel ordering. Spin correlation functions (110) describe two ferromagnetic face-centered cubic (fcc) sublattices with antiferromanetic spin correlations between them. As to the obtained spin configuration (110) in the phase **M** it differs from the classical stripe-ordered state due to the simultaneous turning to zero of the spin-wave spectrum at points $\mathbf{q}_M = (\pi/g)(1,1,0)$, $(\pi/g)(1,0,1)$ and $(\pi/g)(0,1,1)$. Such a structure may be treated as a coherent superposition of the three different classical stripe-ordered states. One of them is constructed by translating 2-D Neel type spin ordering in (**x-y**) plain along **z** axis. It's spin correlation functions correspond to the first term in the right hand side of (110) $a_\mathbf{r}^z = (-1)^{n_x+n_y}$. Other two states may be obtained by Neel type ordering in (**y-z**) and (**z-x**) plains. The resulting superposition leads to the correlation functions (110) which represent four ferromagnetic body-centered cubic (bcc) sublattices with translational vectors $g(1,1,-1)$, $g(1,-1,1)$ and $g(-1,1,1)$. The effective magnetization of each sublattice is $M = \sqrt{\tfrac{9}{2} Q_M}$. The sublattice shortest range spin correlation function $C_\mathbf{f}$, $\mathbf{f} = g(1,1,1)$ is presented in Fig.(14). Between different sublattices the spin correlation functions are antiferromagnetical and at large distances are equal to $-\tfrac{3}{2} Q_M$.

The frustration dependencies of magnetization at $T = 0$ and ground state energy per site are presented on Figs. 15a and 15a respectively. A continuous behaviour of the order parameter M indicates the second order phase transitions between ordered and spin liquid states and leads to a smooth dependence of the energy and spin correlation functions upon frustration (Fig.14).

Let us discuss the limit $p = 1$ when the simple cubic lattice is decoupled into two independent face centered cubic (fcc) sublattices (with NN antiferromagnetic interaction J_2 in each of them). Even at zero temperature there isn't any condensation of spin-wave excitations, leading to LRO. The spin-wave spectrum isn't equal to zero in the whole Brillouine zone except at the two equivalent for the fcc lattice

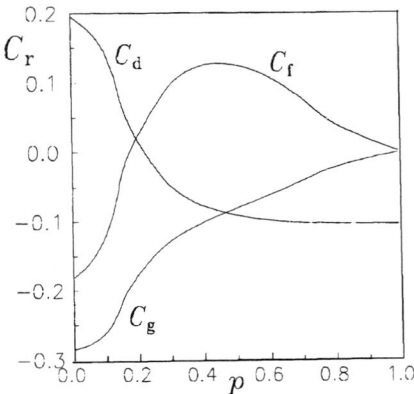

FIGURE 14. The dependence of three short-range correlation functions $C_{\mathbf{r}} = \langle S_i S_{i+r} \rangle$, $\mathbf{r} = \mathbf{g}, \mathbf{d}, \mathbf{f}$ on the frustration parameter p at $T = 0$. $\mathbf{g} = g(1,0,0)$, $\mathbf{d} = g(1,1,0)$ and $\mathbf{f} = g(1,1,1)$, Ref. [14]

points $\mathbf{q} = (0,0,0)$ and $(\pi/g)(1,1,1)$ and it's characteristic feature is the degeneracy line $\mathbf{q} = (\pi/g)(1,y,0)$ (Fig.16 and eq.(107)). The similar degeneration line but with $\omega(\mathbf{q}) = 0$ is in $S = \frac{1}{2}$ four-sublattice AF treated by the traditional Green's function approach [118] or by the linear spin-wave theory [117]. Let us mention that in contrast with our quantum limit $S = \frac{1}{2}$ it results that recent investigations of classical fcc AF [86,45] predict a stabilization of LRO with the ordering vector $\mathbf{q}_X = (\pi/g)(1,0,0)$.

So we have obtained the phase diagram indicating two regions of temperature and frustration parameter $p = J_2/(J_1 + J_2)$ where two different ordered states are possible. The absence of the long range order (LRO) in 3-D lattice even at zero temperature at some values of frustration is emphasized. At zero temperature with increasing frustration the system endures the second order phase transitions from the Neel type LRO state to the stripe type LRO state through the spin liquid state.

In conclusion we want to emphasize once more the difference between our approach and the widely used Schwinger boson mean field (SBMF) theory [9]. As we know the SBMF theory was used by Mila et al [139] to study the 2-D frustrated model and by Sarker et al [168] to treat the 3-D model for simple cubic lattice without frustration. Both the 2-D case $T_N = 0$ and the SBMF theory preserve rotational invariance. But as shown in [139] one needs to introduce different sets of order parameters for the Neel type phase (p close to 0) and the stripe type phase (p close to 1). As a result the 2-D system endures the first order phase transition at $p = 0.36$. We think that such a transition is an artifact of the SBMF approach. In contrast with the SBMF theory our method in 2-D case manifests a second order phase transition [13]. It is obvious that the same difference between two approaches takes place for the 3-D system. Moreover as shown by Sarker et al [168] in a 3-D system the condensation of Schwinger bosons leads to broken rotational symmetry.

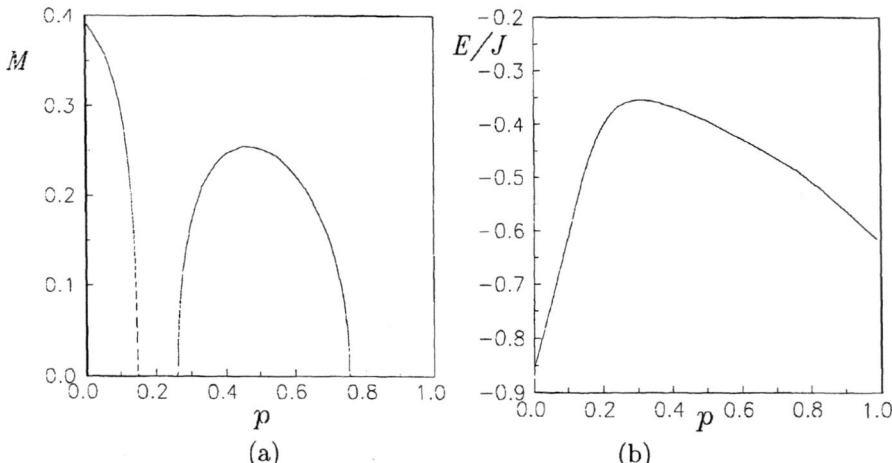

FIGURE 15. a) Magnetization M of **R** and **M** phases at $T = 0$ versus the frustration parameter p. b) Ground state energy E/J per site as a function of the frustration parameter p. Ref. [14].

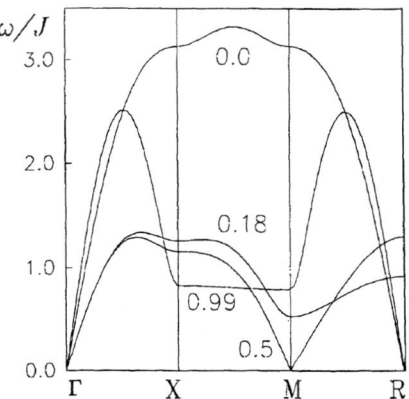

FIGURE 16. Spectrum $\omega(\mathbf{q})$ along the line $\Gamma - X - M - R$ in the Brillouin zone for different values of the frustration parameter p: $0.0, 0.18, 0.5$ and 0.99. Ref. [14].

The SCGF method preserves the rotational symmetry below T_N. Let us also point out that two branches of the Schwinger boson excitation spectrum are symmetric with respect to boundaries of the magnetic Brillouin zone [168]; while in our approach vector $\mathbf{q} = (0,0,0)$ and AF vectors $\mathbf{q}_A$ are not equivalent for the three branches of the spin excitation spectrum (107).

IV COMPLEX LOCAL SPIN POLARON IN A SPHERICALLY SYMMETRIC FRUSTRATED SPIN-BACKGROUND

A Green's functions projection method

In order to determine the spin polaron spectrum $\varepsilon(\mathbf{k})$ the standard Mori-Zwanzig projection technique [141] is often applied (see, for example [29,15,198,197]). Below we give a short outline of this method relatively to our problem. We discuss a variant of the projection technique which is based on the equations of motion for the retarded Green's functions.

Let the Hamiltonian of the square lattice model be determined in each $\mathbf{R}$-cell by Cu spin degrees of freedom and by hole energy levels related to Hubbard operators $Z_\mathbf{R}^{\sigma_1\sigma_2}$ and Fermi operators $c_{\mathbf{R},\sigma}^+$, respectively. Let the Hamiltonian describe the hole hoppings, in-cell interaction between holes and spins and antiferromagnetic interaction between spins. For each cell we introduce a restricted set of one-hole operators (with spin σ) $A_{\mathbf{R},i}^+$, i is the number of operator, $i \leq n$. The projection method approach consists in treating a problem in the subspace of the adopted set of operators. In general the choice of the set is dictated by a physical sense of the problem.

Let each $A_{\mathbf{R},i}^+$ have the following typical symbolical form

$$A_{\mathbf{R},i}^+ = \sum_{\mathbf{R}_1} \alpha(\mathbf{R},\mathbf{R}_1) c_\mathbf{R}^+ Z_{\mathbf{R}_1}, \quad A_{\mathbf{R},j}^+ = \sum_{\mathbf{R}_1,\mathbf{R}_2} \beta(\mathbf{R},\mathbf{R}_1,\mathbf{R}_2) c_\mathbf{R}^+ Z_{\mathbf{R}_1} Z_{\mathbf{R}_2} \quad (111)$$

In order to calculate the spectral function of a bare hole we always take one of the site operators, let it be $A_{\mathbf{R},1}^+$, as a bare hole operator.

Let us introduce the retarded two-time Green's functions $G_{ij}(t,\mathbf{k})$ defined in terms of the Fourier transforms $A_{\mathbf{k},i}$ of the operators $A_{\mathbf{R},i}$:

$$G_{ij}(t,\mathbf{k}) \equiv \left\langle A_{\mathbf{k},i}(t) | A_{\mathbf{k},j}^+(0) \right\rangle = -i\Theta(t) \left\langle \left[A_{\mathbf{k},i}(t), A_{\mathbf{k},j}^+(0) \right] \right\rangle, \quad (112)$$

$$A_{\mathbf{k},j} = \frac{1}{\sqrt{N}} \sum_\mathbf{R} e^{i\mathbf{k}\mathbf{R}} A_{\mathbf{R},j} \quad i,j = 1 \div 10 .$$

The equations of motion for the Fourier transforms of the Green's functions have the form

$$\omega \left\langle A_{\mathbf{k},i} | A_{\mathbf{k},j}^+ \right\rangle_\omega = K_{i,j} + \left\langle B_{\mathbf{k},i} | A_{\mathbf{k},j}^+ \right\rangle_\omega$$
$$K_{i,j} = \left\langle \left[A_{\mathbf{k},i}, A_{\mathbf{k},j}^+ \right] \right\rangle, \quad B_{\mathbf{k},i} = \left[A_{\mathbf{k},i}, H \right] \quad (113)$$

In the projection technique we approximate the new operators $B_{\mathbf{k},i}$ by their projections on the space $\{A_{\mathbf{k},i}\}$ of the basis operators:

$$B_{\mathbf{k},i} \simeq \sum_l \Omega_{i,l}(\mathbf{k}) A_{\mathbf{k},l}, \quad \Omega(\mathbf{k}) = D(\mathbf{k}) K^{-1}, \quad D_{ij}(\mathbf{k}) = \left\langle \left[B_{\mathbf{k},i}, A^+_{\mathbf{k},l} \right] \right\rangle. \quad (114)$$

After we substitute the approximate expressions for the operators $B_{\mathbf{k},i}$ in (114) into the equations of motion (113), the equation system (113) for Green's functions $\left\langle A_{\mathbf{k},i} | A^+_{\mathbf{k},j} \right\rangle_\omega$ becomes closed and can be presented in the matrix form

$$\left(\omega E - D K^{-1} \right) G = K \quad (115)$$

where E is the unit matrix.

The quasiparticle spectrum $\varepsilon(\mathbf{k})$ is determined by the poles of Green's function G and can be derived from the condition

$$det | K \varepsilon(\mathbf{k}) - D | = 0$$

The resulting Green's functions have form

$$G_{i,j}(\omega, \mathbf{k}) = \sum_{l=1}^n \frac{z^{(l)}_{(i,j)}(\mathbf{k})}{\omega - \varepsilon_l(\mathbf{k})}. \quad (116)$$

In particular the value of $z_h(\mathbf{k}) = z^{(1)}_{(1,1)}(\mathbf{k})$ corresponds to the number of bare oxygen holes with the fixed spin σ and the momentum $\mathbf{k}$ in the state $|\mathbf{k}, \sigma\rangle$ of the lowest quasiparticle band $\varepsilon_1(\mathbf{k})$. Let us remind that the spectral weight (residue) $z_{(i,j)}(\mathbf{k})$ satisfies the sum rule $\sum_s z^{(s)}_{(1,1)}(\mathbf{k}) = 1$. This means that in this model in the case of $n > 1$ the Luttinger theorem is not fulfilled and the maximum number of holes per cell is equal to two, despite the presence of n bands.

The calculation of elements of the arrays D and K is usually lengthy and it involves the calculation of complex commutators for the operators $A_{\mathbf{k},i}$ and $B_{\mathbf{k},i}$. Usually some of these commutators can't be expressed explicitly over Green's functions $G_{i,j}$ and some approximations are needed in order to calculate them.

These matrix elements can be considerably simplified for a one-hole problem. In such case they are expressed in terms of two- and multi-site correlation functions of Hubbard operators (in the cases considered below — two-, three-, four- and five-site correlation functions). Taking into account the spherical symmetry of the spin subsystem, the three-site correlators can be reduced to two-cite correlators, and five-cite correlators can be reduced to four-site correlators. Below we represent some typical relations for local correlation functions which we often use to calculate the matrix elements which appear in the projection method:

$$\sum_\sigma Z^{\sigma:\sigma}_\mathbf{R} = 1, \quad Z^{\sigma:\sigma}_\mathbf{R} = \frac{1}{2} + \sigma S^z_\mathbf{R},$$

$$\sum_{\sigma,\gamma=\pm\sigma} Z^{\sigma:\gamma}_{\mathbf{R}_1} Z^{\gamma:\sigma}_{\mathbf{R}_2} = \frac{1}{2} + 2 \mathbf{S}_{\mathbf{R}_1} \mathbf{S}_{\mathbf{R}_2},$$

$$Z_{\mathbf{R}_1}^{\sigma:\sigma} Z_{\mathbf{R}_2}^{\sigma:\sigma} = \frac{1}{4} + \frac{\sigma}{2}(S_{\mathbf{R}_1}^z + S_{\mathbf{R}_2}^z) + S_{\mathbf{R}_1}^z S_{\mathbf{R}_2}^z ,$$

$$\sum_{\gamma=\pm\sigma} Z_{\mathbf{R}_1}^{\sigma:\gamma} Z_{\mathbf{R}_2}^{\gamma:\sigma} = \frac{1}{4} + \frac{\sigma}{2}(S_{\mathbf{R}_1}^z + S_{\mathbf{R}_2}^z) + \mathbf{S}_{\mathbf{R}_1}\mathbf{S}_{\mathbf{R}_2} - i\sigma(S_{\mathbf{R}_1}^x S_{\mathbf{R}_2}^y - S_{\mathbf{R}_1}^y S_{\mathbf{R}_2}^x) ,$$

$$\sum_{\gamma=\pm\sigma} \left\langle Z_{\mathbf{R}_1}^{\sigma:\gamma} Z_{\mathbf{R}_2}^{\gamma:\sigma} \right\rangle = \frac{1}{4} + C_{R_{12}} ,$$

$$\sum_{\lambda=\pm\sigma} \left\langle Z_{\mathbf{R}_1}^{\sigma:\lambda} Z_{\mathbf{R}_2}^{\lambda:\sigma} Z_{\mathbf{R}_3}^{\gamma:\gamma} \right\rangle = \frac{1}{8} + \frac{1}{2} C_{R_{12}} + \frac{1}{6}\sigma\gamma(C_{R_{23}} + C_{R_{31}}) ,$$

$$\sum_{\gamma=\pm\sigma} \left\langle Z_{\mathbf{R}_1}^{\bar\sigma:\gamma} Z_{\mathbf{R}_2}^{\gamma:\sigma} Z_{\mathbf{R}_3}^{\sigma:\bar\sigma} \right\rangle = \frac{1}{3}(C_{R_{23}} + C_{R_{31}}) ,$$

$$\sum_{\lambda,\gamma=\pm\sigma} \left\langle Z_{\mathbf{R}_1}^{\sigma:\gamma} Z_{\mathbf{R}_2}^{\gamma:\lambda} Z_{\mathbf{R}_3}^{\lambda:\sigma} \right\rangle = \frac{1}{8} + \frac{1}{2}(C_{R_{12}} + C_{R_{23}} + C_{R_{31}}) .$$

In the above equations all the sites are supposed to be different and we take the notations $C_{R_{12}} = \langle \mathbf{S}_{\mathbf{R}_1} \mathbf{S}_{\mathbf{R}_2} \rangle$ and $R_{12} = |\mathbf{R}_1 - \mathbf{R}_2|$.

As regards four-site correlators, they are usually expressed over the following form

$$V_{\mathbf{R}_1\mathbf{R}_2\mathbf{R}_3\mathbf{R}_4} = \langle (\mathbf{S}_{\mathbf{R}_1}\mathbf{S}_{\mathbf{R}_2})(\mathbf{S}_{\mathbf{R}_3}\mathbf{S}_{\mathbf{R}_4}) \rangle$$

In order to calculate a four-site correlator an approximation of [193] in often used:

$$V_{\mathbf{R}_1\mathbf{R}_2\mathbf{R}_3\mathbf{R}_4} = C_{R_{12}} C_{R_{34}} + \frac{1}{3} C_{R_{13}} C_{R_{24}} + \frac{1}{3} C_{R_{14}} C_{R_{23}}.$$

The two-site correlation functions can be calculated in the spherically symmetrical approach, described in the previous Section.

B Local spin polaron motion in the effective three-band model on a frustrated spin background

As mentioned in the Introduction, an important issue in developing a microscopic theory of high-T_c superconductivity is the spectrum $\varepsilon(\mathbf{k})$ of hole excitations in the CuO_2 plane. Recent photoemission measurements with angular resolution indicate that there is a nearly flat band near the Fermi level in optimally doped $Bi2212$, $Bi2201$, $Y123$ and $Y124$ cuprates. [192,75,1,44,109] The flat band is located near points $X = \{(\pm\pi, 0),(0, \pm\pi)\}$ and has the shape of an elongated saddle aligned with the $X - \Gamma$ line, where $\Gamma = (0,0)$. The flat band in the spectrum leads to a Van Hove singularity in the density of states near the Fermi level. The proximity of the Van Hove singularity to the Fermi level has been used by various theories to account for the high temperature of superconducting transition. [1,195,138,145] The existence of the optimal doping level is, naturally, related to the coincidence of the Fermi level with the singularity in the density of states. Note that this common property of all the cuprate superconductors, namely, the existence of a large flat region in the electronic spectrum, is hard to interpret in terms of simple band models.

Important photoemission measurements have been performed in $Sr_2CuO_2Cl_2$ [201], whose CuO_2 planes are in an antiferromagnetic dielectric state. These results yield the hole spectrum in a quantum antiferromagnet and present a good test for numerous theoretical approaches to this problem. The measurements indicated [201] that the band bottom is located near point $N = (\pi/2, \pi/2)$ and its width is about $0.3 eV$. Moreover, the minimum at the bottom is nearly isotropic, i.e., the effective masses in the directions $N - \Gamma$ and $N - X$ are approximately equal.

The features of the quasiparticle spectrum discussed above are often ascribed to strong correlations in the hole motion superposed on a two-dimensional antiferromagnetic background [41,33]. The problem usually is treated in terms of generalized $t - J$ and Hubbard models, or the three-band model, [60,199] which is more realistic in the case of the CuO_2 plane because it takes into account both Cu- and O-orbitals. With due account of the antiferromagnetic background and strong correlations between states at lattice sites, the hole quasiparticle spectrum is dominated, even in the simplest approximation, by the component $\varepsilon(\mathbf{k}) \sim (\cos k_x + \cos k_y)^2$ [34,193,23,172], or, more exactly, the spectrum has a strongly anisotropic minimum at point N with lines of equal energy elongated along the boundary of the $X-N-X$ magnetic Brillouin zone. The quasiparticles are hole-spin polarons. The latest theoretical studies indicate that it is important to retain a number of real interactions in these models in order to interpret some "fine" features of experimental spectra detected in experiments. These interactions include, for example, the t'-hoppings, which take into account the possibility of hopping to the next-nearest-neighbour sites of a square lattice, in the generalized $t - J$ and Hubbard models. In the case of the three-band model, it is important to include in the Hamiltonian direct oxygen-to-oxygen hopping of holes. Note that the t'-interaction effectively takes into account direct oxygen-to-oxygen hopping in the CuO_2 plane in the effective three-band model . It turned out that with these terms in both models, the strong anisotropy of the spectrum- near the bottom is eliminated. [15,144,181,52,82,203]

1 Complex local spin polaron structure

Here we shall consider the effect of some real interactions on the spectrum $\varepsilon(\mathbf{k})$ of the local hole-spin polaron in the CuO_2 plane using the three-band Hamiltonian. First of all, the Hamiltonian takes into consideration hopping of an oxygen hole via a copper site with an amplitude τ, including spinflip processes. The Hamiltonian also takes into account direct $O - O$ hopping of holes with an amplitude h. In addition, we analyse the effect of the hopping trajectory dependence of the amplitude τ on the band spectrum. This dependence does not occur when τ is calculated to the lowest order in the $p-d$ hybridization parameter t_{pd} as $\tau = t_{pd}^2/\varepsilon_{pd}$ where $\varepsilon_{pd} = \varepsilon_p - \varepsilon_d$ is the energy difference between the hole levels Cu($d_{x^2-y^2}$) and O(p_x, p_y). But if terms of higher orders in t_{pd}/ε_{pd} are taken into consideration, the amplitude of the hole hopping τ_{x+y} between the oxygen sites $\mathbf{R}-\mathbf{a}_x$ and $\mathbf{R}+\mathbf{a}_y$ will be different from the amplitude τ_{2x} between the sites $\mathbf{R}-\mathbf{a}_x$ and $\mathbf{R}+\mathbf{a}_x$ (from now on $\mathbf{R}$ is the position of the Cu site, $\pm\mathbf{a}_x$ and $\pm\mathbf{a}_y$ are vectors of oxygen sites closest to $\mathbf{R}$). A similar dependence of the amplitude on the hopping trajectory results from including $p-d$ hybridization between remote Cu-O orbitals. [122]

We shall also discuss the effect of temperature and frustration in the spin subsystem on the spectrum $\varepsilon(\mathbf{k})$. It is known that the magnetic correlation length decreases with the doping level. The antiferroimgnelic correlation length ξ_{AFM} also falls with the frustration parameter $\alpha = J_2/J_1$, where J_1 and J_2 are the constants of antiferromagnetic exchange between nearest and next nearest neighbours in the square copper sublattice. The similarity between the doping and frustration was demonstrated by Inui et al. [97] and confirmed by direct calculations of the spin-spin structural factor $S(\mathbf{q})$ on a 4×4 cluster for the doped $t-J$ model and frustrated Heisenberg model. [140] Naturally, the two models are not completely equivalent. For example, the doped $t-J$ model and frustrated J_1-J_2 model produce different results regarding the dynamic spin-spin structural factor and Raman scattering spectrum. [11] Nonetheless, one may assume that frustration mimics the effect of the finite hole density on static spin correlation functions, from which the spin polaron spectrum is derived. Thus, by calculating the spectrum $\varepsilon(\mathbf{k})$ as a function of frustration, we effectively determine the effect of doping on $\varepsilon(\mathbf{k})$.

We shall treat spin excitations in the spherically symmetrical approximation. [13,176,27] This approach allows to investigate a two-dimensional quantum antiferromagnet without invoking the spontaneous symmetry breaking at zero temperature and to describe its paramagnetic state with strong antiferromagnetic correlations, whose ranges are determined by the antiferromagnetic correlation length ξ_{AFM}. This approach seems most adequate for doped HTSC, whose CuO_2 planes have no long-range antiferromagnetic order.

We consider a spin polaron of small radius with spin $S = 1/2$ as an elementary hole excitation in our model. The hole excitation operator perturbs the spin subsystem at two copper sites nearest to the hole. The full basis for such a polaron contains ten local operators from which a Bloch combination is constructed. The spectrum is calculated using the projection technique based on retarded two-time Green's functions.

We shall demonstrate that by taking into consideration oxygen-oxygen hopping and frustration, we can account for the basic features of the hole excitation spectrum mentioned above [15]. Below we shall treat the effective three-band Hubbard model in the frames of the simplified effective Hamiltonian with $U_p = 0, U_d = \infty$. But at the same time we shall also take into account two additional (relative to (53), (52)) interactions, which are important for the explanation of photoemission spectra: direct oxygen-oxygen hoppings between the nearest-neighbour oxygen sites and frustration in the Cu-spin subsystem. e.g. antiferromagnetic exchange between second nearest-neighbours of Cu-sublattice.

Then the effective Hamiltonian has the form [23,182,148,156,204,61]

$$\hat{H} = \hat{T} + \hat{h} + \hat{J} , \qquad (117)$$

where

$$\hat{T} = \tau \sum_{\mathbf{R},\mathbf{a}_1,\mathbf{a}_2,\sigma_1,\sigma_2} Z_{\mathbf{R}}^{\sigma_1\sigma_2} c_{\mathbf{R}+\mathbf{a}_2,\sigma_2}^+ c_{\mathbf{R}+\mathbf{a}_1,\sigma_1} ,$$
$$\hat{h} = -h \sum_{\mathbf{R},\mathbf{a},\mathbf{b},\sigma} c_{\mathbf{R}+\mathbf{a},\sigma}^+ c_{\mathbf{R}+\mathbf{a}+\mathbf{b},\sigma}$$
$$\hat{J} = \hat{J}_1 + \hat{J}_2 \quad \hat{J}_1 = \tfrac{J_1}{2}\sum_{\mathbf{R},\mathbf{g}} \hat{\mathbf{S}}_{\mathbf{R}}\hat{\mathbf{S}}_{\mathbf{R}+\mathbf{g}} \quad \hat{J}_2 = \tfrac{J_2}{2}\sum_{\mathbf{R},\mathbf{d}} \hat{\mathbf{S}}_{\mathbf{R}}\hat{\mathbf{S}}_{\mathbf{R}+\mathbf{d}}$$

The following notations are used in (117): **b** are vectors of the nearest neighbors in the oxygen sublattice, $\mathbf{b} = \pm \mathbf{a}_x \pm \mathbf{a}_y$; $\mathbf{g} = 2\mathbf{a}$ and $\mathbf{d} = 2\mathbf{b}$ are the vectors of the nearest and next nearest neighbours in the copper sublattice.

The hopping amplitude $\tau = \frac{t^2}{\varepsilon}$; the amplitude of direct $O - O$ hopping $h < \tau \sim 0.5\text{eV}$; $J \simeq 0.25\tau$; the typical value of the frustration parameter α is to be $\alpha < 0.35$. [137,96,8].

Before constructing the basis for a complex local polaron we shall represent the initial features of the spectrum when a spin polaron is described in the simplest way, i.e. by a single plaquette operator of Zhang-Rise singlet $A_{\mathbf{R}}^+$ (82). Let us calculate the polaron spectrum by taking into account only the hoping term $\hat{T}$. Then the straightforward calculation leads to the following expression for the spectrum

$$\varepsilon(\mathbf{k}) = \tau(1 + A\gamma_{\mathbf{k}})^{-1}[-3.5 - 8A\gamma_{\mathbf{k}} + B(4\gamma_{\mathbf{k}}^2 - 1)], \qquad (118)$$
$$A = \tfrac{1}{4} + C_g, \ B = \tfrac{1}{8} - C_g + \tfrac{1}{2}C_d, \ \gamma_{\mathbf{k}} = \tfrac{1}{2}[\cos k_x g + \cos k_y g]$$

To obtain this expression we assumed for simplicity that the two-site spin correlation functions between the second- and the third - NN Cu spins coincide, i.e., $C_d = C_{2g}$. This approximation is known to be good at least for the zero value of the frustration parameter $\alpha = J_2/J_1$ [115].

The term with A-factor in the numerator of (118) appears due to the hopping of a polaron to a nearest-neighbour plaquette, the term proportional to B, i.e., determined by hoppings to second and third nearest-neighbours. In the antiferromagnetic regime hops to the nearest-neighbours are suppressed. This is due to the fact, that in this regime C_g is close to -0.25. For the rigid Neel state $A = 0$. For the RVB state C_g is close to -0.32 at $T = 0$ and goes to zero increasing T.

Therefore in the antiferromagnetic regime the main term which determines the spectrum is $B\gamma_{\mathbf{k}}^2$. It is easy to understand the appearance of such a spectrum. Let us discuss the simple one-electron tight-binding case when a carrier moves in a square lattice hopping to a second nearest-neighbour (vector **d**) with amplitude $2t$ and to a third nearest-neighbour (vector **2g**) with amplitude t. Then the carrier will have a spectrum $4t(4\gamma_{\mathbf{k}}^2 - 1)$. We have quite the same situation for the hopping of Zhang-Rice polaron in a strongly antiferromagnetic regime. Firstly, the hopping to a nearest-neighbour is suppressed. Secondly, it can be seen from the Hamiltonian T, that hopping of a Zhang-Rice plaquette **R** in direction **d** is possible in two ways:

a hole jumps over a copper site $\mathbf{R}+\mathbf{g}_x$ or over a site $\mathbf{R}+\mathbf{g}_y$. The hopping to the site $\mathbf{R}+2\mathbf{g}_x$ (third nearest-neighbour) is possible only in one way — over a copper site $\mathbf{R}+\mathbf{g}_x$. So effectively the hopping amplitude of polaron in $\mathbf{d}$ direction is two times greater than in $\mathbf{g}_x$ direction. The spectrum determined by $\gamma_\mathbf{k}^2$ is strongly anisotropic and has a band bottom along the magnetic Brillouin zone boundary (determined by the equation $\gamma_\mathbf{k}^2 = 0$).

As temperature increases and $A \to 1/4$ the term $A\gamma$ becomes the main one and the band bottom shifts to a point $\Gamma = (0,0)$. Let us mention, that even in this paramagnetic case the correct value of the band bottom can't be obtained by averaging the spin operator Z in the Hamiltonian T. This can be explicitly seen by considering one plaquette: the mean-field approach with averaging gives the zero minimal energy and the correct value is close to $-4t$. This shows that even at high temperature one must consider the formation of a local spin polaron.

Now let us consider the technique of constructing the basis from local operators $A_{\mathbf{R},i}^+$ (i is the operator number) generating one hole in the singlet dielectric state $|G\rangle$ of the CuO_2 plane.

It is obvious that all these operators should generate excitations with spin $S = 1/2$ and spin projection $\sigma/2$. This means the state $A_{\mathbf{R},i}^+|G\rangle$ is an eigenstate of the operator $\mathbf{S}_{Tot}$ — the total spin of copper sites and a hole in the plane and its projection S_{Tot}^z:

$$\mathbf{S}_{Tot} = \sum_\mathbf{R} \left(\mathbf{S}_\mathbf{R} + \sum_{\mathbf{a}=\mathbf{a}_x,\mathbf{a}_y} \mathbf{s}_{\mathbf{R}+\mathbf{a}} \right) \tag{119}$$

Here $\mathbf{s}_{\mathbf{R}+\mathbf{a}}$ is the oxygen-hole spin operator.

The first two operators, $A_{\mathbf{R},1(2)}^+$ obviously generate holes at two oxygen sites of an elementary cell without perturbing the copper spin subsystem:

$$A_{\mathbf{R},1}^+ = c_{\mathbf{R}+\mathbf{a}_x,\sigma}^+, \qquad A_{\mathbf{R},2}^+ = c_{\mathbf{R}+\mathbf{a}_y,\sigma}^+. \tag{120}$$

The operator basis can be enhanced in a natural way by taking commutators of the operators $A_{\mathbf{R},1(2)}^+$ with the hopping Hamiltonian $\tilde{T}$. One can check that the result is four new operators $A_{\mathbf{R},3(4)}^+$ and $A_{\mathbf{R},5(6)}^+$ generating hole excitations in the state $|G>$ in the CuO_2 plane:

$$A_{\mathbf{R},3(4)}^+ = \sigma \sum_{\gamma=\pm\sigma} \gamma Z_\mathbf{R}^{\overline{\gamma\sigma}} c_{\mathbf{R}+\mathbf{a}_x(\mathbf{a}_y),\gamma}^+, \quad A_{\mathbf{R},5(6)}^+ = \sigma \sum_{\gamma=\pm\sigma} \gamma Z_{\mathbf{R}+\mathbf{g}_x(\mathbf{g}_y)}^{\overline{\gamma\sigma}} c_{\mathbf{R}+\mathbf{a}_x(\mathbf{a}_y),\gamma}^+ \tag{121}$$

Hereafter we assume that $\overline{\sigma} = -\sigma$.

The operators defined by (121) generate holes at oxygen sites concurrently with a spin excitation at a neighbouring copper site. Note that if the operators in (121) act on a state of a two-site $Cu - O$ system containing one hole at the copper site $\mathbf{R}$ with $S = 1/2$ and $S^z = \sigma/2$, the resulting two-hole state is a singlet. It is obvious that in this sense the linear combination of the operators in (121), i.e.,

$$A^+_{R,3} + A^+_{R,4} + A^+_{R-g_x,5} + A^+_{R-g_y,6}$$

is similar to the Zhang-Rice spin polaron. [205].

The next step in the enlargement of the basis is calculating commutators of the terms $\hat{T}$ and $\hat{J}$ in the Hamiltonian defined by (117) with excitation operators given by (121). From the resulting operators we only select those which are, firstly, linearly independent of those defined by (120) and (121), and secondly, generate excitations at two sites of the copper sublattice nearest to the oxygen site with the hole. As a result, we have four new operators $A_{R,i}$ where $i = 7 \div 10$:

$$A^+_{R,7(8)} = \sigma \sum_{\gamma,\lambda=\pm\sigma} \gamma \left(Z^{\bar{\gamma}:\lambda}_R Z^{\lambda:\bar{\sigma}}_{R+g_x(g_y)} - Z^{\bar{\gamma}:\lambda}_{R+g_x(g_y)} Z^{\lambda:\bar{\sigma}}_R \right) c^+_{R+a_x(a_y),\gamma},$$

$$A^+_{R,9(10)} = (S_R S_{R+g_x(g_y)}) c^+_{R+a_x(a_y),\sigma}$$

(122)

Similar operators of a spin polaron which occupy three sites were discussed in [60,46,198].

The operators $A_{R,i}$ $i = 5 \div 10$ by definition describe excitations $S_{tot} = 1/2$ and $S^z_{tot} = \sigma/2$, since the operator $\mathbf{S}_{tot}$ and its projection S^z_{tot} commute with each term in the Hamiltonian. One can prove that the operators $A_{R,i}$ $i = 1 \div 10$, form a full basis of local operators for a short-range spin polaron (only copper spins closest to the oxygen hole are excited) with $S_{tot} = 1/2$.

Then we use the standard projection method, outlined in the previous Section IV A.

2 Influence of frustration, $O - O$ hoppings and temperature on a local spin polaron spectrum

In this section we only give calculations of the function $\varepsilon(\mathbf{k})$ in the lowest band, i.e., the lowest eigenvalues of the operator DK^{-1} at fixed $\mathbf{k}$. This band determines the conducting properties at low doping levels.

Note an important feature of the Hamiltonian, which should be taken into account when comparing our results to calculations, started from other forms of the Hamiltonian.

In order to make the signs of hopping integrals in $\mathbf{T}$ and $\hat{h}$ in (117) independent of the hopping direction, we perform a unitary transformation changing the phases of the $Cu(d_{x^2-y^2})$ and $O(p)$ wave functions. [133,29,69] This was done by multiplying local wave functions by the factor $\exp(i\mathbf{QR})$ where $\mathbf{R}$ is the radius-vector of the corresponding elementary cell and $\mathbf{Q} = (\pi/g, \pi/g)$ the antiferromagnetic wavevector (below the copper sublattice constant is taken to be $g = 1$). In order to restore the initial translation symmetry of local wave functions, we replace $\varepsilon(\mathbf{k})$ by $\varepsilon(\mathbf{k} + \mathbf{Q})$ and present our results in these variables.

In all plots, except Figs.18c,d, the calculated spectra correspond to the typical value of the exchange integral $J_1 = 0.2\tau$ at two values of the frustration parameter,

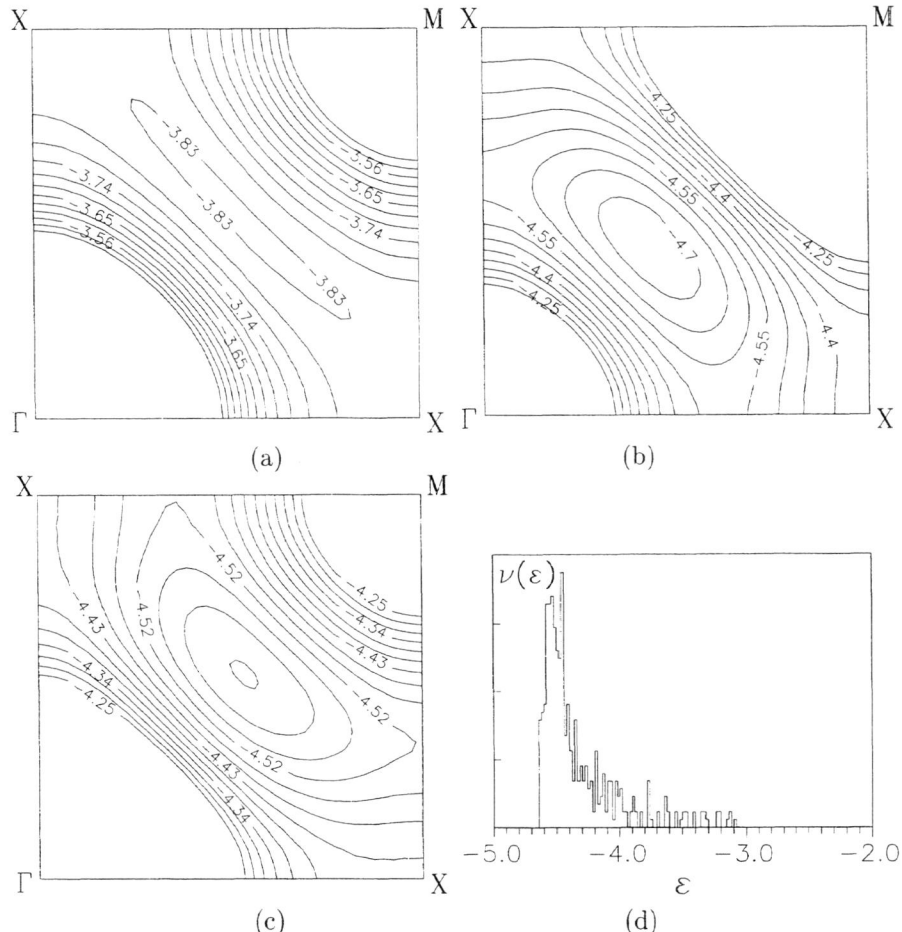

FIGURE 17. Contours of constant energy in the first quadrant of the Brillouin zone. Spectrum $\varepsilon(\mathbf{k})$ at (a) $T = 0$, $\alpha = J_2/J_1 = 0.4$, $h = 0$; (b) $T = 0.0$, $\alpha = J_2/J_1 = 0$, $h = 0.4$; (c) $T = 0$, $\alpha = J_2/J_1 = 0.4$, $h = 0.4$; (d) Density of states $\nu(\varepsilon)$ for $T = 0$, $\alpha = J_2/J_1 = 0.4$, $h = 0.4$. Ref. [13].

$\alpha = 0.0$ and 0.4, and two values of the oxygen-oxygen hopping integral, $h = 0.0$ and 0.4τ. We recall that the energy parameters τ and J_1 and the parameters of the initial model t_{pd} and ε_{pd} are related to each other: [23,182,148,156,204,61]:

$$\tau = \frac{t_{pd}^2}{\varepsilon_{pd}}, \qquad J_1 = \frac{4\tau^2}{\varepsilon_{pd}}. \qquad (123)$$

The calculated spectra are shown by contours of constant energy drawn with equal energy increments in the first quadrant of the first Brillouin zone. The energies of some of the contours are given in the graphs. The density of states $\nu(\varepsilon)$ and curves of $\varepsilon(\mathbf{k})$ along symmetrical directions $\Gamma - X - M - \Gamma$ and $X - N - X$ ($M = (\pi, \pi)$, $N = (\pi/2, \pi/2)$) are also plotted in the graphs. The unit energy is τ.

As was noted earlier, the calculated spectra $\varepsilon(\mathbf{k})$ are interpreted assuming that an increase in the frustration parameter α simulates a higher doping level.

Figure 17a shows the spectrum $\varepsilon(\mathbf{k})$ at $h = 0$, $T = 0$, and at finite frustration parameter $\alpha = 0.4$. The spectrum has a minimum near the point N. In a wide region around the boundary of the antiferromagnetic Brillouin zone $X - N - X$ there is a flat band in which energy is a weak function of $\mathbf{k}$. This portion of the spectrum near the bottom generates a peak in the density of states, whose shape is similar to that of a one-dimensional one with the singularity $\nu(\varepsilon) \sim \varepsilon^{-1/2}$.

At minimum energy the spectrum is highly anisotropic, i.e., the effective masses around point N in the directions $N - M$ and $N - X$ are very different. At a low filling, the Fermi surface is an ellipse and the Hall constant R_H should have the "hole" sign, $R_H > 0$.

We do not show the shape of $\varepsilon(\mathbf{k})$ for $T = 0$, $h = 0$ and $\alpha = 0$. In this case the spectrum has a flat portion around the line $X - N - X$, but the weak modulation of the spectrum near the band bottom is slightly different, so the minimum is on the line $X - \Gamma$ near point X.

The spectrum is radically different at zero frustration factor ($\alpha = 0$), but with oxygen-oxygen hopping. As it can be seen in Fig. 17b at $h = 0.4$ the minimum of the band bottom is still near to point N. But unlike in the case shown in Fig. 17a ($h = 0.0$), this minimum is isotropic. As was noted in the Introduction, this shape of the minimum is observed at zero doping [201] (in our case this corresponds to zero frustration). It may be assumed that the isotropic minimum near the band bottom is due to oxygen - oxygen hopping. Similar results were obtained in [144] and [180].

Another feature of the spectrum in Fig. 17b for $h \neq 0$ which can be seen by comparison with Fig. 17a is a notable modulation around the minimum and a flat portion near point X. As a result, there is a second peak in the density of states. It is interesting that the band portion responsible for the second peak has the shape of an elongated saddle. But this saddle is aligned with the direction $X - M$, as it can be seen in Fig. 17b, whereas the experimentally detected feature is extended along the direction $X - \Gamma$. [192,75,1,44,109].

The case of zero temperature, but with both frustration factors and nonzero oxygen-oxygen hopping amplitude ($\alpha = 0.4$, $h = 0.4$), is shown in Fig. 17c,d. In this case the spectrum $\varepsilon(\mathbf{k})$ is radically different from that in Fig. 17b, since its elongated saddle point is aligned with line $X - \Gamma$, which is in agreement with experiments performed at an optimal doping level (i.e., in our language, at a finite frustration). One can check that at a higher frustration (doping) parameter, this feature and the resulting peak in the density of states are less pronounced. This result is consistent with the general concept that the transition temperature drops in overdoped HTSC if we assume that high-T_c superconductivity is caused by the proximity of the Fermi level to the Van Hove singularity which corresponds to a high peak on the $\nu(\varepsilon)$curve in Fig. 17d.

If the Fermi level coincides with an equal-energy curve near the elongated saddle, for example the line with $\varepsilon = -4.43$ in Fig.17c, the Fermi surface should have a second portion around point M. Note that the Hall constant is directly related to a weighted integral of the curvatures of these lines. [20] One can easily check that the two contours have curvatures of different signs, so the Hall constant may change its sign from plus to minus as the doping level increases.

Note that the comparison of spectra shown in Figs.17b and 17c (($h = 0.4$, $\alpha = 0.0$) and ($h = 0.4$, $\alpha = 0.4$)) demonstrates that the spectrum of the CuO_2plane cannot be described at all in the rigid-band approximation.

Our calculations at finite temperatures indicate that, like the frustration, the temperature effectively changes the spectrum. For example, the spectrum calculated for $h = 0.4$, $\alpha = 0$ and $T = J_1$ fits well to the curves in Fig. 17c for $h = 0.4$, $\alpha = 0.4$ and $T = 0$. This is not surprising because the spectrum is controlled by correlation functions and at moderate frustration parameters $\alpha < 0.5$ an increase in both T and α leads to lower antiferromagnetic correlations (although it is difficult correctly to translate one effect into another).

Now let us turn to the effect of the hopping amplitude on the spectrum. The modified component $\hat{T}$ of the Hamiltonian defined by (117) takes the form

$$\hat{T} = \sum_{\mathbf{R},\mathbf{a}_1,\mathbf{a}_2,\sigma_1,\sigma_2} (\tau + \Delta\tau\delta_{\mathbf{a}_2,-\mathbf{a}_1}) Z_{\mathbf{R}}^{\sigma_1\sigma_2} c_{\mathbf{R}+\mathbf{a}_2,\sigma_2}^{+} c_{\mathbf{R}+\mathbf{a}_1,\sigma_1} . \qquad (124)$$

The Hamiltonian in (124) takes into account that the hole hopping described by the vectors $\pm \mathbf{a}x \pm \mathbf{a}_y$ has amplitude τ, and that described by vectors $\pm 2\mathbf{a}_x$ and $\pm 2\mathbf{a}_y$ has an amplitude $\tau + \Delta\tau$. The physical reason of the difference between these amplitudes has been discussed above. Figures 18a and 18b show spectra calculated for $T = 0$, $h = 0$, $\alpha = 0.4$ and $\Delta\tau/\tau = -0.15$ and $+0.15$ respectively. It can be seen in Fig.18a that the spectrum calculated for $\Delta\tau < 0$ is radically different from all other spectra. The most remarkable feature is that the local minimum at point N disappears. But this situation, apparently, is not realized because calculations by the singlet-triplet model yield an estimate $\Delta\tau \sim 0.1\tau > 0$. The comparison of spectra in Fig.18b ($T = 0$, $\alpha = 0.4$, $h = 0$, and $\Delta\tau/\tau = 0.15$) and Fig.17c ($T = 0$, $\alpha = 0.4$, $h = 0.4$, and $\Delta\tau/\tau = 0$) with that shown in Fig. 17a ($T = 0$, $\alpha = 0.4$, $h = 0$, and $\Delta\tau/\tau = 0$) demonstrates that the inclusion of $\Delta\tau > 0$ is equivalent to a higher amplitude of oxygen-oxygen hopping.

There are few calculations for the three-band model which yield fairly complete information about the spectrum. Dopf et el. [51,49,50] calculated the three-band spectrum by the quantum Monte-Carlo technique, and Putx el al. [158] by finding a self-consistent solution to the Dyson equation to the one-particle Green's function with the account of the interaction between a hole and fluctuations of charge and spin densities. These calculations were based on the three-band model with typical values of parameters: $\varepsilon_{pd} = 4t_{pd}$, $U_d = 6t_{pd}$ the temperature $T = 0$, and the

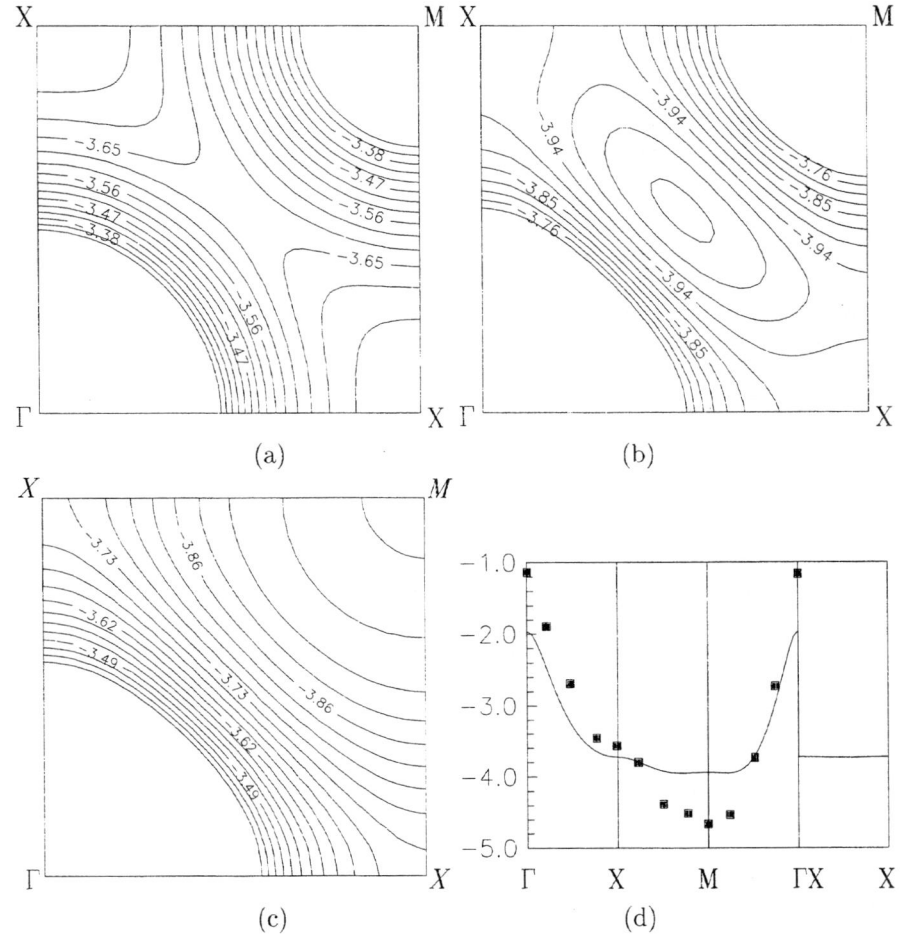

FIGURE 18. (a) Same as Fig.17a, but with the anisotropy of the hopping amplitude $\frac{\Delta\tau}{\tau} = -0.15$. (b) Same as Fig.17a, but with the anisotropy of the hopping amplitude $\frac{\Delta\tau}{\tau} = 0.15$. (c) and (d): Contours of constant energy of the spectrum $\varepsilon(\mathbf{k})$ for $J_1 = 0.25$, $T = 1.6J_1$, $\alpha = J_1/J_2 = 0.4$ and $h = 0.4$, and the spectrum $\varepsilon(\mathbf{k})$ along symmetrical directions– solid trace shows results of the present approach (at $J_1 = 0.25$, $T = 1.6J_1$, $\alpha = J_1/J_2 = 0.4$ and $h = 0.4$) Ref. [13]; dots show Monte Carlo calculations by the three-band model Ref. [158] for $\varepsilon_{pd} = 4t_{pd}$, $U_d = 6t_{pd}$, $T = 0.1t_{pd}$ and $\delta = 0.25$. The energy unit is τ and $t_{pd} = 4\tau$. The energy axis for the Monte Carlo calculations is shifted arbitrarily.

doping level $\delta = 0.25$. Our results are compared to those of [158] in Fig.18c,d which show the spectrum $\varepsilon(\mathbf{k})$ at similar parameters of our effective Hamiltonian, namely $J_1 = 0.25\tau$, $T = 1.6J_1$, $\alpha = 0.4$ and $h = 0.4\tau$. The selection of the latter parameters was based on (123) with $t_{pd}/\varepsilon_{pd} = 0.25$. Note that (123) is valid only for $U_d = \infty$ i.e., this comparison is only intended to show the general trend. Figure 18c demonstrates that the band bottom is at point M because the high temperature breaks antiferromagnetic correlations. Nonetheless, the saddle point near X persists. Figure 18d also shows a Monte-Carlo calculation of the spectrum, which is quite similar to that in [158], and our calculations along symmetrical directions. It can be seen that at high temperature the results are fairly close, although the parameter U_d is different, and the doping $\delta = 0.25$ is simulated by the frustration $\alpha = 0.4$.

In conclusion, let us spell out the basic result of local polaron analysis of the model. There are several important mechanisms, not mutually exclusive, generating singularities in the CuO_2-plane hole spectrum similar to those detected in experiments. The band bottom may become isotropic because of either oxygen-oxygen hopping with amplitude h or the dependence of the hopping amplitude τ on the trajectory. The saddle-like singularity aligned with the $X - \Gamma$ line is generated when, firstly, the frustration (doping) and oxygen-oxygen hopping are taken into account concurrently, secondly, when account is taken of the finite temperature and frustration, and thirdly, when frustration occurs and the hopping amplitude depends on the trajectory. The basic result of our study is the conclusion that with due account of realistic features of the model, one can interpret experimentally observed spectra of hole excitations.

In our opinion, additional interactions, such as Coulomb repulsion of holes at neighbouring copper and oxygen sites, and modifications of the effective Hamiltonian in (117), which apply, strictly speaking, only for $t_{pd} \ll \varepsilon_{pd}$, should not radically change our results.

C The generalized t-J model

It is generally believed that the t-J model qualitatively describes the hole spectrum of the CuO_2 plane in HTSC [144]. There are numerous studies of the hole spectrum in the framework of the t-J model based on the exact diagonalization of small clusters [185,43], the self-consistent Born approximation [108,132] or a "string" ansatz for the hole wave function [55]. One can start either from the two-sublattice Néel-type state [108,132,55,34] or from the spin rotationally invariant spin liquid state [83] and obtain qualitatively the same result: the hole motion occurs mainly on one sublattice, i.e. the dispersion relation is dominated by an effective hopping to the next nearest neighbours and dispersion has minimum at $(\pi/2, \pi/2)$. One finds a flat region of dispersion near point $(\pi, 0)$. This was used to attempt to interpret the experimental data of the optimally doped compounds [42], where a rigid-band picture was assumed (see also [58]). However, the anisotropic minimum of the pure t-J model is in contrast with the experimental result in $Sr_2CuO_2Cl_2$ which shows an isotropic band bottom and a large energy difference between $(\pi/2, \pi/2)$ and $(\pi, 0)$.

In [144,12] it was shown that the qualitative agreement with the experimental results [201] may be improved in the framework of the t_1-t_2-J model, which takes into account the next-nearest-neighbour hopping t_2 of the hole. The inclusion of a t_2-term substantially improves the correspondence of the calculated spectrum near the band bottom with the experimentally observed isotropic minimum of $\varepsilon(k)$ at point $(\pi/2, \pi/2)$. On the other hand, the hole spectrum of the t_1-t_2-J model does not reproduce the flat band region near point $(\pi, 0)$. So, one has the following theoretical problem: the one-hole approach in the pure t-J model can reproduce the flat band region of the optimally doped cuprates [192,75,1,44,109] but not the isotropic band bottom of the insulating compound [201]. And vice versa, one hole in the t_1-t_2-J model leads to an isotropic band bottom but not to a flat band region near $(\pi, 0)$. One possible solution could be that we have to choose different microscopic models for the different compounds. Here, we shall demonstrate the other possibility, whether it is possible to change the hole band by doping so that a flat band region between $(\pi, 0)$ and $(\pi/2, 0)$ arises.

To simulate the effect of doping in a study of the one-hole motion two possibilities were recently compared, namely the influence of frustration and temperature [82]. The inclusion of frustration leads to a t-J_1-J_2 model, where J_1 and J_2 denote the antiferromagnetic exchange interaction between the nearest and next nearest neighbours. The influence of frustration was investigated by two different methods, namely by a variational ansatz where the spin-spin correlations in the frustrated Heisenberg model had been calculated by a spin rotational invariant procedure [13] and by the exact diagonalization of a $4*4$ lattice. It was found that the frustration shifts the minimum of the hole dispersion from $(\pi/2, \pi/2)$ to point $(\pi, 0)$. Both methods had a quite reasonable agreement where the variational method showed the effect in a more pronounced way. Flat dispersion regions were found in the nonfrustrated and frustrated cases. Without frustration, the flat region occurs around $(\pi/2, \pi/2)$, whereas for frustration $J_2/J_1 \gtrsim 0.4$ it occurs between $(\pi, 0)$ and $(\pi, \pi/2)$. Such a flat region is similar to the experiment [192,75,1,44,109], but there it appears between $(\pi, 0)$ and $(\pi/2, 0)$.

The above mentioned investigations demonstrate that the extension of the t-J model by the inclusion of additional hopping terms, frustration or by taking into account a finite temperature may strongly modify the initial features of the hole spectrum. But at the same time, none of these extensions alone can lead to an adequate description of the experimental picture.

In this Section we shall present the investigation of the hole spectrum ([80] and [82]) in the framework of the t_1-t_2-t_3-J_1-J_2 model, e.g. we will take into account the hole hoppings between first (t_1), second (t_2) and third (t_3) nearest neighbours and the frustration in the spin subsystem simultaneously. The additional hoppings t_2 and t_3 naturally appear in the Hamiltonian if one reduces the well known three band Hubbard model of the CuO_2 plane to an extended t-J model [104,31,203]. Throughout the Section t_2 and t_3 are fixed to such values which can be derived from the characteristic parameters of the three band Hubbard model for the CuO_2 plane which were given by Hybertsen et al. [96]. The frustration will be understood to simulate the doping as proposed, for instance, in Ref. [97]. The effect of a finite temperature will also be discussed.

As it will be seen, the structure of the local spin polaron for the $t - J$ model differs from that in the Emery model. Two methods will be used to investigate the problem. The projection Mori-Zwanzig method [141] for two-time retarded Green's functions will be employed to treat the hole excitation as a magnetic polaron of minimal size. Simultaneously we shall use the exact diagonalization of a 4*4 lattice.

1 Spin polaron structure

We consider an extended t-J model on a square lattice with hoppings between first, second and third nearest neighbours and with a frustration term in the exchange interaction. The Hamiltonian of the model is given by

$$H = H_t + H_J = -\sum_{i\sigma l} t_l \, X_i^{\sigma 0} X_{i+l}^{0\sigma} + \frac{1}{2} \sum_{im} J_m \, \mathbf{S}_i \cdot \mathbf{S}_{i+m} , \qquad (125)$$

where

$$t_l = \begin{cases} t_1 > 0 & \text{if } l \text{ is a nearest neighbour vector} \\ t_2 & \text{if } l \text{ is a second nearest neighbour vector} \\ t_3 & \text{if } l \text{ is a third nearest neighbour vector} \\ 0 & \text{else} \end{cases} \qquad (126)$$

and

$$J_m = \begin{cases} J_1 > 0 & \text{if } m \text{ is a nearest neighbour vector} \\ J_2 > 0 & \text{if } m \text{ is a next nearest neighbour vector} \\ 0 & \text{else} \end{cases} \qquad (127)$$

The Hamiltonian is expressed in terms of the Hubbard projection operators which exclude the double occupancy at site i: acting on the vacuum state $X_i^{\sigma 0}$ creates the electron (annihilates the hole) at site i with spin $S = 1/2$ and spin projection $\sigma/2$ ($\sigma = \pm 1$ is a spin index). The following operator equations are fulfilled

$$\begin{aligned} & X_i^{\sigma 0} = c_{i\sigma}^\dagger (1 - n_{i-\sigma}), \quad X_i^{\sigma\sigma} = n_{i\sigma}(1 - n_{i-\sigma}), \quad X_i^{\sigma-\sigma} = c_{i\sigma}^\dagger c_{i-\sigma}, \\ & n_{i\sigma} = c_{i\sigma}^\dagger c_{i\sigma}, \quad S_i^\sigma = X_i^{\sigma-\sigma}, \quad S_i^z = \tfrac{1}{2}\sum_\sigma \sigma X_i^{\sigma\sigma}, \\ & X_i^{00} + \sum_\sigma X_i^{\sigma\sigma} = 1, \quad X_i^{\lambda_1 \lambda_2} X_i^{\lambda_3 \lambda_4} = X_i^{\lambda_1 \lambda_4} \delta_{\lambda_2 \lambda_3}, \end{aligned} \qquad (128)$$

where $\lambda_n = 0$ or σ in the last equation. The hopping parameters t_l in (125) are not free but have to be determined by a reduction procedure from the three band Hubbard model of the CuO_2 plane. Here we use an analytical reduction procedure, the so-called "cell perturbation method" [104,31,203]. For the parameters of the three band Hubbard model we choose the values which were found by Hybertsen et al. [96] for La_2CuO_4 by a constrained density-functional calculation, namely:

$t_{pd} = 1.3\ eV, t_{pp} = 0.65\ eV, \varepsilon_p - \varepsilon_d = 3.6\ eV, U_d = 10.5\ eV, U_p = 4\ eV$ and $U_{pd} = 1.2\ eV$ [96]. Band structure calculations for $Sr_2CuO_2Cl_2$ [147] show only few differences to the La-compound. Therefore, we expect that the parameter values of Hybertsen et al. are also responsible for the Sr-compound. From the reduction procedure (see details in [203]) we found the following hopping parameters: $t_1 = 498\ meV, t_2 = -41\ meV$ and $t_3 = 77\ meV$. They are within the limits which have been found in [31] and also near the results of a numerical mapping in [96]. But differently from Hybertsen et al. we also include the t_3-term. In the following we will choose t_1 as the unit of energy, i.e. $t_2 = -0.08, t_3 = 0.15$. So one can see that the additional hopping terms are rather small, but nevertheless essential as will be shown below. The reduction procedure also gives the possibility to calculate the exchange energy J_1. That is however less important for our purpose, since we shall show that the results for different values of J_1 can be roughly transformed into each other by scaling the bandwidth with J_1.

Let us discuss the one-hole excitations on the background of the half-filled rotationally invariant singlet states $|\Psi\rangle$ of the pure spin system H_J. We are interested in the energetically lowest branch of these excitations and it is known that the bottom of its band is represented by the states with total spin $S_{tot} = 1/2$. We shall treat the problem within the framework of the spin polaron of small radius. That means that we restrict ourselves to spin excitations in the immediate neighbourhood of the hole. It may be seen that the full basis of such excitations is given by the following 5 basis operators $\phi_i^{a\dagger}$ with $a = 0, 1, \ldots 4$:

$$\phi_i^0 = X_i^{\sigma 0} \quad , \quad \phi_i^a = \sum_s X_{i-a}^{\sigma s} X_i^{s0} \quad (a = 1, \ldots, 4), \tag{129}$$

where we shall use the following notation hereafter: the small Latin letters a and b denote either a number between 0 and 4 or the corresponding lattice vector

$$\begin{array}{ll} 0 \leftrightarrow (0,0) & 1 \leftrightarrow (1,0) \quad 3 \leftrightarrow (-1,0) \\ & 2 \leftrightarrow (0,1) \quad 4 \leftrightarrow (0,-1) \end{array} \tag{130}$$

in a synonymous way. Any vector $\phi_i^{a\dagger}|\Psi\rangle$ corresponds to a one hole state with $S_{tot} = 1/2$ and $S_{tot}^z = -\sigma/2$, where S^{tot} and S_{tot}^z are the spin and its projection of the total system.

To obtain the one-hole spectrum $\varepsilon(k)$ we use the two-time retarded matrix Green's function $G^{ab}(t,k)$ for the Fourier transformation of the operators ϕ_i^a:

$$G^{ab}(t,k) = \langle\langle \phi_k^a(t); \phi_k^{b\dagger} \rangle\rangle = -i\Theta(t)\langle\{\phi_k^a(t); \phi_k^{b\dagger}(0)\}\rangle, \tag{131}$$

$$\phi_k^a = \frac{1}{\sqrt{N}} \sum_j e^{ikj} \phi_j^a, \tag{132}$$

where $\{\ldots,\ldots\}$ denotes the anticommutator and where we have used Zubarev's notation [207].

To restrict ourselves to the above chosen set of 5 operators $\{\phi_k^a\}$ (129) we use the standard projection method discussed above.

The corresponding matrix elements of K_{ij} and D_{ij} can be calculated in a straightforward way and we don't give the explicit expressions. These matrix elements are expressed through spin-correlation functions of the spin system for an undoped lattice which is described by the frustrated $S = 1/2$ Heisenberg model. We treat this model in the framework of the spherical symmetric approach which gives values of pair spin-spin correlation functions at any value of frustration parameter J_2/J_1 and temperature.

2 Polaron spectrum

As mentioned above we calculate the spectrum $\varepsilon(k)$ using the hopping parameters t_1, t_2, t_3 which were obtained for La-cuprates and take t_1 as the unit of energy. So we have the values $t_2 = -0.08$ and $t_3 = 0.15$. The calculations within the framework of the magnetic polaron of minimal size are only reliable for J_1 of the order of t_1, since our relevant set of operators is too limited to describe the small J_1 case.

Simultaneously with the Green's function method the Lanczos exact diagonalization scheme is used as a complementary method to calculate the low-lying eigenstates for a square lattice of $4 * 4$ sites (with periodic boundary conditions). These low-lying states are classified by its momentum k and the total spin. Here we shall concentrate on the band with total spin $1/2$. There are only few exceptions where the lowest states have spin $3/2$ and that happens only for higher lying levels with momentum (π, π) or $(0,0)$. The Mori-Zwanzig projection method gives the energy difference between the lowest state with one hole for total spin $1/2$ and fixed momentum and the ground state without any hole. The same energy difference is calculated for the $4 * 4$ lattice.

To clarify the importance of the additional hopping terms and frustration, in Fig. 19 we represent the results for $J_1 = t_1 = 1$ without frustration (Fig.19a) and with $J_2 = 0.4$ (Fig.19b). For convenience we have chosen a finite temperature $T = 0.2$ to calculate the static spin-spin correlation functions [13], but one can check that the differences of the zero temperature case can be neglected. One observes a quite reasonable agreement between the Green's function method and the exact diagonalization data for these parameter values. As it is seen in Fig. 19a, in absence of frustration, the spectrum demonstrates an isotropic band bottom close to point $(\pi/2, \pi/2)$ in accordance to the experimental dispersion of one hole [201]. The same result was obtained in [144] for the t_1-t_2-J model where the following parameter values were taken: $t_2 = -0.35 t_1$, $J = 0.3 t_1$ (see also [12]). Let us mention that the isotropization of the band bottom in comparison with the t-J model may be obtained in a rather wide range of parameters t_2 and t_3. For example, also the values $t_2 = -0.15 t_1$ and $t_3 = 0.1 t_1$ give a dispersion which is similar to the one shown in Fig. 19a. It is important that the spectrum in Fig. 19a does not demonstrate a flat band region near point $(\pi, 0)$.

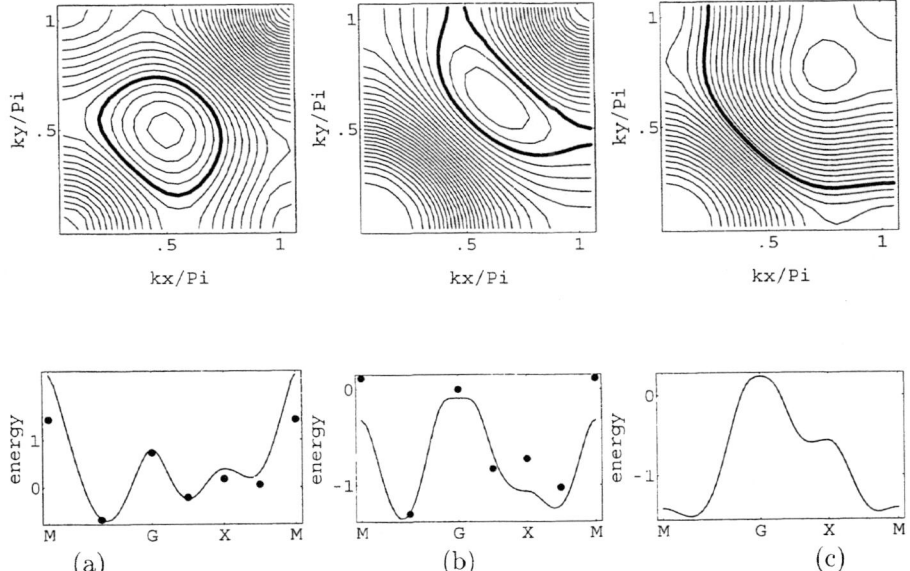

FIGURE 19. Quasiparticle dispersion along the line M-G-X-M and contour plot for $t_1 = 1, t_2 = -0.08, t_3 = 0.15, T = 0.2$ in three cases: a) $J_1 = 1, J_2 = 0, T = 0.2$; b) $J_1 = 1, J_2 = 0.4, T = 0.2$ (i.e. without and with frustration). The circles are the result of the exact diagonalization of a $4*4$ lattice at $T = 0$ with total spin $S_{tot} = 1/2$. The thick line denotes a hypothetical Fermi surface at 20 per cent hole doping where 20 per cent of the BZ are filled with holes (spinless fermions). c) $J_2/J_1 = 0$ and high temperature $T = 1$. The thick line denotes a hypothetical Fermi surface at 20 per cent hole doping where 60 per cent of the BZ are filled with holes (according to Luttinger's rule). Reprinted from Ref. [80], with permission from Springer-Verlag.

Before considering the spectrum of Fig. 19b let us remind that we suppose frustration simulates doping [97]. Of course there is no full equivalence between doping and frustration. For example, the doped t-J model and the frustrated J_1-J_2 model give different results for the dynamical spin-spin structure factor and for the spectrum of Raman scattering [11]. Nevertheless, it is well known that both doping and frustration lead to a decrease of the magnetic correlation length. Furthermore, numerical calculations on finite lattices indicate the equivalence of the mentioned models if we are interested in the static spin-spin correlation functions [140]. Note that this is especially relevant in our present Green's function method where the spectrum $\varepsilon(k)$ is determined by the static spin-spin correlation functions of the spin subsystem.

Fig. 19b represents the spectrum for the same energetical parameters as Fig. 19a, except for the inclusion of a frustration term $J_2 = 0.4J_1$. The comparison of Figs. 19a and 19b indicates that the frustration dramatically changes the spectrum near point $(\pi, 0)$. It leads to the appearance of a flat band region close to that point.

Moreover, this flat band region has the form of an extended saddle point which stretches in the direction $(\pi/2,0)$-$(\pi,0)$. Such a band structure corresponds to the ARPES results for optimally doped cuprates [192,75,1,44,109]. An intuitive explanation for the spectra of Figs.19a and 19b can be given in the following way: the additional hopping terms t_2 and t_3 act in such a way that they suppress motion in the direction of the copper-oxygen bond, but facilitate motion along the diagonal. This explains the isotropic minimum in Fig.19a. Due to frustration, the static spin-spin correlation functions are suppressed. So, the nearest neighbour spin-spin correlation $\langle \vec{S}_i \cdot \vec{S}_{i+x} \rangle$ changes from -0.352 ($J_2 = 0$) to a value of -0.207 for $J_2 = 0.4$. Therefore, a small probability for a transition to a nearest neighbour appears, pushing the energy level at $(\pi,0)$ downwards which leads in the end to the flat band region.

So, we can describe the experimental results for insulating (no frustration, Fig. 19a) and optimally doped compounds (Fig. 19b, the case with frustration) by the same set of hopping parameters. It is important that at the same time our approach demonstrates the non-rigid band behaviour in the framework of a simple and natural mechanism: doping leads to frustration in the spin subsystem and to the variation of spin-spin correlation functions, and correspondingly the alteration of these functions results in the non-rigid behaviour of the spectrum. Or, in other words: the doping changes the state of the magnetic background which strongly determines the form of the spectrum. Let us remind that this mechanism does not take into account the direct interaction between holes which will be of course important in the regime of heavy doping. The reasonable coincidence of the Green's function method with the exact diagonalization data at $J_1 = 1$ is comprehensible since for such large values of J_1 the magnetic energy stabilizes the size of the magnetic polaron and our concept of a polaron of small radius is justified.

Finally, in Fig.19c we represent the results for finite temperature $T = 1$ and zero frustration. As it can be expected, in some sense the inclusion of temperature simulates the effect of doping by decreasing the antiferromagnetic correlation length in a similar way as frustration does. But now, the exact diagonalization method is not applicable. The dispersion in Fig.19c is similar to that of Fig.19b. In particular, the extended saddle point between $(\pi/2,0)$ and $(\pi,0)$ can be observed. There is one important difference that the saddle point in Fig.19c lies much higher with respect to the band bottom than in Fig.19b. This may have consequences on the position of the Fermi level as it will be now discussed.

Before discussing the Fermi surface (FS) let us mention that this question is highly speculative within the present context. Throughout this Section we consider one hole in the extended t-J model which occupies the lowest possible state and there is no FS at all. Nevertheless, we show in Figs. 19 and 19c hypothetical Fermi surfaces (thick lines) corresponding to 20 per cent hole doping, to distinguish two possible scenarios for a transition from a small to a large FS. The first scenario, shown in Fig.19, is connected with the assumption that the holes in an antiferromagnetic background form spinless fermions [108,132] which occupy each

k-point once. Accordingly, 20 per cent of the Brillouin zone (BZ) is enclosed by the FS shown in Fig.19. Without frustration (Fig.19a) we would find small hole pockets around the isotropic minimum $(\pi/2, \pi/2)$, but the change of dispersion due to frustration (Fig.19b) leads to a change of FS the topology. The outer branch in Fig. 19b could be interpreted as the large FS seen in optimally doped cuprates [75,44], but the second branch is not observed in the ARPES experiments. A more correct treatment of the Fermi surface and band filling will be presented in the next Section.

The ARPES experiments show different dispersions in the insulating and optimally doped cuprates. One observes an isotropic band minimum but no flat band in the insulating compound and a flat band region in the optimally doped ones. One can see that the isotropic band minimum of the undoped case may be explained due to additional hopping parameters in a t-J-like Hamiltonian which naturally appear in a proper reduction scheme from the three band Hubbard model. The doping changes the spin background and in the present study we have assumed that it may be simulated by a frustration term in the exchange energy. We have seen that frustration (doping) leads to a non-rigid change of the bands such that a flat band region appears.

Let us emphasize the importance of taking into account the t_2- and t_3-hopping-terms. As mentioned above, these hoppings lead to the isotropization of the band bottom in the non-frustrated case. But these terms are also important for an adequate description of the spectrum near point $(\pi, 0)$. As it was shown in [82] frustration leads to a flat band region with an extended saddle point near point $(\pi, 0)$ also in the case of the t-J_1-J_2 model. But then, in contrast to Fig.19b, this saddle point is extended in the direction $(\pi, 0)$-$(\pi, \pi/2)$ and not in the direction $(\pi/2, 0)$-$(\pi, 0)$ as in the experiment. So, only the simultaneous consideration of t_2-, t_3-hoppings and of frustration leads to the extended saddle point in the direction $(\pi/2, 0)$-$(\pi, 0)$.

V LOCAL POLARON DRESSED BY ANTIFERROMAGNETIC SPIN WAVES

A The case of zero temperature

1 Spin condensate and new polaron operators

Let us examine the formation of complex spin polaron within the framework of the more simple Kondo-lattice model.

Treating this model, Schrieffer [173] replaced the spin subsystem with a classic field with a doubled lattice period and described the elementary excitations as the superposition of electronic states with momenta $\mathbf{p}$ and $\mathbf{p} + \mathbf{Q}$, where $\mathbf{Q} = (\pi, \pi)$ is the antiferromagnetic vector. Such a description is deficient for the two-dimensional $S = \frac{1}{2}$ antiferromagnet because of strong spin fluctuations. Moreover, it becomes impossible if the spin subsystem is found in the homogeneous spherically symmetric state with long range order but with zero value of the average site spin $<S^\alpha> = 0$ (the homogeneous Néel state) [9,176,27,13]. The homogeneous Néel state seems more adequate than the usual two-sublattice Néel state, for treating the ground state of the doped CuO_2 plane.

The distinctive feature of the present investigation consists in considering the one-electron motion on the background of the homogeneous Néel state of the Kondo lattice. This motion is described by a spin polaron with the spectrum periodicity relative to the full Brillouin zone. Let us remind that the conventional two-sublattice spin approach leads to periodicity relative to the magnetic Brillouin zone [156,161,173].

Distinctive feature of this Section consists in treating the spin-polaron as a complex quasiparticle, a coherent superposition of a bare electron, of a local polaron (see the first Section) and of two types of antiferromagnetic delocalized polarons. A local polaron can be attributed to the local singlet (an analogous of the Zhang-Rice polaron in the three-band Hubbard model) and such a polaron represents the lowest band in the limit of the strong Kondo interaction J. As to the antiferromagnetic delocalized polaron it is a bound state of an electron (or of a local polaron) with a spin wave with momentum $\mathbf{Q}$ [124,24]. If the spin subsystem is found in the state with the antiferromagnetic long range order then the amplitude $S_\mathbf{Q}$ of the spin wave with $\mathbf{q} = \mathbf{Q}$ (the Q-wave) has the macroscopic large value and has the properties analogous to the amplitude of a Bose particle with zero momentum in the superfluid Bose-gas. As a result for many problems this amplitude can be treated as a $c-$number. Then the coupling of the Q-wave to local electron states does not represent new states but leads, as mentioned above, to mixing of states with momenta p and $p+Q$. But this treatment fails in the case of the homogeneous Néel state. In this background the average value $<S_Q> = 0$ and only $<S_Q S_Q>$ can be treated as a macroscopic value. Then the coupling of a particle local state to S_Q corresponds to a new delocalized state. It will be shown that it is important to take into account such a quantum nature of the spin Q-wave because the transitions between the local polaron states and the delocalized polaron states essentially determine the spin polaron bands.

The Kondo-lattice Hamiltonian has the form

$$H^{tot} = H_0 + H_1 + H_2, \qquad (133)$$

$$H_0 = \sum_{rg} t_g a^+_{r+g} a_r = \sum_p \epsilon_p a^+_p a_p, \qquad (134)$$

$$H_1 = J \sum_r a^+_r \tilde{S}_r a_r, \qquad (135)$$

$$H_2 = \frac{1}{2} I \sum_{rg} S^\alpha_{r+g} S^\alpha_r. \qquad (136)$$

Here the sums run over the sites r of a square lattice and over the nearest neighbours with the lattice spacing $|g|=1$. For short we miss the spin index for the creation $a^+_{r\sigma}$ and annihilation $a_{r\sigma}$ operators of the Fermi particles (we shall call them electrons) and in the Hamiltonian of the Kondo-interaction H_1 we take the notation $\tilde{S}_r = S^\alpha_r \sigma^\alpha$.

Let us represent the first two equations of the infinite chain of equations for the retarded Green's functions, which describe the motion of one particle in the antiferromagnetic spin background. In the present paper we restrict ourself mainly to the analysis of the spectrum of the quasiparticles. That is why for simplicity of notations we shall simply write operators a instead of the Green's functions $<<a\,|\,b^+>>$ and shall miss the nonhomogeneous terms $<[a,b^+]_+>$. The equations have the form

$$(\omega - \epsilon_p) a_p = J b_p, \quad b_p = N^{-\frac{1}{2}} \sum_r b_r e^{-ipr}, \quad b_r = \tilde{S}_r a_r, \qquad (137)$$

$$\omega \tilde{S}_{r+R} a_r = [\tilde{S}_{r+R} a_r, H_0 + H_1 + H_2] = \\ = \sum_g t_g \tilde{S}_{r+R} a_{r+g} + J \tilde{S}_{r+R} \tilde{S}_r a_r + I i e_{\alpha\beta\gamma} \sum_g \sigma_\alpha S^\beta_{r+R+g} S^\gamma_{r+R} a_r. \qquad (138)$$

In Ref. [173] the mean-field approach was taken for the Néel lattice and the spins were treated as the classical vectors:

$$S^\alpha_r = \delta_{\alpha z} S_0 e^{iQr}, \quad S_0 = const, \quad Q = (\pm\pi, \pm\pi). \qquad (139)$$

In this approximation the Kondo-interaction Hamiltonian (135) takes the form of the potential energy with the doublet period and the equation (137) has a closed form:

$$(\omega - \epsilon_p)a_p = \sigma J S_0 a_{p-Q}. \tag{140}$$

Here p is in the Brillouin zone with the periods $(2\pi, 0), (0, 2\pi)$, but the spectrum has the periodicity of the magnetic Brillouin zone:

$$E_{ap} = \pm\sqrt{\epsilon_p^2 + \Delta^2}, \quad \Delta = S_0 J. \tag{141}$$

In (141) and below we suppose the spectrum model with the "nesting":

$$\epsilon_p = -\epsilon_{p+Q} = -2t(cos p_x + cos p_y) \tag{142}$$

In the case of the $S = \frac{1}{2}$ spin system the quantum fluctuations are important and the equation (138) must be used in order to find the last term in Eq.(137). In one's turn the equation (138) hasn't a closed form. In order to close the chain of equations we use the standard Mori-Zwanzig projection technique [141]. In our case this means that we must approximate the last two terms in the right-hand side of (138) by their projections on the restricted space of bases operators. The choice of the restricted set of basis operators must be dictated by the physics of the problem under discussion.

For a paramagnetic spin subsystem state the simplest set of basis operators may be taken as two operators that appear in the first equation (137), that are the operator of a "bare" electron a_r and the annihilation operator of a one-site spin polaron b_r (see the first Section).The local polaron excitations in the Kondo lattice were also discussed in Refs. [156,161]

Now let the spin subsystem be in the homogeneous Néel state with the tenser parameter of the long range order:

$$<S_r^\alpha> = 0, \tag{143}$$

$$C_R^{\alpha\beta} = <S_r^\alpha S_{r+R}^\beta> = M^{\alpha\beta} e^{iQR}, \quad |R| \gg 1. \tag{144}$$

Then it will be consecutive to introduce the operators which take into account the correlation of the electron and the local polaron with the antiferromagnetic Q-wave

$$c_r = \tilde{Q}_r a_r, \quad c_p = \tilde{Q}_0 a_{p+Q}, \tag{145}$$

$$\tilde{Q}_r = N^{-1} \sum_R e^{iQR} \tilde{S}_{r+R} = e^{iQR} \tilde{Q}_0 : . \tag{146}$$

$$d_r = \tilde{Q}_r \tilde{S}_r a_r, \quad d_p = \tilde{Q}_0 b_{p+Q}. \tag{147}$$

Note, that the operators c_r and d_r allow distant correlations between the spin and the electron ($|R| \gg 1$). So, in a sense our approach is alternative to the widely used t-J model investigations based on the decoupling of an on-site Fermi operator into a spinless fermion and an antiferromagnetic magnon operator [170,131,108,132,123,120].

It is substantial that the adopted set of the basis operators is closed relative to the Kondo-interaction Hamiltonian:

$$[a_r, H_1] = Jb_r,$$
$$[b_r, H_1] = J(\tfrac{3}{4}a_r - b_r),$$
$$[c_r, H_1] = Jd_r,$$
$$[d_r, H_1] = J(\tfrac{3}{4}c_r - d_r).$$
(148)

Taking into account Eq.(138) this gives, in particular,

$$(\omega - \epsilon_{p+Q})c_p = Jd_p. \tag{149}$$

In order to get a closed form for Green's functions equations we shall project the corresponding commutators on the following orthonormal set of basis operators:

$$B_{1r} = a_r,$$
$$B_{2r} = f_2^{-1/2}b'_r, \quad b'_r = b_r - c_r,$$
$$f_2 = <[b'_r, b'^{+}_r]_+> = 3/4 - M, \quad M = M^{\alpha\alpha},$$
$$B_{3r} = f_3^{-1/2}c_r,$$
$$f_3 = <[c_r, c^{+}_r]_+> = M,$$
$$B_{4r} = f_4^{-1/2}d'_r, \quad d'_r = d_r - Ma_r + Mf_2^{-1}(b_r - c_r),$$
$$f_4 = <[d'_r, d'^{+}_r]_+> = M(f_2 - Mf_2^{-1}).$$
(150)

The coefficients of the projection relations

$$[B_i, H] = \sum_j a_{ij} B_j \tag{151}$$

are determined as

$$a_{ij} = <[[B_i, H], B^{+}_{jr}]_+>. \tag{152}$$

Taking into account the equations (137), (149) and (152), we obtain in momentum representation a simple system of four equations

$$\omega B_{i,p} = \sum_j a_{ij} B_{j,p} \tag{153}$$

The matrix elements of the spectral matrix a_{ij} are

$$a(1,1) = \epsilon_p, \quad a(1,2) = J\sqrt{f_2}, \quad a(1,3) = J\sqrt{M}, \quad a(1,4) = 0,$$
$$a(2,2) = f_2^{-1}[(C+M)\epsilon_p - J(f_2 - M) - 4IC],$$
$$a(2,3) = -J\sqrt{Mf_2}^{-1},$$
$$a(2,4) = [Mf_2^{-1}(C+M)\epsilon_p - Jf_4 - 2IM(1+2Cf_2^{-1})]/\sqrt{f_2 f_4},$$
$$a(3,3) = -\epsilon_p, \quad a(3,4) = J\sqrt{f_4 M^{-1}},$$
$$a(4,4) = Mf_4^{-1}[(C+M)Mf_2^{-2} - (C+\tfrac{M}{3})]\epsilon_p - 3J/4f_2$$
$$+IMf_4^{-1}[8M/3 - 4C(1+Mf_2^{-2}) - 4Mf_2^{-2}].$$
(154)

Here $C = C_{R=1}^{\alpha\alpha} < 0$. To obtain these expressions we used the following approximation of Takahashi [193] for four different-site spin correlation functions

$$< S_{r_1}^i S_{r_2}^j S_{r_3}^k S_{r_4}^l > = C_{r_1-r_2}^{ij} C_{r_3-r_4}^{kl} + C_{r_1-r_3}^{ik} C_{r_2-r_4}^{jl} + C_{r_1-r_4}^{il} C_{r_2-r_3}^{jk}, \quad r_1 \neq r_2 \neq r_3 \neq r_4. \tag{155}$$

Hence, in the adopted approximation the electron motion in the antiferromagnetic background is described by the quasiparticle that is a coherent superposition of four Fermi fields - the field of the bare electron a_p, the field of the delocalized polaron c_p and two fields of the localized polaron b_p and d_p, which are hybridized mainly due to the Kondo-interaction. The system of equations (153) can be treated as the system of the Schroedinger equations where the matrix elements a_{ij} reproduce the amplitude of the transitions from state j to state i. The presence of nondiagonal matrix elements underlines the quantum nature of the spin S and the Q-waves. The eigenfunctions and the eigenvalues of Eqs. (153) describe the elementary excitations which generate four bands. The form of the bands depends on the state of the magnetic subsystem (the quantities M and C) and the relations between the energetic parameters t, J, I. Below we take the following typical values for the quantities M and C: $M = 0.1$ and $C = -0.335$ [128,176,27,13].

In the model adopted all the spectra depend on momentum only through ϵ_p. Let us mention that calculating the matrix elements (152) we took the low density approximation of Fermi particles, i.e., $< a_p^+ a_p > \to 0$.

Four branches of the quasiparticle spectrum $E^{(i)}(p)$, $i = 1-4$ for the case $J \gg t, I$ (strongly correlated limit) are represented in Fig.20a. In Fig.20b we also represent the values of the electron Green's function residues $Z_p^{(i)}$ which correspond to the poles $E^{(i)}(p)$

$$<< a_{p\sigma} | a_{p\sigma}^+ >> = << B_{1,p} | B_{1,p}^+ >> = \sum_{i=1}^{4} \frac{Z_p^{(i)}}{\omega - E^{(i)}(p)}. \tag{156}$$

In the low density limit the residues $Z_p^{(i)}$ are determined by the solution of the equations of motion for Green's functions $<< B_{i,p} | B_{1,p}^+ >>$ when the nonhomogeneous terms are equal to $W_{i,1} = < [B_{i,p}, B_{1,p}^+]_+ > = \delta_{i,1}$. The $Z_p^{(i)}$ values represent one-particle spectral function $A(p,\omega) = \sum_i Z_p^{(i)} \delta[\omega - E^{(i)}(p)]$. Each of these values $Z_p^{(i)}$ characterizes the contribution (weight) of a bare particle state a_p to the quasiparticle state with energy $E^{(i)}(p)$. For this reason $\sum_i Z_p^{(i)} = 1$.

Fig. 20A describes case $J \gg t, I$. The lowest bands (with the center at $E^{(s)} \cong -\frac{3}{2}J$) correspond to the motion of the local polaron in a singlet state $b^{(s)} = b - \frac{1}{2}a$ and to the motion of such a polaron coupled to the Q-wave. The upper bands (centered at $E^{(t)} \cong \frac{1}{2}J$) describe the motion of the local polaron in a triplet state $b^{(t)} = b + \frac{3}{2}a$ and also its coupling to the Q-wave. The "singlet" and "triplet" terminology suppose, that acting on the spin subsystem state, the

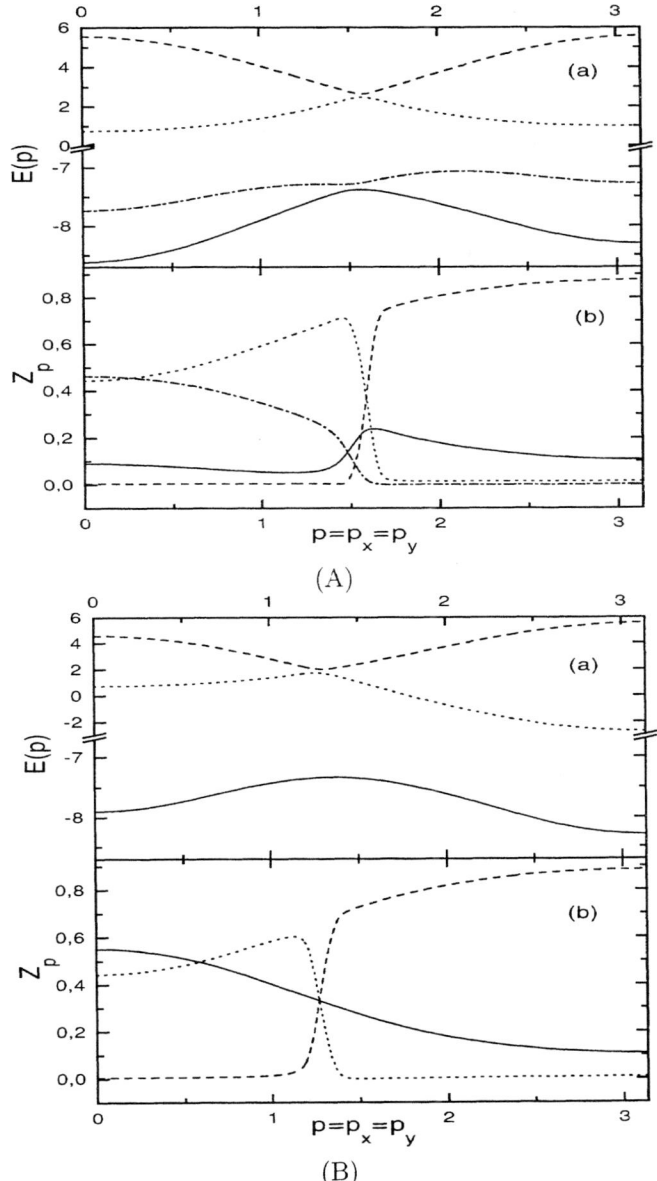

FIGURE 20. A). The quasipartical spectra $E^{(i)}(p)$ for $J = 5 \gg t, I, t = 1, I = 0.1$ along the symmetry line $p = p_x = p_y$ (a) and (b) $<< a_{p\sigma} \mid a_{p\sigma}^+ >>$ residues $Z_p^{(i)}$ for the poles $E^{(i)}(p)$. Different line types represent four bands and residues. B).The same as in case A, but for truncated basis of three operators a_p, b_p, c_p Ref. [124].

operators $b_r^{(s)+}, b_r^{(t)+}$ create accordingly a singlet or triplet spin-electron pair in the site r.

It is essential that the resulting four bands are only approximately symmetric relative to the magnetic Brillouin zone boundary. In the momentum space the distance between two quasiparticle states with the equal energy can differ from the antiferromagnetic vector Q. This means that dictated by these bands the "shadow band" effect can be other than the usual one displaced by the Q vector in the two-sublattice Néel antiferromagnet [107].

The bottom of the spectrum of the elementary excitations turns out to be substantially determined by the b_p and d_p-states. At the same time, as it may be seen from Fig.20A(b) (the solid line), the possible filling of each p state by a particle with a fixed spin is not too small for the lowest band and is close to 0.1. Let us remind that such a strong deviation of filling from 1 must lead to a relatively large Fermi surface even at small filling.

Let us underline the importance of taking into account the Q-polaron d_p. In order to illustrate this circumstance, in Fig.20B we show the spectrum of the elementary excitations calculated for the same values of energetic parameters as in Fig.20A, but using the basis of the first three operators a_p, b_p, c_p. As it may be seen by comparison of Figs.20A and 20B, this leads to the shift of the lowest band bottom from point $\mathbf{p} = (0,0)$ to point $\mathbf{p} = (\pi,\pi)$. At the same time the value of the residue Z_p for the band bottom is close to 0.1 as in Fig.20.

If three energetic parameters t, I, J are of the same order of magnitude then the elementary excitation in any of the four bands is a coherent superposition of the states a, b, c, d and these states have relatively the same weight.

Let us stress, that ignoring the Q-polarons, i.e., the states c and d, leads to the disappearance of bands, and can essentially change the remaining bands. The importance of taking into account Q-polarons in such models as $t - J$ and the three-band Hubbard model, where the particle motion is strongly coupled to the spin subsystem, will be demonstrated in the next Section.

Finally, let us mention once more that our calculations are performed in the limit of low filling. Nevertheless, the calculated upper bands have physical meaning if we study the electron transitions. In a general case it may be seen that the one-particle spectral density strongly depends on the filling.

2   Effective three-band model

In the present Section we shall study the local polaron, dressed in an antiferromagnetic spin wave, for a more realistic model, described by effective three-band Hamiltonian (it was introduced above).

$$\hat{H} = \hat{T} + \hat{h} + \hat{J}, \qquad \hat{T} = \tau \sum_{\mathbf{R},\mathbf{a}_1,\mathbf{a}_2,\sigma_1,\sigma_2} Z_{\mathbf{R}}^{\sigma_1\sigma_2} c^+_{\mathbf{R}+\mathbf{a}_2,\sigma_2} c_{\mathbf{R}+\mathbf{a}_1,\sigma_1}$$
$$\hat{h} = -h \sum_{\mathbf{R},\mathbf{a},\mathbf{b},\sigma} c^+_{\mathbf{R}+\mathbf{a},\sigma} c_{\mathbf{R}+\mathbf{a}+\mathbf{b},\sigma}, \qquad \hat{J} = \tfrac{J}{2} \sum_{\mathbf{R},\mathbf{g}} \hat{\mathbf{S}}_{\mathbf{R}} \hat{\mathbf{S}}_{\mathbf{R}+\mathbf{g}}. \qquad (157)$$

Here the CuO_2 plane is described by the square sublattice with lattice constant g and two O sites per a Cu unit cell; $\mathbf{R}$ — the vectors of Cu sites, $\mathbf{R}+\mathbf{a}$ — are four vectors of O sites nearest to Cu site $\mathbf{R}$, $\mathbf{a} = \pm\mathbf{a}_x, \pm\mathbf{a}_y$, $\mathbf{a}_x = g(\frac{1}{2},0)$, $\mathbf{a}_y = g(0,\frac{1}{2})$. We assume $g = 1$. In (157) we take the notations: $\mathbf{b}$ are the nearest neighbour (NN) vectors for the oxygen sublattice; $\mathbf{g} = 2\mathbf{a}$ and $\mathbf{d} = 2\mathbf{b}$ are the first- and second- NNs for the Cu sublattice; the operators c_σ^+ and $Z^{\sigma 0}$ create a hole with spin $S = \frac{1}{2}$ and spin projection $\frac{\sigma}{2}$ ($\sigma = \pm 1$) at the O and Cu sites respectively and $Z_\mathbf{R}^{\sigma_1 \sigma_2}$ are the Hubbard projection operators which are convenient for excluding doubly occupied Cu sites.

The first term $\hat{T}$ in (157) describes the effective hole hopping with amplitude from O to O sites through the intervening Cu sites. The term $\hat{J}$ corresponds to the AFM interaction between Cu sites. $\hat{h}$ represents the direct O–O NN hopping.

In this Section we take the zero temperature $T = 0$, and the spin subsystem has a nonzero effective magnetization M^2.

Let us discuss the hole excitations with spin $S = \frac{1}{2}$ and the spin projection $\frac{\sigma}{2}$. We shall restrict ourselves to a finite number of site operators $A_{\mathbf{R},j}$, ($1 \le j \le 12$), in each unit cell $\mathbf{R}$.

In order to treat the hole excitations in the framework of the local spin-polaron concept we introduce six site operators for each unit cell $\mathbf{R}$ which describe the polaron of small radius.

$$A^+_{\mathbf{R},\sigma,1(2)} = c^+_{\mathbf{R}+\mathbf{a}_x(\mathbf{a}_y),\sigma}, \qquad A^+_{\mathbf{R},\sigma,3(4)} = \sigma \sum_{\gamma=\pm 1} \gamma X^{\bar{\gamma}:\bar{\sigma}}_\mathbf{R} c^+_{\mathbf{R}+\mathbf{a}_x(\mathbf{a}_y),\gamma}$$
$$A^+_{\mathbf{R},\sigma,5(6)} = \sigma \sum_{\gamma=\pm 1} \gamma X^{\bar{\gamma}:\bar{\sigma}}_{\mathbf{R}+\mathbf{g}_x,(\mathbf{g}_y)} c^+_{\mathbf{R}+\mathbf{a}_x,(\mathbf{a}_y),\gamma}, \qquad \bar{\sigma} = -\sigma \tag{158}$$

$$A^+_{\mathbf{k},\sigma,i} = \frac{1}{\sqrt{N}} \sum_\mathbf{R} e^{i\mathbf{k}\mathbf{R}} A_{\mathbf{R},\sigma,i}$$

$A^+_{\mathbf{k},\sigma,i}$ are the Fourier transforms of $A^+_{\mathbf{R},\sigma,i}$.

As it was shown before, this basis of local spin-polaron operators gives the proper description of the experimentally observed important features of the CuO_2 plane hole spectrum: extended saddle point and isotropic band bottom.

In this Section we shall demonstrate the role of the delocalized spin polarons which correspond to the coupling of the local polarons to the AFM spin wave with momentum $\mathbf{Q} = (\pi,\pi)$, the so called Q-polarons.

As before we consider the spin system to be in a spherically symmetric state. At $T = 0$ the average value $\langle S_\mathbf{Q}\rangle = 0$ but $\langle S_\mathbf{Q}S_\mathbf{Q}\rangle$ can be treated as a macroscopic value. Then the coupling of local polaron states to $S_\mathbf{Q}$ corresponds to new delocalized states. In order to take these states into account we introduce the additional six operators based on the basis of (158):

$$A^+_{\mathbf{R},\sigma,j} = \sigma \sum_{\gamma=\pm 1} \gamma Q_\mathbf{R}^{\bar{\gamma}\bar{\sigma}} A^+_{\mathbf{R},\gamma,i}, \qquad A^+_{\mathbf{k},\sigma,j} = \sigma \sum_\gamma \gamma A^+_{\mathbf{k}+\mathbf{Q},\gamma,i} S_\mathbf{Q}^{\bar{\gamma}\bar{\sigma}},$$
$$Q_\mathbf{R}^{\bar{\gamma}\bar{\sigma}} \equiv e^{i\mathbf{Q}\mathbf{R}} S_\mathbf{Q}^{\bar{\gamma}\bar{\sigma}} = N^{-1} \sum_{\mathbf{R}_1} e^{i\mathbf{Q}(\mathbf{R}+\mathbf{R}_1)} X_{\mathbf{R}_1}^{\bar{\gamma}\bar{\sigma}}, \qquad j = i+6, \; i = (1 \div 6). \tag{159}$$

In order to determine the spin polaron spectrum $\varepsilon_i(\mathbf{k})$ of 12 quasiparticle bands we use the two-time retarded matrix Green's functions $G_{i,j}(t,\mathbf{k})$ for the operators $A_{\mathbf{k},\sigma,i}$:

$$G_{i,j}(t,\mathbf{k}) \equiv \left\langle A_{\mathbf{k},i}(t) \mid A^+_{\mathbf{k},j}(0) \right\rangle = -i\Theta(t) \left\langle \left\{ A_{\mathbf{k},i}(t), A^+_{\mathbf{k},j}(0) \right\} \right\rangle, \qquad (160)$$

We solve the system of the equations of motion for $G_{i,j}(\omega, \mathbf{k})$ by using the standard Mori-Zwanzig projection technique and restricting ourselves to the above chosen basis of operators $\{A_{\mathbf{k},\sigma,i}\}$ (158,159).

Note that the dependence of the spin polaron excitations spectrum on the long-range correlation function $\langle S_Q S_Q \rangle$ appears only due to the treatment of the Q-polaron states [124,24]. Below we take the following numerical values of the spin-correlation functions: $\langle S_R S_{R+g} \rangle = -0.3521$, $\langle S_R S_{R+d} \rangle = 0.229$, $\langle S_R S_{R+2g} \rangle = 0.2$, $M^2 = 0.0914$. The matrix elements for matrices D and K were calculated in the low hole doping limit, $n \ll 1$, where n is the total number of oxygen holes per unit cell. This approximation is justified by the fact that we are mainly interested in the lowest band. As it will be seen below, the maximum filling of the lowest band is $n \approx 0.26$.

The Green's functions have the form

$$G_{i,j}(\omega, \mathbf{k}) = \sum_{l=1}^{12} \frac{Z^{(l)}_{(i,j)}(\mathbf{k})}{\omega - \varepsilon_l(\mathbf{k})}. \qquad (161)$$

In particular the value of $Z_h(\mathbf{k}) = Z^{(1)}_{(1,1)}(\mathbf{k}) + Z^{(1)}_{(2,2)}(\mathbf{k})$ corresponds to the number of bare oxygen holes with the fixed spin σ and the momentum $\mathbf{k}$ in the state $|\mathbf{k}, \sigma\rangle$ of the lowest quasiparticle band $\varepsilon_1(\mathbf{k})$. Let us remind that spectral weight (residue) $Z_{(i,j)}(\mathbf{k})$ satisfy the sum rule $\sum_s Z^{(s)}_{(i,i)}(\mathbf{k}) = 1, i = 1, 2$. This means that in this model the Luttinger theorem is not fulfilled and the maximum number of holes per cell is equal to four despite the presence of twelve bands.

The results of the most interesting two lowest bands $\varepsilon_1(\mathbf{k})$ and $\varepsilon_2(\mathbf{k})$ for realistic values of model parameters $J_1 = 0.2\tau$, $h = 0.4\tau$ (below all energetical parameters are expressed in units τ) are presented in Fig.21a. In the same figure the spectrum of the lowest of six bands $\underline{\varepsilon}_1(\mathbf{k})$, calculated in the approximation of six operators (158) is shown. In Fig.21b the spectrum $\varepsilon_1(\mathbf{k})$ is presented by the equal-energy lines $\varepsilon_1(\mathbf{k}) = \text{const}$. Let us remind that the Q-polarons may lead to a rather complex form of the Fermi surface. This circumstance can give a nontrivial behaviour of the Hall effect on doping, and even cause the inversion of the Hall constant if the Fermi energy is close to $\varepsilon_1(\mathbf{k}) = -4.5$.

As it is seen from Fig.21a, the inclusion of the Q-wave qualitatively leads to the decoupling of the lowest band of the local small polaron excitations and $\varepsilon_1(\mathbf{k})$ is close to $\underline{\varepsilon}_1(\mathbf{k})$. This means that the main features of the lowest band excitations previously calculated in the local polaron approximation [15] are preserved.

The importance of the treatment of the Q-wave polarons may be seen if we discuss the filling of the lowest band by bare holes. In Fig.21c the filling of the $\varepsilon_1(\mathbf{k})$ and $\underline{\varepsilon}_1(\mathbf{k})$ are shown, i.e., the spectral weights (residues) $Z_h(\mathbf{k})$ and $\underline{Z}_h(\mathbf{k})$ (the underlined values correspond to local polaron approximation). One can see that the introduction of the Q-polarons leads to the essential decrease of hole filling along

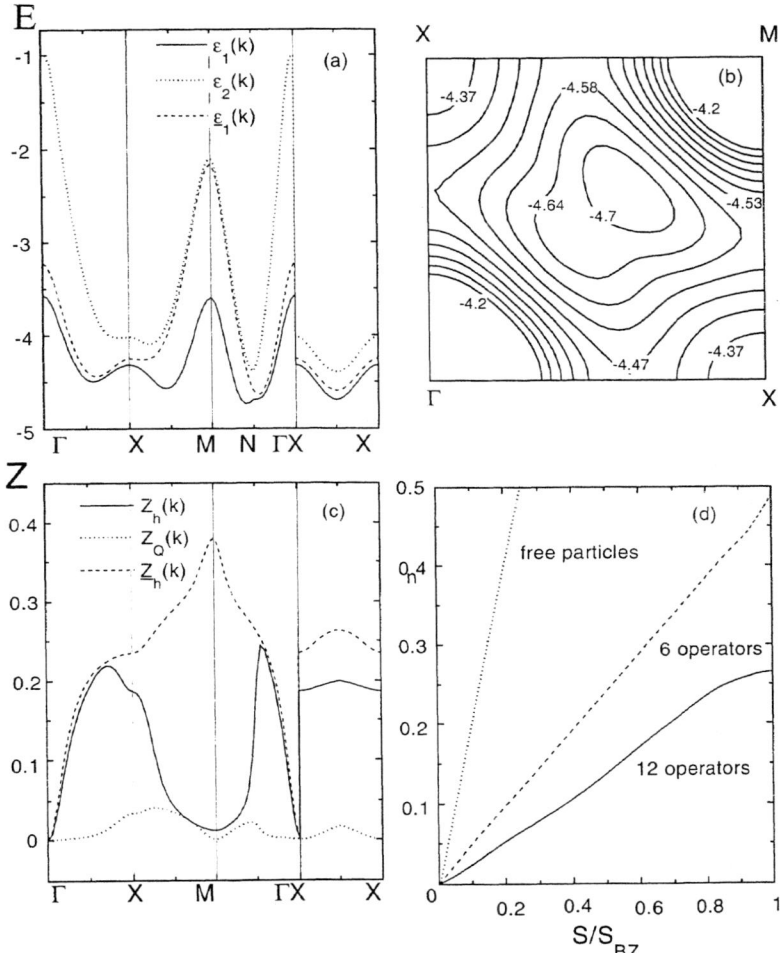

FIGURE 21. a) The spectra $\varepsilon_1(\mathbf{k})$ and $\varepsilon_2(\mathbf{k})$ calculated for the basis (158) and (159), and the spectrum $\underline{\varepsilon}_1(\mathbf{k})$ calculated for the basis (158). The spectra are given along symmetry directions $\Gamma - X - M - N - \Gamma$ and $X - N - X$; $\Gamma = (0,0)$; $X = (\pi,0), (0,\pi)$, $M = (\pi,\pi)$; $N = (\pi/2, \pi/2)$. b) Spectrum equal-energy lines $\varepsilon_1(\mathbf{k}) = const.$ c) $Z_h(\mathbf{k})$ and $\underline{Z}_h(\mathbf{k})$ — the number of bare holes (the spectral weights (residues) of corresponding Green's functions) in the quasiparticle excitations for the spectra $\varepsilon_1(\mathbf{k})$ and $\underline{\varepsilon}_1(\mathbf{k})$; $Z_Q(\mathbf{k})$ is the residue of the lowest pole $\varepsilon_1(\mathbf{k})$ for the Green's functions $G_{11,11}(\omega, \mathbf{k}) + G_{12,12}(\omega, \mathbf{k})$, which characterize the "shadow band" effect. d) The dependence of the number of holes n per unit cell on the value of the Fermi surface area S (S_{BZ} - the area of the first Brillouin zone): solid line — for the spectrum $\varepsilon_1(\mathbf{k})$; dashed line — for the spectrum $\underline{\varepsilon}_1(\mathbf{k})$; dotted straight line — the case of the noninteracting particle filling. Ref. [28].

the lines X-M and M-N. This redistribution of the bare hole spectrum weight explains the results of photoemission experiments when the "flat band region" is observed along the direction X-Γ and is invisible along the line X-M. Fig.21d demonstrates that the local polaron concept (six operators (158)) leads to a strong decrease of the filling under the Fermi surface. This means the violation of the Luttinger theorem approximately in four times (the analogous effect was found in [85]). But the inclusion of Q-polaron states (159) leads to the additional essential reduction of the filling, approximately in 1.5 times. The maximum $\varepsilon_1(\mathbf{k})$ -band filling is equal to $n = 0.22$, and the smallness of this value justifies our low density approximation.

Formula (159) points out that the Q-polaron $A^-_{\mathbf{k},\sigma,7(8)}$ contains a bare hole state $c^+_{\mathbf{k}+\mathbf{Q},\sigma}$. It means that the residual $Z_\mathbf{Q}$ of the corresponding Green's function $G_\mathbf{Q}(\mathbf{k}) = G_{7,7} + G_{8,8}$ is responsible for the "shadow band" effect [107]. The close results may be obtained in the case of the $t - J$ model [79].

We are to note that the Q-polaron scenario reproduces the essential decrease of the lowest band width with the decrease of the AFM constant J. Usually such an effect is obtained only in the self-consistent Born approximation [132].

B Influence of temperature and frustration (doping)

1 Polaron structure at $T \neq 0$

In Section V A the splitting of the local polaron band due to coupling with the antiferromagnetic spin wave $S_\mathbf{Q}$ at $T = 0$ was described. The description of this splitting was based on the nonzero macroscopic value $\langle S_\mathbf{Q} S_\mathbf{Q} \rangle$. It is obvious, that the spectrum must change continuously with the increase of temperature and the mentioned splitting cannot disappear, even though at any finite temperature $\langle S_\mathbf{Q} S_\mathbf{Q} \rangle$ is no longer macroscopic. Here we shall present a method for describing the evolution of spectrum on temperature. The method is based on the introduction of a more complex structure of a spin polaron: superposition of a local spin polaron, a local spin polaron dressed by a continuum of spin waves $S_\mathbf{q}$ with $\mathbf{q}$ close to $\mathbf{Q}$, and additionally a local spin polaron dressed by $S_\mathbf{Q}$ if $T = 0$. We illustrate the reconstruction of the spin polaron structure by an example of the Kondo-lattice model with hoppings to the first-, second-, and third-nearest neighbours. This model is more realistic compared to a simple model with hoppings only to the first nearest neighbors, because it gives a minimum of the lowest band close to point $\mathbf{k} = (\pi/2, \pi/2)$.

The Hamiltonian has the following form

$$\hat{H} = \hat{J} + \hat{T} + \hat{I} \qquad (162)$$

$$\begin{aligned}
\hat{J} &= J \sum_\mathbf{r} a^+_\mathbf{r} \tilde{\mathbf{S}}_\mathbf{r} a_\mathbf{r}, \\
\hat{T} &= \tau_g \sum_{\mathbf{r},\mathbf{g}} a^+_{\mathbf{r}+\mathbf{g}} a_\mathbf{r} + \tau_d \sum_{\mathbf{r},\mathbf{d}} a^+_{\mathbf{r}+\mathbf{d}} a_\mathbf{r} + \tau_{2g} \sum_{\mathbf{r},\mathbf{g}} a^+_{\mathbf{r}+2\mathbf{g}} a_\mathbf{r}, \\
\hat{I} &= \tfrac{I}{2} \sum_{\mathbf{r},\mathbf{g}} \hat{\mathbf{S}}_\mathbf{r} \hat{\mathbf{S}}_{\mathbf{r}+\mathbf{g}}
\end{aligned}$$

Here $\mathbf{g} = \pm\mathbf{g}_x \pm \mathbf{g}_y$ are the vectors of nearest neighbours; $\mathbf{d}$ and $2\mathbf{g}$ are the vectors of the second- and the third-nearest neighbours. The Fermi operator $a^+_{\mathbf{r},\sigma}$ creates a hole with spin $S = 1/2$ and spin projection $\sigma/2$ ($\sigma = \pm 1$). The notation $\tilde{S}_\mathbf{r} = S_\mathbf{r}^\alpha \sigma^\alpha$ is used in the Kondo interaction Hamiltonian $\hat{J}$.

To study elementary excitations on the basis of the spin-polaron approach, we introduce the following basis of site operators:

$$A_{\mathbf{r},1} = a_\mathbf{r}, \qquad A_{\mathbf{r},2} = \tilde{S}_\mathbf{r} a_\mathbf{r}. \tag{163}$$

To take into account an infinite-radius polaron at $T = 0$, we also include the operators:

$$A_{\mathbf{r},j} = \tilde{Q}_\mathbf{r} A_{\mathbf{r},i}, \qquad \tilde{Q}_\mathbf{r} = N^{-1} \sum_\rho e^{i\mathbf{Q}(\mathbf{r}+\rho)} \tilde{S}_{\mathbf{r}+\rho}, \quad j = i+2, \; i = (1 \div 2) \tag{164}$$

which describe the pairing of local polaron with spin excitations with long-range order. We note that for $M = \langle \mathbf{S}_\mathbf{Q} \mathbf{S}_\mathbf{Q} \rangle = 0$ the contribution from these operators vanishes. Finally, we introduce the operators

$$A_{\mathbf{r},j} = \tilde{Q}^{(1)}_\mathbf{r} A_{\mathbf{r},i}, \qquad \tilde{Q}^{(1)}_\mathbf{r} = N^{-1} \sum_{\rho, \mathbf{q} \in \Omega} e^{i\mathbf{q}(\mathbf{r}+\rho)} \tilde{S}_{\mathbf{r}+\rho}, \quad j = i+4, \; i = (1 \div 2). \tag{165}$$

Here Ω is a small region of the Brillouin zone near the (π, π), presented below in Fig. 23. The summation does not include $\mathbf{q} = \mathbf{Q}$.

The operators $A_{\mathbf{r},5}$ and $A_{\mathbf{r},6}$ describe an intermediate-radius polaron, i.e. the coupling of a local polaron with spin waves, whose momenta belong to square regions around $\mathbf{Q}$ with linear dimensions L.

To determine the spin-polaron spectrum $\varepsilon_i(\mathbf{k})$ (here i is the band number) we shall employ the matrix of two-time retarded Green's functions $G_{i,j}(t, \mathbf{k})$ for the operators $A_{\mathbf{k},\sigma,i}$.

We shall solve the system of equations of motion for $G_{i,j}(\omega, \mathbf{k})$ using the standard Mori-Zwanzig projection method [141] and we shall confine ourselves to the previously chosen basis of operators $\{A_{\mathbf{k},\sigma,i}\}$. Corresponding matrix elements are rather cumbersome and we do not present them here.

At $T = 0$ (case 1) we study a basis consisting of six operators $i = 1 - 6$. At finite temperature (case 2) the operators $A_{\mathbf{k},\sigma,3}$ and $A_{\mathbf{k},\sigma,4}$ are absent in the set of operators since they describe an infinite-radius polaron which exists only in the presence of long-range order.

As a result, the Green's function has the form

$$G_{i,j}(\omega, \mathbf{k}) = \sum_{l=1}^{12} \frac{Z^{(l)}_{(i,j)}(\mathbf{k})}{\omega - \varepsilon_l(\mathbf{k})}. \tag{166}$$

Specifically, the quantity $Z^{(l)}_{(1,1)}(\mathbf{k})$ refers to the number of bare holes with fixed spin σ and momentum $\mathbf{k}$ in the state $|\mathbf{k}, \sigma\rangle$ of the quasiparticle band $\varepsilon_l(\mathbf{k})$. We note that the spectral weights (residues) $Z_{(i,j)}(\mathbf{k})$ satisfy the sum rule $\sum_s Z^{(s)}_{(1,1)}(\mathbf{k}) = 1$.

We note that the dependence of the spin-polaron excitations spectrum ε_l on the correlation function $M = \langle S_Q S_Q \rangle$ is manifested only when the Q-polaron states $A_{r,3}$ and $A_{r,4}$ are taken into account. To determine the spectrum $\varepsilon(\mathbf{k})$ it is necessary to know the spin correlation function $\langle S_{R_1} S_{R_2} \rangle$. We calculated them in the same manner as in [27,13].

Specifically, the following numerical values of the spin correlation function were obtained at $T = 0$ $\langle S_R S_{R+g} \rangle = -0.332$, $\langle S_R S_{R+d} \rangle = -0.145 + M$ and $\langle S_R S_{R+2g} \rangle = -0.144 + M$, where $M = 0.0577$. The following values of the model parameters were used: $\tau = 1, \tau_{2g} = 0.5\tau, \tau_d = \tau_{2g}, \tau_g = -\tau_{2g}, J = 5\tau_{2g}$ and $I = 0.5\tau_{2g}$, which were chosen so that the minimum of the lower band would lie near the point $(\pi/2, \pi/2)$. The value of parameter L, which gives for the operators $A_{\mathbf{q},5(6)}$ the region Ω was chosen by a variational method for each $\mathbf{k}$ in the Brillouin zone. It was found that region Ω strongly depends on $\mathbf{k}$. For example, $L(0,0) = 0.05\pi$ and $L(\pi,\pi) = 0.35\pi$. The calculations were performed in the light-doping limit $n << 1$, n being the total number of carriers per cell. Case 2 is represented by temperature $T = 0.1I$, which corresponds to a correlation length $\xi = 40$ lattice constants.

The results for the most interesting lower bands $\varepsilon_l(\mathbf{k})$ for $\mathbf{k}$ along the symmetry direction from $\Gamma = (0,0)$ to $M = (\pi,\pi)$ are presented in Figs. 22A and 22B, respectively, for cases 1 and 2. Fig. 23 displays the constant-energy surfaces of the first band. As one can see from Figs. 22A(a) and 22B(a) the minimum of the lower band in both cases lies near point $(\pi/2, \pi/2)$, which is characteristic in the Emery model, and corresponds to the experimentally observed spectrum [201]. As one can see from these figures, cases $T = 0$ and $T \neq 0$ are similar, with the exception of the vanishing of the second band in case 2, where there is no long-range order.

Figures 22A(b) and 22B(b) display bare-carrier weights $Z_{1,1}$ as function of $\mathbf{k}$ for the cases $T = 0$ and $T \neq 0$, respectively. The second band for the case $T = 0$ has negligible residues, and except for its presence the figures are similar. Therefore the set of operators $A_{1,2,3,4}$ at $T = 0$ can be adequately replaced by the set of operators $A_{1,2,5,6}$ when long-range order vanishes.

We call attention to the fact that the result calculations on the basis of the Mori-Zwanzig method for the lower band are qualitatively identical to the computation results for the coherent part of the spectral function $A(\mathbf{q}, \omega) = -\frac{1}{\pi} \mathrm{Im} G_{11}$ by the self-consistent Born approximation method [114], which will be presented in the last Section. Specifically, both methods describe the "shadow-band" effect [107,3]. According to this effect, if the motion of the electron occurs against the background of a Néel state with a doubled lattice period, then each coherent peak of the spectral function will have a corresponding peak with the same energy but with wave vector $\mathbf{q} + \mathbf{Q}$. In the case of a spherically symmetric spin state, the distance between the two peaks is only approximately equal to $\mathbf{Q}$, and, most importantly, the heights of the peaks (i.e., the corresponding residues) differ in magnitude (in Fig.23 the residue at point P_1 is 0.4, while the residue at point P_2 is 0.2). It is important that the shadow-band effect does not vanish at $T \neq 0$.

We also note that the differences of the residues of the lower band in Fig. 22A(b) and 22B(b) from 1 have the effect that the Luttinger theorem is strongly violated.

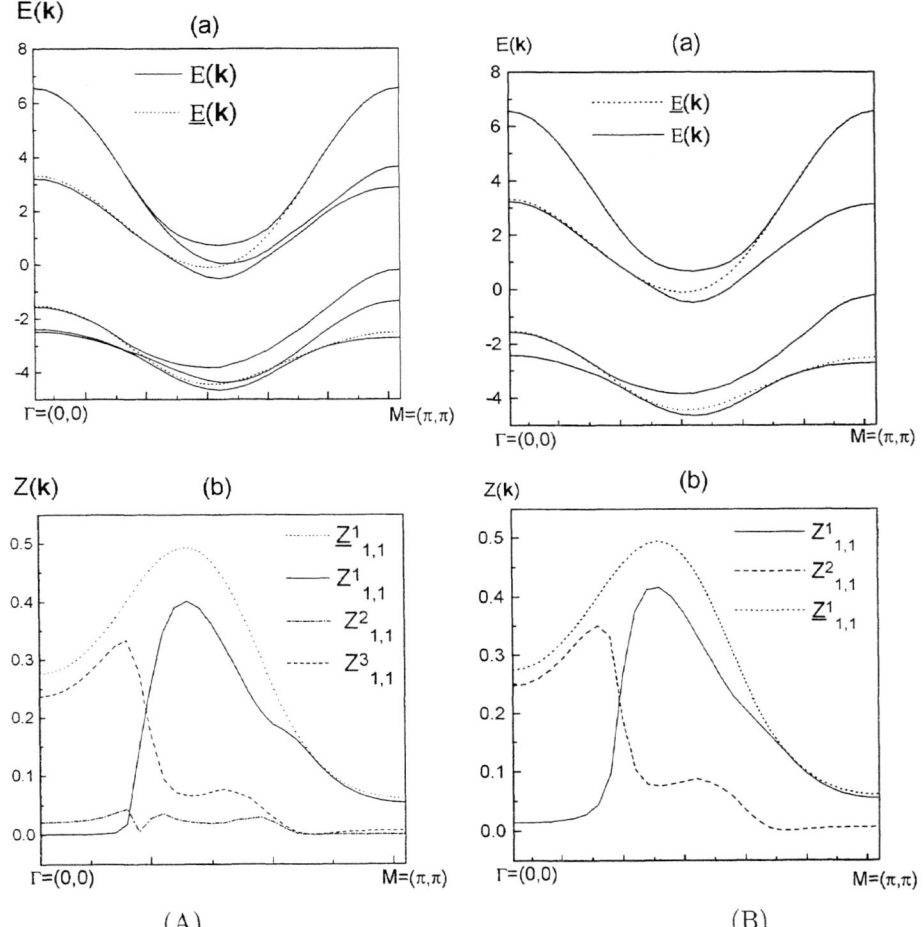

FIGURE 22. A).(a) Spectrum $E^i(k)$ at $T=0$ along $k=k_x=k_y$. $\underline{E}^{(i)}(k)$ – is the local-polaron approximation. (b) Residues $Z_{1,1}^{(i)}(k)$ of $<<a_{k\sigma} \mid a_{k\sigma}^+>>$ which correspond to the poles of $E^{(i)}(k)$. B).(a) Spectrum $E^i(k)$ of quasiparticle excitations at $T=0.1I$ along $k=k_x=k_y$. $\underline{E}^{(i)}(k)$ – is the local-polaorn approximation. (b) Residues $Z_{1,1}^{(i)}(k)$ of $<<a_{k\sigma} \mid a_{k\sigma}^+>>$ which correspond to poles of $E^{(i)}(k)$. $\underline{Z}_{1,1}^{(1)}(k)$ corresponds to $\underline{E}^{(1)}(k)$ in the local-polaron approximation. Ref. [28].

For example, if in our model there are 0.16 states below the Fermi level $\varepsilon = -4.0$ (see Fig. 23), then for the case of the same one-electron spectrum in the tight-binding model there are 0.50 states.

It is interesting that, as our calculations show, the form of the bands remains qualitatively the same right up to temperatures at which the spin correlation length ξ decreases to 3-6 lattice constants.

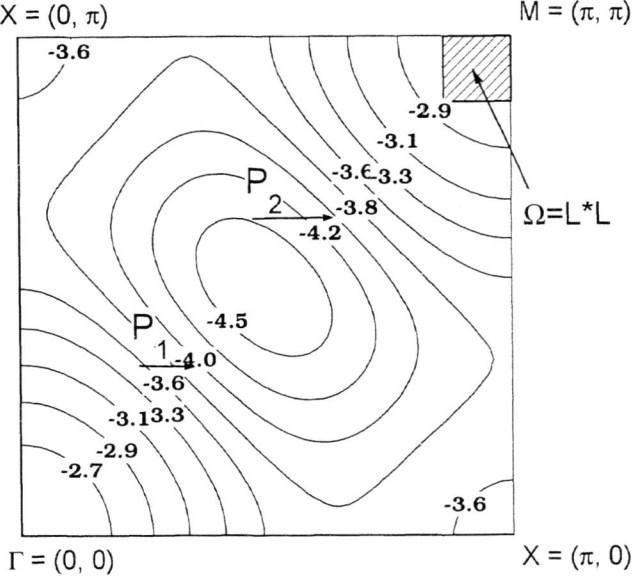

FIGURE 23. Constant-energy surfaces of the lower band $E^{(1)}(k)$ at $T = 0.1I$ Ref. [28].

2 From underdoped to overdoped regime

In the present Section it will be shown that many of the aspects of APRES experiments on 2D AFM mentioned in the Introduction can be explained qualitatively in the framework of the spin-polaron concept if one takes into account a complex structure of a spin polaron and the frustration of the spin subsystem [16]. Let us remind that it is generally believed that the frustration in the spin subsystem is governed mainly by doping (see for example [97]).

In our discussion below namely the frustration in the spin subsystem modifies the spin polaron spectrum. That is why when we speak about the evolution of the spectrum on doping we suppose some qualitative equivalence between the doping and frustration in the spin subsystem. Note, that frustration can take place even for undoped 2D AFM if one takes into account a realistic structure of CuO_2 plane interactions [8].

The problem is studied on the basis of the Kondo lattice Hamiltonian with electron hopping to first, second and third nearest neighbours (such a Hamiltonian gives partial mapping for the Emery model [156,160,161]).

The Hamiltonian has the form:

$$H^{tot} = H_0 + H_1 + H_2,$$
$$H_0 = \sum_{rg} t_g a^+_{r+g} a_r + \sum_{rd} t_d a^+_{r+d} a_r + \sum_{r2g} t_{2g} a^+_{r+2g} a_r, \qquad (167)$$
$$H_1 = J \sum_r a^+_r \tilde{S}_r a_r, \quad H_2 = \tfrac{1}{2} I_1 \sum_{rg} S^\alpha_{r+g} S^\alpha_r + \tfrac{1}{2} I_2 \sum_{rd} S^\alpha_{r+d} S^\alpha_r.$$

Here $\mathbf{g} = \pm\mathbf{g_x} \pm \mathbf{g_y}$ are the vectors of the nearest neighbours; $\mathbf{d}$ and $2\mathbf{g}$ are the vectors of the second- and third-nearest neighbours. The Fermi operator $a_{\mathbf{r},\sigma}^+$ creates a hole with spin $S = 1/2$ and spin projection $\sigma/2$ at a lattice site $\mathbf{r}$. We omit spin indexes and for the on-site Kondo interaction $\hat{J}$ we use the notation $\tilde{S}_\mathbf{r} = S_\mathbf{r}^\alpha \sigma^\alpha$. The Hamiltonian $\hat{T}$ describes hole hopping between first, second and third-nearest neighbours with amplitudes t_g, t_d, t_{2g}. The exchange Hamiltonian $\hat{I}$ corresponds to the AFM frustrated interaction between the spins, p ($0 \le p \le 1$) — the frustration parameter, $I_1 = (1-p)I$ and $I_2 = pI$ are the exchange constants for the first and second nearest neighbours.

Treating this model in the two sublattice spin structure, Schrieffer [173] pointed out the crucial importance of taking into account certain coherence factors. We shall treat the system following the spherically symmetric approach [13]. In order to take into account the mentioned coherent factor at finite temperature one must treat the spin polaron operator as the superposition of the local spin polaron states $A_{\mathbf{r},1}$, $A_{\mathbf{r},2}$ and $A_{\mathbf{r},3}$, $A_{\mathbf{r},4}$ — two local polarons dressed by spin waves with momenta q which are close to the antiferromagnetic vector $\mathbf{Q} = (\pi, \pi)$, where spin susceptibility is sharply peaked. This operators have the form:

$$A_{\mathbf{r},1} = a_\mathbf{r}; \quad A_{\mathbf{r},2} = \tilde{S}_\mathbf{r} a_\mathbf{r}; \quad A_{\mathbf{r},3} = \tilde{Q}_\mathbf{r}^{(1)} a_\mathbf{r}; \quad A_{\mathbf{r},4} = \tilde{Q}_\mathbf{r}^{(1)} \tilde{S}_\mathbf{r} a_\mathbf{r}; \tag{168}$$

$$\tilde{Q}_\mathbf{r}^{(1)} = N^{-1} \sum_{\rho, q \in \Omega} e^{i\mathbf{q}(\mathbf{r}+\rho)} \tilde{S}_{\mathbf{r}+\rho}, \quad \Omega = \{\mathbf{q}, |\pm\pi - q_{x,y}| < L\} \tag{169}$$

As it was shown in [24,28] the two operators $A_{\mathbf{r},3}$, $A_{\mathbf{r},4}$ (spin polarons of intermediate radius) are important for the description of the splitting of the local polaron bands. We take region Ω to be square regions around $\mathbf{Q}$ (described by q which satisfies inequality (169)) with linear dimension L (four squares $\Omega = L \times L$ related to corners of the first Brillouin zone (BZ)). Below we take $L = 0.5\pi$.

To obtain the one-hole spectrum $\varepsilon(\mathbf{k})$ we use the two-time retarded matrix Green's function $G_{ij}(\omega, \mathbf{k})$ for the Fourier transforms of the operators $A_{\mathbf{r},i}$. We use the standard projection method to solve the Green function equations and treat the problem in the limit of low doping (we will consider doping $x < 0.25$). The projection method matrix elements have a cumbersome form and are expressed over a static spin-spin correlation functions. These spin–spin correlation functions were calculated in a spherically symmetric approach taking frustration into account. As a result the Green's functions have the form:

$$G_{ij}(\omega, \mathbf{k}) = \sum_{l=1}^{4} \frac{Z_{ij}^{(l)}(\mathbf{k})}{\omega - \varepsilon_l(\mathbf{k})}. \tag{170}$$

In particular the value of $Z_h(\mathbf{k}) = Z_{1,1}^{(1)}(\mathbf{k})$ gives the weight of bare oxygen holes with the fixed spin projection σ and momentum $\mathbf{k}$ in the state $|\mathbf{k}, \sigma\rangle$ of the lowest quasiparticle band $\varepsilon_1(\mathbf{k})$. Let us remind that spectral weight (residue) $Z_{i,j}(\mathbf{k})$

satisfies the sum rule $\sum_s Z_{11}^{(s)}(\mathbf{k}) = 1$. This means that in this model the Luttinger theorem is not fulfilled and the maximum number of holes per cell is equal to two despite the presence of four bands.

The qualitative reduction of the effective three band model (see [23,15]) to the model under discussion leads to the following set of Hamiltonian parameters: $t_g = 0.5\tau; t_d = 0.25\tau; t_{2g} = 0.2\tau; J = 3\tau; I = 0.4\tau$ where $\tau = t^2/\varepsilon_{pd}$. Below we put $\tau = 1$. We present the results of spin-polaron spectrum calculation for temperature $T = 0.3I$. Let us mention that for the frustration value parameter $p > 0.1$ the spectrum has a weak temperature dependence up to temperature $T \sim 0.5I$.

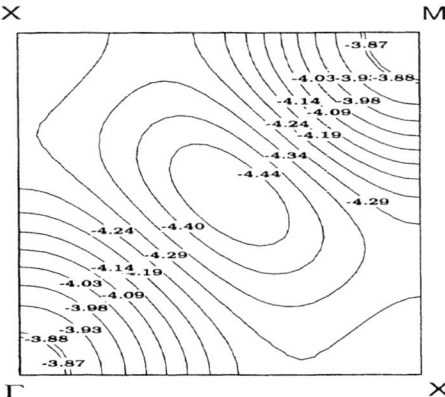

FIGURE 24. Spectrum of the lowest band for $p = 0.1, T = 0.3I$. Equal-energy lines $\varepsilon_1(\mathbf{k}) = $ const, line $\varepsilon_f = -4.44$ correspond to the Fermi surface for doping $x = 0.032$. Ref. [16].

In Fig. 24 the spectrum $\varepsilon_1(\mathbf{k})$ is presented by the equal-energy lines $\varepsilon_1(\mathbf{k}) = $ const for the frustration parameter $p = 0.1$. We suppose that this value of frustration correspond to a small doping case. As it is seen from Fig. 24 the minimum of $\varepsilon_1(\mathbf{k})$ is close to $(\pi/2; \pi/2)$ and the spectrum is rather isotropic near the band bottom. The dispersion along the directions $\Gamma(0;0)-M(\pi;\pi)$, and $\Gamma-X(\pi;0)-M$ reproduces the ARPES results (compare for example Fig. 24 with the dispersion in Fig. 3 in Ref. [130]). The band width is also close to ARPES results if we take a realistic value of parameter $\tau = 0.4$ eV. The spin-polaron spectrum in Fig. 24 has a symmetry close to the symmetry of magnetic BZ but the residues $Z_{1,1}^{(1)}(\mathbf{k})$ have the symmetry of the initial BZ (for example $Z_{1,1}^{(1)}(\mathbf{k} = (0;0)) = 0.1 \neq Z_{1,1}^{(1)}(\mathbf{k} = (\pi;\pi)) = 0.21$). The residues $Z_{1,1}^{(1)}(\mathbf{k})$ are close to 0.3 near the band bottom. If we suppose that the frustration value parameter $p = 0.1$ is related to the doping $x = 0.032$ then the Fermi surface corresponds to the equienergetical line $\varepsilon_f = -4.44$ given in Fig. 24. If we will determine a pseudogap δ as $\delta = \varepsilon_1(\pi, 0) - \varepsilon_f$ ([166]) then the value of the pseudogap is equal to 0.19.

In Figs. 25a and 25b we give correspondingly the spin-polaron spectrum and the

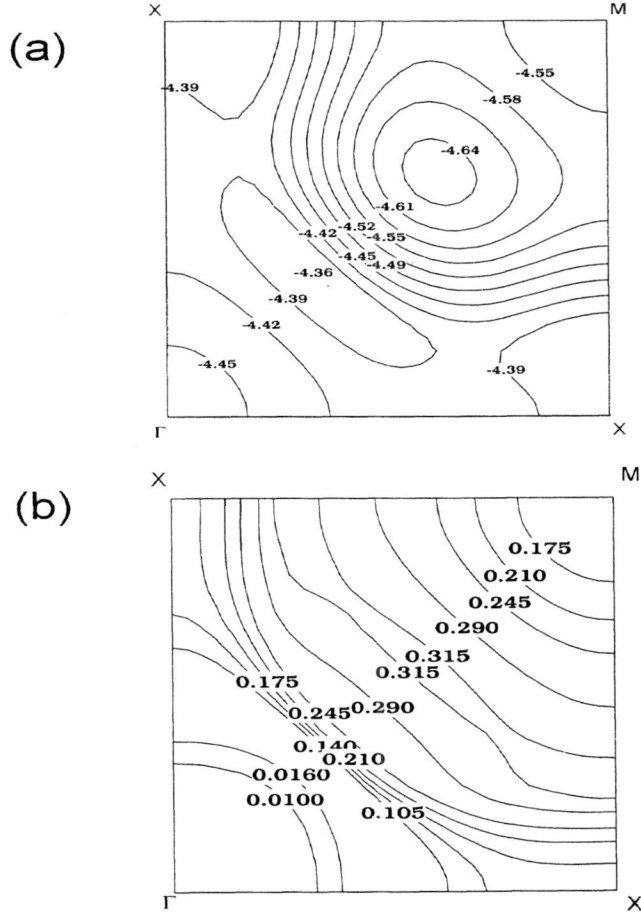

FIGURE 25. Spectrum of the lowest band for $p = 0.3, T = 0.3I$. (a) Equal-energy lines $\varepsilon_1(\mathbf{k}) = $ const, line $\varepsilon_f = -4.437$. correspond to the Fermi surface for doping $x = 0.25$. (b) Residues contour lines $Z_{1,1}^{(1)}(\mathbf{k}) = $ const for $\varepsilon_1(\mathbf{k})$ $p = 0.3$.

residues $Z_{1,1}^{(1)}(\mathbf{k})$ for the frustration parameter value $p = 0.3$ which we relate to the doping $x = 0.25$. The evolution of the spectrum on frustration leads to the Fermi surface which corresponds to $\varepsilon_f = -4.437$. The corresponding equienergetical line

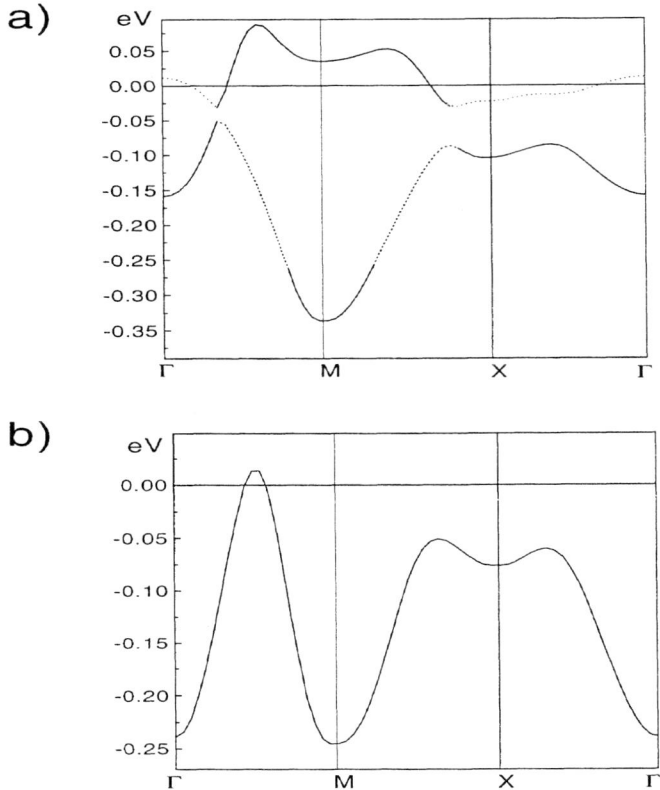

FIGURE 26. The electronic spectrum in eV along symmetrical lines Zero energy corresponds to the Fermi level. (a) The first and second bands for $p = 0.3, T = 0.3I$. The part of the both bands which has considerable residues is plotted by the solid line, the part with small residues – by the dotted line. (b) The first band for $p = 0.1, T = 0.3I$ Ref. [16].

crosses the $X - M$ boundary. Such a crossing is given by ARPES experiments for optimally doped (or overdoped) HTSC ([121,166,130]). The pseudogap in this case is $\delta = 0.04$. Let us mention that the residues $Z_{1,1}^{(1)}(\mathbf{k})$ have a sudden drop on the $(\pi/2; \pi/2)$ to $(0; 0)$ cut. We don't reproduce the second band, but the residues in the second band $Z_{1,1}^{(2)}(\mathbf{k})$ are not small in the region close to Γ and we think that namely this band is given by ARPES for the mentioned area. Another important feature of the calculated spectrum is the existence of a flat band region close to point X.

In Figs. 26a and 26b we represent the spectrum along the symmetric directions of BZ, correspondingly for $p = 0.3$ and for $p = 0.1$. In this figure for $p = 0.3$ we take into account the second band (dashed line) in the region close to Γ and miss a part of the $\varepsilon_1(\mathbf{k})$ spectrum where the residues are small. The comparison with the ARPES experiment (see Fig. 3 in [130]) shows that the spin-polaron approach qualitatively reproduces the evolution of the hole spectrum from an undoped to an optimally doped region.

VI SELF-CONSISTENT BORN APPROXIMATION FOR A LOCAL SPIN-POLARON IN THE EFFECTIVE THREE-BAND MODEL

In recent years much work has been devoted to treating the hole motion in the two-dimensional antiferromagnet in the framework of the self-consistent Born approximation (SCBA) [33]. Mainly this work was devoted to the 2D t-J model [41,170,108,55,132] treating the antiferromagnet as a two-sublattice Néel state. It was shown that the hole motion is characterized by a quasiparticle band of bandwidth $\sim 2J$ that is separated from the incoherent part [41,170,108,55,132].

The dispersion energy minimum is found at momentum $(\pi/2, \pi/2)$ and is very close to the value of the energy at $(\pi, 0)$. Such a result was obtained by several methods. One of them is the self-consistent Born approximation for a reduced version of the t-J model consisting of spinless holes coupled to spinons. [170,108,132,120,152].

This method has become especially popular since it is sufficiently simple and allows the calculation of the spectral function. It gives rise to a dispersion and also a spectral function which is symmetric with respect to the magnetic Brillouin zone (BZ). Similar results were obtained for the Emery model. [56,106,181].

In a remarkable experiment [201] the quasiparticle dispersion of such a spin polaron was measured directly by creating a photohole in the antiferromagnetic insulator $Sr_2CuO_2Cl_2$. Whereas the width of the dispersion was in reasonable agreement with the theoretical predictions, its shape had some deviations especially at point $(\pi, 0)$ in momentum space. Afterwards it was shown that the deviations can be diminished by taking into account hopping terms to second and third neighbors in the t-J model. [181,144,12] These additional terms correspond to a proper treatment of direct oxygen-oxygen hopping in the Emery model. [181,203] However, some deviations between theory and experiment remain. So, the measurement shows no quasiparticle weight at the Γ-point $(0,0)$ while the spin polaron in the t-J model has a definite weight. Next, one observes quite a remarkable damping especially away from the minimum. Some theories (especially the usual SCBA) are symmetric with respect to the antiferromagnetic (AFM) BZ, whereas the experimental data show a strong asymmetry: after crossing the AFM BZ boundary (coming from the Γ-point) the quasiparticle looses its weight nearly abruptly. This deficiency, however, seems to be merely a property of the usual SCBA and not of the t-J model itself. [57,62]

In the usual SCBA there is no dispersion of the bare spinless hole; the dispersion appears only due to the coupling to the spinons. Therefore, one might expect large vertex corrections. However, the low order vertex corrections cancel in the pure t-J model as it was noted in Refs. [132,120]. Such a cancelation is absent in the more realistic Emery model. But there are many indications (see also Ref. [200]), and we shall add some arguments below, that the photoemission data cannot be explained by the t-J model alone, at least not throughout the whole BZ.

In this Section we concentrate on the spin-fermion model (the reduced form of the three-band Emery model) [60,61] for the CuO$_2$ plane and present a new scheme [18,114] to calculate the spectral function starting from the mean-field description [23,15] of the local spin polaron (not a bare hole). One can expect that the fluctuations around the mean field spin polaron are smaller than around the dispersionless spinless hole. The mean-field description of the spin polaron in the Emery model was presented in previous Sections.

Below we present a technique to calculate the spectral function of quasiparticles with a complicated operator structure. The technique combines the Mori-Zwanzig projection method with a self-consistent calculation of the self-energy in terms of irreducible Green's functions (Tserkovnikov technique [194]). Such approach gives the possibility to introduce a simultaneous description of a spin polaron and a bare hole, and investigate the effects of direct oxygen-oxygen hopping which only gives us the possibility to reach a realistic description. Furthermore, we also discuss the upper part of the spectrum in the spin-fermion model whose lowest branch corresponds to the spin polaron. Numerical results for the spectral function show that one can distinguish three regions in **k**-space — where the spin polaron exists as a quasiparticle with infinite lifetime, with finite lifetime or where it is strongly overdamped. The discussed method leads to several qualitatively new features which are not present in the SCBA of spinless holes coupled to spinons. But these features may be observed in the experimental data for Sr$_2$CuO$_2$Cl$_2$. We are concentrating here on a qualitative discussion rather than on fine tuning the parameters of the Emery model in its spin-fermion form to obtain maximal agreement with experiment.

A   Basis operators and projection method

As it was mentioned above, the main features of the hole motion in the CuO$_2$ plane are described by the model: [60,61]

$$\hat{H} = \hat{\tau} + \hat{J} + \hat{h} , \tag{171}$$

$$\hat{\tau} = 4\tau \sum_{\mathbf{R}} p_{\mathbf{R}}^{\dagger} \left(\frac{1}{2} + \tilde{S}_{\mathbf{R}}\right) p_{\mathbf{R}} , \qquad \hat{J} = \frac{J}{2} \sum_{\mathbf{R},\mathbf{g}} S_{\mathbf{R}}^{\alpha} S_{\mathbf{R}+\mathbf{g}}^{\alpha} , \tag{172}$$

$$\hat{h} = -h \sum_{\mathbf{R}} \left[c_{\mathbf{R}+\mathbf{a}_x}^{\dagger} \left(c_{\mathbf{R}+\mathbf{a}_y} + c_{\mathbf{R}-\mathbf{a}_y} + c_{\mathbf{R}+\mathbf{g}_x+\mathbf{a}_y} + c_{\mathbf{R}+\mathbf{g}_x-\mathbf{a}_y} \right) + h.c. \right] , \tag{173}$$

with

$$p_{\mathbf{R}} = \frac{1}{2} \sum_{\mathbf{a}} c_{\mathbf{R}+\mathbf{a}}, \quad \tilde{S}_{\mathbf{R}} \equiv S_{\mathbf{R}}^{\alpha} \hat{\sigma}^{\alpha}, \quad \left\{p_{\mathbf{R}}, p_{\mathbf{R}'}^{\dagger}\right\} = \delta_{\mathbf{R},\mathbf{R}'} + \frac{1}{4} \sum_{\mathbf{g}} \delta_{\mathbf{R},\mathbf{R}'+\mathbf{g}} , \tag{174}$$

where $\mathbf{a}_{x,y} = \frac{1}{2}\mathbf{g}_{x,y}$, $\mathbf{g} = \pm\mathbf{g}_x, \pm\mathbf{g}_y$.

Here and below the summation over repeated indexes is understood everywhere; $\{\ldots,\ldots\}$, $[\ldots,\ldots]$ stand for anticommutator and commutator respectively; $\mathbf{g}_{x,y}$ are basis vectors of a copper square lattice ($|\mathbf{g}| \equiv 1$), $\mathbf{R} + \mathbf{a}$ are four vectors of O sites nearest to the Cu site $\mathbf{R}$; the operator $c^{\dagger}_{\mathbf{R}+\mathbf{a}}$ creates a hole predominantly at the O site (the spin index is dropped in order to simplify the notations); $\hat{\sigma}^{\alpha}$ are the Pauli matrices; the operator $\mathbf{S}$ represents the localized spin on the copper site. As before (see Chap.2) we do not introduce the explicit relative phases of p- and d-orbitals since they can be transformed out by redefining the operators with phase factors $\exp(i\mathbf{q}_0\mathbf{R})$, $\mathbf{q}_0 = (\pi, \pi)$. In order to compare our results with other authors and with experiment we restore these phases by changing in the final results $\mathbf{k} \to \mathbf{k}' = \mathbf{k} - \mathbf{q}_0$. The parameter τ is the hopping amplitude of oxygen holes that take into account the coupling of the hole motion with the copper spin subsystem, J is the constant of the nearest neighbour AFM exchange between the copper spins. Throughout the Section for the parameter values we take $\tau \sim 0.5\text{eV}$, $J = 0.2\tau$ and $h = 0.3\tau$.

From the very beginning let us properly take into account the local correlations $\hat{\tau}$ without the violation of spin commutation relations. For the spin-liquid state with the spin rotational symmetry for the copper subsystem the spin-spin correlation functions satisfy the relation $C_r \equiv \langle S^{\alpha}_{\mathbf{R}} S^{\alpha}_{\mathbf{R}+\mathbf{r}} \rangle = 3 \langle S^{x(y,z)}_{\mathbf{R}} S^{x(y,z)}_{\mathbf{R}+\mathbf{r}} \rangle$. We choose a set of three basis operators. The first two of them constitute the Zang-Rice (ZR) polaron

$$B_{1,\mathbf{k}} = \frac{1}{\beta_{\mathbf{k}}\sqrt{N}} \sum_{\mathbf{R}} e^{-i\mathbf{k}\mathbf{R}} p_{\mathbf{R}}, \quad B_{2,\mathbf{k}} = \frac{1}{\nu_{\mathbf{k}}\sqrt{N}} \sum_{\mathbf{R}} e^{-i\mathbf{k}\mathbf{R}} \tilde{S}_{\mathbf{R}} p_{\mathbf{R}}, \tag{175}$$

where the factors $\beta_{\mathbf{k}}$ and $\nu_{\mathbf{k}}$ arise due to the orthonormalization

$$\beta_{\mathbf{k}} = \sqrt{1 + \gamma_{\mathbf{k}}}, \quad \nu_{\mathbf{k}} = \sqrt{\frac{3}{4} + C_g \gamma_{\mathbf{k}}}, \tag{176}$$

and $\gamma_{\mathbf{k}} \equiv \frac{1}{2}(\cos k_x + \cos k_y)$. In fact, the basis operators (175) can be combined in two ways corresponding to the ZR singlet and part of the ZR triplet state. Only the lower lying singlet combination builds the spin polaron quasiparticle. We can also write

$$B_{1,\mathbf{k}} = \frac{1}{\beta_{\mathbf{k}}} \left[\cos\left(\frac{k_x}{2}\right) c_{\mathbf{k},x} + \cos\left(\frac{k_y}{2}\right) c_{\mathbf{k},y} \right] \tag{177}$$

where

$$c_{\mathbf{k},x,y} = \frac{1}{\sqrt{N}} \sum_{\mathbf{R}} e^{-i\mathbf{k}(\mathbf{R}+\mathbf{a}_{x,y})} c_{\mathbf{R}+\mathbf{a}_{x,y}}. \tag{178}$$

Now it is easy to see that one has to introduce the operator

$$B_{3,\mathbf{k}} = \frac{1}{\beta_{\mathbf{k}}} \left[\cos\left(\frac{k_y}{2}\right) c_{\mathbf{k},x} - \cos\left(\frac{k_x}{2}\right) c_{\mathbf{k},y} \right] \tag{179}$$

in order to represent the full set of bare hole operators. That is important if we want to find the hole spectral weight of the spin polaron band.

In a first step we construct the three eigenoperators $\mathcal{B}_{i,\mathbf{k}}$ and the corresponding bands $\Omega_{\mathbf{k}}^{(i)}$ in the mean field approach:

$$\mathcal{B}_{i,\mathbf{k}} = \alpha_j^{(i)}(\mathbf{k}) B_{j,\mathbf{k}}, \quad \mathcal{B}_{\mathbf{k}} \equiv \mathcal{B}_{1,\mathbf{k}}, \quad \Omega_{\mathbf{k}} \equiv \Omega_{\mathbf{k}}^{(1)}, \tag{180}$$

where we introduced a simplified notation for the most interesting lowest spin polaron band $\Omega_{\mathbf{k}}$ with the polaron annihilation eigenoperator $\mathcal{B}_{\mathbf{k}}$. We use the projection method of Mori and Zwanzig [141,72] and introduce the retarded Green's functions (GF):

$$G_{ij}(\mathbf{k},\omega) = \langle B_{i,\mathbf{k}} | B_{j,\mathbf{k}}^\dagger \rangle_\omega \equiv -\imath \int_{t'}^\infty dt e^{\imath\omega(t-t')} \langle \{B_{i,\mathbf{k}}(t), B_{j,\mathbf{k}}^\dagger(t')\} \rangle. \tag{181}$$

Here we use Zubarev's notations. [207] Within the Mori-Zwanzig projection method (PM) the equations of motion for the GF (181) are projected onto the subspace spanned by the operators $B_{i,\mathbf{k}}$ which leads to the following eigenvalue problem to determine $\alpha_j^{(n)}$ and $\Omega_{\mathbf{k}}^{(n)}$:

$$\omega G_{ij}^{(0)}(\mathbf{k},\omega) = \delta_{ij} + \mathcal{L}_{il} G_{lj}^{(0)}(\mathbf{k},\omega), \quad \left(\mathcal{L}_{ij} - \Omega_{\mathbf{k}}^{(n)} \delta_{ij}\right) \alpha_j^{(n)}(\mathbf{k}) = 0 \tag{182}$$

where

$$\mathcal{L}_{ij} \equiv \langle\{[B_{i,\mathbf{k}},\hat{H}], B_{j,\mathbf{k}}^\dagger\}\rangle; \quad \langle\{B_{i,\mathbf{k}}, B_{j,\mathbf{k}}^\dagger\}\rangle = \delta_{ij}. \tag{183}$$

An explicit calculation expresses $\mathcal{L}_{ij}$ in terms of the two-site spin-spin correlation functions $C_{\mathbf{r}}$ and gives:

$$\begin{aligned}
\mathcal{L}_{11} &= 2\tau\beta_{\mathbf{k}}^2 - 8h\pi_{\mathbf{k}}^2\beta_{\mathbf{k}}^{-2}, \quad \mathcal{L}_{12} = \mathcal{L}_{21} = 4\tau\beta_{\mathbf{k}}\nu_{\mathbf{k}}, \quad \mathcal{L}_{13} = \mathcal{L}_{31} = -4h\pi_{\mathbf{k}}\delta_{\mathbf{k}}\beta_{\mathbf{k}}^2, \\
\mathcal{L}_{22} &= \tau\nu_{\mathbf{k}}^{-2}\left[-9/8 + C_{\mathbf{g}}(1-4\gamma_{\mathbf{k}}) + 8^{-1}\sum_{\mathbf{g}_1\neq\mathbf{g}_2} e^{-\imath\mathbf{k}(\mathbf{g}_1-\mathbf{g}_2)}C_{\mathbf{g}_1-\mathbf{g}_2}\right] \\
&\quad + J\nu_{\mathbf{k}}^{-2}C_{\mathbf{g}}(\gamma_{\mathbf{k}}-4) - h\nu_{\mathbf{k}}^{-2}\left[3/2 + 4C_{\mathbf{g}}\gamma_{\mathbf{k}} + 2C_{\mathbf{g}_x+\mathbf{g}_y}(\gamma_{\mathbf{k}}^2 - \delta_{\mathbf{k}}^2)\right], \\
\mathcal{L}_{23} &= \mathcal{L}_{32} = 0, \quad \mathcal{L}_{33} = 8h\pi_{\mathbf{k}}^2\beta_{\mathbf{k}}^{-2}
\end{aligned} \tag{184}$$

where

$$\delta_{\mathbf{k}} \equiv \frac{1}{2}(-\cos k_x + \cos k_y), \quad \pi_{\mathbf{k}} \equiv \cos\left(\frac{k_x}{2}\right)\cos\left(\frac{k_y}{2}\right). \tag{185}$$

The projected equation of motion (182) defines the mean field Green's functions $G_{ij}^{(0)}$ with three bands $\Omega_{\mathbf{k}}^{(1)}, \ldots, \Omega_{\mathbf{k}}^{(3)}$ corresponding to the ZR singlet, the nonbonding oxygen and the triplet bands. The mean field spectrum of all three bands is shown in Fig. 27a. In the following we shall mainly concentrate on the eigenoperator of the lowest band, the polaron eigenoperator $\mathcal{B}_{\mathbf{k}} = \mathcal{B}_{1,\mathbf{k}}$. The GF of the polaron quasiparticle is defined as:

$$G_p(\mathbf{k}, \omega) = \langle \mathcal{B}_{\mathbf{k}} | \mathcal{B}_{\mathbf{k}}^{\dagger} \rangle_{\omega} . \quad (186)$$

One has to distinguish between the polaron GF defined in (186) and the GF $G_h(\mathbf{k}, \omega)$ which gives the number of holes in a unit cell and is introduced as:

$$G_h(\mathbf{k}, \omega) = \langle c_{\mathbf{k},x} | c_{\mathbf{k},x}^{\dagger} \rangle_{\omega} + \langle c_{\mathbf{k},y} | c_{\mathbf{k},y}^{\dagger} \rangle_{\omega} = \langle B_{1,\mathbf{k}} | B_{1,\mathbf{k}}^{\dagger} \rangle_{\omega} + \langle B_{3,\mathbf{k}} | B_{3,\mathbf{k}}^{\dagger} \rangle_{\omega} . \quad (187)$$

Below we shall suppose that the intensity given by the photoemission experiments can be roughly compared with the hole spectral function which is dictated by $G_h(\mathbf{k}, \omega)$. In the mean field approximation $G_h^{(0)}(\mathbf{k}, \omega)$ has the form

$$G_h^{(0)}(\mathbf{k}, \omega) = \sum_i \frac{\left|\alpha_1^{(i)}(\mathbf{k})\right|^2 + \left|\alpha_3^{(i)}(\mathbf{k})\right|^2}{\omega - \Omega_{\mathbf{k}}^{(i)}} = \sum_i \frac{Z^{(i)}(\mathbf{k})}{\omega - \Omega_{\mathbf{k}}^{(i)}} , \quad (188)$$

which defines the pole strength $Z^{(i)}(\mathbf{k})$ of all three bands. If we restrict ourselves to the vicinity of the lowest pole $\Omega_{\mathbf{k}} = \Omega_{\mathbf{k}}^{(1)}$ the hole GF can be approximated by

$$G_h^{(0)}(\mathbf{k}, \omega) \approx Z(\mathbf{k}) G_p^{(0)}(\mathbf{k}, \omega) , \quad (189)$$

with

$$Z(\mathbf{k}) = Z^{(1)}(\mathbf{k}) , \quad \text{and} \quad G_p^{(0)} = \frac{1}{\omega - \Omega_{\mathbf{k}}} . \quad (190)$$

Let us first discuss the mean field spectrum of all three bands (see Fig.27a). First of all, one observes a well separated singlet band between 1 and 2 eV below the nonbonding one. The singlet-triplet splitting is maximal at (π, π) and roughly 3 eV (assuming τ to be 0.5 eV). That agrees quite reasonably with the splitting between the ZR singlet and triplet states of 3.4 eV that were found in calculations for a CuO_4 cluster. [63] Experimental indications for the triplet state were found in a recent photoemission measurement of the substance $Ba_2Cu_3O_4Cl_2$. [169] It is also interesting to note that the singlet band dispersion nearly obeys the antiferromagnetic symmetry (with only small deviations due to the spin rotational invariant ground state) but the triplet band does not. The nonbonding band receives a dispersion due to h. It has the maximum (which would correspond to a binding energy maximum in a spectroscopic measurement) at (π, π). One should note, however, that in our model we did not include all the three oxygen $2p$ orbitals at a given site. There are also other nonbonding oxygen bands and especially the one with the lowest binding energy at (π, π) (seen experimentally in Ref. [155]) is not present in our model.

Now, we discuss the lowest spin polaron band $\Omega_{\mathbf{k}}$. From Fig.27b we see that the pole strength $Z(\mathbf{k})$ vanishes at the Γ point (with $\mathbf{k} = (0,0)$). This result is the

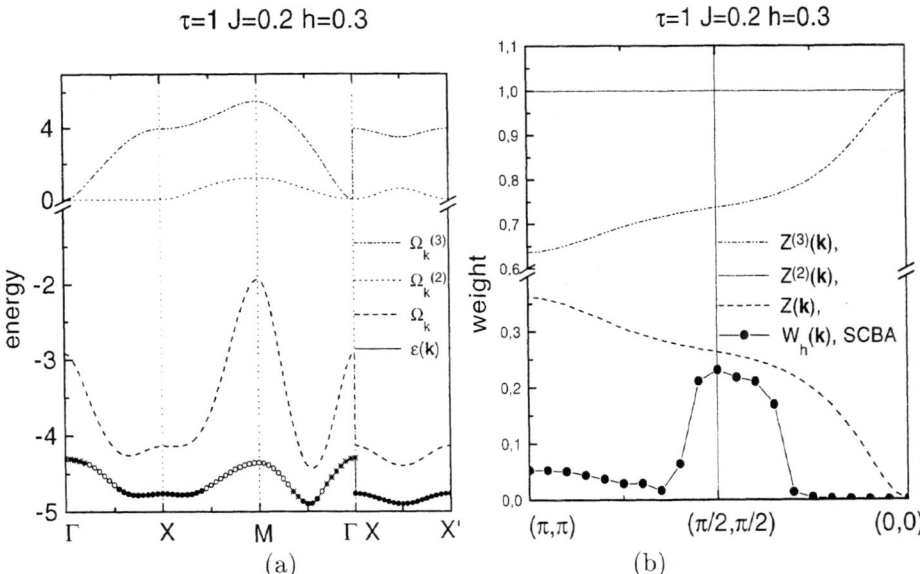

FIGURE 27. a) The dispersion of the lowest QP band $\varepsilon(\mathbf{k})$ and the mean field (MF) dispersions $\Omega_\mathbf{k}^{(i)}$ along high symmetry lines in the Brillouin zone for $J = 0.2$, $\tau = 1$ and $h = 0.3$. Filled and open circles mark quasiparticles with infinite and finite lifetime. Stars correspond to the lowest peak in the spectral density which is strongly overdamped. b) The QP weights $Z^{(i)}(\mathbf{k})$ of all three bands in the MF and the area under the lowest peak of the hole spectral function $W_h(\mathbf{k})$.

consequence of a vanishing hybridization between oxygen and copper states at this point. Furthermore, the triplet and the nonbonding oxygen bands are degenerated here. Consequently, all the orbitals which are incorporated into our model give rise to only one peak in the photoemission spectra at the Γ point. And indeed, the spin polaron is not visible in the photoemission spectra there. [201,155] Note that the vanishing of hybridization, and consequently the vanishing of the spin polaron is ignored in the course of Emery model reduction to the one-band Hamiltonian (see Refs. [203,205] and references therein), and thus the hole spectral weight is nonzero at $\mathbf{k} = (0,0)$ in the t-J and other one-band models. Even the theories for the three-band Emery model which start from the Néel order [106,181] cannot obtain zero spectral weight at $\mathbf{k} = (0,0)$ since they always deal with a linear combination of states with momentum $\mathbf{k}$ and $\mathbf{k} + \mathbf{q}_0$ ($\mathbf{q}_0 = (\pi,\pi)$). But if we go beyond the mean-field approximation in the mentioned models, spectral weight may be small when damping is large at Γ-point [153,154].

We also note an apparent similarity of the polaron branch $\Omega_\mathbf{k}$ of the present spectrum with the calculations based on the Néel state and the linear spin wave theory for t-J [170,108,55,132,152] and three-band [56,106,181] models. However, our results show a slight deviation from the symmetry of the magnetic Brillouin zone $\Omega_{\mathbf{k}} \neq \Omega_{\mathbf{k}+\mathbf{q}_0}$ due to the spin singlet state of the magnetic background. We see from Fig. 27 that the mean field polaron bandwidth $w \sim \tau$ ($w = 2.4\tau$). That is larger than expected due to the following reason: in the limit of small $J/\tau \ll 1$ a state $|\mathcal{B}_{\mathbf{k}}\rangle = \mathcal{B}_{\mathbf{k}}^{\dagger}|\mathcal{G}\rangle$ for a momentum $\mathbf{k}$ near $(0,0)$ or (π, π) far from the band bottom has an energy $\Omega_{\max} \sim \Omega_{\min} + w$. This energy may be much greater than that of the states

$$|\mathcal{Y}_{\mathbf{k},\mathbf{q}}\rangle = \tilde{S}_{\mathbf{q}} \mathcal{B}_{\mathbf{k}-\mathbf{q}}^{\dagger}|\mathcal{G}\rangle \qquad (191)$$

where we defined

$$\tilde{S}_{\mathbf{q}} = \frac{1}{\sqrt{N}} \sum_{\mathbf{R}} e^{i\mathbf{q}\mathbf{R}} \tilde{S}_{\mathbf{R}}. \qquad (192)$$

The energy of the states $|\mathcal{Y}_{\mathbf{k},\mathbf{q}}\rangle$ is in the order of $\Omega_{\min} + J$ for $\mathbf{k} - \mathbf{q}$ near the band bottom. It means that $|\mathcal{B}_{\mathbf{k}}\rangle$ is unstable with respect to the decay into $|\mathcal{Y}_{\mathbf{k},\mathbf{q}}\rangle$ states and analogous states, which contain more spinwaves. These states contain spin distortions that are situated far from the hole. In order to take them into account within the PM we should extend the basis set up to infinity. On the other hand the effect of these states can be described by treating the scattering of the spin polaron in terms of irreducible GF, also known as the Tserkovnikov technique. As we shall see, this repairs the too large polaron bandwidth in the simple projection method.

B Local polaron scattering

We treat the polaron operator $\mathcal{B}_{\mathbf{k}}^{\dagger} = \mathcal{B}_{1,\mathbf{k}}^{\dagger}$ (180) as a candidate for the elementary excitation and calculate the corresponding two-time retarded GF $G_p(\mathbf{k},\omega)$ which is defined in (186). The Dyson equation for G_p has the form

$$G_p^{-1}(\mathbf{k},\omega) = \left[G_p^{(0)}\right]^{-1} - \Sigma(\mathbf{k},\omega); \quad \Sigma(\mathbf{k},\omega) = \langle \mathcal{R}_{\mathbf{k}} | \mathcal{R}_{\mathbf{k}}^{\dagger} \rangle_\omega^{(irr)}, \qquad (193)$$

with

$$G_p^{(0)} = (\omega - \Omega_{\mathbf{k}})^{-1}; \quad \mathcal{R}_{\mathbf{k}} = [\mathcal{B}_{\mathbf{k}}, H] \qquad (194)$$

and where we used the irreducible GF

$$\langle \mathcal{R}_{\mathbf{k}} | \mathcal{R}_{\mathbf{k}}^{\dagger} \rangle_\omega^{(irr)} = \langle \mathcal{R}_{\mathbf{k}} | \mathcal{R}_{\mathbf{k}}^{\dagger} \rangle_\omega - \langle \mathcal{R}_{\mathbf{k}} | \mathcal{B}_{\mathbf{k}}^{\dagger} \rangle_\omega \frac{1}{\langle \mathcal{B}_{\mathbf{k}} | \mathcal{B}_{\mathbf{k}}^{\dagger} \rangle_\omega} \langle \mathcal{B}_{\mathbf{k}} | \mathcal{R}_{\mathbf{k}}^{\dagger} \rangle_\omega \qquad (195)$$

in accordance with the definition given by Tserkovnikov. [194] The results for $G_p(\mathbf{k},\omega)$ can be used to calculate $G_h(\mathbf{k},\omega)$ for low energies

$$G_h(\mathbf{k},\omega) \approx Z(\mathbf{k}) G_p(\mathbf{k},\omega) . \qquad (196)$$

It should be underlined that Eq. (193) coincides only formally with the Dyson equation for the causal Green's function. The "self-energy" $\Sigma(\mathbf{k},\omega)$ in (193) has no diagrammatic representation and it is represented by more complex Green's functions on the right hand side of Eq. (195). The derivation of Eq. (195) and its relation to the conventional projection technique [141,72,68] are represented in [114,154]. As follows from (193) Re Σ gives the renormalization of the energy of the polaron $B_\mathbf{k}^\dagger$ and Im Σ gives its damping when $|\text{Im }\Sigma(\mathbf{k},\omega)| \ll |\text{Re }\Sigma(\mathbf{k},\omega)|$. In a general case the elementary excitations should be investigated self-consistently. The self-energy $\Sigma(\mathbf{k},\omega)$ is dictated by the interaction of the polaron with the spin subsystem, i.e., by the polaron scattering on the spin waves. For this reason the main problem of the present technique consists in calculating the irreducible Green's function (195).

To determine $\mathcal{R}_\mathbf{k}$ we have to calculate the commutator with the Hamiltonian. That gives in detail

$$\begin{aligned}
\left[B_{1,\mathbf{k}}, \hat{H}\right] &= \left(2\tau\beta_\mathbf{k}^2 - 8h\pi_\mathbf{k}^2\beta_\mathbf{k}^{-2}\right) B_{1,\mathbf{k}} + 4\tau\beta_\mathbf{k}\nu_\mathbf{k} B_{2,\mathbf{k}} - 4h\pi_\mathbf{k}\delta_\mathbf{k}\beta_\mathbf{k}^{-2} B_{3,\mathbf{k}}\,, \\
\left[B_{2,\mathbf{k}}, \hat{\tau}\right] &= 3\tau\beta_\mathbf{k}\nu_\mathbf{k}^{-1} B_{1,\mathbf{k}} - \tau 2 B_{2,\mathbf{k}} \\
&+ \tau 2 N^{-1/2}\nu_\mathbf{k}^{-1}\sum_\mathbf{q}\gamma_{\mathbf{k}-\mathbf{q}}\left(\beta_{\mathbf{k}-\mathbf{q}} Y_{1,\mathbf{k},\mathbf{q}} + 2\nu_{\mathbf{k}-\mathbf{q}} Y_{2,\mathbf{k},\mathbf{q}}\right)\,, \\
\left[B_{2,\mathbf{k}}, \hat{J}\right] &= 4JN^{-1/2}\nu_\mathbf{k}^{-1}\sum_\mathbf{q}\gamma_\mathbf{q}\sqrt{2/3}\nu_{\mathbf{k}-\mathbf{q}} Y_{J,\mathbf{k},\mathbf{q}}\,, \\
\left[B_{2,\mathbf{k}}, \hat{h}\right] &= -2h B_{2,\mathbf{k}} + hN^{-1/2}\nu_\mathbf{k}^{-1}\sum_\mathbf{q}\left[-\left(8\pi_{\mathbf{k}-\mathbf{q}}^2 - 2\beta_{\mathbf{k}-\mathbf{q}}^2\right)\beta_{\mathbf{k}-\mathbf{q}}^{-1} Y_{1,\mathbf{k},\mathbf{q}} \right. \\
&\left. - 4\pi_{\mathbf{k}-\mathbf{q}}\delta_{\mathbf{k}-\mathbf{q}}\beta_{\mathbf{k}-\mathbf{q}}^{-1} Y_{3,\mathbf{k},\mathbf{q}}\right]\,,
\end{aligned} \quad (197)$$

where we introduced the notation

$$Y_{i,\mathbf{k},\mathbf{q}} = \tilde{S}_\mathbf{q} B_{i,\mathbf{k}-\mathbf{q}}\,. \quad (198)$$

In the scattering term of the exchange energy $\hat{J}$ arises

$$Y_{J,\mathbf{k},\mathbf{q}} \equiv S_\mathbf{q}^\alpha\left[\frac{1}{\nu_{\mathbf{k}-\mathbf{q}}\sqrt{\frac{2}{3}N}}\sum_\mathbf{R} e^{-\imath(\mathbf{k}-\mathbf{q})\mathbf{R}}\left(\imath\epsilon_{\alpha\beta\gamma} S_\mathbf{R}^\beta \hat{\sigma}^\gamma p_\mathbf{R}\right)\right]\,. \quad (199)$$

which has a different structure ($\epsilon_{\alpha\beta\gamma}$ is the antisymmetric tensor). Therefore, it has to be projected

$$\left[B_{2,\mathbf{k}}, \hat{J}\right] \approx 4J\frac{1}{\nu_\mathbf{k}\sqrt{N}}\sum_\mathbf{q}\gamma_\mathbf{q}\left(\frac{2}{3}\right)\nu_{\mathbf{k}-\mathbf{q}} Y_{2,\mathbf{k},\mathbf{q}}, \quad (200)$$

using

$$\left\langle\left\{Y_{J,\mathbf{k},\mathbf{q}}, Y_{2,\mathbf{k},\mathbf{q}}^\dagger\right\}\right\rangle = \frac{2}{3}C_\mathbf{q}; \quad C_\mathbf{q} = \sum_\mathbf{R} e^{-\imath\mathbf{q}(\mathbf{n}-\mathbf{R})}\left\langle S_\mathbf{n}^\alpha S_\mathbf{R}^\alpha\right\rangle\,. \quad (201)$$

Finally one finds

$$\left[B_{3,\mathbf{k}}, \hat{H}\right] = -\frac{4h\pi_{\mathbf{k}}\delta_{\mathbf{k}}}{\beta_{\mathbf{k}}^2}B_{1,\mathbf{k}} + \frac{8h\pi_{\mathbf{k}}^2}{\beta_{\mathbf{k}}^2}B_{3,\mathbf{k}} . \tag{202}$$

After carrying out the projection (200), all terms $R_{i,\mathbf{k}} \equiv \left[B_{i,\mathbf{k}}, \hat{H}\right]$ can be represented in unique form

$$R_{i,\mathbf{k}} = \lambda_{ij} B_{j,\mathbf{k}} + \frac{1}{\sqrt{N}} \sum_{\mathbf{q}} g_{ij,\mathbf{k},\mathbf{q}} Y_{j,\mathbf{k},\mathbf{q}} \tag{203}$$

with coefficients λ_{ij} and $g_{ij,\mathbf{k},\mathbf{q}}$ that can be simply derived from (197,200) and (202). One should note that the scattering has its origin mainly in the second basis operator $B_{2,\mathbf{k}}$ (175) since only coefficients $g_{2j,\mathbf{k},\mathbf{q}}$ are different from zero. We find from Eq. (203):

$$\mathcal{R}_{\mathbf{k}} = \alpha_i^{(1)}(\mathbf{k}) R_{i,\mathbf{k}} = \alpha_i^{(1)}(\mathbf{k}) \lambda_{ij} B_{j,\mathbf{k}} + \check{R}_{\mathbf{k}} , \tag{204}$$

where we define

$$\check{R}_{\mathbf{k}} = \frac{1}{\sqrt{N}} \sum_{\mathbf{q}} \alpha_i^{(1)}(\mathbf{k}) g_{ij,\mathbf{k},\mathbf{q}} Y_{j,\mathbf{k},\mathbf{q}} = \frac{1}{\sqrt{N}} \sum_{\mathbf{q}} \tilde{S}_{\mathbf{q}} \left(\alpha_i^{(1)}(\mathbf{k}) g_{ij,\mathbf{k},\mathbf{q}} B_{j,\mathbf{k}-\mathbf{q}} \right) . \tag{205}$$

We see from Eqs. (193,195), that the self-energy $\Sigma(\mathbf{k}, \omega)$ accounting for interaction effects is expressed through higher order Green's functions. One should notice that the terms linear in $\mathcal{B}_{\mathbf{k}} \equiv B_{1,\mathbf{k}}$ in Eq. (204) do not contribute to the irreducible Green's function (195) for $\Sigma(\mathbf{k}, \omega)$. The terms $\propto \mathcal{B}_{j,\mathbf{k}}$, j= 2, 3, in Eq. (204) are orthogonal to $\mathcal{B}_{\mathbf{k}}$ and give nonvanishing contribution to the irreducible Green's function (195) only in the energy region of the upper polaron bands. If the lowest polaron band may be regarded as isolated (i.e. if it is energetically well separated from the other bands), we can neglect the interband scattering and retain only the intraband one. This is the case in our problem, since the energy gap (1.3τ for $h = 0$ and 1.9τ for $h = 0.3$, see Fig. 27) between the two lowest bands is about half of the mean field bandwidth of the polaron. Then we project the operator $[\alpha_i^{(1)}(\mathbf{k}) g_{ij,\mathbf{k},\mathbf{q}} B_{j,\mathbf{k}-\mathbf{q}}]$ onto $\mathcal{B}_{\mathbf{k}-\mathbf{q}}$ and finally obtain

$$\check{R}_{\mathbf{k}} \approx \frac{1}{\sqrt{N}} \sum_{\mathbf{q}} \Gamma(\mathbf{k}, \mathbf{q}) \tilde{S}_{\mathbf{q}} \mathcal{B}_{\mathbf{k}-\mathbf{q}} \tag{206}$$

where

$$\Gamma(\mathbf{k}, \mathbf{q}) = \left\langle \left\{ \alpha_i^{(1)}(\mathbf{k}) g_{ij,\mathbf{k},\mathbf{q}} B_{j,\mathbf{k}-\mathbf{q}}, \alpha_l^{(1)}(\mathbf{k}-\mathbf{q}) B_{l,\mathbf{k}-\mathbf{q}}^\dagger \right\} \right\rangle = \alpha_i^{(1)}(\mathbf{k}) g_{ij,\mathbf{k},\mathbf{q}} \alpha_j^{(1)*}(\mathbf{k}-\mathbf{q}) . \tag{207}$$

In the following we shall use $\check{R}_{\mathbf{k}}$ instead of $\mathcal{R}_{\mathbf{k}}$ in Eq. (195). Then the lowest order self-energy contribution is provided by the first term on the right hand side of the expression (195), which is responsible for the scattering of polarons, while the second term leads to higher order corrections. One should mention that the operators $Y_{i,\mathbf{k},\mathbf{q}}$ are orthogonal among each other. However, unlike the basis operators (175, 179) only up to terms of order $1/N$. Therefore, one may obtain artificial terms in the vertex $\Gamma(\mathbf{k},\mathbf{q})$ if one carries out the calculation in a changed order than the one presented here.

In [18] only one operator in the PM was used. The present method generalizes the recipe of vertex calculation from Ref. [18] to the more complex case of several operators in the PM basis set.

The self-energy (195) with $\check{R}_\mathbf{k}$ (206) provides the opportunity to apply the mode-coupling approximation in terms of an independent propagation of the polaron and spin excitations. It consists in the proper decoupling procedure for the two time correlation function

$$\langle \check{R}_{\mathbf{k},\sigma}(t)\check{R}^\dagger_{\mathbf{k},\sigma}(t')\rangle \simeq \frac{1}{N}\sum_\mathbf{q} \Gamma^2(\mathbf{k},\mathbf{q}) \langle \mathcal{B}_{\mathbf{k}-\mathbf{q},\sigma}(t)\mathcal{B}^\dagger_{\mathbf{k}-\mathbf{q},\sigma}(t')\rangle \langle \mathbf{S}_{-\mathbf{q}}(t)\mathbf{S}_\mathbf{q}(t')\rangle, \qquad (208)$$

$$\langle \mathbf{S}_{-\mathbf{q}}(t)\mathbf{S}_\mathbf{q}(t')\rangle = \frac{1}{N}\sum_{\mathbf{r},\mathbf{r}'} e^{i\mathbf{q}\cdot(\mathbf{r}'-\mathbf{r})} \langle \mathbf{S}_\mathbf{r}(t)\mathbf{S}_{\mathbf{r}'}(t')\rangle. \qquad (209)$$

In the frameworks of such a mode-coupling approximation the second term on the right hand side of Eq. (195) may be neglected. The first term $\langle \check{R}_\mathbf{k}|\check{R}^+_\mathbf{k}\rangle_\omega$ of (195) may be expressed in terms of the Fourier transform of the above two-time correlation function (208). The presence of the polaron-polaron correlation function $\langle \mathcal{B}_{\mathbf{k}-\mathbf{q}}(t)\mathcal{B}^\dagger_{\mathbf{k}-\mathbf{q}}(t')\rangle$ in the right hand side of Eq. (208) allows the self-consistent calculation of $G_p(\mathbf{k},\omega)$. Note that in the case of the t-J model, investigated in terms of spinless holes, the analogous decoupling procedure for the irreducible Green's function of the form (195) is equivalent to SCBA in a usual diagrammatic technique. [152–154].

The spin-spin correlation function $\langle \mathbf{S}_{-\mathbf{q}}(t)\mathbf{S}_\mathbf{q}(t')\rangle$ in (208) is calculated from the spin excitation Green's function $D(\mathbf{q},\omega)$. We treat the spin subsystem following the spherically symmetric approach, [111,176,27,13] (see Section III). Green's function $D(\mathbf{q},\omega)$ has the form

$$D(\mathbf{q},\omega) = \langle S^\alpha_{-\mathbf{q}}|S^\alpha_\mathbf{q}\rangle_\omega = -8JC_\mathbf{g}\frac{1-\gamma_\mathbf{q}}{\omega^2-\omega^2_\mathbf{q}}; \qquad (210)$$

$$\omega^2_\mathbf{q} = -32J\alpha_1\,(C_\mathbf{g}/3)\,(1-\gamma_\mathbf{q})(2\Delta+1+\gamma_\mathbf{q}). \qquad (211)$$

We neglect the influence of doped holes on the copper spin dynamics and take the spin spectrum parameters calculated in Ref. [27,13]: $\Delta = 0$, for $T = 0$, the vertex correction $\alpha_1 = 2.35$, the spin excitations condensation part $m^2 = 0.09$. Note that the spin excitation spectrum $\omega_\mathbf{q}$ has the magnetic BZ symmetry at $\Delta = 0$, but Green's function $D(\mathbf{q},\omega)$ has the symmetry of the full BZ due to the numerator $(1-\gamma_\mathbf{q})$.

As a result of decoupling (208) we come to the integral equation for Green's function

$$G_p(\mathbf{k}, \omega) = \frac{1}{\omega - \Omega_\mathbf{k} - \Sigma(\mathbf{k}, \omega)}, \qquad (212)$$

where

$$\Sigma(\mathbf{k}, \omega) = \frac{1}{N} \sum_\mathbf{q} M^2(\mathbf{k}, \mathbf{q}) G_p(\mathbf{k} - \mathbf{q}, \omega - \omega_\mathbf{q}), \qquad (213)$$

$$M^2(\mathbf{k}, \mathbf{q}) = \Gamma^2(\mathbf{k}, \mathbf{q}) \frac{(-4C_\mathbf{g})(1 - \gamma_\mathbf{q})}{\omega_\mathbf{q}}. \qquad (214)$$

$\Gamma(\mathbf{k}, \mathbf{q})$ corresponds to the bare vertex for the coupling between a spin polaron and a spin wave. It is known [173,38] that this vertex is substantially renormalized for $\mathbf{q}$ close to the antiferromagnetic vector $\mathbf{q}_0 = (\pi, \pi)$. This renormalization is due to the strong interaction of a polaron with the condensation part of spin excitations that must be taken into account from the very beginning. As a result, the renormalized vertex $\tilde{\Gamma}(\mathbf{k}, \mathbf{q})$ must be proportional to [173] $\left[(\mathbf{q} - \mathbf{q}_0)^2 + L_s^{-2}\right]^{1/2}$, L_s being the spin-spin correlation length, $L_s \to \infty$ in our case of a long range order state of the spin subsystem. Below, this renormalization is taken into account empirically by the substitution

$$\Gamma(\mathbf{k}, \mathbf{q}) \to \tilde{\Gamma}(\mathbf{k}, \mathbf{q}) = \Gamma(\mathbf{k}, \mathbf{q}) \sqrt{(1 + \gamma_\mathbf{q})}. \qquad (215)$$

The introduced vertex correction is proportional to $|\mathbf{q} - \mathbf{q}_0|$ for $\mathbf{q}$ close to $\mathbf{q}_0$. Let us mention that the bare vertex leads to a dramatic decrease of the QP bandwidth.

C Results for polaron and hole spectral densities

In this section we present the results for the low energy part of the one-particle retarded Green's function $G_h(\mathbf{k}, \omega)$ (187). In our treatment we ignore the scattering of a polaron to the upper bands. In this approximation the low energy part of $G_h(\mathbf{k}, \omega)$ is related to the polaron Green's function $G_p(\mathbf{k}, \omega)$ through Eq. (196). We are mostly interested in the spectral function of the polaron or hole GF, respectively, which are given by

$$A_{p/h}(\mathbf{k}, \omega) = -\frac{1}{\pi} \text{Im}\, G_{p/h}(\mathbf{k}, \omega). \qquad (216)$$

The spectral function of a bare hole A_h is roughly proportional to the intensity in angle resolved photoemission (ARPES) experiments. The self-consistent equation (212) was solved by means of the recursive procedure that provides step by step the coefficients a_n, b_n of the continued fraction expansion of $G_p(\mathbf{k}, \omega)$.

$$G_p(\mathbf{k},\omega) = \cfrac{b_0^2}{\omega - a_0 -} \cfrac{b_1^2}{\omega - a_1 -} \cdots \cfrac{b_n^2}{\omega - a_n -} \cdots$$

$$a_n = a_n(\mathbf{k}), \quad b_n = b_n(\mathbf{k})$$

The details of the calculation can de found in [18]. The two-dimensional Simpson's rule was used for the integration in Eq. (213) over 231 points in the irreducible part of the Brillouin zone. The prominent feature of the expansion is the fast convergence of coefficients a_n, b_n to the asymptotic behaviour which is characterized by a linear dependence on n. The slope for a_n is twice as large as the slope for b_n. This feature allows to use the incomplete gamma function as an analytic terminator [77,78] for the continued fraction. We applied the terminator after the calculation of 30 pairs of coefficients. Our calculations for various models of strongly correlated electrons indicate that the linear growth of continued fraction coefficients seems to be a common feature of such models.

The solid line in Fig. 27 shows the dispersion $\varepsilon(\mathbf{k})$ of the lowest peak in the spectral density. First, we may see that the overall shape of the dispersion curve has not changed in comparison with the mean field result. On the other hand, we may note that the energy renormalization for the top of the band is larger than for the bottom resulting in substantial narrowing of the band. Although the band width of the polaron in mean field $\Omega_\mathbf{k}$ is proportional to the hopping amplitude τ, it is reduced to a band width proportional to J due to the scattering of the polaron at spin fluctuations which are inherent in $\varepsilon(\mathbf{k})$. Let us note that the polaron mean field energy $\Omega_\mathbf{k}$ represents the center of gravity of the polaron spectral function $A_p(\mathbf{k},\omega)$ that is

$$\Omega_\mathbf{k} = \int_{-\infty}^{\infty} \omega A_p(\mathbf{k},\omega)\, d\omega. \tag{217}$$

On the other hand, we see from the explicit expressions for the Liouvillean matrix elements (183) that $\Omega_\mathbf{k}$ is unambiguously related to the nearest neighbour and next-nearest neighbor static spin-spin correlation functions of the undoped system. The latter is described by the Heisenberg Hamiltonian (172) on the square lattice, and the correlation functions are known at present time with a very high accuracy from various analytic [176,30] and numerical [115,150] approaches. That means the asymmetry of $\Omega_\mathbf{k}$ (with respect to the magnetic Brillouin zone) is not only a consequence of the particular approximations made in the present work, but already a property of the polaron spectral function itself.

Fig. 28 gives the polaron and hole spectral densities for three characteristic momenta $\mathbf{k}$ to distinguish three regimes. In addition, we show the real and imaginary parts of the self-energy $\Sigma(\mathbf{k},\omega)$ which are approximately identical for polaron or hole GF according to Eq. (196). Let us recall that we have made the change $\mathbf{k} \to \mathbf{k}' = \mathbf{k} - \mathbf{q}_0$ in the final results, so that the quasimomenta in the figures correspond to the actual Brillouin zone of the CuO_2 plane, in contrast with work. [18] We find a sharp quasiparticle peak at the bottom of the spectra for $\mathbf{k}_a = (\pi/2, \pi/2)$

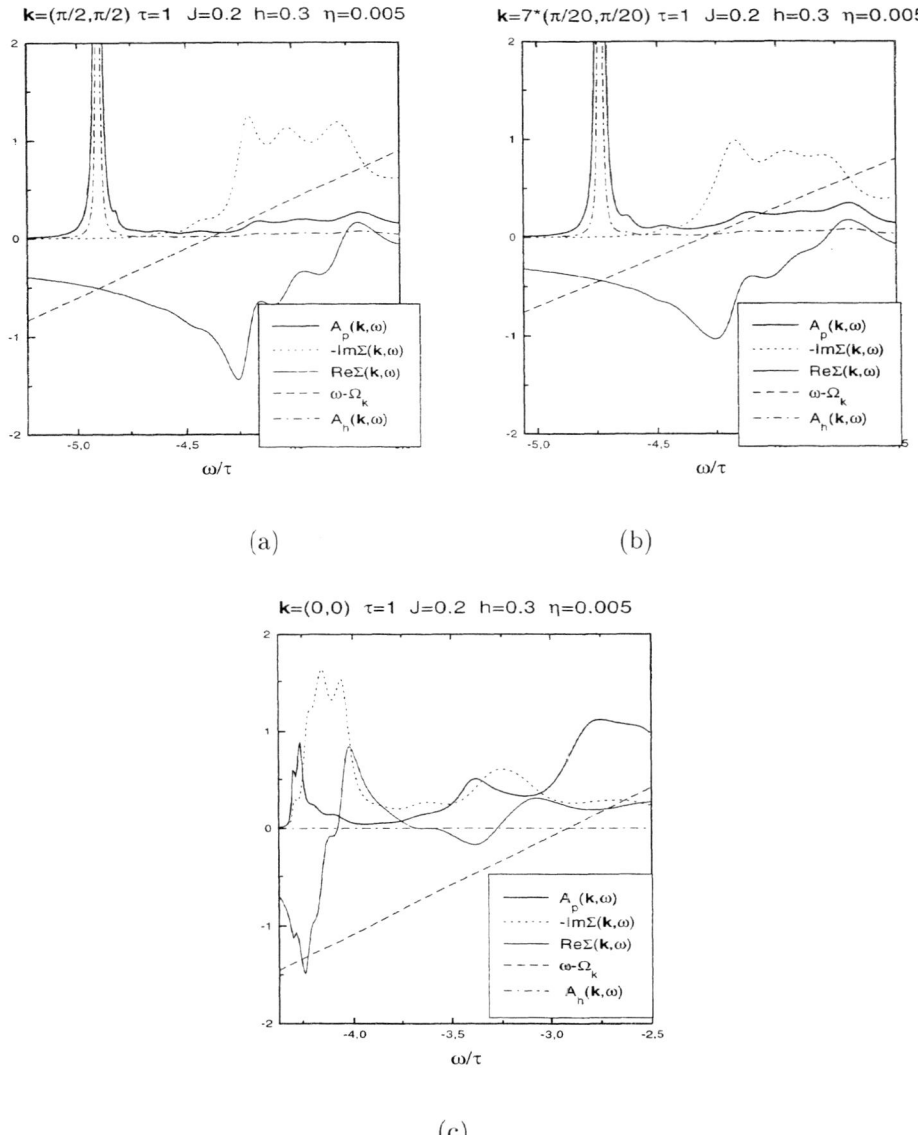

FIGURE 28. Polaron and hole spectral densities $A_{p/h}(\mathbf{k},\omega)$ around the lowest peak for the momenta: a) $\mathbf{k} = (\pi/2, \pi/2)$, b) $\mathbf{k} = 7(\pi/20, \pi/20)$ and c) $\mathbf{k} = (0,0)$. Also shown the imaginary part of the self-energy and the crossing of the curves $\omega - \Omega_{\mathbf{k}}$ and $\mathrm{Re}\Sigma(\mathbf{k},\omega)$. The spectral functions are broadened by a small imaginary part $\eta = 0.005$ in ω. Ref. [114].

and $\mathbf{k}_b = (7\pi/20, 7\pi/20)$. The position of the quasiparticle peak corresponds to the condition Re $G_p^{-1}(\mathbf{k}, \omega) = 0$, i.e. the point where we have a crossing of functions $\omega - \Omega_\mathbf{k}$ and Re $\Sigma(\mathbf{k}, \omega)$, see Figs. 28a and 28b. Quite a different shape has the spectral density $A_p(\mathbf{k}_c, \omega)$ at $\mathbf{k}_c = (0, 0)$. Its lowest peak is not related to the zero of Re $G_p^{-1}(\mathbf{k}, \omega)$ and is due to abrupt changes in the self-energy. This leads to a large imaginary part of $\Sigma(\mathbf{k}, \omega)$ at the position of the lowest peak in the spectral function. We may say that there is no quasiparticle excitation for this quasimomentum value.

We may distinguish three qualitatively different kinds of spectral function $A_p(\mathbf{k}, \omega)$ behaviour:

(i) The lowest peak tends to a delta function $W_p \delta[\omega - \varepsilon(\mathbf{k})]$, when Im $\omega \to 0$. We have a quasiparticle, characterized by the excitation energy $\varepsilon(\mathbf{k}) = \Omega_\mathbf{k} + $ Re $\Sigma[\mathbf{k}, \varepsilon(\mathbf{k})]$ with infinite lifetime τ_l, $(1/\tau_l) \equiv -$Im $\Sigma[\mathbf{k}, \varepsilon(\mathbf{k})] = 0$. These k-points are situated near the band minimum $\mathbf{k}_{\min}$. They are marked by solid circles in Fig. 27.

(ii) The lowest peak has an approximate Lorentzian form

$$A_p(\mathbf{k}, \omega) = -\frac{1}{\pi} \text{Im } G_p(\mathbf{k}, \omega) \approx W_p \frac{(1/\tau_l)}{[\omega - \varepsilon(\mathbf{k})]^2 + (1/\tau_l)^2} + A_{incoh} \qquad (218)$$

and corresponds to a quasiparticle with energy $\varepsilon(\mathbf{k}) = \Omega_\mathbf{k} + $ Re $\Sigma[\mathbf{k}, \varepsilon(\mathbf{k})]$, and finite lifetime $(1/\tau_l) \approx -$Im $\Sigma[\mathbf{k}, \varepsilon(\mathbf{k})] \ll |\varepsilon(\mathbf{k}) - \varepsilon(\mathbf{k}_{\min})|$. The corresponding k-values are marked by open circles in Fig. 27.

(iii) The spectral density is completely incoherent and is of the same order of magnitude as the imaginary part of the self-energy at all ω (see Fig. 28c). These k-points are marked by stars.

In order to understand the nature of the polaron damping ($1/\tau_l$ in (218)), let us consider a particular k-value and estimate the imaginary part of the self-energy at a particular ω. Changing the summation variable in (213) we can calculate the imaginary part of the self-energy as follows

$$-\text{Im } \Sigma(\mathbf{k}, \omega) = \pi \frac{1}{N} \sum_{\mathbf{q}'} M^2(\mathbf{k}, \mathbf{k} - \mathbf{q}') A_p(\mathbf{q}', \omega - \omega_{\mathbf{k}-\mathbf{q}'}) . \qquad (219)$$

Now we suppose that the lowest peak at all $\mathbf{q}' = \mathbf{k} - \mathbf{q}$ may be described approximately by Lorentzian (218). Then we see from Eq. (219) that large damping, characterized by a large value of $-$Im $\Sigma(\mathbf{k}, \omega)$ may only arise when the condition

$$\omega - \varepsilon(\mathbf{q}') - \omega_{\mathbf{k}-\mathbf{q}'} = 0 \qquad (220)$$

is satisfied for some values of $\mathbf{q}'$ (see also Ref. [203]). On the contrary, if Eq. (220) does not hold even at the lowest edge of the spectral function $\omega = \varepsilon(\mathbf{k})$, damping is absent. In Fig. 29 we plot $\varepsilon(\mathbf{q}')$ together with the curve $\varepsilon(\mathbf{k}) - \omega_{\mathbf{k}-\mathbf{q}'}$ for various values of $\mathbf{k}$ (for $h = 0.3\tau$ and where we choose only one direction for both $\mathbf{k}$, and $\mathbf{q}'$). We may see that for $\mathbf{k}_{\min}=(\pi/2, \pi/2)$ the condition (220) holds

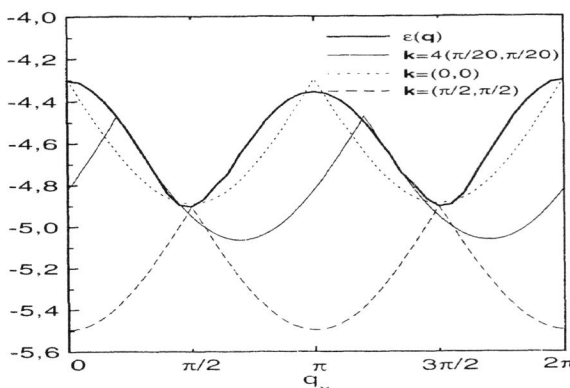

FIGURE 29. Crossing of the curves $\varepsilon(\mathbf{q}')$ and $\varepsilon(\mathbf{k}) - \omega_{\mathbf{k}-\mathbf{q}'}$ for three different values of $\mathbf{k}$ along the diagonal of the BZ for $\mathbf{q}' = (q'_x, q'_x)$: there are only trivial crossing points ($\mathbf{q} = \mathbf{k} - \mathbf{q}'$ equal to Γ or M) for $\mathbf{k} = (\pi/2, \pi/2)$ but nontrivial crossings for $\mathbf{k} = (4\pi/20, 4\pi/20)$ and $\mathbf{k} = (0,0)$.

only for the trivial values $\mathbf{q} = \mathbf{k} - \mathbf{q}' = (0,0)$ and $\mathbf{q} = \mathbf{q}_0$, where, however, the vertex $M^2(\mathbf{k}, \mathbf{q})$ vanishes. It demonstrates that it is impossible to scatter from $\mathbf{k}_{\min}$ into other states. The same situation occurs for some finite regions of $\mathbf{k}$ values around the band bottom. These are exactly those momenta with a quasiparticle of infinite lifetime. Different is the situation for $\mathbf{k}=(0,0)$ and $\mathbf{k} =(4\pi/20, 4\pi/20)$ where several nontrivial intersections take place.

Fig. 30 shows the low energy part of the hole spectral density $A_h(\mathbf{k}, \omega)$ for $\mathbf{k}$ along high symmetry directions. Let us recall that the intensity of the lowest peak of $A_h(\mathbf{k}, \omega)$ is governed by two circumstances: the $\mathbf{k}$-dependence of the mean field residue $Z(\mathbf{k})$ (see Eq.(188) and the discussion about Fig.27b) and the intensity of the lowest polaron peak of $A_p(\mathbf{k}, \omega)$. One can clearly observe a well defined quasiparticle peak near the bottom of the spectrum at $(\pi/2, \pi/2)$, everywhere along the line $(0, \pi)$-$(\pi, 0)$ and also near $(\pi, 0)$. At the Γ-point $(0,0)$, there is no hole spectral density due to the vanishing residue for the hole Green's function in the lowest mean field polaron band. At the same time one observes intensity at (π, π). A clear asymmetry of the peak intensity is seen along the diagonal of the BZ. The abrupt drop of the peak intensity in the region $(\pi/2, \pi/2)$ -(π, π) is related to the strong polaron damping there. In other words, most of the mean field hole spectral weight ($Z(\mathbf{k})$ in Fig. 27b) goes into the incoherent part of the spectrum (situated at higher energies) due to the strong coupling of the polaron $\mathcal{B}_{\mathbf{k}}$ with spin excitations. An analogous behaviour of the quasiparticle peak intensity was obtained in Ref. [57] within a variational ansatz for the t-J model (see also [154]).

Let us compare the results with those of the usual SCBA of spinless holes in

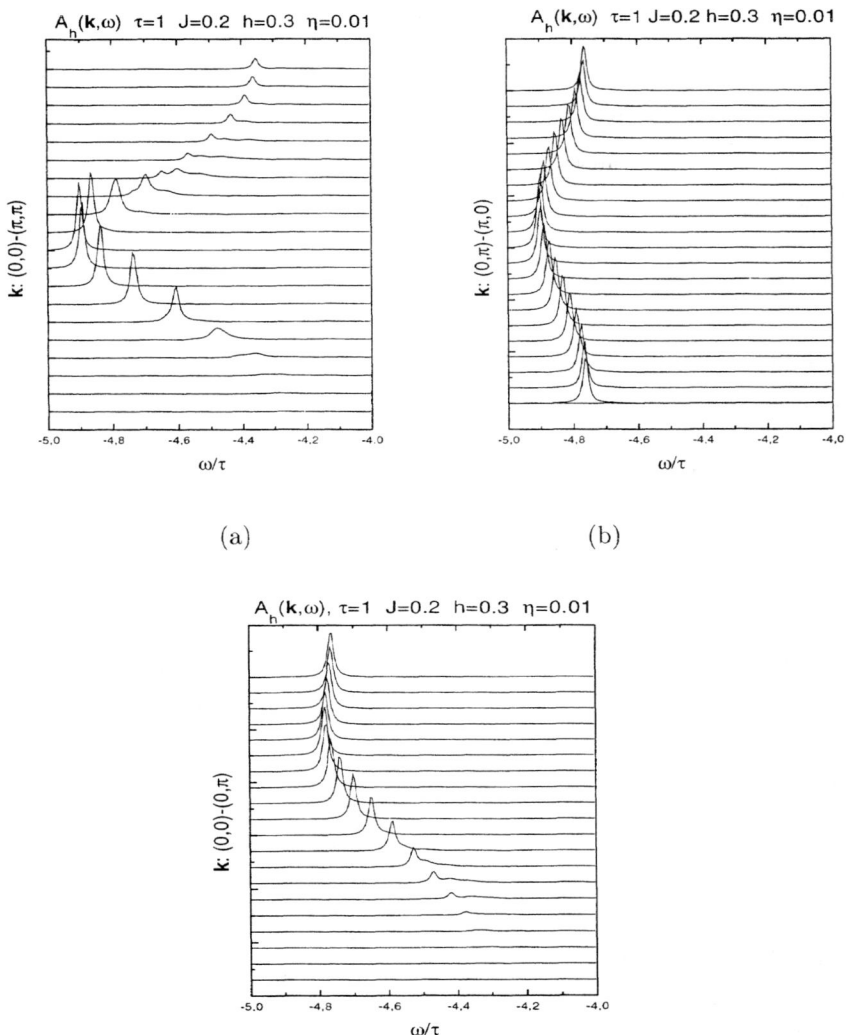

FIGURE 30. Contour plot of the hole spectral function $A_h(\mathbf{k},\omega)$ for $J = 0.2$, $\tau = 1$, $h = 0.3$ and $\eta = 0.01$ along high symmetry lines in the BZ a) $(0,0)$-(π,π), b) $(\pi,0)$-$(0,\pi)$ and c) $(0,0)$-$(\pi,0)$. Ref. [114].

the pure t-J model. Here, one finds quasiparticles with infinite lifetime and finite weight everywhere in **k**-space. Additionally, the spectral function has the symmetry of the magnetic BZ. Introducing additional hopping terms (corresponding to direct oxygen-oxygen hopping in the Emery model) leads to a scattering mechanism such that the upper parts of the spectrum loose their quasiparticle character. [12,203] On the contrary, in our approach, the damping of the spin polaron is already present without direct oxygen-oxygen hopping. If we only consider the dispersion relation $\varepsilon(\mathbf{k})$ we observe a remarkable similarity between the present calculation and earlier results using the SCBA of spinless holes or other methods. In fact, the deviations from the magnetic BZ symmetry in Fig. 27 are not very large. On the other hand, we find some qualitatively new features of the spectral function which are absent in the numerical results of Refs. [132,152]. Besides the strong polaron scattering away from the band bottom, we especially note: (i) the absence of the polaron quasiparticle at the Γ-point and (ii) the asymmetry of the peak intensity with respect to the magnetic BZ.

Our results can also be summarized in such a way that a well pronounced polaron quasiparticle peak exists only around the bottom of the band (See Figs. 30a-c). In the direction $(\pi/2, \pi/2)$-$(\pi, 0)$ it is more clearly seen than perpendicular to it. The recent experimental finding [155] shows, indeed, that the Zhang-Rice singlet can be observed in $Sr_2CuO_2Cl_2$ only in a similar region of the BZ. And also the higher peak intensity going from $(\pi/2, \pi/2)$ to the Γ-point in comparison with the opposite direction is in agreement with the experiment. So we see that those details in which our calculation differs from the standard one [170,108,132,120,152] are essential for a better understanding of the experiment.

Let us emphasize that for a detailed comparison to a real experiment there are a considerable number of complications which have not been taken into account in the present work: first, in general a photohole can be generated also on copper, and there will be (in general **k**-dependent) interference between the photoholes created on different atoms. An operator $c^\dagger_{\mathbf{R}+\mathbf{a}}$ that enters the spin-fermion Hamiltonian (171) creates a hole in a state that is an antibonding (in hole notation) combination of oxygen p_σ state and singly occupied copper $d_{x^2-y^2}$ state at adjacent sites. In a more realistic theory one should deal with a complete set of bare oxygen and copper hole creation operators. This may be expected to have some influence on the spectral weight. Next, one should take into account interference terms like $<p_x|p_y>$, because the photoemission operator is in general a linear combination of p_x and p_y. Thereby the relative phase between p_x and p_y depends on the momentum transfer, and is in general even different whether one is in the 1st or 2nd Brillouin zone etc. Clearly this will give additional momentum dependence of the weight. Moreover, it is not even a priori clear that p_x and p_y have equal weight in the PES-operator. Rather this depends in a relatively complicated way, e.g., on the polarization of the incoming X-ray photons. It is in fact experimentally well known that the spectra depend quite sensitively on the X-ray polarization in several oxychlorides like $Sr_2CuO_2Cl_2$ or $Ba_2Cu_3O_4Cl_2$. [169,76] At the present stage of many-body theory it is impossible to take all these complications into account. Thus, our comparison with experiment has only a qualitative character. Nevertheless, we point out that according to our results, many features of the experimental spectra may originate from the peculiarities of the correlated ground state of the material rather than from the above complications.

We may consider now what happens if we begin to dope our system. First, we suppose that for extremely small doping the Fermi surface consists of hole pockets centered at $(\pi/2, \pi/2)$. The small spectral weight of the quasiparticle poles in this region means that the surface enclosed inside the Fermi surface may be much larger than the number of holes. Note that due to the strong weight asymmetry the experimental observation of the Fermi surface parts facing M point with $\mathbf{k} = (\pi, \pi)$ point may be very hard to do. In Ref. [130] the observation of weak features in photoemission spectra in this region of k-space was reported, but the authors can not unambiguously interpret them as Fermi surface cuts. With the increase of doped holes the antiferromagnetic correlations in the spin subsystem are weakened resulting in a deformation of the hole dispersion. [15,83,80] The minimum of the hole spectrum should shift to the M point and a 'large' Fermi surface develops. Such a shift on frustration was demonstaterd in previous Sections (an analogous scenario was also developed in Ref. [38]).

So the discussed method allows to calculate the spectral function of the Emery model in its spin-fermion form combining the Mori-Zwanzig projection method with a representation of the self-energy by irreducible GF. Self-energy effects reduce the mean field bandwidth of the spin polaron to a realistic bandwidth in the order of $2J$, but do not change the overall shape of the dispersion. Direct oxygen-oxygen hopping leads to a more isotropic band minimum around $(\pi/2, \pi/2)$. The quasiparticle weight is maximal (about 0.25) around the minimum of dispersion. Despite some similarities, we observed several new features of the spectral function which are not present in the usual SCBA [170,108,132,120,152] for spinless holes coupled to spinons in the pure t-J model, namely the absence of quasiparticle weight at the Γ point, the strong damping away from the minimum and the asymmetry with respect to the antiferromagnetic BZ (where the asymmetry is present, however, in other approaches to the t-J model) [57,62,154]. Qualitatively, these features can be observed in the ARPES experiments on $Sr_2CuO_2Cl_2$.

VII SUMMARY

We have studied the spin polaron motion in different models of 2-D antiferromagnet using the two-time retarded Green's functions. Here we shall mention once more some important (from our point of view) results which give a spin-polaron concept.

From the discussion of cluster states in Section II it is clear that namely a start from a small radius polaron represents a good approximation for the true quasiparticle low-energy excitations in a strongly correlated limit. Such a description at the same time immediately leads to a strong deviation from Fermi's statistic for bare carries. The Kondo-lattice case gives a most explicit example of such a situation. Another interesting point is that the investigation of a two-particle small polaron state decay into two one-particle states leads to the dynamic mechanism of hole pairing (see Sections II A and II C).

As shown above even a small radius spin polaron at the mean-field level gives an important features of ARPES experiments: the isotropic bottom of the lowest carrier band and the existence of saddle-like singularity aligned with the $(\pi, 0)$ – $(0, 0)$ line. It is important that in order to get these results one has to take into account the realistic features of the models. Such features are: 1) the presence of the oxygen-oxygen hopping in the three-band model and the hopping to the second and third nearest neighbours in the generalized t-J (these hoppings in both models are responsible for the dispersion along the line $(0, \pi) - (\pi, 0)$ and the isotropic bottom); 2) the presence of the spin frustration and its increase with doping (frustration is responsible for the existence of saddle-like singularity).

As the movement of the carries on the spin background strongly depends on the features of the background an adequate spin-subsystem treatment is important in physically interesting situations of finite temperature and frustration. Here we represented a spin-rotation-invariant approach which preserves the periodicity relative to a full Brillouin zone (in contrast with a Neel-type two-sublattice description).

The important moment in the extension of the spin-polaron approach consists in the construction of complex polaron states in order to take into account the coupling of a local polaron with such spin waves which dictate a long or intermediate antiferromagnetic range order. Presented in cases of the effective three-band model and the Kondo-lattice model, this complex polaron approach leads to the decoupling of the lowest local polaron band and to a very strong violation of the Luttinger theorem. As a result the lowest polaron band reproduces such features of ARPES experiments as unusual evolution of the Fermi surface with doping, the possibility of the "shadow band" effect and a sudden drop of intensity (the bare hole residues) at the $(\pi/2, \pi/2)$ to $(0, 0)$ cut.

The results for the mean field complex spin-polaron approach and the local polaron (not bare hole) SCBA (in case of the effective three-band model) demonstrate that the mean field lowest band qualitatively reproduces the features of the SCBA quasiparticle band. This can be explicitly seen by the comparison of the lowest quasipartical band and the bare hole residue $Z(\mathbf{k})$ in Fig.27– SCBA and in Fig.21– complex mean-field polaron dressed by AFM waves (see direction $\Gamma - M$): in both cases we see the same narrowing of the band (relative to a local polaron band) and the same nontrivial dependence of weights $Z(\mathbf{k})$.

In Fig.1 we also compared local spin polaron results with the previous studies [106] which started from the bare hole. We see that the small polaron mean-field energy $\Omega_{\mathbf{k}}$ lies much lower than the QP pole obtained from SCBA for the bare hole. Since $\Omega_{\mathbf{k}}$ determines the center of gravity for the Green's function spectral density, the actual QP pole position (at least for the band bottom) should lie deeper in energy than $\Omega_{\mathbf{k}}$. This means that in a strongly correlated regime the important local correlations should be taken into account in zero approximation and small spin polaron should be constructed. The polaron scattering on spin waves will then be of less importance and may be treated with perturbation methods.

As the complex spin-polaron concept gives a rather simple description of the important lowest excitation band we strongly believe that in the near future this concept will be useful for theoretical investigations on such a phenomena as superconductivity and normal kinetic properties of HTSC and other strongly correlated systems.

We conclude that the low-energy physics of high-T_c superconductors should be considered in terms of spin polaron dynamics. In particular, the problem of superconducting hole pairing must be treated as pairing of these quasiparticles rather than pairing of bare holes.

VIII ACKNOWLEDGMENTS

This work was supported, in part, by the INTAS (project No. 97-11066) and by RFFS (projects No. 98-02-17187 and No. 98-02-16730)

REFERENCES

1. Abrikosov, A. A. *et al.*, 1993, Physica **C214**, 73.
2. Aebi, P. *et al.*, 1994, Phys. Rev. Lett. **72**, 2757.
3. Aebi, P. *et al.*, 1995, Phys. Rev. Lett. **74**, 1885.
4. Aksenov, V. L. *et al.*, 1987, J. Phys. C: Solid State Phys. **20**, 375.
5. Alcantara Bonfim, O. F. de, and G. F. Reiter, 1991, in *Proc. of the Univ. of Miami Workshop on Electronic Structure and Mechanisms for High Temperature Superconductivity*, ed. J. Ashkenazi. Plenum, N.Y.
6. Anderson, P. W., 1987, Science, **235**, 1196.
7. Anderson, P. W., G. Baskaran, Z. Zou, and T. Hsu, 1987, Phys. Rev. Lett. **58**, 2790.
8. Annet, J. F. *et al.*, 1989, Phys. Rev. **B40**, 2620.
9. Arovas, D. P., and A. Auerbach, 1988, Phys. Rev. **B38**, 316.
10. Auerbach, A., and D. P. Arovas, 1988, Phys. Rev. Lett. **61**, 617.
11. Bacci, S., E. Gagliano, and F. Nori, 1991, Int. J. Mod. Phys. **B5**, 325.
12. Bala, J., A. Ole's, and J. Zaanen, 1995, Phys. Rev. **B52**, 4597.
13. Barabanov, A. F., and V. M. Beresovsky, 1994, Zh. Eksp. Teor. Fiz. **106**, 1156. [1994, JETP **79**, 627]; 1994, J. Phys. Soc. Jpn. **63**, 3974; 1994, Phys. Lett. **A186**, 175.
14. Barabanov, A. F, V. M. Beresovsky, and E. Žasinas, 1995, Phys. Rev. **B52**, 10177.
15. Barabanov, A. F. *et al.*, 1996, Zh. Eksp. Teor. Fiz. **110**, 1480 [JETP **83**, 819]; 1991, J. Phys. Cond. Matter **3**, 9129; 1995, Physica **C252**, 308; 1993, Physica **C212**, 375.
16. Barabanov, A. F., A. A. Kovalev, O. V. Urazaev, and A. M. Belemouk, 2000, Phys. Lett. **A**, to be published.
17. Barabanov, A. F., and R. O. Kuzian, 1991, Sol. St. Commun. **78**, 963.
18. Barabanov, A. F. *et al.*, 1998, Zh. Eksp. Teor. Fiz. **113**, 1758 [1998, JETP **86**, 959]; 1997, Phys. Rev. **B55**, 4015.
19. Barabanov, A. F., and L. A. Maksimov, 1995, Phys. Lett. **A207**, 390.
20. Barabanov, A. F., L. A. Maksimov, and A. V. Mikheyenkov, 1989, J. Phys:Cond. Matt. **1**, 10143; 1991, Sverkhprovodimost': Fizika, Khimiya, Teknollogiya **4**, 3.
21. Barabanov, A. F. *et al.*, 1990, Int. J. Mod. Phys. **B4**, 2319; 1990, J. Phys:Cond. Matt. **2**, 8925.
22. Barabanov, A. F., L. A. Maksimov, and G. V. Uimin, 1988, Pisma v Zh. Eksp. Teor. Fiz. **47**, 532. [1988, JETP Lett. **47**, 622].
23. Barabanov, A. F., L. A. Maksimov, and G. V. Uimin, 1989, Zh. Eksp. Teor. Fiz. **96**, 655. [JETP **69**, 371 (1989); 1988, Pisma v Zh. Eksp. Teor. Fiz. **47**, 532. [1988, JETP Lett. **47**, 622].
24. Barabanov, A. F. *et al.*, 1997, Pisma v Zh. Eksp. Teor. Fiz. **66**, 173. [1997, JETP Lett. **66**, 182].
25. Barabanov, A. F., and A. V. Mikheyenkov, 1986, Fiz. Tverd. Tela **28**, 998.
26. Barabanov, A. F., and O. A. Starykh, 1990, Pis'ma Zh. Eksp. Teor. Fiz. **51**, 271 [1990, JETP Lett. **5**, 311].
27. Barabanov, A. F., and O. A. Starykh, 1992, J. Phys. Soc. Jpn. **61**, 704.
28. Barabanov, A. F. *et al.*, 1998, Pis'ma Zh. Eksp. Teor. Fiz. **68**, 386 [JETP Lett. **68**

, 412]; 1999, Doc. Akad. RAN **336**, 188 [1999, Doklady **44**].
29. Becker, K. W., W. Brenig, and P. Fulde, 1990, Z. Phys. **B81**, 163.
30. Becker, K. W., H. Won, and P. Fulde, 1989, Z. Phys. **B75**, 335.
31. Belinicher, V. I., A. L. Chernyshev, and V. A. Shubin, 1996, Phys. Rev. **B53**, 335; 1994, Phys. Rev. **B49**, 9746.
32. Birgenau, R. J., and G. Shirane, 1989, in *Physical Properties of High Temperature Superconductors*, ed.D. M. Ginsberg, World Scientific, Singapore, Vol.1, p.151.
33. Brenig, W., 1995, Phys. Rep. **251**, 4.
34. Carneiro, C. E., M. J. De Oliveira, S. R. A. Salinas, and G. V. Uimin, 1990, Physica **C166**, 206.
35. Chakravarty, S., B. I. Halperin, and D. R. Nelson, 1988, Phys. Rev. Lett. **60**, 1057.
36. Chakravarty, S., B. I. Halperin, and D. R. Nelson, 1989, Phys. Rev. **B39**, 2344.
37. Chandra, P., and B. Doucot, 1988, Phys. Rev. **B39**, 9335.
38. Chubukov, A. V., and D. K. Morr, 1997, cond-mat/9701196.
39. Chubukov, A. V., and T. Jolicoeur, 1991, Phys. Rev. **B44**, 12050.
40. Dagotto, E., 1991, Int. J. Mod. Phys. **B5**, 77, .
41. Dagotto, E., 1994, Rev. Mod. Phys. **66**, 763.
42. Dagotto, E., A. Nazarenko, and M. Boninsegni, 1994, Phys. Rev. Lett. **73**, 728.
43. Dagotto, E. *et al.*, 1990, Phys. Rev. **B41**, 9049.
44. Dessau, D. S. *et al.*, 1993, Phys. Rev. Lett. **71**, 278.
45. Diep, H. T., and H. Kawamura, 1989, Phys. Rev. **B40**, 7019.
46. Ding, H., G. Lang, and W. Goddard, 1992, Phys. Rev. **B46**, 14317.
47. Ding, H. *et al.*, 1997, Phys. Rev. Lett. **78**, 2628.
48. Doniach, S., M. Inui, V. Kalmeyer, and M. Gabay, 1988, Europhys. Lett. **6**, 663.
49. Dopf, G., A. Muramatsu, and W. Hanke, 1992, Phys. Rev. Lett. **68**, 353.
50. Dopf, G., A. Muramatsu, and W. Hanke, 1992, Europhys. Lett. **17**, 559.
51. Dopf, G., J. Wagner, P. Dieterich *et al.*, 1992, Phys. Rev. Lett. **68**, 2082.
52. Duffi, D., and A. Moreo, 1995, Phys. Rev. **B52**, 15607.
53. Dyson, F. J., 1956, Phys. Rev. **102**, 1217.
54. Dyson, F. J., 1956, Phys. Rev. **102**, 1230.
55. Eder, R., and K. Becker, 1990, Z. Phys. **B78**, 219.
56. Eder, R., and K. Becker, 1990, Z. Phys. **B79**, 333.
57. Eder, R., and K. Becker, 1991, Phys. Rev. **B44**, 6982.
58. Eder, R., Y. Ohta, and T. Shimozato, 1994, Phys. Rev. **B50**, 3350.
59. Einarsson, T., P. Fröjdh, and H. Johanesson,. 1991, Preprint ITP-91-60, Goteborg.
60. Emery, V. J., 1987, Phys. Rev. Lett. **58**, 2794.
61. Emery, V. J., and G. Reiter, 1988, Phys. Rev. **B38**, 4547.
62. Eskes, H., and R. Eder, 1996, Phys. Rev. **B54**, R14226.
63. Eskes, H., L. H. Tjeng, and G. A. Sawatzky, 1990, Phys. Rev. **B41**, 288.
64. Eskes, H., M. B. J. Meinders, and G. A. Sawatzky, 1991, Phys. Rev. Lett. **67**, 1035.
65. Eskes, H., and G. A. Sawatzky, 1991, Phys. Rev. **B43**, 119.
66. Fan, Y., and M. Ma, 1988, Phys. Rev. **B37**, 1820.
67. Fazekas, P., and P. W. Anderson, 1974, Phil. Mag., **30**, 432.
68. Forster D., 1975, *Hydrodynamic Fluctuations, Broken Symmetry, and Correlation Functions* (Reading, MA: Benjamin).

69. Frenkel, D. M., R. J. Gooding, B. I. Shraiman, and E. D. Siggia, 1990, Phys. Rev. **B41**, 350.
70. Friedel, J., 1987, J. Physique **49**, 1435.
71. Friedel, J., 1987, J. Physique **48**, 1787.
72. Fulde, P., 1995, *Electronic correlations in Molecules and Solids*, Springer, Berlin.
73. Glazman, L. I., and A. S. Iosilevich, 1988, Pis'ma v Zh. Eksp. Teor. Fiz. **47**, 464 [1988, JETP Lett. **47**, 547].
74. Gochev, I. G., 1994, Phys. Rev. **B49**, 9594.
75. Gofron, K. *et al.*, 1993, J. Phys. Chem. Solids **54**, 1193.
76. Golden, M. S. *et al.*, 1997, Phys. Rev. Lett. **78**, 4107.
77. Haydock, R., 1980, in *Solid State Physics*, ed. H. Ehrenreich, F. Seitz, and D. Turnbull, Academic, N.Y.
78. Haydock, R., and C. M. M. Nex, 1985, J. Phys. C: Solid State Phys. **18** 2235.
79. Hayn, R., A. F. Barabanov, and L. A. Maksimov, Physica **B259-261**, 749.
80. Hayn, R., A. F. Barabanov, and J. Schulenburg, 1997, Z. Phys. **B102**, 359.
81. Hayn, R., A. F. Barabanov, J. Schulenburg, and J. Richter, 1996, Phys. Rev. **B53**,11714.
82. Hayn, R., A. F. Barabanov, J. Schulenburg, and J. Richter, 1996, Phys. Rev. **B53**, 1179.
83. Hayn, R., J. -L. Richard, V. Yu. Yushankhai, 1995, Solid State Commun. **93**, 127.
84. Hayn, R., V. Yushankhai, and A. F. Barabanov, Physica **B230-232**, 903.
85. Hayn, R., V. Yushankhai, and S. Lovtsov, 1993, Phys. Rev. **B47**, 5253.
86. Heinila, M. T., and A. S. Oja, 1993, Phys. Rev. **B48**, 16514.
87. Hirsh, J. E., 1987, Phys. Rev. **B35**, 8726.
88. Hirsh, J. E., and S. Tang, 1989, Phys. Rev. **B39**, 2887.
89. Hirsh, J. E., and S. Tang, 1990, Phys. Rev. **B40**, 4769.
90. Hirsh, J. E., S. Tang, E. Loh, and D. J. Salapino, 1988, Phys. Rev. Lett. **60**, 668.
91. Horsch., P., and W. Stephan, 1989, *Proc. of the NATO Advanced Research Workshop on Interacting Electrons in Reduced Dimensions*, ed. D.Baeriswyl and D.Campbell, Plenum, N.Y.
92. Hubbard, J., 1963, Proc. R. Soc. **A276**, 238.
93. Hüfner, S., 1995, *Photoemission Spectroscopy*, Springer, Berlin.
94. Huse, D. A., and V. Elser, 1988, Phys. Rev. Lett. **60**, 2531.
95. Hybertsen, M. S., M. Schlüter, and N. E. Christensen, 1989, Phys. Rev. **B39**, 9028.
96. Hybertsen, M. S. *et al.*, 1990, Phys. Rev. **B41**, 11068.
97. Inui, M., S. Doniach, and M. Gabay, 1988, Phys. Rev. **B38**, 6631.
98. Ioffe, L. B., and A. I. Larkin, 1988, Int. J. Mod. Phys. **B2**, 203, .
99. Iordanskii, S. V., and A. V. Smirnov, 1980, Zh. Eksp. Teor. Fiz. **79**, 1942 [1980, JETP **52**, 981].
100. Irkhin, V. Yu., A. A. Katanin, and M. I. Katsnelson, 1992, J. Phys. Cond. Matt. **4**, 5227.
101. Ivanov, N. B., and J. Richter, 1994, J. Phys. Cond. Matt. **6**, 3785.
102. Izyumov, Yu. A., 1997, Usp. Fiz. Nauk **167**, 465.
103. Izyumov, Yu. A., 1999, Usp. Fiz. Nauk **169**, 225.
104. Jefferson, J. H., H. Eskes, and L. F. Feiner, 1992, Phys. Rev. **B45**, 7959.

105. Johnston, D. C., 1989, Phys. Rev. Lett. **62**, 957.
106. Kabanov, V. V., and A. Vagov, 1993, Phys. Rev. **B47**, 12134.
107. Kampf, A. P., and J. R. Schrieffer, 1990, Phys. Rev. **B42**, 7967.
108. Kane, C. L., P. A. Lee, and N. Read, 1989, Phys. Rev. **B39**, 6880.
109. King, D. M. *et al.*, 1994, Phys.Rev. Lett. **73**, 3298.
110. Kivelson, S., 1987, Phys. Rev. **B36**, 7237.
111. Kondo, J., and K. Yamaji, 1972, Prog. Theor. Phys. **47**, 807.
112. Kopayev, V. Yu., 1988, Pisma v Zh. Eksp. Teor. Fiz. **47**, 628. [1988, JETP Lett. **47**, 726].
113. Kozlov, A. N., L. A. Maksimov, and A. F. Barabanov, 1992, Physica **C200**, 183.
114. Kuzian, R. O., R. Hayn, A. F. Barabanov, L. A. Maksimov, 1998, Phys. Rev. **B58**, 6194.
115. Liang, R. O., B. Doucot, and P. W., Anderson, 1988, Phys. Rev. Lett. **61**, 365.
116. Lieb, E. H., and F. Y. Wu, 1966, Phys. Rev. Lett. **20**, 1445.
117. Lines, M. E., 1963, Proc. Roy. Soc. London **A271**, 105.
118. Lines, M. E., 1964, Phys. Rev. **135**, A1336.
119. Littlewood, P. B., C. M. Varma, and E. Abrahams, 1987, Phys. Rev. Lett **60**, 379.
120. Liu, Z., and E. Manousakis, 1992, Phys. Rev. **B45**, 2425.
121. Loeser, A. G. *et al.*, 1996, Science **273**, 325.
122. Lovtsov, S. V., and V.Yu.Yushankhai, 1991, Physica **C179**, 159.
123. Lui, Z., and E. Manousakis, 1991, Phys. Rev. **B44**, 2414.
124. Maksimov, L. A., A. F. Barabanov, and R. O. Kuzian, 1997, Phys. Lett. **A232**, 286.
125. Maksimov, L. A., R. Hayn, and A. F. Barabanov, 1998, Phys. Lett. **A 238**, 288.
126. Maleev, S. V., 1958, Zh. Eksp. Teor. Fiz. **33**, 1010. [Sov. Phys. JETP **6** , 776 (1958)].
127. Mancini, F., *et al.*, 1996, Phys. Lett. **A210**, 429.
128. Manousakis, E., 1991. Rev. Mod. Phys. **63**, 1.
129. Markiewicz, R. S., 1990, J. Phys. Cond. Matter **2**, 6223.
130. Marshall, D. S.*et al.*, 1996, Phys. Rev. Lett. **76**, 4841.
131. Marsiglio, F., *et al.*, 1991, Phys. Rev. **B43**, 10882.
132. Martinez, G., and P. Horsch, 1991, Phys. Rev. **B44**, 317.
133. Matsukawa, H., and H. Fukuyama, 1989, J. Phys. Soc. Jap. **58**, 2845.
134. McMahan, A. K., J. F. Annett and R. M. Martin, 1990, Phys. Rev. **B42**, 6268.
135. McMahan, A. K., R. M. Martin, and S. Satpathy, 1988, Phys. Rev. **B38**, 6650.
136. Mermin, N. D., and Wagner, H., 1966, Phys. Rev. Lett. **17**, 1133.
137. Mila, F., 1988, Phys. Rev. **B38**, 11358.
138. Mila, F., 1990, Phys. Rev. **B42**, 2677.
139. Mila, F., D. Poilblanc, and C. Bruder, 1991, Phys. Rev. **B43**, 7891.
140. Moreo, A., E. Dagotto, T. Jolicoeur, J. Riera, 1990, Phys. Rev. **B42**, 6283.
141. Mori, H., 1965, Prog. Theor. Phys. **33**, 423.
142. Nagaev, E. L., 1968, Zh. Eksp. Teor. Fiz. **54**, 228. [1968, JETP **27**, 122].
143. Nagaoka, Y., 1966, Phys. Rev. **147**, 392.
144. Nazarenko, A. *et al.*, 1995, Phys. Rev. **B51**, 8676.
145. Newns, D. M., P. C. Pattnaik, and C. C. Tsuei, 1991, Phys. Rev. **B43**, 3075.

146. Nishimori, H., and Y. Saika, 1990, J. Phys. Soc. Jap. **59**, 4454.
147. Novikov, D. L., A. J. Freeman, and J. D. Jorgensen, 1995, Phys. Rev. **B51**, 6675.
148. Ogata, M., and H. Shiba, 1988, J. Phys. Soc. Jpn **57**, 3074.
149. Oguchi, T., and H. Kitatani, 1990, J. Phys. Soc. Jap. **59**, 3322.
150. Oitmaa, J., and D. D. Betts, 1978, Can. J. Phys **56**, 897.
151. Park., K. T. et al., 1988, J. Phys. Soc. Jpn. **57**, 3445.
152. Plakida, N. M., V. S. Oudovenko, and V. Yu. Yushankhai, 1994, Phys. Rev. **B50**, 6431.
153. Plakida, N. M. et al., 1997, Phys. Rev. B55, R11997; 1995, Phys.Rev. **B51**, 16599.
154. Plakida, N. M., and V. S. Oudovenko, 1999, Phys. Rev. **B59**, 11949.
155. Pothuizen, J. J. M. et al., 1997, Phys. Rev. Lett. **78**, 717.
156. Prelovshek, P., 1988, Phys. Lett. **A126**, 287.
157. Press, W. H., B. P. Flannery, S. A. Teukolsky, and W. T. Vetterling, 1981, "Numerical Recipes", Cambridge University Press.
158. Putz, R., R. Preuss, A. Muramatsu and W. Harke, 1996, Phys. Rev. **B53**, 5133.
159. Raimondi, R., Jefferson, and J. H., Feiner, L. F., 1996, Phys. Rev. **B53**, 8774.
160. Ramsak, A., and P. Prelovshek, 1989, Phys. Rev. **B40**, 2239.
161. Ramsak, A., and P. Prelovshek, 1990, Phys. Rev. **B42**, 10415.
162. Read, N., and S. Sachdev, 1989, Phys. Rev. Lett. **62**, 1694.
163. Read, N., and S. Sachdev, 1991, Phys. Rev. Lett. **66**, 1773.
164. Richter, J., 1993, Phys. Rev. **B47**, 5794.
165. Richter, J., N. B. Ivanov, K. Retzlaff, 1995, J. Mag. Mag. Mat. **140-144**, 1609.
166. Ronning, F. et al., 1998, Science **282**, 2067.
167. Sachdev, S., N. Read, 1991, Int. J. Mod. Phys. **B5**, 219, .
168. Sarker, S. et al., 1989, Phys. Rev. **B40**, 5028.
169. Schmelz, H. C. et al., 1998, Phys. Rev. **B57**, 10936.
170. Schmitt-Rink R. O., C. M. Varma, and A. E. Ruckenstein, 1988, Phys. Rev. Lett. **60**, 2793.
171. Schrieffer, J. R. and P. A. Wolf, 1966, Phys.Rev. **149**, 491.
172. Schrieffer, J. R., 1989, Phys.Rev. **B39**, 11663.
173. Schrieffer, J. R., 1995, J. Low. Temp. Phys. **99**, 397.
174. Schulz, H. J., and T. A. L. Ziman, 1992, Europhysics Letters **18**, 355.
175. Shen, Z. X., and D. S. Dussau, 1995, Phys. Rep. **253**, 1.
176. Shimahara, H., and S. Takada, 1991, J. Phys. Soc. Jpn. **60**, 2394.
177. Shimahara, H., and S. Takada, 1992, J. Phys. Soc. Jap. **61**, 989.
178. Shirane, G. et al., 1987, Phys. Rev. Lett. **59**, 1613.
179. Sinha, S. K. et al., 1988, J. Appl. Phys. **63**, 4015.
180. Starykh, O. A., O. F. A. Bonfim, and G. F. Reiter, 1995, Phys. Rev. **B52**, 12534.
181. Starykh, O. A., O. F. de Alcantara Bonfim, and G. F. Reiter, 1995, Phys. Rev. **B52**, 12534.
182. Stechel, E. B., and D. R. Jennison, 1988, Phys. Rev. **B38**, 4632.
183. Stechel, E. B., and D. R. Jennison, 1988, Phys. Rev. **B38**, 8873.
184. Stephan, W., and P. Horsch, 1990, Phys. Rev. Lett. **66**, 2258.
185. Szczepanski von, K. J., Horsch, P., Stephan, W., Ziegler, M., 1990, Phys. Rev. **B41**, 2017.

186. Tohyama, T., and Mayekava, 1992, Physica **C191**, 193.
187. Takahashi, M., 1989, J. Phys. Soc. Jap. **58**, 1524.
188. Takahashi, M., 1989, Phys. Rev. **B40**, 2494.
189. Takahashi, Y., 1988, Z. Phys. **B71**, 425.
190. Tang, S., M. E. Lazzouni, and J. E. Hirsh, 1989, Phys. Rev. **B40**, 5000.
191. Thurston, T. R., et al., 1989, Phys. Rev. **B40**, 4585.
192. Tobin, J. G. et al., 1992, Phys. Rev. **B45**, 5563.
193. Trugman, S., 1988, Phys.Rev. **B37**, 1597.
194. Tserkovnikov, A., 1981, Teor. Mat. Fiz. **49**, 219 [1982, Theor. Math. Phys., **49**, 993].
195. Tsuei, C. C., D. M. Newns, C. C. Chi, and P. C. Pattnaik, 1990, Phys. Rev. Lett. **65**, 2724.
196. Tyč, S. and B. I. Halperin, 1990, Phys.Rev. **B42**, 2096.
197. Unger., P., and P. Fulde, 1993, Phys.Rev. **B47**, 8947.
198. Unger., P., and P. Fulde, 1993, Phys.Rev. **B48**, 16607.
199. Varma, C. M., S. Schmitt-Rink, and E. Abrahams, 1987, Solid. State. Commun. **62**, 681.
200. Vos, K. J. E., and R. J. Gooding, 1996, Z. Phys. **B101** , 79.
201. Wells, B. O., Z. -X. Shen, A. Matsuura, D. M. King, M. A. Kastner, M. Greven, and R. J. Birgeneau, 1995, Phys. Rev. Lett. **74**, 964.
202. Xu, J., and C. Ting, 1990, Phys. Rev. **B42**, 6861.
203. Yushankhai, V. Yu., V. S. Oudovenko, and R. Hayn, 1997, Phys. Rev. **B55**, 15562.
204. Zaanen, J., and A. M. Ole's, 1988, Phys. Rev. **B37**, 9423.
205. Zhang, F. C., and T. M. Rice, 1988, Phys. Rev. **B37**, 3759.
206. Zhang, F. C., and T. M. Rice, 1990, Phys. Rev. **B41**, 7243.
207. Zubarev, D. N., 1960, Usp. Fiz. Nauk **71**, 71. [1960, Sov. Phys. Uspekhi **3**, 320].

Ferromagnetism and Electronic Correlations

W. Nolting

Humboldt-Universität zu Berlin, Institut für Physik,
Lehrstuhl Festkörpertheorie, Invalidenstraße 110
10 115 Berlin, Germany

Abstract. Ferromagnetism belongs to the oldest phenomena in solid state physics being, however, even today not yet fully understood. The main shortcoming for an understanding of this basic phenomenon is the lack of a general, unified theory. The different types of magnetic materials need for their description rather different theoretical models. each of them with a fairly restricted range of validity. To avoid misunderstandings it is therefore recommendable to start the discussion with a simple but clear classification of magnetism and to address the respective theoretical models to the appropriate classes of materials (Sect.I). The most prominent models are inspected in detail in the following sections.

For an at least qualitative description of ferromagnets with itinerant magnetic moments (*bandferromagnets*) the single-band Hubbard model is considered a good starting point. It poses a highly non-trivial many-body problem. Basic concepts and methods of many-body theory are therefore shortly introduced (Sect.II) as far as they are vital for the understanding of the following: Green functions, spectral densities, selfenergies, quasiparticles, quasiparticle densities of states, quasiparticle bandstructures,

Starting with the Hubbard-Hamiltonian and several exactly solvable limiting cases are discussed (Sect.II). The latter are important for the construction of reliable approaches to the not rigorously tractable many-body problem. Examples are the zero-bandwidth limit, the strong coupling behaviour of the spectral density, weak coupling perturbational treatments, and high-energy expansions.

We try a systematic improvement in the evaluation of the Hubbard-Hamiltonian by a series of analytic approximations fulfilling certain sum rules of the spectral density and the above-mentioned exact limiting cases (Sect.III). Starting from two-pole approaches (Hubbard I, spectral density approach (SDA), ...) via different alloy analogies and some applications of dynamical mean field theory we come to a conclusion for a magnetic phase diagram. The calculated Curie temperatures are compared with numerically essentially exact, recent Quantum-Monte Carlo calculations for infinite-dimensional lattices. The role of physically decisive correlation functions, which lead to bandshifts and bandwidth corrections, and the quasiparticle damping on the stability of (ferro) magnetism is investigated.

While the Hubbard model gives a frame for the itinerant moment systems the so-called s-f model is used to describe magnetic materials (insulators, semiconductors, metals) which take their magnetic properties from localized moments. It traces back the characteristic features of these systems to an interband exchange interaction between localized *magnetic* electrons (f) and itinerant conduction electrons (s). It is in principle identical to the Kondo-lattice and the double-exchange model. As a typical consequence of the s-f exchange a new quasiparticle appears which is called the *magnetic polaron*. It can be calculated exactly for a non-trivial special case. A selfenergy approach combined with a modified RKKY-theory is proposed to determine selfconsistently the electronic and the magnetic properties of the exchange-coupled s and f electrons.

I MODELS OF MAGNETISM

The phenomenon of *collective magnetism* belongs to the oldest and most fundamental problems of solid state physics. However, only by reading the content of a modern journal of solid state physics one realizes that this *old* phenomenon continues to be a very hot topic of current research. That can be understood only by the fact that *collective magnetism* is not yet fully understood. The main shortcoming is the lack of a unified theory which would be able to explain the rich variety of magnetic features within the framework of one and the same theoretical model. The various types of magnetic materials require for their description a full set of rather different models, each of them with a fairly restricted range of validity. When speaking about magnetism one should therefore start with a clear specification of which theoretical model is to be applied to which magnetic material. To present a rough but unambiguous classification of magnetism and to address respective theoretical models is the intention of this introductory section.

A Preparations

Magnetism is bound to the existence of magnetic moments. Which kind of Hermitean operator represents the observable *magnetic moment*?

1 Magnetic moment operator

Using the correspondence principle we derive from classical electrodynamics the quantum mechanical observable *magnetic moment*:

$$\hat{\mathbf{m}} = -\nabla_{\mathbf{B}} \hat{H} \tag{1.1}$$

$\hat{H}$ is the Hamiltonian of the system and $\nabla_{\mathbf{B}}$ means the gradient with respect to the external field $\mathbf{B} = \mu_0 \mathbf{H}$. While $\hat{\mathbf{m}}$ is a vector-**operator** the *magnetization* $\mathbf{M}(\mathbf{r})$ is a normal vector with real components. It can be interpreted as the average magnetic moment per volume:

$$\mathbf{M}(\mathbf{r}) = \frac{1}{v(\mathbf{r})} \sum_{j=1}^{N(v)} \langle \hat{\mathbf{m}}_j \rangle \tag{1.2}$$

The system shall consist of single particles (atoms, ions, molecules, clusters,...) numbered by the index j. The summation runs over all particles in the microscopic volume $v(\mathbf{r})$ around $\mathbf{r}$. The bracket $\langle ... \rangle$ denotes quantum statistical average

$$\langle \hat{\mathbf{m}} \rangle = \frac{1}{Z} Tr(e^{-\beta \hat{H}} \hat{\mathbf{m}}) \qquad (1.3)$$

Z is the canonical partition function,

$$Z = Tr(e^{-\beta \hat{H}}); \qquad \beta = \frac{1}{k_B T}. \qquad (1.4)$$

The main goal in the theory of *collective magnetism* is the determination of the *spontaneous* magnetization for a system of permanent magnetic moments.

2   Bohr van Leeuwen theorem

Magnetism is a purely quantum mechanical effect. According to a strictly classical calculation the magnetization of an arbitrary macroscopic particle system in thermodynamic equilibrium vanishes in a finite magnetic field. This is the statement of the Bohr-van Leeuwen theorem [1,2]. It is thus recommendable to discuss a magnetic problem from the very beginning on the basis of quantum mechanics.

The proof of the theorem is very simple. Starting point is a system of N particles (mass m_i, charge q_i) in the Volume V. $H(\mathbf{r}, \mathbf{p})$ is the classical Hamilton function and Z the classical partition function

$$Z = \frac{1}{h^{3N} N!} \int \ldots \int d^3 r_1 \ldots d^3 r_N d^3 p_i \ldots d^3 p_N \, exp(-\beta H(\mathbf{r}, \mathbf{p})) \qquad (1.5)$$

Z is used for the determination of the *classically* averaged magnetic moment

$$\langle \hat{\mathbf{m}} \rangle = \frac{1}{\beta Z} \nabla_\mathbf{B} Z \qquad (1.6)$$

in a finite magnetic field

$$\mathbf{B} = rot \mathbf{A}$$

The Hamilton function reads

$$H = \sum_{i=1}^{N} \frac{1}{2m_i} (\mathbf{p}_i - q_i \mathbf{A}(\mathbf{r}_i))^2 + U(\mathbf{r}_i, \ldots, \mathbf{r}_N) \qquad (1.7)$$

$U(\mathbf{r})$ represents the potential energy. Inserting H into (1.5) and using the substitution

$$p_{i\alpha} - q_i A_\alpha(\mathbf{r}_i) \equiv u_{i\alpha}(\mathbf{r}_i); \quad \alpha = x, y, z \qquad (1.8)$$

allows to perform the momenta integrations. Defining

$$\Lambda(T) = \prod_{i=1}^{N} \lambda_i(T)$$

with

$$\lambda_i(T) = h(2\pi m_i k_B T)^{-1/2} \qquad (1.9)$$

as the thermal de Broglie-wave length we get for the partition function

$$Z = \frac{1}{N!\, \Lambda^3(T)} \int \cdots \int d^3 r_1 \ldots d^3 r_N \, exp(-\beta U(\mathbf{r}_1, \ldots, \mathbf{r}_N)) \qquad (1.10)$$

In particular we have found

$$Z \neq Z(\mathbf{B}) \qquad (1.11)$$

and therefore according to (1.6):

$$\langle \mathbf{m} \rangle = \mathbf{0} \qquad (1.12)$$

That proves the theorem. The result appears rather surprising because there do exist various aspects of magnetism which are successfully explained by semiclassical vector models. However, these models must contain certain inconsistencies, i. e. more or less hidden quantum mechanical elements. For getting a first classification of magnetism we discuss in the next section a very simple but illustrative example.

3 Single particle in a homogeneous magnetic field

We use the following simplified model to derive the magnetic moment operator of a particle consisting of a positively charged nucleus and N electrons:

1. *Fixed position of the nucleus, no influence on the homogeneous external field:*

$$\mathbf{B} = \mathrm{rot}\,\mathbf{A} = (0, 0, B) \qquad (1.13)$$

We use the Coulomb gauge:

$$\mathrm{div}\,\mathbf{A} = 0 \qquad (1.14)$$

Both relations can be fulfilled by

$$\mathbf{A} = \frac{1}{2}\mathbf{B} \times \mathbf{r} = \frac{B}{2}(-y, x, 0) \qquad (1.15)$$

Because of (1.14) the electron momentum $\mathbf{p}_i$ commutes with the site-dependent vector potential

$$\hat{\mathbf{p}}_i \cdot \mathbf{A}(\hat{\mathbf{r}}_i) = \mathbf{A}(\hat{\mathbf{r}}_i) \cdot \hat{\mathbf{p}}_i \tag{1.16}$$
$$= \frac{1}{2} B(-\hat{y}_i \hat{p}_{ix} + \hat{x}_i \hat{p}_{iy})$$
$$= \frac{1}{2} \mathbf{B} \cdot \mathbf{l}_i$$

$\mathbf{l}_i = \hat{\mathbf{r}}_i \times \hat{\mathbf{p}}_i$ is the orbital angular momentum of electron i.

2. *Kinetic energy:*

$$\hat{T} = \frac{1}{2m} \sum_{i=1}^{N} (\hat{\mathbf{p}}_i + e\mathbf{A}(\hat{\mathbf{r}}_i))^2 \tag{1.17}$$
$$= \hat{T}_0 + \frac{\mu_B}{\hbar} \mathbf{B} \cdot \mathbf{L} + \frac{e^2 B^2}{8m} \sum_{i=1}^{N} (\hat{x}_i^2 + \hat{y}_i^2)$$

Here we have introduced the total angular momentum

$$\mathbf{L} = \sum_{i=0}^{N} \mathbf{l}_i \tag{1.18}$$

$\hat{T}_0$ gathers all field-independent terms

$$\hat{T}_0 = \frac{1}{2m} \sum_{i=1}^{N} \hat{\mathbf{p}}_i^2 \tag{1.19}$$

$\mu_B = \frac{e\hbar}{2m}$ is the Bohr magneton.

3. *Phenomenological introduction of the spin, no spin-orbit coupling*
The total spin

$$\mathbf{S} = \sum_{i=1}^{N} \mathbf{s}_i \tag{1.20}$$

is connected with a magnetic moment which gives rise to an additional term in the Hamiltonian:

$$\hat{H}_S = 2\frac{\mu_B}{\hbar} \mathbf{B} \cdot \mathbf{S} \tag{1.21}$$

The complicated electron-electron and electron-nucleus interactions are assumed not to be dramatically influenced by the magnetic field. Together with

$\hat{T}_0$ they are absorbed by the field-independent partial operator $\hat{H}_0$ of the total Hamiltonian $\hat{H}$:

$$\hat{H} = \hat{H}_0 + \frac{\mu_B}{\hbar}(\mathbf{L} + 2\,\mathbf{S}) \cdot \mathbf{B} + \left(\frac{e^2}{8m}\sum_{i=1}^{N}(\hat{x}_i^2 + \hat{y}_i^2)\right)\mathbf{B}^2 \qquad (1.22)$$

By (1.1) we get as the magnetic moment of the considered particle:

$$\hat{\mathbf{m}} = -\frac{\mu_B}{\hbar}(\mathbf{L} + 2\,\mathbf{S}) - \left(\frac{e^2}{4m}\sum_{i=1}^{N}(\hat{x}_i^2 + \hat{y}_i^2)\right)\mathbf{B} \qquad (1.23)$$

We can use this result for a first rough classification of magnetic materials. The second term in (1.23) obviously represents an induced magnetic moment disappearing when the field is switched off. As a pure induction effect it is antiparallel to the external field **B** giving rise to what is called

diamagnetism.

Normally, however, this is only a tiny contribution to magnetism being overcompensated by the first term in (1.23). A material is therefore classified as a *diamagnet* only if the first term in (1.22) vanishes (closed electron shells). For not completely filled electron shells

$$\hat{\mathbf{m}}_p = -\frac{\mu_B}{\hbar}(\mathbf{L} + 2\mathbf{S}) \qquad (1.24)$$

represents a permanent magnetic moment. If there is no remarkable interaction between the moments they will be disordered in zero field for all temperatures T. There is no spontaneous order of the moments. Only a magnetic field can force the moments into a collective alignment. This phenomenon is called

paramagnetism.

If an exchange interaction creates an ordered phase of the moments below a critical temperature T^* then one speaks of *collective magnetism*, which can manifest itself as

ferro-, ferri-, antiferromagnetism.

Para- and diamagnetism are effectively single-particle problems being considered as rather well understood. We shall concentrate ourselves therefore exclusively on the collective magnetism. To be specific we discuss in detail *ferromagnetism*, only. What is the reason for a spontaneous low-temperature magnetic order? ($T < T_C$; T_C : Curie temperature). The simplest concept is that of an internal magnetic field that forces the moments into a ferromagnetic order.

4 Exchange field

The non-interacting moments of a paramagnet can be forced into a certain order by an external magnetic field. So one might suppose that the spontaneous moment order in a ferromagnet is due to an *internal exchange field* $\mathbf{B}_{\text{ex}}$. Before investigating the origin of this hypothetical field let us estimate its order of magnitude. That can be done by use of the free energy

$$F = U - TS \quad (1.25)$$

There are two competing effects. The exchange field $\mathbf{B}_{\text{ex}}$ takes care for an ordering tendency thus minimizing the internal energy U. Temperature T provokes a disorder thus maximizing the entropy. Both tendencies lower the free energy. At low temperatures the ordering effect of $\mathbf{B}_{\text{ex}}$ will dominate (ferromagnetic phase), at high temperatures the disordering effect of the entropy (paramagnetic phase).

At the transition temperature T_C thermal energy and magnetic field energy should be comparable. Therewith we can approximate the *exchange energy*

$$E_{\text{ex}} \approx \mu_B B_{\text{ex}} \approx k_B T_C \quad (1.26)$$

Using the experimental T_C-values we can exploit this formula to estimate the exchange field $\mathbf{B}_{\text{ex}}$. What comes out are extremely large fields compared to normal laboratory fields ($B < 10 \ldots 20\text{T}$)

TABLE 1.1: Estimation of the hypothetical exchange field from measured T_C-values

	T_C [K]	$k_B T$ [meV]	B_{ex} [T]
Fe	1043	89.9	1552.8
Co	1393	120.1	2073.9
Ni	631	54.4	939.4
Gd	292	25.0	431.7
EuO	69.3	6.0	103.2

or to fields created by conventional dipole-dipole interaction. We have to conclude, since such large fields are unimaginable, that the simple concept of an internal exchange field cannot be the explanation for the origin of ferromagnetism. We have to look after different theories.

However, we have to face the fact that there does not exist a unified theory. This lack forces us to apply to the different types of ferromagnetic materials a full set of rather different theoretical models with strongly restricted ranges of validity. It is thus absolutely necessary to check which model is appropriate for a given material. For this purpose let us start with a simple classification of magnetism that uses the fundamental preconditions of ferromagnetism: A solid is needed and must contain permanent magnetic moments. The solids are roughly decomposed into insulators and metals, while the moments can be localized or itinerant. This gives four possibilities of combinations permitting a fairly reasonable classification of the magnetic materials which we shall follow in the next sections.

We start with a discussion of the insulators which take their magnetism from localized moments. This subclass is considered as best understood.

B   Insulators with localized moments (Heisenberg model)

The permanent and localized magnetic moments must stem from any inner only partially filled electron shell. Typically these are (3d, 4f, 5f)-shells, in particular the 4f-shell of rare earths and their compounds. Prototypes are the Europium chalcogenides with an exactly half-filled 4f-shell. The oxide and sulfide are ferromagnets, while the telluride is an antiferromagnet. The reason why these materials are considered as well understood rests on the fact that there obviously exists a theoretical model which is able to describe at least the purely magnetic properties of these insulators in a highly realistic manner. It is the well-known Heisenberg model [3]

$$H = -\sum_{i,j} J_{ij}\, \mathbf{S}_i \cdot \mathbf{S}_j \tag{1.27}$$

which represents an exchange-coupled system of localized spins. J_{ij} are the respective exchange coupling constants. In spite of its rather simple operator structure the respective many-body problem could not yet be solved for the general case. The elaborated approaches are, however, so convincing that one does not expect very much additional information from the unknown exact solution. Many experimental data are astonishingly well explained by the model [4]: Blochs $T^{\frac{3}{2}}$-law, the Curie-Weiß-law, the concept of spin waves and magnons, critical phenomena and so on. In spite of its great successes one must not forget that the Heisenberg operator is an effective Hamiltonian which should not be misinterpreted as if ferromagnetism were a consequence of a moment-moment interaction. We have already found out that this is not the case. H simply simulates in the finite dimensional spin space the effects of the exclusively responsible Coulomb interaction between charged particles. How this works we are going to inspect in the next subsections.

1 Direct exchange

Starting point is the Hamiltonian

$$H = H_0 + H_1 \tag{1.28}$$

The interaction part H_1 contains the Coulomb-interaction between electrons or other charged particles. It is easy to demonstrate how the Pauli principle can lead to magnetic effects without any spin part in the interaction term H_1. We demonstrate this for the simplest system, a two-electron system

$$H = \sum_{i=1}^{2} \frac{\mathbf{p}_i^2}{2m} + V(\mathbf{r}_1, \mathbf{r}_2) \tag{1.29}$$

The two electrons have spins $\mathbf{S}_1, \mathbf{S}_2$ ($S_1 = S_2 = \tfrac{1}{2}$) which either couple to an antisymmetric singlet

$$|00\rangle^{(-)} = \frac{1}{\sqrt{2}}(|\uparrow\downarrow\rangle - |\downarrow\uparrow\rangle) \tag{1.30}$$

or to a symmetric triplet:

$$|11\rangle^{(+)} = |\uparrow\uparrow\rangle \tag{1.31}$$

$$|10\rangle^{(+)} = \frac{1}{\sqrt{2}}(|\uparrow\downarrow\rangle + |\downarrow\uparrow\rangle) \tag{1.32}$$

$$|1-1\rangle^{(+)} = |\downarrow\downarrow\rangle \tag{1.33}$$

The eigenstate $|\Psi\rangle$ of H for the two identical fermions must be antisymmetric with respect to a particle-interchange. Since H is spin-independent the eigenstate $|\Psi\rangle$ factorizes into a spin- and a space-part. Both parts must carry a definite symmetry character to guarantee that the full state is antisymmetric:

$$|\Psi_+\rangle = |q\rangle^{(+)}|00\rangle^{(-)} \qquad\qquad |\Psi_-\rangle = |q\rangle^{(-)}|1M_S\rangle^{(+)} \tag{1.34}$$

The antisymmetric spin-singlet $|00\rangle^{(-)}$ couples to the symmetric space part $|q\rangle^{(+)}$ and the symmetric spin-triplet $|1M_S\rangle^{(+)}$ ($M_S = 0, \pm 1$) to the antisymmetric space part $|q\rangle^{(-)}$. The Hamiltonian H acts only on $|q\rangle^{(\pm)}$:

$$H|\Psi_+\rangle = (H|q\rangle^{(+)})|00\rangle^{(-)} = E_+|\Psi_+\rangle \tag{1.35}$$

$$H|\Psi_-\rangle = (H|q\rangle^{(-)})|1M_S\rangle^{(+)} = E_-|\Psi_-\rangle \tag{1.36}$$

We now search for an effective Hamiltonian H_{eff} which acts exclusively on the spin-part but with the same consequences as in (1.35) and (1.36):

$$H_{\text{eff}} |\Psi_+\rangle = |q\rangle^{(+)} (H_{\text{eff}} |00\rangle^{(-)}) \stackrel{!}{=} E_+ |\Psi_+\rangle \tag{1.37}$$

$$H_{\text{eff}} |\Psi_-\rangle = |q\rangle^{(-)} (H_{\text{eff}} |1M_S\rangle^{(+)}) \stackrel{!}{=} E_- |\Psi_-\rangle \tag{1.38}$$

This can indeed be achieved by the following choice:

$$H_{\text{eff}} = J_0 - J_{12} \, \mathbf{S}_1 \cdot \mathbf{S}_2 \tag{1.39}$$

J_0 is an unimportant constant

$$J_0 = \frac{1}{4} (E_+ + 3 E_-) \tag{1.40}$$

Decisive is the coupling constant:

$$\hbar^2 J_{12} = E_+ - E_- \tag{1.41}$$

H_{eff} (1.39) has indeed the form of the Heisenberg model Hamiltonian (1.27) provided that J_{12} is really unequal zero. May it be possible that $E_+ \neq E_-$? This can be demonstrated by a special two-electron system as, e. g., realized in the H_2-molecule [5]:

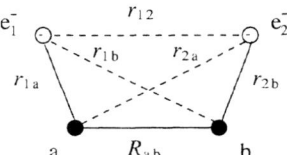

FIGURE 1.1. Electron-electron, proton-proton and electron-proton distances in the H_2-molecule

$$H = H_0 + H_1 \tag{1.42}$$

Because of their big masses, compared to those of the electrons, the proton positions can be considered as fixed at the distance R_{ab}.

H_0 represents two independent hydrogen problems (r_{1a}, r_{2b}) realized by an infinite distance $(R_{ab} \to \infty)$ between the two protons a and b. The correctly symmetrized eigenstates read,

$$|q_0\rangle^{(\pm)} = \frac{1}{\sqrt{2}} (|\phi_a^{(1)}\rangle |\phi_b^{(2)}\rangle \pm |\phi_a^{(2)}\rangle |\phi_b^{(1)}\rangle) \tag{1.43}$$

where $\phi_{a,b}^{(1,2)}$ are the well-known solutions of the hydrogen problem. Both shall be the respective groundstates of the atoms a and b:

$$H_0 |q_0\rangle^{(\pm)} = 2E_0 |q_0\rangle^{(\pm)} \qquad (1.44)$$

In this limiting case we have $E_+^{(0)} = E_-^{(0)}$, i. e. no spin coupling J_{12} and therewith no spontaneous spin order.

The *perturbation* H_1 incorporates all the other interactions ($R_{ab}, r_{12}, r_{2a}, r_{1b}$) coming into play for a finite R_{ab}. The mutual influence of the two hydrogen atoms prevents an exact solution of the problem. We use the Ritz variational principle to estimate the ground state energy. $|q_0\rangle^{(\pm)}$ is a plausible choice for the variational state, guaranteeing the correct symmetry behaviour:

$$E_\pm = \frac{{}^{(\pm)}\langle q_0 | H |q_0\rangle^{(\pm)}}{{}^{(\pm)}\langle q_0 | q_0\rangle^{(\pm)}} = 2E_0 + \frac{C_{ab} \pm A_{ab}}{1 \pm |L_{ab}|^2} \qquad (1.45)$$

According to (1.41) this means for the exchange coupling constant J_{12}:

$$J_{12} = \frac{1}{\hbar^2}(E_+ - E_-) = -\frac{2}{\hbar^2} \frac{C_{ab}|L_{ab}|^2 - A_{ab}}{1 - |L_{ab}|^4} \qquad (1.46)$$

J_{12} is determined by several overlap-integrals of different type. One of them is:

$$L_{ab} = \langle \phi_a^{(i)} | \phi_b^{(i)} \rangle \quad ; i = 1, 2 \qquad (1.47)$$

The ground state wave functions $\phi_a^{(i)}$ and $\phi_b^{(i)}$ are centred at different nucleus positions being therefore non-orthogonal with a finite overlap. C_{ab} is the *Coulomb integral* which can be understood as the interaction between two different charge densities:

$$C_{ab} = \langle \phi_a^{(1)} | \langle \phi_b^{(2)} | H_1 | \phi_a^{(1)} \rangle | \phi_b^{(2)} \rangle = \langle \phi_a^{(2)} | \langle \phi_b^{(1)} | H_1 | \phi_b^{(1)} \rangle | \phi_a^{(2)} \rangle \qquad (1.48)$$

Still more decisive is the *exchange integral*

$$A_{ab} = \langle \phi_a^{(1)} | \langle \phi_b^{(2)} | H_1 | \phi_a^{(2)} \rangle | \phi_b^{(1)} \rangle = \langle \phi_a^{(2)} | \langle \phi_b^{(1)} | H_1 | \phi_a^{(1)} \rangle | \phi_b^{(2)} \rangle \qquad (1.49)$$

which is classically not understandable. Using the ground state wave functions of the hydrogen atoms located at $\mathbf{R}_a$ and $\mathbf{R}_b$ one finds for *normal* distances R_{ab}:

$$|L_{ab}| \ll 1 \ , \ |A_{ab}| \gg |L_{ab}|, C_{ab} \ , A_{ab} < 0$$

That means for the exchange coupling:

$$J_{12} \approx \frac{2}{\hbar^2} A_{ab} < 0 \qquad (1.50)$$

The result is an antiferromagnetic coupling between the two electron spins. If we postulate that a generalization to more than two electrons really works then we get an effective Hamiltonian of type (1.27). However, we have to bear in mind

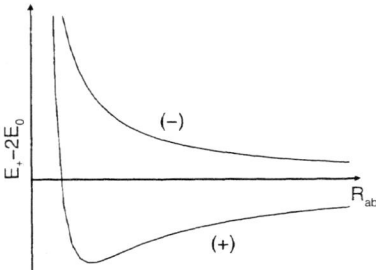

FIGURE 1.2. Ground state energies in the H_2-molecule

that it is an effective Hamiltonian that simulates the complicated Coulomb interactions in the finite-dimensional spin space. Furthermore, it has been shown that the above-procedure does not converge (*non-orthogonality catastrophe*). Even for only three electrons one is running into complications. When applying the Heisenberg-Hamiltonian (1.27) to an actual problem one is obliged to consider the exchange constants J_{ij} as parameters rather than as given by (1.50). However, it is sure that the J_{ij} consist of overlap-integrals so that a spontaneous moment order is excluded if there is no wave-function overlap. The consequence is that J_{ij} decreases exponentially with the distance $R_{ij} = |\mathbf{R}_i - \mathbf{R}_j|$. For most of the magnetic materials a *direct* exchange as discussed in this chapter is therefore unlikely. Indirect couplings as the *superexchange*, which is illustrated in the next section, appear to be more realistic. They lead, however, to the same operator structure as in (1.27).

2 Superexchange

This kind of indirect exchange is realized in systems like

$$MnO, MnF_2, FeF_2, \ldots$$

for which the magnetic moment distances are too large to allow for a direct exchange. It is commonly accepted that an indirect coupling between two magnetic transition metal ions mediated by the diamagnetic O^{2-} or $(F_2)^{2-}$-ion is responsible for a collective order of the localized d-moments. The simplest explanation of the

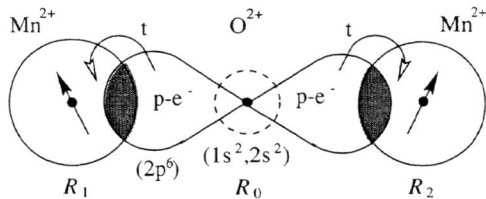

FIGURE 1.3. Model of superexchange mechanism with respect to MnO

superexchange mechanism uses a semiclassical model [6,7]:

1. The half-filled d-shells of the Mn^{2+}-ions are represented by classical spins of variable relative orientation

$$\mathbf{S}_1 \cdot \mathbf{S}_2 = \hbar^2 S^2 \cos\theta \quad \left(S = \frac{5}{2}\right) \tag{1.51}$$

2. The p electrons of the diamagnetic ion may hop to the neighbouring magnetic Mn^{2+}-ions because of a finite overlap of the respective p- and d-wave functions.

Since the 3d-shell of the Mn^{2+} is with five electrons exactly half-filled the p-electron can enter the Mn^{2+}-ion only spin antiparallel to the localized 3d-spin. That leads to an effective coupling between the Mn^{2+} spins. One can construct the Hamiltonian matrix by considering the following four states as a complete set:

$|1\rangle$: both p-electrons at the diamagnetic ion

$$E_1 = \langle 1| H |1\rangle = \epsilon$$

$|2\rangle$: one of the two p-electrons hops to the *left* Mn^{2+}:

$$E_2 = \langle 2| H |2\rangle = \epsilon + U$$

U measures the additional Coulomb interaction between the p electron and the Mn^{2+}-d electrons

$|3\rangle$: one of the two p-electrons hops to the *right* Mn^{2+}:

$$E_3 = \langle 3| H |3\rangle = \epsilon + U$$

$|4\rangle$: one of the p electrons hops to the *right*, the other to the *left* Mn^{2+}:

$$E_4 = \langle 4| H |4\rangle = \epsilon + U + V$$

V is a further Coulomb contribution.

Neglecting higher order processes one gets the Hamiltonian matrix

$$\hat{H} \equiv \begin{pmatrix} \epsilon & t & t & 0 \\ t & \epsilon + U & 0 & t\sin\frac{\theta}{2} \\ t & 0 & \epsilon + U & t\sin\frac{\theta}{2} \\ 0 & t\sin\frac{\theta}{2} & t\sin\frac{\theta}{2} & \epsilon + U + V \end{pmatrix} \tag{1.52}$$

t is the transfer matrix element. The transitions are virtual so we can assume: $t \ll V, U$. Diagonalizing the matrix one finds for the ground state energy

$$E_0 = E_{00} + 2 \left(\frac{t^2}{U}\right)^2 \frac{\cos\theta}{U+V} + \mathcal{O}(\sqcup)^{\prime} \qquad (1.53)$$

Expressing $\cos\theta$ by the scalar product $\mathbf{S}_1 \cdot \mathbf{S}_2$ (1.51) one can define a model Hamiltonian of type (1.27) with an effective exchange coupling:

$$J_{12}^{SA} = -\frac{2}{\hbar^2 S^2} \frac{t^4}{U^2(U+V)} \qquad (1.54)$$

Again we have found an antiferromagnetic coupling.

C Insulators with itinerant moments

It sounds a bit contradictory, but, there exist, in caertain sense, magnetic materials which can be considered as insulators with itinerant moments. These are the so-called *Mott-Hubbard insulators* [8]. Prototypes can be found among the transition metal oxides, in particular the monoxides. Most intensively investigated is surely the antiferromagnetic NiO. To the keyword *theory* we have to offer essentially question marks since it is up to now not really clear which theoretical model might be able to describe the extraordinary physical properties of the Mott-Hubbard insulators:

CuO, NiO, CoO, FeO, MnO, Fe_2O_3, Mn_2O_3,...

What should be explained by a properly chosen theoretical model?

a) All these materials are insulators what is really astonishing because of the following reason. It is surely allowed to consider the oxygen as O^{2-} ion. But then it is not difficult to check that the number of 3d-electrons per transition metal ion n_{3d} is smaller than 10. Following elementary band theory all these materials should therefore be metallic. *First principles* bandstructure calculation indeed predict metallic ground states.

Inspecting the surprising insulating behaviour in more detail teaches us that it is obviously a necessary condition that n_{3d} is an integer between 5 and 10:

$$5 \leq n_{3d} \leq 10 \qquad (1.55)$$

The mentioned systems become immediately metallic if n_{3d} is non-integer, e. g. , because of non-stoichiometry. An interesting speculation in this context is that valence mixing, too, may be the reason for metallic behaviour: Fe_3O_4 is composed by FeO and Fe_2O_3. FeO and Fe_2O_3 are both Mott-insulators. However, the iron is divalent in FeO and trivalent in Fe_2O_3. So the average 3d-number is non-integer. Maybe that is the reason why Fe_3O_4 is a metal. Metallic behaviour is even observed for an integer n_{3d}, but below 5. VO with $n_{3d} = 3$ is a paramagnetic metal.

b) All the Mott-Hubbard insulators are antiferromagnets and that cannot be only by chance. It must have a deeper reason. There is even to be seen a certain systematics in the magnetic properties:

TABLE 1.2: Magnetic properties of transition metal monoxides.

	NiO	CoO	MnO
n_{3d}	8	7	5
T_N [K]	523	289	122
μ/μ_B	1.6-1.9	3.8	4.8

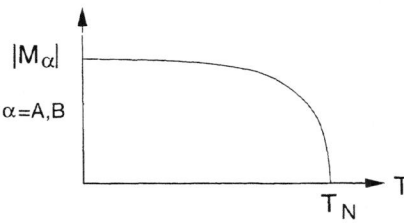

FIGURE 1.4. Sublattice-magnetization as a function of temperature.

As to the purely magnetic properties they are almost ideal Heisenberg-antiferromagnets. The sublattice magnetization curves are well-described by a Brillouin function. The moments are stable for arbitrary high temperatures.

c) A lot of photoemission experiments have sketched a rather clear picture of the density of states. Obviously there are three not connected regions with predominantely 3d-character. The lower two are occupied, the upper one is empty.

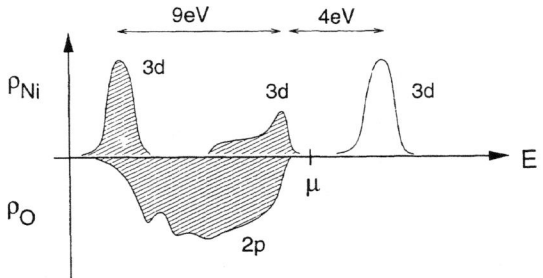

FIGURE 1.5. Schematic plot of the quasiparticle density of states of NiO.

Between the middle and the uppermost structure there is the insulating gap which amounts to some 4eV in NiO. The gap remains stable for all temperatures showing no reaction at the phase transition at $T = T_N$. Obviously strong electron correlations split the 3d-band into several quasiparticle subbands, each of them either fully occupied or completely empty. So the system is an insulator but an insulator with a permanent magnetic moment since the original 3d-band is only partially filled. The moment is due to itinerant electrons in 1...2eV broad subbands. Therefore we can speak of insulators with itinerant moments.

D Local moment metals

A ferromagnetic metal is characterized by two striking properties: A high electrical conductivity and a spontaneous magnetic order. If these two properties are carried by two different groups of electrons then the permanent magnetic moment are necessarily localized. They stem from any inner only partially filled electron shell, as, e. g., the 4f-shell of the Rare Earth. Prototype is the ferromagnetic Gd with an exactly half-filled 4f-shell:

$$\text{Gd:} \quad J = S = \tfrac{7}{2}, \quad T_C = 293.2\,\text{K}, \quad \mu(T=0) = 7.63\,\mu_B$$

According to Hunds rule this leads to a pure spin moment. The $T = 0$ moment is slightly higher than $7\mu_B$ pointing to a certain contribution of the *a priori* non-magnetic (5d-6s) conduction electrons which occupy a rather broad conduction band being responsible for the electrical conductivity. Other similar examples are Dy, Tb. To this class also belong interesting compounds like $Eu_{1-x}Gd_xS$ and Eu-rich EuO. Many striking properties of these local moment metals can be traced back to an intimate correlation between the two well-defined electronic subsystems and as far as such cross-effects are meant there exists a reliable theoretical model, the so-called s-f model [9], sometimes also denoted as (ferromagnetic) Kondo-lattice model. It explains the decisive features by an intraatomic exchange, i. e. by a local interaction between the conduction electron spin σ_i and the localized f-spin $\mathbf{S}_i$. The model-Hamiltonian reads:

$$H = H_s + H_{sf} \tag{1.56}$$

H_s refers to uncorrelated conduction electrons

$$H_s = \sum_{\mathbf{k}\sigma} \epsilon(\mathbf{k}) c^+_{\mathbf{k}\sigma} c_{\mathbf{k}\sigma} = \sum_{ij\sigma} T_{ij} c^+_{i\sigma} c_{j\sigma} \tag{1.57}$$

c^+ and c are, respectively, creation and annihilation operators of a band electron which is specified by the lower indices. $\mathbf{k}$ is a wave-vector from the first Brillouin zone, $\sigma = \uparrow$ or $\downarrow$ denotes the spin projection. The indices i and j number the lattice sites. The single-electron Bloch energies $\epsilon(\mathbf{k})$ are connected by Fourier transformation to the hopping integrals T_{ij}:

$$T_{ij} = \frac{1}{N} \sum_{\mathbf{k}} \epsilon(\mathbf{k}) e^{i\mathbf{k}\cdot(\mathbf{R}_i - \mathbf{R}_j)} \tag{1.58}$$

The characteristic model properties are due to the mentioned intra-atomic exchange interaction between the conduction electron spin σ and the localized spin $\mathbf{S}$

$$H_{sf} = -J \sum_i \sigma_i \cdot \mathbf{S}_i \tag{1.59}$$

J is the corresponding coupling constant. Using for the conduction electron spin σ_i the formalism of second quantization,

$$\sigma_i^z = \frac{\hbar}{2}(n_{i\uparrow} - n_{i\downarrow}) = \frac{\hbar}{2}\sum_\sigma z_\sigma n_{i\sigma} \quad ; \quad n_{i\sigma} = c_{i\sigma}^+ c_{i\sigma} \tag{1.60}$$

$$z_\uparrow = +1 \quad ; \quad z_\downarrow = -1 \tag{1.61}$$

$$\sigma_i^+ = \hbar c_{i\uparrow}^+ c_{i\downarrow} \quad ; \quad \sigma_i^- = \hbar c_{i\downarrow}^+ c_{i\uparrow} \tag{1.62}$$

the s-f interaction can be brought into a more useful form:

$$H_{sf} = -\frac{1}{2}J\hbar \sum_i \left(S_i^z(n_{i\uparrow} - n_{i\downarrow}) + S_i^+ c_{i\downarrow}^+ c_{i\uparrow} + S_i^- c_{i\uparrow}^+ c_{i\downarrow} \right) \tag{1.63}$$

This exchange interaction gives rise to a striking mutual influence of the itinerant conduction electrons and the localized magnetic moments.

1 The "red shift" effect

The ferromagnetism of the local moment system induces a remarkable temperature-dependence of the conduction band structure. In simple terms this can easily be understood. Below T_C the 4f-moments order themselves, building there-

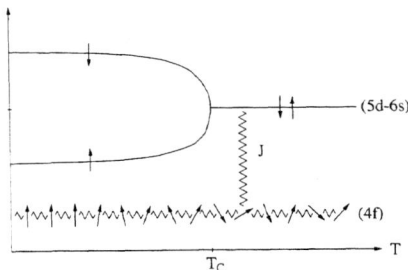

FIGURE 1.6. Schematic plot of the *red shift*-effect

with something like an internal magnetic field which removes by an inner Zeeman effect the spin-degeneracy in the conduction band. The subsequent temperature-shift of the conduction band states has been first observed some 35 years ago as *red shift* of the optical absorption edge in EuO being verified in the meantime for all ferromagnetic local moment systems [10]. Performing on the model-Hamiltonian (1.56) the simplest mean field approach,

$$H \longrightarrow \sum_{ij\sigma} \left(T_{ij} - \frac{1}{2} J\hbar z_\sigma \langle S_i^z \rangle \delta_{ij} \right) c_{i\sigma}^+ c_{j\sigma} \qquad (1.64)$$

the *red shift* proportional to the 4f magnetization directly appears via an effective temperature-dependent magnetic field. One has tried to exploit the *red shift* effect for constructing a perfect electron spin filter by field emission through a thin EuS layer deposited on a tungsten tip [11]. Without external field the EuS conduction

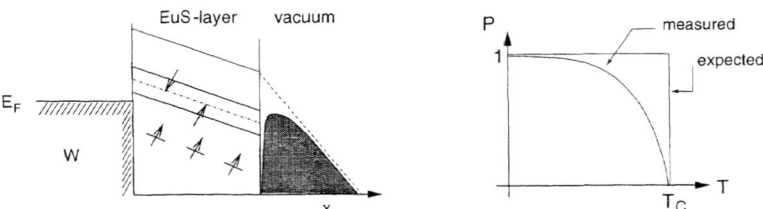

FIGURE 1.7. Spin-filter experiment on an EuS-layer deposited on a tungsten tip.

band lies clearly above the Fermi level of the tungsten. An external electric field, being able to penetrate the insulator EuS, tilts the energy bands of the EuS in such a way that electrons from the W-Fermi level can tunnel through an inner and an outer barrier into vacuum (s. Fig. 1.7). Field strength and film thickness are chosen just so that tunneling of 4f electrons is insignificant. Only electrons from the conduction band come into question when they have overcome the inner barrier between the W-Fermi level and the EuS-conduction band. The conditions of the experiment take care of a vanishing current at room temperature. If indeed the spin polarized band splittings happen for $T < T_C$, then only ↑ electrons may be able to tunnel because of a decreasing internal barrier, which increases, on the other hand, for ↓ electrons. For the electron spin polarization $P(T)$ one therefore expects something like a step function jumping from zero for $T > T_C$ onto the ideal value 1 for $T < T_C$. The achieved degrees of polarization have been indeed rather high, the expected ideal value, however, was never reached. This means that there is indeed a strong reaction of the conduction band states on the ferromagnetism of the local moment system. The mean field description (1.64), however, strongly oversimplifies the situation.

A further striking effect which is due to the mentioned induced temperature-dependence of the band states is a metal-insulator transition observed in Eu-rich EuO [12]. The Eu-richness manifests itself in oxygen vacancies. With each vacancy there are two Eu-electrons being no longer needed for the binding. They will be trapped, however, by the (2+)-charged vacancy. The idea is that a single electron is tightly bound to the vacancy while in case that both electrons are at the same vacancy the Coulomb repulsion takes care for an impurity level fairly close to the lower band edge. Upon cooling below T_C the band edge crosses the impurity level freeing therewith the excess electrons. A conductivity jump of up to 14 orders of magnitude has been observed. This spectacular insulator-metal transition, which takes place at a temperature $T^* < T_C$, can be understood in simple terms again by the mean-field solution (1.64) of the s-f model.

Further effects result from the interaction of the band electron with collective excitations of the spin system. One of them is the appearance of a new quasiparticle, which is called *magnetic polaron*. It will be discussed in detail in IV. Another effect concerns peculiarities in the electrical resistivity at the magnetic phase transition known as *spin disorder resistivity*.

The influence of the s- and f-subsystems is of course of mutual character. That means, that a finite bandoccupation has observable consequences for the spin system, too. The ferromagnetic order in Gd, e. g. , can be explained only by an indirect exchange interaction between the localized moments mediated by a spin polarization of the conduction electrons. This is known as RKKY-interaction [2], which shall be discussed in the next section. The RKKY-mechanism can excellently be monitored by inspecting the alloy $Eu_{1-x}Gd_xS$. Replacing in EuS the divalent Eu^{2+}-ion by a trivalent Gd^{3+}-ion leads in a definite manner to a population of the conduction band without diluting the moment system ($S_{4f}(Gd^{3+}) = S_{4f}(Eu^{2+}) = \frac{7}{2}$). So one can study directly the density-dependence of the Curie temperature: $T_C = T_C(n)$ [13].

2 RKKY interaction

The so-called RKKY interaction [14–16] is the characteristic coupling mechanism between localized magnetic moments in metallic systems like Gd. It is an indirect moment coupling via a spin polarization of the *a priori* non-magnetic conduction electrons. Starting point for an explanation of this interaction type is the

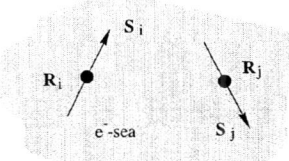

FIGURE 1.8. Two localized spins (moments) in a homogeneous electron sea.

model-Hamiltonian (1.56). The *uncorrelated* conduction electrons are described by the partial operator H_s in (1.57). It is assumed that there is no direct exchange interaction between the local moments. The phenomenon is fully ascribed to the intraatomic exchange interaction (1.59) between the localized moments (spins) and the conduction electron spins. It turns out to be convenient for the following to perform a partial Fourier transformation of the exchange interaction operator (1.63):

$$H_{sf} = -\frac{1}{2}J\frac{\hbar}{N}\sum_i\sum_{kq}e^{-iq\cdot R_i}\{S_i^z\left(c^+_{k+q\uparrow}c_{k\uparrow} - c^+_{k+q\downarrow}c_{k\downarrow}\right) + \quad (1.65)$$
$$+S_i^+c^+_{k+q\downarrow}c_{k\uparrow} + S_i^-c^+_{k+q\uparrow}c_{k\downarrow}\}$$

The conventional RKKY-interaction is the result of a second order perturbation theory where H_{sf} acts as perturbation. The goal is to approach the ground state energy.

The *unperturbed* ground state ($H_{sf} = 0$) can be formulated as

$$|0F\rangle \equiv |0\rangle|F\rangle \quad (1.66)$$

$|0\rangle$ represents the Slater-determinant of the unpolarized, completely filled conduction electron *Fermi sphere*. $|F\rangle$ symbolizes a spin state of the localized moment system.

$$H_s|0F\rangle = E_0^{(0)}|0F\rangle \quad (1.67)$$
$$E_0^{(0)} = \frac{3}{5}N_e\epsilon_F \quad (1.68)$$

N_e is the number of electrons and ϵ_F the *free* Fermi energy. In first order perturbation theory there is no energy correction,

$$\Delta E_0^{(1)} = \langle 0F|H_{sf}|0F\rangle = 0 \quad (1.69)$$

because H_{sf} creates particle-hole pairs (1.65). In second order we have to evaluate

$$\Delta E_0^{(2)} = \sum_{(a,F')}^{\neq (0,F)} \frac{|\langle 0F|H_{sf}|aF'\rangle|^2}{E_0^{(0)} - E_a^{(0)}} \quad (1.70)$$

$|a\rangle$ is an excited state of the Fermi sphere with one electron-hole pair. That means

$$|a\rangle: \qquad E_0^{(0)} - E_0^{(a)} = \epsilon(k+q) - \epsilon(k) \quad (1.71)$$

It remains to determine

$$\langle 0F|H_{sf}|aF'\rangle\langle aF'|H_{sf}|0F\rangle$$

incorporating matrix elements of the following form:

$$\langle 0|c^+_{k+q\sigma_1}c_{k\sigma_2}|a\rangle\langle a|c^+_{k'\sigma_2'}c_{k'+q'\sigma_1'}|0\rangle \longrightarrow \Theta_{k,k+q}\delta_{kk'}\delta_{\sigma_2\sigma_2'}\delta_{qq'}\delta_{\sigma_1\sigma_1'}$$

$\Theta_{k,k+q}$ is a product of step functions:

$$\Theta_{\mathbf{k},\mathbf{k}+\mathbf{q}} = \Theta(k_F - |\mathbf{k}+\mathbf{q}|)\,\Theta(|\mathbf{k}| - k_F) \qquad (1.72)$$

The second order energy correction then reads

$$\Delta E_0^{(2)} = \frac{1}{4} J^2 \frac{\hbar^2}{N^2} \sum_{i,j} \sum_{\mathbf{k},\mathbf{q}} e^{-i\mathbf{q}\cdot(\mathbf{R}_i-\mathbf{R}_j)} \frac{\Theta_{\mathbf{k},\mathbf{k}+\mathbf{q}}}{\epsilon(\mathbf{k}+\mathbf{q})-\epsilon(\mathbf{k})} \times$$
$$\times \sum_{F'} \langle F| \{ S_i^z |F'\rangle\langle F'| S_j^z + (-S_i^z)|F'\rangle\langle F'|(-S_j^z)$$
$$+ S_i^+ |F'\rangle\langle F'| S_j^- + S_i^- |F'\rangle\langle F'| S_j^+ \} |F\rangle$$

Using the completeness relation for the spin states and the definition

$$J_{ij}^{\text{RKKY}} = \frac{1}{2} J^2 \frac{\hbar^2}{N^2} \sum_{i,j} \sum_{\mathbf{k},\mathbf{q}} \frac{\Theta_{\mathbf{k},\mathbf{k}+\mathbf{q}}}{\epsilon(\mathbf{k}+\mathbf{q})-\epsilon(\mathbf{k})} e^{-i\mathbf{q}\cdot(\mathbf{R}_i-\mathbf{R}_j)} \qquad (1.73)$$

the second order energy correction can be written as:

$$\Delta E_0^{(2)} = -\sum_{i,j} J_{ij}^{\text{RKKY}} \langle F| \mathbf{S}_i \cdot \mathbf{S}_j |F\rangle \qquad (1.74)$$

$\Delta E_0^{(2)}$ can be interpreted as the expectation value of an "effective" Heisenberg-Hamiltonian in the spin space:

$$H_{\text{RKKY}} = -\sum_{i,j} J_{ij}^{\text{RKKY}} \mathbf{S}_i \cdot \mathbf{S}_j \qquad (1.75)$$

This represents the indirect coupling between localized spins mediated by spin polarizable conduction electrons.

The exchange integrals J_{ij}^{RKKY} show an oscillatory behaviour as function of the distance $R_{ij} = |\mathbf{R}_i - \mathbf{R}_j|$. Within the *Sommerfeld model* of free conduction electrons one gets from (1.73):

$$J_{ij}^{\text{RKKY}} = \alpha J^2 \frac{n_e^2}{\epsilon_F} F(2k_F R_{ij}) \qquad (1.76)$$

The function $F(x)$ takes care for a damped oscillation

$$F(x) = \frac{\sin(x) - x\cos(x)}{x^4} \qquad (1.77)$$

$n_e = N_e/V$ is the electron density and α a constant:

$$\alpha = \frac{9}{8}\pi\hbar^2 \frac{V^2}{N^2}$$

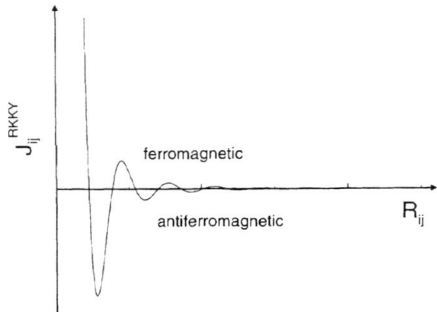

FIGURE 1.9. Distance-dependence of the RKKY-exchange coupling

N is the number of sites. In dependence of the distance the indirect spin interaction changes between ferro- and antiferromagnetic coupling. It is a long range interaction,

$$J_{ij}^{\text{RKKY}} \sim \frac{1}{R_{ij}^3} \tag{1.78}$$

For comparison the direct exchange (1.46) decreases exponentially with the distance. The RKKY mechanism is a second order effect,

$$J_{ij}^{\text{RKKY}} \sim J^2 \tag{1.79}$$

and happens only in metallic local moment systems. Because of $\epsilon_F = \frac{\hbar^2}{2m}(3\pi^2 n)^{\frac{2}{3}}$ it holds

$$J_{ij}^{\text{RKKY}} \sim \frac{n^2}{\epsilon_F} \sim n^{\frac{4}{3}} \tag{1.80}$$

E Itinerant moment metals

If in a ferromagnetic metal electrical conductivity and spontaneous magnetization are provoked by one and the same group of electrons then the permanent magnetic moments are necessarily itinerant. This situation is called *bandferromagnetism* [17]. Archetypal representatives are the classical ferromagnets Fe, Co, Ni. Below a critical temperature T_C (*Curie temperature*) a spontaneous spin-dependent shift of the density of states of a partially filled energy band takes care for $N_{e\uparrow} > N_{e\downarrow}$ and therewith for a finite total moment. To understand bandferromagnetism thus means to understand the origin of the spontaneous exchange splitting as a consequence of strong electron correlations. The exchange splitting is of course energy- and wave vector-dependent. It is, however, surprising that the average $(T = 0)$-values differ by a full order of magnitude for Fe and Ni:

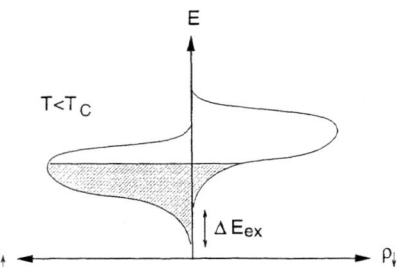

FIGURE 1.10. Stoner-splitting in bandferromagnets.

$$\Delta E_{\text{ex}}(T=0) \approx \begin{cases} 0.2\text{eV} : \text{Ni} \\ 2.0\text{eV} : \text{Fe} \end{cases} \quad (1.81)$$

Another striking difference is the temperature-dependence of ΔE_{ex}. For $T \longrightarrow T_C$ the exchange splitting exhibits a collapsing (*Stoner-like*) behaviour in Ni, while in Fe the splitting seems to persist in the paramagnetic phase (*non-collapsing*). This has to be explained by a proper theory. This theory should also reproduce more or less quantitatively magnetic key-data, as for instance the Curie temperature,

$$T_C = \begin{cases} 631\text{K} : \text{Ni} \\ 1044\text{K} : \text{Fe} \end{cases} \quad (1.82)$$

or the $T=0$ moment:

$$\mu(T=0) = \begin{cases} 0.56\mu_B : \text{Ni} \\ 2.2\mu_B : \text{Fe} \end{cases} \quad (1.83)$$

Last but not least important details such as the *Ni-6eV satellite* should be interpreted.

A first qualitative explanation of bandferromagnetism seems to be rather simple. Consider for simplicity a non-degenerate energy band. Each mechanism that keeps the equally charged band electrons at distance lowers the potential energy because of the Coulomb repulsion. Such a mechanism is provided by the Pauli principle that hinders electrons with parallel spin projection to enter the same lattice site. So a high degree of electron spin polarization is convenient with respect to the interaction energy. However, since the spin-polarized electrons have to be brought into empty single-particle states, the spin polarization appears inconvenient with respect to the single-electron energy. Whether or not spin polarization, say bandferromagnetism, is energetically favoured is due to a compromise. A high density of states around the Fermi edge would allow to flip the spins without too big an increase in single-particle energy. A strong and a strongly screened Coulomb interaction would result in a high gain of interaction energy by keeping the spin-polarized electrons at distance. These findings are gathered in the well-known *Stoner criterion*:

$$U\rho(\epsilon_F) > 1 \quad (1.84)$$

where U denotes the intraatomic Coulomb matrix element, and $\rho(\epsilon)$ is the *free* Bloch-density of states. The minimum set of ingredients for a reasonable model of bandferromagnetism is therefore:

Pauli principle,

strong and strongly screened Coulomb interaction,

narrow band, high density of states at the Fermi edge,

kinetic energy.

The simplest model which incorporates these points is the *Hubbard model* [18], which shall be inspected in detail in Sects. II and III:

$$H = \sum_{i,j,\sigma} T_{ij} c_{i\sigma}^+ c_{j\sigma} + \frac{1}{2} U \sum_{i,\sigma} n_{i\sigma} n_{i-\sigma} \qquad (1.85)$$

The notation is clear from (1.57) to (1.60). The three main model simplifications,

one orbital per atom (s band),

intraatomic Coulomb interaction,

one atom per unit cell,

of course prevent from the very beginning a direct application of model results to any real bandferromagnet. Strictly speaking, it was not clear until very recently whether or not ferromagnetism is at all possible within the frame of the possibly oversimplified Hubbard model. As we are going to prove in Sects. II and III we can indeed assume that the Hubbard model provides a good qualitative insight into the basic phenomenon *bandferromagnetism*. Decisive model parameters, which fix the detailed properties of the *correlated Hubbard electrons*, are

Coulomb coupling U/W (W: width of the *free* Bloch-band)

band occupation $n = \sum_\sigma \langle n_\sigma \rangle$ ($0 \le n \le 2$)

lattice structure

In terms of these entities we shall inspect in the following two sections the physics of the Hubbard model.

F Conclusions

While dia- and paramagnetism are basically single-particle phenomena, and therefore comparatively well understood, ferromagnetism is a consequence of strong electron correlations, and thus a highly complicated many-body problem. The main shortcoming for the present understanding of ferromagnetism is the lack of a unified theory. The different types of ferromagnetic materials require for their description a full set of theoretical models, the most important ones we have commented on in this introductory section. It turns out to be reasonable first to classify the ferromagnetic materials into insulators and metals as well as into those with respectively itinerant and localized magnetic moments.

The class of materials considered as best understood are the insulators with localized magnetic moments (EuO, EuS). They are excellently described by the Heisenberg model as far as the purely magnetic properties are meant. The general problem is not exactly solvable, but fully convincing approximate results do exist. The temperature-dependence of the spontaneous magnetization (Brillouin-function like), the Curie-Weiß law for the static susceptibility, the Bloch-$T^{\frac{3}{2}}$-law, the concept of spin waves and magnons as well as critical phenomena are well-known successful examples.

Mott-Hubbard insulators such as NiO, CoO, FeO take their moment from a completely filled quasiparticle subband with finite bandwidth. Therefore, they can be interpreted in a certain sense as insulators with itinerant (!) moments. A commonly accepted model to derive and explain the striking physical properties does not exist up to now. They represent a very hot topic of current research.

In so-called *local-moment* metals (Gd, Dy, Tb) electrical conductivity and spontaneous magnetism are provoked by different and well-defined groups of electrons. Many interesting features of these materials can be traced back to an intimate correlation between these two subsystems. In IV we inspect in detail what can be said about the *local moment* metals by use of the s-f model *Kondo-lattice model*.

If conductivity and magnetization are both determined by the electrons of one and the same partially filled energy band then one speaks of bandferromagnetism (Fe, Co, Ni). In Sects. II,III we investigate the Hubbard model as a candidate for a qualitative understanding of the *itinerant-moment* ferromagnetism.

II ITINERANT MOMENT SYSTEMS: HUBBARD MODEL, BASIC FEATURES

We first inspect ferromagnetic systems which base their magnetism on itinerant magnetic moments. Electrical conductivity as well as spontaneous magnetization are provoked by one and the same group of bandelectrons. That is what is called *bandferromagnetism*. We consider as an acceptable starting point for a qualitative understanding of this phenomenon the Hubbard-Hamiltonian (1.85) [18]

$$H = H_0 + H_1 = \sum_{i,j,\sigma}(T_{ij} - \mu\delta_{ij})c_{i\sigma}^+ c_{j\sigma} + \frac{1}{2}U\sum_{i,\sigma} n_{i\sigma}n_{i-\sigma} \quad (2.1)$$

$n_{i\sigma} = c_{i\sigma}^+ c_{i\sigma}$ is the occupation number operator and μ the chemical potential. The hopping integrals T_{ij} are connected with the Bloch energies $\epsilon(\mathbf{k})$ as given in (1.58). Very often, not necessarily, the hopping is restricted to next neighbours:

$$T_{ij} = \begin{cases} T_0 & \text{if } i = j \\ -t & \text{if } i,j \text{ next neighbours} \\ 0 & \text{otherwise} \end{cases} \quad (2.2)$$

T_0 is the centre of gravity of the free Bloch-band:

$$\rho_0(E) = \frac{1}{N}\sum_{\mathbf{k}} \delta(E - \epsilon(\mathbf{k})) \quad (2.3)$$

The model ingredients, the model parameters as well as the model restrictions are detailedly listed up in Sect. I E. The Hubbard model plays a central role for an understanding of correlated electrons on a lattice. It is considered to give an at least qualitative insight into the electronic properties of narrow band solids (3d transition metals and their compounds), in particular what concerns bandferromagnetism (Fe, Co, Ni). Furthermore, since recently it is treated as a candidate for the description of high-T_C-superconductors in their *normal conductor* phase. In spite of its simple operator structure the Hubbard-Hamiltonian defines a highly non-trivial many-body problem. In the following sections we extract some basic features of the not exactly solvable Hubbard model in order to have the right means to develop in Sect. III some reliable approaches to the fairly complicated many-body problem.

A The Many-Body Problem

Practically all informations we are interested in can be derived from the retarded single-electron Green function [19], [20], [21]:

$$G_{ij\sigma}(E) = \langle\langle c_{i\sigma}; c_{j\sigma}^+ \rangle\rangle_E$$
$$= -i \int_0^\infty dt\, e^{\frac{i}{\hbar}Et} \langle\, [c_{i\sigma}(t), c_{j\sigma}^+(0)]_+ \,\rangle \tag{2.4}$$

$$G_{\mathbf{k}\sigma}(E) = \langle\langle c_{\mathbf{k}\sigma}; c_{\mathbf{k}\sigma}^+ \rangle\rangle_E$$
$$= \frac{1}{N} \sum_{i,j} G_{ij\sigma}(E)\, e^{-i\mathbf{k}(\mathbf{R}_i - \mathbf{R}_j)} \tag{2.5}$$

The construction operators on the right-hand side of (2.4) are written in their time-dependent Heisenberg representation. Using the eigenvalue equation

$$\mathcal{H}\,|E_n\rangle = E_n\,|E_n\rangle \tag{2.6}$$

one easily gets the Lehmann-representation of the Green function,

$$G_{\mathbf{k}\sigma}(E) = \frac{\hbar}{\Xi} \sum_{n,m} |\langle E_n|c_{\mathbf{k}\sigma}^+|E_m\rangle|^2 \frac{e^{-\beta E_n} + e^{-\beta E_m}}{E - (E_n - E_m)} \tag{2.7}$$

that tells us that the poles of the Green function coincide with the exact excitation energies of the system. Ξ is the grandcanonical partition function. Note that

$$\mathcal{H} = H - \mu \hat{N};\; [H, \hat{N}]_- = 0$$

$$\mathcal{H}\,|E_n\rangle = (H - \mu\hat{N})\,|E_n\rangle = (E_n(N) - \mu N)\,|E_n\rangle$$
$$= E_n\,|E_n\rangle$$

A first attempt to find the fundamental Green function can start from the equation of motion, which in turn follows directly from that of the time-dependent Heisenberg operator $c_{i\sigma}(t)$ in (2.4):

$$(E + \mu - \epsilon(\mathbf{k}))\, G_{\mathbf{k}\sigma}(E) = \hbar + \langle\langle\, [c_{\mathbf{k}\sigma}, H_1]_-\,;\, c_{\mathbf{k}\sigma}^+ \,\rangle\rangle_E \tag{2.8}$$

The *higher* Green function on the right-hand side prevents a direct solution, but the decomposition

$$\langle\langle\, [c_{\mathbf{k}\sigma}, H_1]_-\,;\, c_{\mathbf{k}\sigma}^+ \,\rangle\rangle_E \stackrel{!}{=} \Sigma_{\mathbf{k}\sigma}(E)\cdot G_{\mathbf{k}\sigma}(E) \tag{2.9}$$

at least does permit a formal solution:

$$G_{\mathbf{k}\sigma}(E) = \frac{\hbar}{E + \mu - \epsilon(\mathbf{k}) - \Sigma_{\mathbf{k}\sigma}(E)} \qquad (2.10)$$

The *selfenergy* $\Sigma_{\mathbf{k}\sigma}(E)$ incorporates all interaction effects being therefore a very fundamental and normally unknown term. In general it is a complex function:

$$\Sigma_{\mathbf{k}\sigma}(E) = R_{\mathbf{k}\sigma}(E) + iI_{\mathbf{k}\sigma}(E) \qquad (2.11)$$

The Green function is not directly observable in the experiment but the *single-electron spectral density*

$$S_{\mathbf{k}\sigma}(E) = -\frac{1}{\pi} \mathrm{Im} G_{\mathbf{k}\sigma}(E + i0^+) \qquad (2.12)$$

representing the bare line shape of an angle- and spin-resolved (direct or inverse) photoemission experiment:

$$S_{\mathbf{k}\sigma}(E) = -\frac{\hbar}{\pi} \frac{I_{\mathbf{k}\sigma}(E)}{(E + \mu - \epsilon(\mathbf{k}) - R_{\mathbf{k}\sigma}(E))^2 + I_{\mathbf{k}\sigma}^2(E)} \qquad (2.13)$$

If real and imaginary part of the selfenergy are sufficiently well-behaved functions of E then one can expect more or less pronounced Lorentzian-like peaks at the zeros $E_{i\sigma}(\mathbf{k})$ of the bracket in (2.13):

$$E_{i\sigma} + \mu - \epsilon(\mathbf{k}) - R_{\mathbf{k}\sigma}(E_{i\sigma}) \stackrel{!}{=} 0 \qquad i = 1, 2, \ldots \qquad (2.14)$$

This leads to the classical *quasiparticle picture*. Assume that near such a *quasiparticle-peak*

$$I_{\mathbf{k}\sigma}(E) \approx I_{\mathbf{k}\sigma}(E_{i\sigma}(\mathbf{k})) \equiv I_{\mathbf{k}\sigma}^{(i)} \qquad (2.15)$$

then (2.13) can be reformulated by a Taylor expansion around the *quasiparticle-peak*

$$S_{\mathbf{k}\sigma}^{(i)}(E) \approx \frac{\hbar}{\pi} z_{\mathbf{k}\sigma}^{(i)} \frac{|z_{\mathbf{k}\sigma}^{(i)} I_{\mathbf{k}\sigma}^{(i)}|}{(E + \mu - E_{i\sigma}(\mathbf{k}))^2 + (z_{\mathbf{k}\sigma}^{(i)} I_{\mathbf{k}\sigma}^{(i)})^2} \qquad (2.16)$$

$z_{\mathbf{k}\sigma}^{(i)}$ is called *spectral weight* or *quasiparticle weight*:

$$z_{\mathbf{k}\sigma}^{(i)} = \left| 1 - \frac{dR_{\mathbf{k}\sigma}}{dE} \right|_{E=E_{i\sigma}(\mathbf{k})-\mu}^{-1} \qquad (2.17)$$

The Fourier transform $S_{\mathbf{k}\sigma}^{(i)}(t - t')$ represents in first approximation a damped oscillation, which makes reasonable the definition of a *quasiparticle lifetime*:

$$\tau_{\mathbf{k}\sigma}^{(i)} = \frac{\hbar}{|z_{\mathbf{k}\sigma}^{(i)} I_{\mathbf{k}\sigma}^{(i)}|} \qquad (2.18)$$

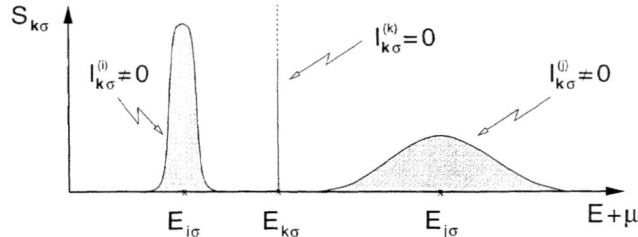

FIGURE 2.1. Schematic plot of the single-electron spectral density

The lifetime is mainly determined by the imaginary part of the selfenergy being therewith connected to the widths of the *quasiparticle peaks* in the spectral density. Vanishing imaginary part leads to a δ-function in the spectral density representing the special case of a quasiparticle with infinite lifetime.

We see that three characteristics qualify the quasiparticle: the energy $E_{i\sigma}(\mathbf{k})$, the lifetime $\tau_{\mathbf{k}\sigma}^{(i)}$ and the spectral weight $z_{\mathbf{k}\sigma}^{(i)}$. Later we shall use the peak-positions to define the *quasiparticle bandstructure*. Summing the spectral density (2.12) over all wave-vectors of the first Brillouin zone yields the *quasiparticle density of states* [21]:

$$\rho_\sigma(E) = \frac{1}{N\hbar} \sum_{\mathbf{k}} S_{\mathbf{k}\sigma}(E - \mu) , \qquad (2.19)$$

which refers to an angle-averaged photoemission experiment. As to its very definition in general $\rho_\sigma(E)$ can carry an explicit (T, n, σ) - dependence.

The macroscopic thermodynamics is accessible via the spectral theorem, e.g.:

$$n_\sigma = \frac{1}{N} \sum_i \langle n_{i\sigma} \rangle = \int_{-\infty}^{\infty} dE \, f_-(E) \rho_\sigma(E) \qquad (2.20)$$

We only consider paramagnetic and ferromagnetic configurations and are therefore allowed to assume translational symmetry ($\langle n_{i\sigma} \rangle \equiv n_\sigma \;\forall i$). $f_-(E)$ is the Fermi function:

$$f_-(E) = f(E - \mu); f(E) = (e^{\beta E} + 1)^{-1} \qquad (2.21)$$

A non-vanishing spontaneous magnetic moment,

$$m = \mu_B (n_\uparrow - n_\downarrow) \qquad (2.22)$$

indicates the appearance of ferromagnetism.

B Zero-Bandwidth Limit

Ferromagnetism is certainly a strong coupling phenomenon. So it might be instructive to inspect the extreme limit of the *zero-bandwidth* $W \to 0$:

$$T_{ij} \to T_0 \delta_{ij} \; ; \; \epsilon(\mathbf{k}) \to T_0 \; \forall \mathbf{k} \tag{2.23}$$

The conduction band is shrunk to an N-fold degenerate level T_0. That means for the Hubbard-Hamiltonian (1.85):

$$H_{W=0} = (T_0 - \mu) \sum_{i,\sigma} n_{i\sigma} + \frac{1}{2} U \sum_{i,\sigma} n_{i\sigma} n_{i-\sigma} \tag{2.24}$$

The general first equation of motion of the Green function,

$$(E + \mu) G_{ij\sigma} = \hbar \delta_{ij} + \sum_m T_{im} G_{mj\sigma}(E) + U \Gamma_{iii,j\sigma}(E) \; , \tag{2.25}$$

with

$$\Gamma_{imn,j\sigma}(E) \equiv \langle\langle c^+_{i-\sigma} c_{m-\sigma} c_{n\sigma} \; ; \; c^-_{j\sigma} \rangle\rangle_E \tag{2.26}$$

simplifies to

$$(E + \mu - T_0) G_{ii\sigma}(E) = \hbar + U \Gamma_{iii,i\sigma}(E) \tag{2.27}$$

which cannot be solved directly because of Γ. The equation of motion of Γ, however, decouples exactly, since for Fermions $n^2_{i-\sigma} = n_{i-\sigma}$ can be exploited:

$$(E + \mu - T_0 - U) \Gamma_{iii,i\sigma}(E) = \hbar n_{-\sigma} \tag{2.28}$$

The last two equations lead to the rigorous solution:

$$G_{ii\sigma}(E) = \hbar \left\{ \frac{1 - n_{-\sigma}}{E + \mu - T_0} + \frac{n_{-\sigma}}{E + \mu - T_0 - U} \right\} \tag{2.29}$$

The original atomic level T_0 splits into two quasiparticle levels:

$$E_{1\sigma} = T_0 \; ; \quad E_{2\sigma} = T_0 + U \tag{2.30}$$

with formally spin-dependent weights,

$$z_{1\sigma} = 1 - n_{-\sigma} \; ; \quad z_{2\sigma} = n_{-\sigma} \tag{2.31}$$

which can be interpreted as probabilities for an excited σ-electron to enter a site being already preoccupied by a $(-\sigma)$-electron ($z_{2\sigma}$) or not ($z_{1\sigma}$). Spectral density and quasiparticle density of states both have a two-peak structure:

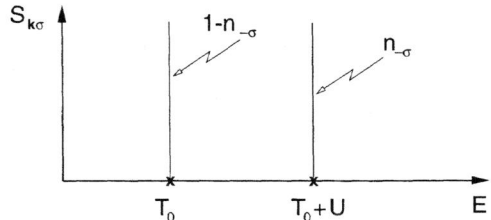

FIGURE 2.2. Quasiparticle density of states of the Hubbard model in the zero bandwidth limit.

$$\rho_\sigma(E) \equiv S_{ii\sigma}(E-\mu) \equiv S_{\mathbf{k}\sigma}(E-\mu) = \sum_{j=1}^{2} z_{j\sigma}\delta(E-E_{j\sigma}) \quad (2.32)$$

Eqs. (2.30) to (2.32) imply the following selfenergy structure:

$$\Sigma_\sigma^{W=0}(E) = Un_{-\sigma}\frac{E+\mu-T_0}{E+\mu-T_0-U(1-n_{-\sigma})} \quad (2.33)$$

For a complete solution we have to determine $n_{-\sigma}$ by use of the spectral theorem (2.20):

$$n_{-\sigma} = \frac{f_-(T_0)}{1-f_-(T_0+U)+f_-(T_0)} = \frac{1}{2}n \quad (2.34)$$

There is no spin-dependence, a spontaneous magnetization is therefore excluded. Ferromagnetism is thus impossible in the zero-bandwidth limit!

C Strong coupling regime

If we now switch on the *electron hopping*, but so that

$$W \ll U \quad (2.35)$$

holds that the two degenerate levels (2.30) get a real dispersion, slightly shifting their energetic positions and possibly modifying their spectral weights. Nevertheless, some exact statements can be done as first demonstrated by Harris and Lange [22]. For $W=0$ the double occupation of lattice sites remains conserved,

$$[n_{i\sigma}n_{i-\sigma}, H]_- = 0 \ ,$$

which allows to decompose the construction operators,

$$c_{i\sigma} = c_{i\sigma}n_{i-\sigma} + c_{i\sigma}(1-n_{i-\sigma}) \ ,$$

into operators which change the number of doubly occupied sites in a definite manner. Harris and Lange [22] developed for $W \ll U$ a unitary transformation, $c = U^+ \hat{c} U$, to new quasiparticles, the hopping of which retains the number of doubly occupied sites. The procedure can be iterated up to any desired order in W/U. The decisive result is that there are essentially two charge-excitation peaks in the spectrum close to T_0 and $T_0 + U$, respectively. Additional satellite peaks at $T_0 + p \cdot U$ ($p = -1, -2, \ldots; p = 2, 3, \ldots$) acquire a weight of order $(W/U)^4$ or even less and can thus be neglected in the strong-coupling regime. We do not know the

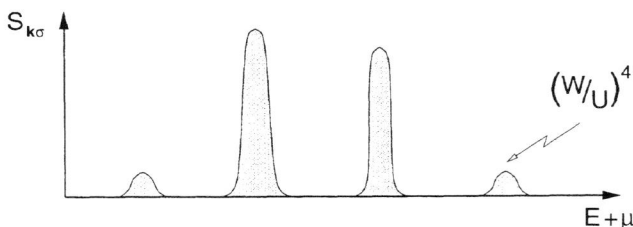

FIGURE 2.3. Single-electron spectral density of the Hubbard model in the strong coupling regime ($W \ll U$).

exact shape of the two main peaks, but their centres of gravity,

$$T_{1\sigma} = (1 - n_{-\sigma})\epsilon(\mathbf{k}) + n_{-\sigma} B_{\mathbf{k}-\sigma} + \mathcal{O}(\tfrac{W}{U}) \tag{2.36}$$

$$T_{2\sigma} = U + n_{-\sigma}\epsilon(\mathbf{k}) + (1 - n_{-\sigma}) B_{\mathbf{k}-\sigma} + \mathcal{O}(\tfrac{W}{U}) \tag{2.37}$$

as well as the spectral weights that refer to the areas under the main peaks:

$$z_{1\sigma}(\mathbf{k}) \equiv 1 - n_{-\sigma} + \mathcal{O}(\tfrac{W}{U}) = 1 - z_{2\sigma}(\mathbf{k}) \tag{2.38}$$

These expressions contain a higher correlation function,

$$B_{\mathbf{k}-\sigma} = B_{-\sigma} + F_{\mathbf{k}-\sigma} , \tag{2.39}$$

which will turn out to be of decisive importance in particular with respect to the stability of spontaneous ferromagnetism. Because of

$$B_{\mathbf{k}-\sigma} \xrightarrow[W \to 0]{} T_0 \tag{2.40}$$

the zero-bandwidth limit (2.30), (2.31) is correctly fulfilled. The k-independent part of the higher correlation involves a *correlated electron hopping*,

$$B_{-\sigma} = T_0 + \frac{1}{n_{-\sigma}(1 - n_{-\sigma})} \frac{1}{N} \sum_{i,j}^{i \neq j} T_{ij} \langle c^+_{i-\sigma} c_{j-\sigma}(2n_{i\sigma} - 1) \rangle \tag{2.41}$$

which may provoke a spin-dependent shift of the band centres of gravity being therefore called *band shift*. We shall elaborate that the explicit spin-dependence of $B_{-\sigma}$ will be a decisive precondition for a spontaneous band electron spin order.

The k-dependent part of the higher correlation function (2.39),

$$F_{\mathbf{k}-\sigma} = \frac{1}{n_{-\sigma}(1-n_{-\sigma})} \frac{1}{N} \sum_{i,j}^{i \neq j} T_{ij} \, e^{-i\mathbf{k}(\mathbf{R}_i - \mathbf{R}_j)} *$$
$$*\{ \langle n_{i-\sigma} n_{j-\sigma}\rangle - n_{-\sigma}^2 - \langle c_{j\sigma}^+ c_{-\sigma}^+ c_{i-\sigma} c_{i\sigma}\rangle - \langle c_{j\sigma}^+ c_{i-\sigma}^+ c_{j-\sigma} c_{i\sigma}\rangle \} \qquad (2.42)$$

does not directly influence the centres of gravity,

$$\frac{1}{N} \sum_{\mathbf{k}} F_{\mathbf{k}-\sigma} = 0 \;, \qquad (2.43)$$

but may lead to a spin-dependent *band width correction*. It is built up by *density-density*, *double hopping* and *spin flip* correlation terms. Obviously it vanishes in the zero-bandwidth limit. Eqs. (2.36) and (2.37) exhibit the *Hubbard splitting* of the quasiparticle subbands by about U and the bandnarrowing by the factors $(1-n_{-\sigma})$ and $n_{-\sigma}$.

For further evaluations it will show up decisive that the *band shift* $B_{-\sigma}$ can exactly be expressed by the single-electron spectral density:

$$n_{-\sigma}(1-n_{-\sigma})(B_{-\sigma} - T_0) = \frac{1}{N\hbar} \sum_{\mathbf{k}} \int_{-\infty}^{\infty} dE \, f(E) \, (\epsilon(\mathbf{k}) - T_0) *$$
$$* \left[\frac{2}{U}(E + \mu - \epsilon(\mathbf{k})) - 1 \right] S_{\mathbf{k}-\sigma}(E) \qquad (2.44)$$

In case of a local, i. e. k-independent selfenergy this relation can still be reformulated as follows:

$$n_{-\sigma}(1-n_{-\sigma})(B_{-\sigma} - T_0) = -\frac{1}{\pi\hbar} Im \int_{-\infty}^{\infty} dE \, f(E) *$$
$$* [(E + \mu - T_0 - \Sigma_{-\sigma}(E))G_{ii-\sigma}(E) - \hbar] \, (\frac{2}{U}\Sigma_{-\sigma}(E) - 1) \qquad (2.45)$$

This version of the spin-dependent band shift very often turns out to be convenient.

D Spectral moments

For controlling unavoidable approximations or even for constructing reliable approaches the moments of the spectral density can be of great importance:

$$M^{(n)}_{ij\sigma} = \frac{1}{\hbar} \int_{-\infty}^{\infty} dE \cdot E^n S_{ij\sigma}(E) \qquad (2.46)$$

$$n = 0, 1, 2, \ldots$$

In principle, they can be calculated rigorously via the equivalent expression [21]:

$$M^{(n)}_{ij\sigma} = \langle [\underbrace{[\ldots[c_{i\sigma}, \mathcal{H}]_-, \ldots, \mathcal{H}]_-}_{(n-p)-fold}, \underbrace{[\mathcal{H}, \ldots [\mathcal{H}, c^+_{j\sigma}]_- \ldots]_-}_{p-fold}]_+ \rangle \qquad (2.47)$$

$$0 \leq p \leq n$$

In practice, however, the moments are useful only for low order n. The limitation is due to the fact that with increasing n eq. (2.47) involves expectation values of higher and higher order being usually unknown.

Some very general, but nevertheless useful statements are possible:

(a) *Centres of Gravity*

The spin-dependent centres of gravity of the total weighted spectrum,

$$T_\sigma = \frac{1}{\hbar N} \sum_{\mathbf{k}} \int_{-\infty}^{\infty} dE \cdot E S_{\mathbf{k}\sigma}(E - \mu) \;, \qquad (2.48)$$

are correctly given, if the first two moments

$$M^{(n)}_{\mathbf{k}\sigma} \;\; ; \;\; n = 0, 1$$

are fulfilled.

(b) *Hubbard-quasiparticle bands*

A necessary condition for the evolution of the so-called *Hubbard-bands* (see eqs. (2.36), (2.37) and Sects. III B 1 and III B 2) being separated by an energy gap of order U ($U \gg W$) [23] is the consistency with

$$M^{(n)}_{\mathbf{k}\sigma} \;\; ; \;\; n = 0, 1, 2$$

The condition is not sufficient. The self-consistent second order perturbation theory in U (Sect. II G) for infinite lattice dimension ($d = \infty$, Sect. III I) predicts the correct moments up to $n = 2$ but does not yield the Hubbard splitting for large U/W.

(c) $W/U \ll 1$-*regime*

Consistency with the W/U-perturbation theory (Sect. II C) requires two conditions. Firstly, the above-mentioned *Hubbard-bands* must exist in the strong-coupling region. This is fulfilled by any approach that recovers the zero-bandwidth limit (Sect. II B). Secondly, the first four spectral moments

$$M_{\mathbf{k}\sigma}^{(n)} \; ; \; n = 0, 1, 2, 3$$

must correctly be reproduced. These requirements guarantee the right centres of gravity (2.36), (2.37) and spectral weights (2.38) of the two main peaks of the strong coupling single-electron spectral density, a possible spin-dependent peak shift included.

For the Hubbard model the first four spectral moments read as [24,25]:

$$M_{\mathbf{k}\sigma}^{(0)} = 1 \tag{2.49}$$

$$M_{\mathbf{k}\sigma}^{(1)} = \hat{\epsilon}(\mathbf{k}) + U n_{-\sigma} \tag{2.50}$$

$$M_{\mathbf{k}\sigma}^{(2)} = \hat{\epsilon}^2(\mathbf{k}) + 2\hat{\epsilon}(\mathbf{k}) U n_{-\sigma} + U^2 n_{-\sigma} \tag{2.51}$$

$$M_{\mathbf{k}\sigma}^{(3)} = \hat{\epsilon}^3(\mathbf{k}) + 3\hat{\epsilon}^2(\mathbf{k}) U n_{-\sigma} + \hat{\epsilon}(\mathbf{k}) U^2 n_{-\sigma}(2 + n_{-\sigma})$$
$$+ U^3 n_{-\sigma} + U^2 n_{-\sigma}(1 - n_{-\sigma})(B_{\mathbf{k}-\sigma} - \mu) \tag{2.52}$$

The $n = 3$-moment contains the *higher* correlation function (2.39), which as already mentioned plays a decisive role for spontaneous ferromagnetism. For brevity we have written:

$$\hat{\epsilon}(\mathbf{k}) = \epsilon(\mathbf{k}) - \mu \tag{2.53}$$

E High-energy expansion

There is a close connection between the spectral moments and the high-energy behaviour of the Green functions. Starting point is the spectral representation:

$$G_{\mathbf{k}\sigma}(E) = \int_{-\infty}^{\infty} dE' \frac{S_{\mathbf{k}\sigma}(E')}{E - E'} \tag{2.54}$$

That can be written as

$$G_{\mathbf{k}\sigma}(E) = \frac{1}{E} \int_{-\infty}^{\infty} dE' \frac{S_{\mathbf{k}\sigma}(E')}{1 - \frac{E'}{E}} =$$

$$= \frac{1}{E} \sum_{n=0}^{\infty} \int_{-\infty}^{\infty} dE' \left(\frac{E'}{E}\right)^n S_{\mathbf{k}\sigma}(E')$$

According to (2.46) that means

$$G_{\mathbf{k}\sigma}(E) = \hbar \sum_{n=0}^{\infty} \frac{M_{\mathbf{k}\sigma}^{(n)}}{E^{n+1}} \qquad (2.55)$$

A similar expression is useful for the selfenergy:

$$\Sigma_{\mathbf{k}\sigma}(E) = \sum_{m=0}^{\infty} \frac{C_{\mathbf{k}\sigma}^{(m)}}{E^m} \qquad (2.56)$$

which can be derived from the general Dyson-equation (2.9)

$$E\, G_{\mathbf{k}\sigma}(E) = \hbar + (\hat{\epsilon}(\mathbf{k}) + \Sigma_{\mathbf{k}\sigma}(E))\, G_{\mathbf{k}\sigma}(E) \qquad (2.57)$$

Inserting the high-energy expansions (2.55) and (2.56),

$$\sum_{n=0}^{\infty} \frac{\hbar}{E^n} M_{\mathbf{k}\sigma}^{(n)} = \hbar + \hat{\epsilon}(\mathbf{k}) \sum_{n=0}^{\infty} \frac{\hbar}{E^{n+1}} M_{\mathbf{k}\sigma}^{(n)} + \sum_{m,n=0}^{\infty} \frac{\hbar}{E^{m+n+1}} C_{\mathbf{k}\sigma}^{(m)} M_{\mathbf{k}\sigma}^{(n)},$$

one gets by comparison of the coefficients to each power $\frac{1}{E^n}$:

$$C_{\mathbf{k}\sigma}^{(0)} = M_{\mathbf{k}\sigma}^{(1)} - \hat{\epsilon}(\mathbf{k}) \qquad (2.58)$$

$$C_{\mathbf{k}\sigma}^{(1)} = M_{\mathbf{k}\sigma}^{(2)} - (M_{\mathbf{k}\sigma}^{(1)})^2 \qquad (2.59)$$

$$C_{\mathbf{k}\sigma}^{(2)} = M_{\mathbf{k}\sigma}^{(3)} - 2 M_{\mathbf{k}\sigma}^{(2)} M_{\mathbf{k}\sigma}^{(1)} + (M_{\mathbf{k}\sigma}^{(1)})^3 \qquad (2.60)$$

$$C_{\mathbf{k}\sigma}^{(3)} = M_{\mathbf{k}\sigma}^{(4)} - 2 M_{\mathbf{k}\sigma}^{(3)} M_{\mathbf{k}\sigma}^{(1)} + 3 M_{\mathbf{k}\sigma}^{(2)} (M_{\mathbf{k}\sigma}^{(1)})^2$$
$$- (M_{\mathbf{k}\sigma}^{(2)})^2 - (M_{\mathbf{k}\sigma}^{(1)})^4 \qquad (2.61)$$

In case of the Hubbard model one finds with the moments (2.50) to (2.52):

$$C_{\mathbf{k}\sigma}^{(0)} = U n_{-\sigma} \qquad (2.62)$$

$$C_{\mathbf{k}\sigma}^{(1)} = U^2 n_{-\sigma}(1 - n_{-\sigma}) \qquad (2.63)$$

$$C_{\mathbf{k}\sigma}^{(2)} = U^2 n_{-\sigma}(1 - n_{-\sigma})(B_{\mathbf{k}-\sigma} - \mu) +$$
$$+ U^3 n_{-\sigma}(1 - n_{-\sigma})^2 \qquad (2.64)$$

Any analytical (approximate) expression for the selfenergy can easily be checked against these rigorous results simply by expanding in powers of $1/E$. Provided that expansion coefficients turn out to be correct up to $m = 2$, the moments of the resulting spectral density are fulfilled up to $m = 3$. In the strong coupling regime this ensures complete consistency with the W/U perturbational results of ref. [22].

Let us try to construct the simplest ansatz consistent with the W/U perturbational results [26]:

$$\Sigma_{k\sigma}(E) = C_{k\sigma}^{(0)} + \frac{C_{k\sigma}^{(1)}}{E} + \frac{C_{k\sigma}^{(2)}}{E^2} + \ldots$$

$$= C_{k\sigma}^{(0)} + \frac{C_{k\sigma}^{(1)}}{E}\left(1 + \frac{C_{k\sigma}^{(2)}}{C_{k\sigma}^{(1)} \cdot E} + \ldots\right)$$

$$\approx C_{k\sigma}^{(0)} + \frac{C_{k\sigma}^{(1)}}{E - \frac{C_{k\sigma}^{(2)}}{C_{k\sigma}^{(1)}}}$$

$$= Un_{-\sigma} + \frac{U^2 n_{-\sigma}(1-n_{-\sigma})}{E + \mu - B_{k-\sigma} - U(1-n_{-\sigma})}$$

That means:

$$\Sigma_{k\sigma}(E) \approx Un_{-\sigma}\frac{E+\mu-B_{k-\sigma}}{E+\mu-B_{k-\sigma}-U(1-n_{-\sigma})} \qquad (2.65)$$

The structure is the same as for the $W = 0$-limit (2.33) except for the replacement $T_0 \to B_{k-\sigma}$. The result (2.65) will reappear at a later stage of our investigation as the selfenergy of the *spectral density approach* (SDA), that proves to give an excellent qualitative picture of bandferromagnetism. The approach becomes especially attractive if evidence for a local approximation ($\Sigma_{k\sigma}(E) \to \Sigma_\sigma(E)$, $B_{k-\sigma} \to B_{-\sigma}$, $F_{k-\sigma} \approx 0(2.42)$) is available as, e. g., in the case of infinite lattice dimensions ($d \to \infty$) (see Sect. III I). Then the correlation functions $n_{-\sigma}$ and $B_{-\sigma}$ in (2.65) are rigorously expressable by the spectral density and by the selfenergy, respectively, via eqs. (2.44) and (2.45). The resulting closed system of equations can then be solved self-consistently. A convincing determination of the non-local part $F_{k-\sigma}$ is only possible with further approximations which, however, can plausibly be motivated (see Sect. III A 2) [25].

F The exactly half-filled band in the strong-coupling regime

In the simplified Hubbard model *exactly half-filled band* means that the number of band electrons equals the number of lattice sites:

$$n = 1 \; ; \; W \ll U \qquad (2.66)$$

In the zero-bandwidth limit (Sect. II B) the ground state refers to the situation with exactly one (localized) electron at each lattice site, i. e. there is no double occupancy of any site. The only variable is then the electron spin ($\sigma = \uparrow$ or $\downarrow$). The ground state is obviously 2^N-fold degenerate. If we switch on the electron hopping but with $W \ll U$ the electrons will remain highly localized and double occupancies are very unlikely. Virtual hoppings from site to site, however, provide an indirect coupling between the electron spins at different sites. Such a situation is described by the

Heisenberg model. The equivalence of the Hubbard - with the Heisenberg model for the situation (2.66) can indeed be shown by a simple perturbation theory.

The electron hopping represents the *perturbation*

$$H = H_0 + H_1 \tag{2.67}$$

$$H_0 = T_0 \sum_{i,\sigma} n_{i\sigma} + \frac{1}{2} U \sum_{i,\sigma} n_{i\sigma} n_{i-\sigma} \tag{2.68}$$

$$H_1 = \sum_{\substack{i,j,\sigma \\ i \neq j}} T_{ij} c_{i\sigma}^{+} c_{j\sigma} \tag{2.69}$$

We are exclusively interested in the ground state. All eigenstates and eigenvalues are characterized by the number d of doubly occupied sites

$$H_0 |d\alpha\rangle^{(0)} = E_d^{(0)} |d\alpha\rangle^{(0)} = (N T_0 + d U) |d\alpha\rangle^{(0)} \tag{2.70}$$

Because of the electron spin the eigenstates are highly degenerated, distinguished by the index α. Perturbation theory of first order requires the solution of the secular equation

$$\det \left[{}^{(0)}\langle 0\alpha' | H_1 | 0\alpha \rangle^{(0)} - E_0^{(1)} \delta_{\alpha\alpha'} \right] \stackrel{!}{=} 0 \tag{2.71}$$

However, all matrix elements vanish, since (2.69) leads to

$${}^{(0)}\langle d\alpha' | H_1 | 0\alpha \rangle^{(0)} \sim \delta_{d1} \tag{2.72}$$

Non-trivial results thus require second order perturbation theory:

$$\sum_\alpha c_\alpha \left\{ \sum_{d,\gamma}^{d \neq 0} {}^{(0)}\langle 0\alpha' | H_1 | d\gamma \rangle^{(0)} \, {}^{(0)}\langle d\gamma | H_1 | 0\alpha \rangle^{(0)} \frac{1}{E_0^{(0)} - E_d^{(0)}} - E_0^{(2)} \delta_{\alpha\alpha'} \right\} \stackrel{!}{=} 0 \tag{2.73}$$

This can be interpreted as eigenvalue equation of an *effective Hamiltonian*

$${}^{(0)}\langle 0\alpha' | H_1 \sum_{d,\gamma}^{d \neq 0} \frac{|d\gamma\rangle^{(0)} \, {}^{(0)}\langle d\gamma|}{E_0^{(0)} - E_d^{(0)}} H_1 |0\alpha\rangle^{(0)} =$$

$$\stackrel{\left(\substack{2.72 \\ 2.70}\right)}{=} -\frac{1}{U} {}^{(0)}\langle 0\alpha' | H_1 \left(\sum_{d,\gamma} |d\gamma\rangle^{(0)} \, {}^{(0)}\langle d\gamma| \right) H_1 |0\alpha\rangle^{(0)}$$

The term $d = 0$ can formally be included because it does not lead to any contribution as can be seen by (2.72). The effective Hamiltonian can therefore be read as:

156

$$H_{eff} = P_0 \left(-\frac{H_1^2}{U}\right) P_0 \qquad (2.74)$$

P_0 : projection operator onto the $d = 0$-subspace

Inserting (2.69)

$$H_{eff} = -\frac{1}{U} P_0 \left(\sum_{ij\sigma}^{i\neq j} \sum_{mn\sigma'}^{m\neq n} T_{ij} T_{mn}\, c_{i\sigma}^+ c_{j\sigma} c_{m\sigma'}^+ c_{n\sigma'} \right) P_0$$

one recognizes that only the terms $i = n$, $j = m$ contribute:

$$H_{eff} = -\frac{1}{U} P_0 \left(\sum_{\substack{ij \\ \sigma\sigma'}}^{i\neq j} T_{ij} T_{ji}\, c_{i\sigma}^+ c_{j\sigma} c_{j\sigma'}^+ c_{i\sigma'} \right) P_0$$

$$= -\frac{1}{U} P_0 \left(\sum_{\substack{ij \\ \sigma\sigma'}}^{i\neq j} |T_{ij}|^2\, c_{i\sigma}^+ c_{i\sigma'} (\delta_{\sigma\sigma'} - c_{j\sigma'}^+ c_{j\sigma}) \right) P_0$$

$$= -\frac{1}{U} P_0 \left(\sum_{ij\sigma}^{i\neq j} |T_{ij}|^2 (n_{i\sigma} - n_{i\sigma} n_{j\sigma} \right.$$

$$\left. - c_{i\sigma}^+ c_{i-\sigma}\, c_{j-\sigma}^+ c_{j\sigma}) \right) P_0 \qquad (2.75)$$

We now introduce spin operators:

$$S_i^z = \frac{\hbar}{2} \sum_\sigma z_\sigma n_{i\sigma} \qquad (z_\uparrow = +1,\ z_\downarrow = -1) \qquad (2.76)$$

$$S_i^\sigma = \hbar c_{i\sigma}^+ c_{i-\sigma} \qquad (S_i^\uparrow \equiv S_i^+;\ S_i^\downarrow \equiv S_i^-) \qquad (2.77)$$

Therewith it is easy to show

$$P_0 \left(\sum_\sigma n_{i\sigma} n_{j\sigma} \right) P_0 = P_0 \left(\frac{2}{\hbar^2} S_i^z S_j^z + \frac{1}{2} \right) P_0 \qquad (2.78)$$

when we exploit

$$P_0 \left(\sum_\sigma n_{i\sigma} \right) P_0 \equiv P_0\, \mathbb{1}\, P_0 \qquad (2.79)$$

what of course only holds for the half-filled band. Directly from the definition (2.77) we get

$$P_0 \left(\sum_\sigma c_{i\sigma}^+ c_{i-\sigma} c_{j-\sigma}^+ c_{j\sigma} \right) P_0 = P_0 \left(\frac{1}{\hbar^2} \sum_\sigma S_i^\sigma S_j^{-\sigma} \right) P_0$$

$$\stackrel{i\neq j}{=} \frac{1}{\hbar^2} P_0 \left(2 S_i^x S_j^x + 2 S_i^y S_j^y \right) P_0 \qquad (2.80)$$

The effective Hamiltonian has indeed the structure of the Heisenberg-Hamiltonian. Eqs. (2.78) and (2.80) inserted into (2.75) yield:

$$H_{eff} = P_0 \left\{ \eta - \sum_{i,j}^{i \neq j} J_{ij}\, \mathbf{S}_i \cdot \mathbf{S}_j \right\} P_0 \tag{2.81}$$

η is an unimportant constant:

$$\eta = -\frac{1}{2U} \sum_{i,j}^{i \neq j} |T_{ij}|^2 \tag{2.82}$$

The effective *exchange integrals*

$$\hbar^2 J_{ij} = -2\frac{|T_{ij}|^2}{U} < 0 \tag{2.83}$$

are always negative favouring an antiferromagnetic ordering of the electron spins. For the half-filled band ($n = 1$) we should therefore expect antiferromagnetism in the Hubbard model.

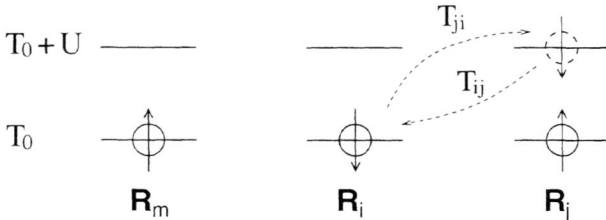

FIGURE 2.4. Intersite exchange interaction mediated by virtual electron hopping.

Interpretation: Eq. (2.83) describes virtual transitions from $\mathbf{R}_i$ to $\mathbf{R}_j$ and back. According to second-order perturbation theory (Ritzs variational principle) that leads to an energy-decrease. The transitions are forbidden in a ferromagnet because of the Pauli principle. The T_{ij} are short-range so that the antiferromagnetic order appears even more convenient than the paramagnetic disorder. The more virtual hopping processes possible, the larger is the gain in energy.

G   Weak-coupling regime

Up to now only the strong coupling regime was discussed, which at first glance seems to be the decisive region for spontaneous ferromagnetism. The opposite limit,

$$U/W \ll 1 \;,$$

is of course also tractable by a perturbational approach. We use the *diagrammatic language* [19–21]:

$-i\, G_{ij\sigma}(iE_n)$ ⇒⟵⇒ *full propagator*
E_n

$-i\, G^{(0)}_{ij\sigma}(iE_n)$ ⟵ *free propagator*
E_n

$-\dfrac{1}{\hbar}\Sigma_{ij\sigma}(iE_n)$ ⌢Σ⌢ *selfenergy*
E_n

E_n means the *fermionic Matsubara energies*.

$$E_n = (2n+1)\dfrac{\pi}{\beta}\ ;\quad n=\ldots,-1,0,1,\ldots \tag{2.84}$$

The *Dyson equation* (2.9) reads

$$G_{ij\sigma}(iE) = G^{(0)}_{ij\sigma}(iE) + \sum_{l,m} G^{(0)}_{il\sigma}(iE)\,\dfrac{1}{\hbar}\Sigma_{lm\sigma}(iE)\,G_{mj\sigma}(iE) \tag{2.85}$$

One can show [19–21] that the selfenergy is the sum of all *dressed skeleton diagrams*. A *skeleton-diagram* is a selfenergy diagram being built up only by (free) propagators which do <u>not</u> contain any selfenergy insertion. If in such diagrams the *free* propagators are additionally replaced by full propagators, then the skeleton is *dressed*.

(a) *Hartree-Fock approximation (HF)*
In lowest order one has to evaluate for the selfenergy:

(*Hartree*) + (*Fock*)

The *Fock-contribution* disappears in the case of the Hubbard model. The *Hartree-term* yields:

$$\Sigma_{ij\sigma}^{HF} \equiv \delta_{ij} U \langle n_{i-\sigma} \rangle \qquad (2.86)$$

In this approximation the Hubbard model is identical to the *older* Stoner model, which is known to overestimate ferromagnetism in a highly unrealistic manner. It fulfils the first two spectral moments being therefore correct for the spin-dependent centres of the total energy spectrum (2.48). Note, however, that (2.86) is not at all an analytical solution since $\langle n_{i-\sigma} \rangle = n_{-\sigma}$ remains to be determined by the *full* spectral density via the spectral theorem (2.20). For $U \to 0$ eq. (2.86) becomes exact.

(b) *Self-consistent second order perturbation theory (SSOPT)*

We have to evaluate in the next step:

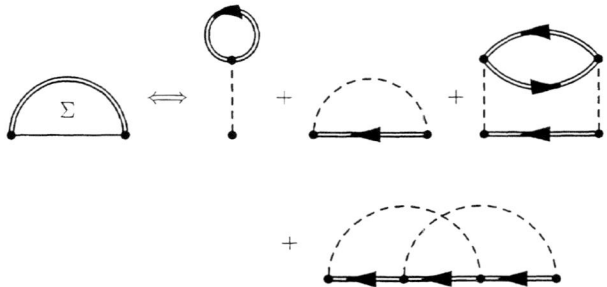

It is easy to verify that in the special case of the Hubbard model the second and the fourth diagram vanish. New is the second order contribution (SOC):

$$\Sigma_{ij\sigma}^{(SSOPT)}(E) = \delta_{ij} U n_{-\sigma} + \Sigma_{ij\sigma}^{(SOC)}(E) \qquad (2.87)$$

Tedious but straightforward and standard exploitation of the respective diagram yields the following expression:

$$\Sigma_{ij\sigma}^{(SOC)}(E) = \frac{U^2}{\hbar^3} \int\!\!\!\int\!\!\!\int_{-\infty}^{+\infty} dx\, dy\, dz\, \frac{S_{ij\sigma}(x) S_{ji-\sigma}(y) S_{ij-\sigma}(z)}{E + i0^+ - x + y - z} *$$
$$* \left\{ f(x)f(-y)f(z) + f(-x)f(y)f(-z) \right\} \qquad (2.88)$$

Note that $f(-x) = 1 - f(x)$. The second order contribution turns out to be non-local, energy-dependent and complex. Several alternatives are thinkable:

(1) $S \to S^{(0)}$

The spectral density $S_{ij\sigma}$ is replaced by the free one. This represents the *conventional perturbation theory* (CPT). Note however, that the replacement attacks only (2.88). The first-order diagram must remain dressed to account for contributions like

(2) $S \to S^{(1)} = S^{(0)}(E - Un_{-\sigma})$

We call it *perturbation theory around Hartree-Fock* (SOC-HF) [27,28]. At a later stage (Sect. III) we shall come back to this case.

All variants are equivalent up to order U^2! A further very important result can be derived for the case (a). A low-energy expansion of (2.88) shows:

$$-\frac{1}{\pi} Im\Sigma_{ij\sigma}^{(SOC)}(E) = \left(\frac{U^2}{\hbar^3} S_{ij\sigma}^{(0)} S_{ji-\sigma}^{(0)} S_{ij-\sigma}^{(0)}\right)_{|E=0} \cdot E^2 + \mathcal{O}(E^4) \qquad (2.89)$$

This means that the imaginary part of the selfenergy vanishes like E^2 at the Fermi edge. The generalization of this important result has been proven by Luttinger [29].

$$Im\Sigma_{ij\sigma}^{(n)}(E) \sim E^{2n-2} \quad \text{for } E \to 0 \qquad (2.90)$$

$\Sigma_{ij\sigma}^{(n)}$ denotes the n-th order perturbation theory for the selfenergy. So the above statement for the second order contribution to the selfenergy holds for the full function.

H Fermi-liquid behaviour

The interacting electron system becomes a *free Fermi gas* when the interaction is switched off. The properties of the Fermi gas are well-known from elementary *Quantum Statistics*:

$$\chi \sim const , \ \rho \sim T^2 , \ c_V \sim T , \ \ldots \tag{2.91}$$

(χ: magnetic susceptibility, ρ: electrical resistivity, c_V: heat capacity). For only weakly interacting particles these properties should not be dramatically modified. In particular there should be a *one-to-one correspondence* to the excitations of the free system. In such a case we call the interacting particle system a *Fermi liquid*.

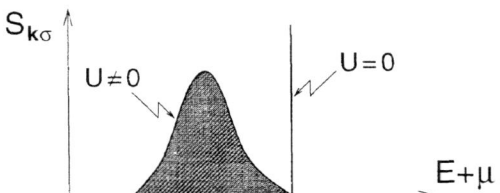

FIGURE 2.5. Schematic plot of the Fermi liquid ($U \neq 0$) and Fermi gas ($U = 0$) single-electron spectral density.

This shall be inspected in more detail by use of the results of the preceding section. Above all that means that we presume the perturbation series in any case to converge. This cannot be proven on a general basis but will be the only assumption for the following.

Let us restrict ourselves to

$$T = 0 \ ; \ low\text{-}energy \ region$$

Then we use from the considerations in the last section for the real and imaginary part of the selfenergy:

$$R_{\mathbf{k}\sigma}(E) = a_{\mathbf{k}\sigma} + b_{\mathbf{k}\sigma} E + \mathcal{O}(E^2) \tag{2.92}$$
$$I_{\mathbf{k}\sigma}(E) = d_{\mathbf{k}\sigma} E^2 + \mathcal{O}(E^4) \tag{2.93}$$

As schematically indicated in Fig. 2.5 Fermi-liquid behaviour means in particular that the quasiparticle equation (2.14) has only one solution:

$$E_\sigma - \epsilon(\mathbf{k}) - R_{\mathbf{k}\sigma}(E_\sigma - \mu) \stackrel{!}{=} 0 \tag{2.94}$$

In analogy to the free system we define tentatively a

$$Fermi \ surface \ : \ \{\mathbf{k} | E_\sigma(\mathbf{k}) \stackrel{!}{=} \mu\} \tag{2.95}$$

We have to inspect in the following whether or not this definition is reasonable. Eq. (2.95) is equivalent to

$$\mu \stackrel{!}{=} \epsilon(\mathbf{k}) + R_{\mathbf{k}\sigma}(0) = \epsilon(\mathbf{k}) + \Sigma_{\mathbf{k}\sigma}(0) , \qquad (2.96)$$

where we have used (2.94) and (2.93). We now reformulate the spectral density (2.13) for $(\mathbf{k}, E)$-points close to the Fermi surface. Applying (2.92) and (2.94) one easily finds:

$$E + \mu - \epsilon(\mathbf{k}) - R_{\mathbf{k}\sigma}(E) \approx (E - E_\sigma(\mathbf{k}) + \mu)(1 - b_{\mathbf{k}\sigma})$$

Using the definition (2.17) the *quasiparticle weight* close to the Fermi surface reads:

$$z_{\mathbf{k}\sigma} \longrightarrow (1 - b_{\mathbf{k}\sigma})^{-1} \qquad (2.97)$$

For the imaginary part of the selfenergy eq. (2.93) can be inserted into (2.13):

$$S_{\mathbf{k}\sigma}(E) \approx -\frac{\hbar}{\pi} z_{\mathbf{k}\sigma} \frac{z_{\mathbf{k}\sigma} d_{\mathbf{k}\sigma} E^2}{(E + \mu - E_\sigma(\mathbf{k}))^2 + (z_{\mathbf{k}\sigma} d_{\mathbf{k}\sigma})^2 E^4} \qquad (2.98)$$

This expression replaces in the Fermi-liquid region the cruder formula (2.16). It shows up the expected one-peak structure (see Fig. 2.5). The quasiparticle peak is located at $E = E_\sigma(\mathbf{k}) - \mu$ with a width of

$$\Delta_{\mathbf{k}\sigma} = |d_{\mathbf{k}\sigma}| z_{\mathbf{k}\sigma} (E_\sigma(\mathbf{k}) - \mu)^2 \qquad (2.99)$$

representing a measure of *quasiparticle damping*. According to (2.18) the reciprocal damping defines the *quasiparticle lifetime*:

$$\tau_{\mathbf{k}\sigma} = \frac{\hbar}{|d_{\mathbf{k}\sigma}| z_{\mathbf{k}\sigma} (E_\sigma(\mathbf{k}) - \mu)^2} \qquad (2.100)$$

Accordingly, quasiparticles at the Fermi surface have an infinite lifetime.

$$S_{\mathbf{k}\sigma}(E) \xrightarrow[E \to 0]{} \hbar z_{\mathbf{k}\sigma} \delta(E + \mu - E_\sigma(\mathbf{k})) \qquad (2.101)$$

Because of the quadratic energy dependence they are well-defined only very close to the Fermi surface. For larger energetic distances the presumptions of the above-theory, however, are no longer valid. The quasiparticle picture breaks down.

Let us finally inspect the momentum distribution at $T = 0K$:

$$\langle n_{\mathbf{k}\sigma} \rangle = \frac{1}{\hbar} \int_{-\infty}^{0} dE\, S_{\mathbf{k}\sigma}(E) =$$

$$= \int_{-\epsilon}^{0} dE\, z_{\mathbf{k}\sigma} \delta(E + \mu - E_\sigma(\mathbf{k})) + \langle \tilde{n}_{\mathbf{k}\sigma} \rangle \qquad (2.102)$$

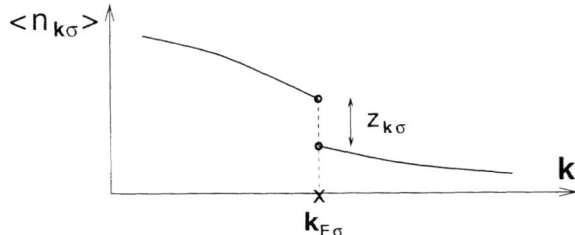

FIGURE 2.6. k-dependence of the momentum distribution of a Fermi liquid.

ϵ is a small positive number:

$$\langle n_{\mathbf{k}\sigma} \rangle = \langle \tilde{n}_{\mathbf{k}\sigma} \rangle + z_{\mathbf{k}\sigma}\, \Theta(\mu - E_\sigma(\mathbf{k})) \tag{2.103}$$

The jump at $\mathbf{k}_{F\sigma}$ makes the concept of a Fermi surface reasonable. We use the above results to define *normal Fermi liquids*

1. Existence of a Fermi surface.

2. One-to-one correspondence with the Fermi gas, well defined quasiparticles as low-energy excitations.

3. $Im\Sigma_{\mathbf{k}\sigma}(E) \sim E^2$, $\tau_{\mathbf{k}\sigma}^{-1} \sim E^2$

4. $\langle n_{\mathbf{k}\sigma} \rangle$-discontinuity at the Fermi surface.

We remember that all these results are based on the validity of the diagram perturbational expansion. Further-going considerations in the same spirit lead to the

Luttinger theorem [30]

(a) The number of (conduction) electrons equals the Fermi volume divided by the volume per **k**-state in the reciprocal space.

(b) The Fermi volume (volume in **k**-space enclosed by the Fermi surface) is invariant with respect to changes in the interaction strength, being identical with that of the Fermi gas. The shape of the Fermi surface, however, may change because of the **k**-dependence of the selfenergy $\Sigma_{\mathbf{k}\sigma}(0)$:

$$N_e = \sum_{\mathbf{k}\sigma} \Theta\left(\mu - \epsilon(\mathbf{k}) - \Sigma_{\mathbf{k}\sigma}(0)\right) \tag{2.104}$$

I Infinite lattice dimension

Exact statements about the many-body problem of the Hubbard model are possible for infinite lattice dimensions $d \to \infty$ [31]. However, to keep the physical problem reasonable and non-trivial the hopping integral t has to be rescaled. With $t = const$ the kinetic energy per particle would increase up to infinity for $d \to \infty$ while the potential energy would remain constant. As a consequence the physical properties would be those of the free Fermi gas. From this reason we require

$$\sqrt{\frac{1}{N}\sum_{\mathbf{k}} \epsilon^2(\mathbf{k})} \underset{d\to\infty}{\overset{!}{=}} \text{finite} \qquad (2.105)$$

That means for a hypercubic lattice, e. g. , for which each site has $2d$ next neighbours:

$$\frac{1}{N}\sum_{\mathbf{k}} \epsilon^2(\mathbf{k}) = \sum_{j} T_{ij}T_{ji} = 2d\,t^2 \underset{d\to\infty}{\longrightarrow} \text{finite}$$

This is possible only if

$$t \sim \frac{1}{\sqrt{d}}$$

Let us use this fact to estimate the Bloch-density of states

$$\rho_0(E) = \frac{1}{N}\sum_{\mathbf{k}} \delta(E - \epsilon(\mathbf{k}))$$

$$= \frac{a^d}{(2\pi)^d} \int d^d\mathbf{k}\, \delta(E - \epsilon(\mathbf{k})) \qquad (2.106)$$

a is the lattice constant and $\mathbf{k}$ the d-dimensional wave-vector:

$$\epsilon(\mathbf{k}) = -2t\sum_{i=1}^{d} \cos k_i a \qquad (2.107)$$

$$\mathbf{k} = (k_1, k_2, \ldots, k_d) \;;\; -\pi \le k_i a \le +\pi$$

Inserting (2.107) into (2.106) one gets after Fourier-transformation of the δ-function:

$$\rho_0(E) = \left(\prod_{i=1}^{d} \int_{-\pi}^{\pi} \frac{dk_i a}{2\pi}\right) \delta\left(E + 2t\sum_{j=1}^{d} \cos k_j a\right)$$

$$= \frac{1}{2\pi\hbar} \int_{-\infty}^{\infty} d\tau\, e^{\frac{i}{\hbar}E\tau} \prod_{i=1}^{d} \left(\int_{-\pi}^{\pi} \frac{dx_i}{2\pi} e^{2\frac{i}{\hbar}t \cos x_i \cdot \tau}\right)$$

The x_i-integral can be estimated as follows: $(t \sim d^{-\frac{1}{2}}, d \to \infty)$

$$\int_{-\pi}^{\pi} \frac{dx_i}{2\pi} e^{2\frac{i}{\hbar}t\cos x_i \cdot \tau} = 1 - \frac{t^2}{\hbar^2}\tau^2 + \mathcal{O}(t^4)$$

$$\approx \exp(-\frac{t^2}{\hbar^2}\tau^2)$$

Therefore we have

$$\rho_0(E) \approx \frac{1}{2\pi\hbar} \int_{-\infty}^{\infty} d\tau \, e^{\frac{i}{\hbar}E\tau - \frac{t^2}{\hbar^2}\tau^2 d}$$

eventually arriving at

$$\rho_0^{(\infty)}(E) = \frac{1}{\sqrt{2\pi}t^*} \exp(-\frac{E^2}{2t^{*2}}) \qquad (2.108)$$

To get a non-trivial density of states we have to postulate

$$t^* = t \cdot \sqrt{2d} \stackrel{!}{=} const \qquad (2.109)$$

t^* is then identical to the variance (2.105). By this scaling of the hopping integral the competition between kinetic and potential energy as well as the physically interesting correlation effects are retained.

Let us now discuss the consequences of this scaling. The kinetic energy per particle is by construction a finite quantity:

$$\frac{1}{N}\langle H_0 \rangle = -t \sum_{\substack{(i,j) \\ \sigma}} \langle c_{i\sigma}^+ c_{j\sigma} \rangle^{(0)} \stackrel{!}{=} finite$$

The sum runs over the $2d$ next neighbours of the site i. Because of (2.109) it must therefore hold:

$$\langle c_{i\sigma}^+ c_{j\sigma} \rangle^{(0)} = \mathcal{O}(\frac{1}{\sqrt{d}}) \ ; \ i \neq j \qquad (2.110)$$

Via spectral theorem this fact transfers to the spectral density and the Green function:

$$S_{ij\sigma}^{(0)} = \mathcal{O}(\frac{1}{\sqrt{d}}) \ ; \ G_{ij\sigma}^{(0)} = \mathcal{O}(\frac{1}{\sqrt{d}}) \quad i \neq j \qquad (2.111)$$

The consequences for the selfenergy are obvious [31] as can be seen from the following diagrammatic example:

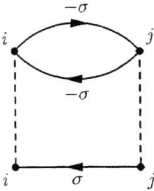

The three propagators which appear in the sketched selfenergy diagram of second order contribute

$$\sum_j G^{(0)}_{ij\sigma}(E)\, G^{(0)}_{ij-\sigma}(E)\, G^{(0)}_{ji-\sigma}(E) =$$
$$= \mathcal{O}(d \cdot (\frac{1}{\sqrt{d}})^3) = \mathcal{O}(\frac{1}{\sqrt{d}})$$

For $d \to \infty$ it vanishes. That holds for all off-diagonal terms [32]:

$$\Sigma_{ij\sigma}(E) = \Sigma_\sigma(E)\, \delta_{ij} \quad ; \quad d \to \infty \tag{2.112}$$

This exact result can be exploited in many fruitful respects. A special important application is represented in the next section.

J Effective impurity problem

Starting point for the following considerations is the locality of the selfenergy (2.112), being exact for $d \to \infty$ [33]. There is some evidence that for finite d it represents a reasonable *local approach*. It leads to a simplified Dyson equation for the local propagator:

$$G_{ii\sigma}(E) = G^{(0)}_{ii\sigma}(E) + \sum_j G^{(0)}_{ij\sigma}(E)\, \frac{1}{\hbar}\Sigma_\sigma(E)\, G_{ji\sigma}(E) \tag{2.113}$$

In terms of diagrams that reads:

$$\tag{2.114}$$

It is easy to show that the following alternative Dyson equation exists,

$$G_{ii\sigma}(E) = F_{ii\sigma}(E) + F_{ii\sigma}(E) \frac{1}{\hbar}\Sigma_\sigma(E) G_{ii\sigma}(E) \qquad (2.115)$$

if we define:

$$F_{ii\sigma}(E) = G_{ii\sigma}^{(0)}(E) + \sum_j^{\neq i} G_{ij\sigma}^{(0)}(E) \frac{1}{\hbar}\Sigma_\sigma(E) F_{ji\sigma}(E) \qquad (2.116)$$

It differs from eq. (2.113) by leaving out the diagonal (local) terms. Representing $F_{ii\sigma}(E)$ by a wavy line,

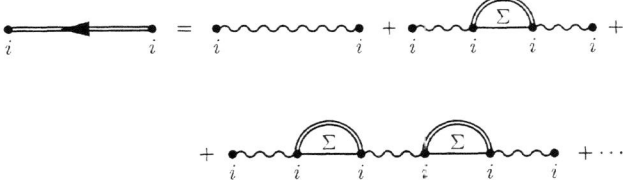

the local propagator reads as:

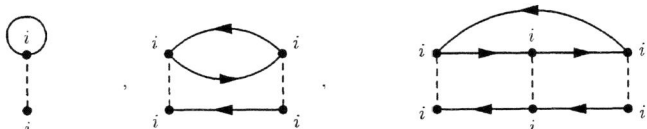

Let us now switch off all scattering centres except that at site $\mathbf{R}_i$ to get the *impurity selfenergy diagrams* (ISD)

They define a special diagram class C_σ:

$$\Sigma_\sigma^{ISD}(E) = C_\sigma\left[\{G_{ii\sigma}^{(0)}(E)\}, U\right] \qquad (2.117)$$

C_σ is independent of the special realization of the diagrams by the *free* propagator $G_{ii\sigma}^{(0)}(E)$.

Let us now look which diagrams contribute to the Hubbard-$d = \infty$-selfenergy. In first place there are all the *local diagrams*, which are defined as those where all site-indices are equal. These are just the ISD. But we have to count also certain non-local diagrams which get their non-locality by a selfenergy-insertion. This is illustrated by the following example:

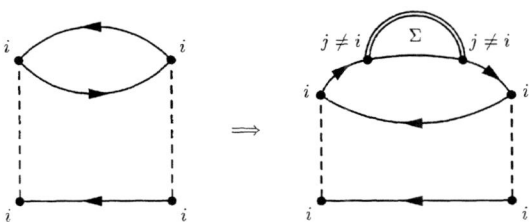

They do not disappear for $d \to \infty$ since the two non-local Green functions

$$G^{(0)}_{ij\sigma} \cdot G^{(0)}_{ji\sigma} \sim \mathcal{O}(\frac{1}{d})$$

are outweighed by the additional summation

$$\sum_\gamma \longrightarrow \mathcal{O}(d)$$

Obviously we get all these diagrams by replacing in the above ISD the *free* propagator by the *modified* propagator $F_{ii\sigma}(E)$ (2.116)

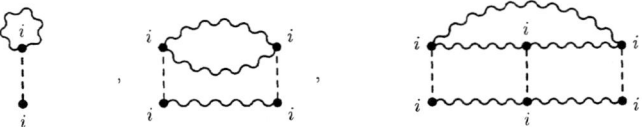

According to (2.117) this means:

$$\Sigma_\sigma(E) = C_\sigma \left[\{ F_{ii\sigma} \}, U \right] \tag{2.118}$$

Therewith we have expressed the $d = \infty$ Hubbard-selfenergy by the diagrammatic functional C_σ of the impurity scattering. This is the basis for the *dynamical mean field theory* (DMFT) which is discussed in detail in section III D

III FERROMAGNETISM IN ITINERANT MOMENT SYSTEMS

In this Section we investigate the possibility of bandferromagnetism in the framework of the Hubbard model. Starting point is therefore the Hamiltonian (2.1). Almost all informations we are interested in is conveyed by the single-electron Green function (2.4). Its equation of motion reads

$$(E + \mu)G_{ij\sigma}(E) = \hbar\delta_{ij} + \sum_{m} T_{im}G_{mj\sigma}(E) + U\Gamma_{iii,j\sigma}(E) \tag{3.1}$$

The *higher* Green function on the right-hand side,

$$\Gamma_{ilm,j\sigma}(E) = \langle\langle c^+_{i-\sigma}c_{l-\sigma}c_{m\sigma}; c^+_{j\sigma}\rangle\rangle_E \tag{3.2}$$

prevents a direct solution of the many-body problem. To get a first insight we first apply the simplest decoupling procedure. The mean-field decoupling of the *higher* Green function in (3.2),

$$\Gamma_{iii,j\sigma}(E) \longrightarrow \langle n_{i-\sigma}\rangle G_{ij\sigma}(E) \tag{3.3}$$

is equivalent to a replacement of the Hubbard interaction (2.1) by an effective exchange field

$$B^{\text{ex}}_\sigma = \frac{U}{\mu_\text{B}} n_{-\sigma} \tag{3.4}$$

which, because of $n_{-\sigma}$, has to be fixed selfconsistently. The effective-field ansatz defines the Stoner model, the idea of which is very much older than the Hubbard model [34]. The result is a $(\mathbf{k}, E)$-independent selfenergy:

$$\Sigma^\text{S}_\sigma(E) = Un_{-\sigma} \tag{3.5}$$

corresponding to a first order perturbation theory (2.86). The quasiparticle density of states is easily found to be

$$\rho_\sigma(E) = \rho_0(E - Un_{-\sigma}) \tag{3.6}$$

which can be used to express $n_{-\sigma}$ via (2.20). It is an instructive exercise to prove that ferromagnetism ($n_\uparrow \neq n_\downarrow$) is possible as soon as the *Stoner criterion* (1.84) is fulfilled. The spin asymmetry is due to a rigid but spin-dependent shift of the energy bands.

$$E_\downarrow(\mathbf{k}) - E_\uparrow(\mathbf{k}) = U(n_\uparrow - n_\downarrow) \tag{3.7}$$

The detailed evaluation of the Stoner model yields a strong overestimation of collective ferromagnetism that manifests itself in unrealistically high values for the Curie temperature. The reason for the failure of the simple Stoner model lies in the suppression of typical electron correlation effects.

A Strong coupling approaches

1 Hubbard-I-decoupling

A first proposal for an approach that incorporates electron correlation effects is due to Hubbard in his pioneering paper [18]. He starts with the equation of motion of the *higher* Green function (3.2)

$$(E + \mu - U)\Gamma_{iii,j\sigma}(E) = \hbar\delta_{ij}n_{-\sigma} + \sum_m T_{im}(\Gamma_{iim,j\sigma}(E) \qquad (3.8)$$
$$+ \Gamma_{imi,j\sigma}(E) - \Gamma_{mii,j\sigma}(E))$$

An approximate solution is offered by a mean-field decoupling of the Γ-functions on the right-hand site:

$$\Gamma_{iim,j\sigma}(E) \xrightarrow{i \neq m} n_{-\sigma} G_{mj\sigma}(E) \qquad (3.9)$$

$$\Gamma_{imi,j\sigma}(E) \xrightarrow{i \neq m} \langle c^+_{i-\sigma} c_{m-\sigma} \rangle G_{ij\sigma}(E) \qquad (3.10)$$

$$\Gamma_{mii,j\sigma}(E) \xrightarrow{i \neq m} \langle c^+_{m-\sigma} c_{i-\sigma} \rangle G_{ij\sigma}(E) \qquad (3.11)$$

Exploiting translational symmetry

$$\sum_m T_{im}(\langle c^+_{i-\sigma} c_{m-\sigma} \rangle - \langle c^+_{m-\sigma} c_{i-\sigma} \rangle) = 0 \qquad (3.12)$$

the equation of motion is solved by Fourier transformation,

$$G_{\mathbf{k}\sigma}(E) = \frac{\hbar}{E + \mu - \epsilon(\mathbf{k}) - \Sigma^{HI}_{\mathbf{k}\sigma}} \qquad (3.13)$$

where the resulting selfenergy is identical to that of the zero-bandwidth limit (2.33):

$$\Sigma^{HI}_{\mathbf{k}\sigma}(E) \equiv \Sigma^{W=0}_{\sigma}(E) \qquad (3.14)$$

It is easy to demonstrate that the solution (3.13) fulfils the trivial *band limit* $U = 0$, the zero-bandwidth limit $W = 0$ and the first three spectral moments $M^{(0,1,2)}_{\mathbf{k}\sigma}$. It violates the fourth moment $M^{(3)}_{\mathbf{k}\sigma}$ being therefore unable to reproduce the correct strong-coupling behaviour (Sects. II C and II D).

The selfenergy (3.14) is k-independent and real, therewith describing quasiparticles with infinite lifetimes. The quasiparticle density of states (2.19) can be expressed by the Bloch-density of states

$$\rho_\sigma(E) = \rho_0\left(E - \Sigma^{W=0}_{\sigma}(E)\right) \qquad (3.15)$$

The selfenergy has a singularity at $E_{0\sigma} = E + \mu = T_0 + U(1 - n_{-\sigma})$. As schematically

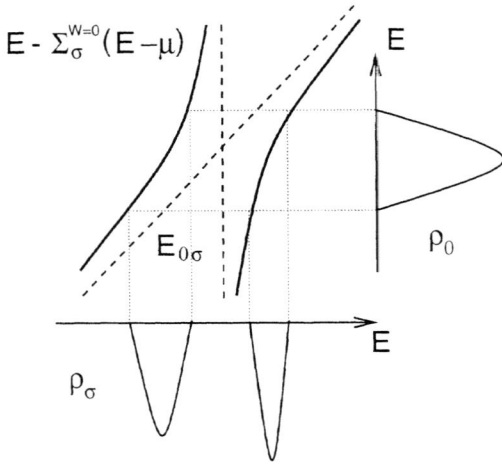

FIGURE 3.1. Schematic plot of the energy band splitting into two quasiparticle subbands according to the Hubbard-I-solution

plotted in Fig. 3.1 that leads to a splitting of the Bloch band ($\rho_0(E)$) into two quasiparticle subbands (*Hubbard bands*).

The interesting question is whether or not there does appear under certain circumstances an additional spin splitting (spin asymmetry) of the quasiparticle subbands. Giving rise to a finite spontaneous magnetic moment it would indicate a ferromagnetic phase.

The above-mentioned quasiparticle splitting manifests itself in the respective quasiparticle dispersion (zeros of the denominator of eq. (3.13)), that reads in the strong-coupling region:

$$E_{1\sigma}(\mathbf{k}) = (1 - n_{-\sigma})\epsilon(\mathbf{k}) + n_{-\sigma}T_0 + \mathcal{O}\left(\frac{W}{U}\right) \qquad (3.16)$$

$$E_{2\sigma}(\mathbf{k}) = U + n_{-\sigma}\epsilon(\mathbf{k}) + (1 - n_{-\sigma})T_0 + \mathcal{O}\left(\frac{W}{U}\right) \qquad (3.17)$$

The corresponding spectral weights are

$$\alpha_{1\sigma}(\mathbf{k}) \approx 1 - n_{-\sigma} \quad ; \quad \alpha_{2\sigma}(\mathbf{k}) \approx n_{-\sigma} \qquad (3.18)$$

The partial densities of states of the two *Hubbard bands*,

$$\rho_\sigma^{\text{lower}}(E) \approx \rho_0\left(\frac{E - n_{-\sigma}T_0}{1 - n_{-\sigma}}\right) \qquad (3.19)$$

$$\rho_\sigma^{\text{upper}}(E) \approx \rho_0\left(\frac{E - U - (1 - n_{-\sigma})T_0}{n_{-\sigma}}\right) \qquad (3.20)$$

have spin-independent centres of gravity:

$$T_\sigma^{\text{upper}_{\text{lower}}} \equiv \int dE\, E\, \rho_\sigma^{\text{upper}_{\text{lower}}}(E)$$

$$T_\uparrow^{\text{lower}} = T_\downarrow^{\text{lower}} = T_0 \quad;\quad T_\uparrow^{\text{upper}} = T_\downarrow^{\text{upper}} = T_0 + U \tag{3.21}$$

There may result a spin-dependent bandwidth modification but no shift of the

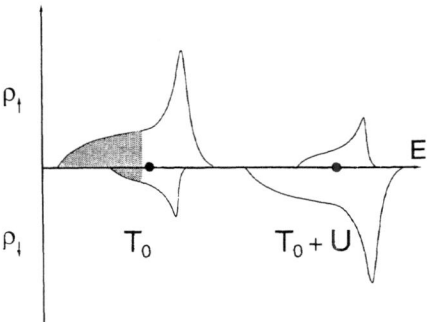

FIGURE 3.2. Schematic plot of the Hubbard-bands in the Hubbard-I-approximation

centres of gravity of the quasiparticle subbands. This makes the appearance of ferromagnetism as a selfconsistent solution of the Hubbard-I-theory fairly unlikely. A rather simple consideration leads to the following criterion for ferromagnetism:

$$(U \gg W\,;\, n < 1\,;\, T = 0) \tag{3.22}$$

$$1 - n + (E_F - T_0)\rho_0 \left(\frac{2E_F - nT_0}{2 - n}\right) \leq 0$$

E_F is the Fermi energy of the interacting system. Because of , $(1 - n) > 0$, $\rho_0 > 0$ the inequality requires $E_F < T_0 = T_\sigma^{\text{lower}}$. Ferromagnetism in the Hubbard-I-approach therefore needs a high density of states at the lower band edge, in particular a highly asymmetric DOS, and a relatively low band occupation.

2 *Spectral Density Approach (SDA)*

It is easy to show that the Hubbard-I-solution correctly fulfills only the first three spectral moments. According to Sect. II D a direct consequence is that the strong coupling behaviour (Sect. II C) is not reproduced by this theory, surely a serious disadvantage with respect to spontaneous ferromagnetism. An improvement in this respect can be found by the *spectral density approach* [35,24,25], which represents a very simple, non-perturbational approach consisting of three steps: First, one tries to guess the general structure of the spectral density guided by rigorous spectral representations, exactly solvable limiting cases, sum rules as well as *plausibility* and *intuition*. According to what we have found in Sects. II B and II C a two-pole ansatz suggests itself:

$$S_{\mathbf{k}\sigma}(E) = \hbar \sum_{j=1}^{2} z_{j\sigma}(\mathbf{k})\delta\left(E + \mu - E_{j\sigma}(\mathbf{k})\right) \qquad (3.23)$$

This ansatz neglects from the very beginning quasiparticle damping, the importance of which with respect to magnetic stability will be discussed in subsequent sections. It goes without saying that Fermi-liquid behaviour (Sect. II H) will also be violated by (3.23).

The second step is the calculation of the first four moments of the Hubbard model by use of the rigorous relation (2.47). The results are given in eqs. (2.49) to (2.52). These moments are used in the third step to fit the parameters in the ansatz (3.23) via (2.46). As a result of this simple procedure we find a selfenergy which exactly agrees with the expression (2.65) derived from a high-energy expansion. By construction the SDA is exact in the zero-bandwidth limit (2.33) and fulfils the first four moments guaranteeing therewith the right strong-coupling behaviour (Sect. II D). The selfenergy (2.65) differs from that of the Hubbard-I-approach by the replacement

$$T_0 \to B_{\mathbf{k}-\sigma}$$

while otherwise maintaining the same formal structure. $B_{\mathbf{k}-\sigma}$ is the higher correlation function defined in (2.39). It turns out to be decisive with respect to ferromagnetism. The local part $B_{-\sigma}$ (*bandshift*) in $B_{\mathbf{k}-\sigma}$ can be rigorously expressed by the required spectral density $S_{\mathbf{k}\sigma}(E)$ (see (2.44)). The **k**-dependent part $F_{\mathbf{k}-\sigma}$ appears a little bit more complicated. Assuming translational invariance and hopping between nearest neighbours only, the **k**-dependence can be separated [25]:

$$n_{-\sigma}(1 - n_{-\sigma})F_{\mathbf{k}-\sigma} = (\epsilon(\mathbf{k}) - T_0)\sum_{i=1}^{3} F_{-\sigma}^{(i)} \qquad (3.24)$$

$$F_{-\sigma}^{(1)} = \langle n_{i-\sigma} n_{j-\sigma}\rangle - n_{-\sigma}^2 \qquad \text{density correlation} \qquad (3.25)$$

$$F_{-\sigma}^{(2)} = -\langle c_{j\sigma}^{+} c_{j-\sigma}^{+} c_{i-\sigma} c_{i\sigma}\rangle \qquad \text{double hopping correlation} \qquad (3.26)$$

$$F_{-\sigma}^{(3)} = -\langle c_{j\sigma}^{+} c_{i-\sigma}^{+} c_{j-\sigma} c_{i\sigma}\rangle \qquad \text{spin flip correlation} \qquad (3.27)$$

$$(3.28)$$

i and j are numbering next neighbours. The method used to determine the higher correlations $F_{-\sigma}^{(i)}$ shall shortly be exemplified for $F_{-\sigma}^{(3)}$.

First we rewrite $F_{-\sigma}^{(3)}$ as

$$F_{-\sigma}^{(3)} = -\sum_{l} \delta_{jl} \langle c_{l\sigma}^{+} c_{j+\Delta-\sigma}^{+} c_{j-\sigma} c_{j+\Delta\sigma} \rangle \quad (3.29)$$

where the index Δ corresponds to the lattice vector which connects two neighbouring lattice sites $\mathbf{R}_i$ and $\mathbf{R}_j$. Because of translational invariance $F_{-\sigma}^{(3)}$ does not depend on the explicit value of Δ. We now introduce a *higher* spectral density $S_{\mathbf{k}\sigma}^{(3)}(E)$ as the $(\mathbf{k}, E)$-dependent Fourier transform of

$$S_{jl\sigma}^{(3)}(t, t') = \frac{1}{2\pi} \langle [c_{j+\Delta-\sigma}^{+} c_{j-\sigma} c_{j+\Delta\sigma}(t), c_{l\sigma}^{+}(t')]_{+} \rangle \quad (3.30)$$

The spectral theorem yields

$$F_{-\sigma}^{(3)} = -\frac{1}{N} \sum_{\mathbf{k}} \int_{-\infty}^{+\infty} dE\, f_{-}(E) S_{\mathbf{k}\sigma}^{(3)}(E - \mu) \quad (3.31)$$

According to the definition (3.30) the poles of $S_{\mathbf{k}\sigma}^{(3)}$ belong to the single-electron excitations of the Hubbard system. Inspecting the spectral representation of $S_{\mathbf{k}\sigma}^{(3)}$ and comparing it with that of the single-electron spectral density $S_{\mathbf{k}\sigma}(E)$ suggests to use again a two-pole ansatz

$$S_{\mathbf{k}\sigma}^{(3)}(E) = \hbar \sum_{j=1}^{2} \hat{z}_{j\sigma}(\mathbf{k}) \delta(E + \mu - E_{j\sigma}(\mathbf{k})) \quad (3.32)$$

where the quasiparticle energies $E_{j\sigma}(\mathbf{k})$ are the same as those in eq. (3.23), so that the spectral weights $\hat{z}_{j\sigma}(\mathbf{k})$ are the only unknown parameters. They are fixed by the first two spectral moments of $S_{\mathbf{k}\sigma}^{(3)}$. A lengthy but straight forward evaluation then leads via (3.31) to an explicit expression for $F_{-\sigma}^{(3)}$ [25].

In an analogous manner the correlation terms $F_{-\sigma}^{(1)}$ and $F_{-\sigma}^{(2)}$ are determined by two-pole ansatze like (3.9) for properly chosen *higher* spectral densities $S_{\mathbf{k}\sigma}^{(1)}$, $S_{\mathbf{k}\sigma}^{(2)}$. For details the reader is referred to ref. [25]. We note in passing that the same procedure can of course also be applied to the term $\langle n_{i\sigma} c_{i-\sigma}^{+} c_{j-\sigma} \rangle$ appearing in the band shift (2.41) yielding then the exact result (2.44). That can be considered as strong support of the described concept.

Fig. 3.3 shows a typical result of the SDA applied to a strongly coupled electron system ($\frac{U}{W} = 2.5$) on a bcc lattice. Plotted is the Q-DOS as function of energy, respectively, for various band occupations n at $T = 0$ (left column) and for various temperatures with a fixed particle density of $n = 0.65$ (right column). Let us start with the n-dependence of $\rho_\sigma(E)$ at $T = 0$. For $n \leq 0.55$ there is no selfconsistent ferromagnetic solution and the QDOS consists of two quasiparticle subbands. The lower band refers to an electron hopping mainly over lattice sites

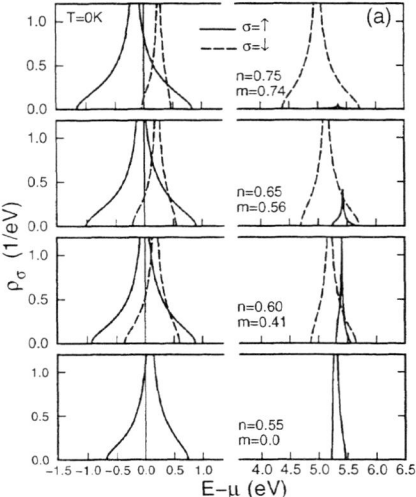
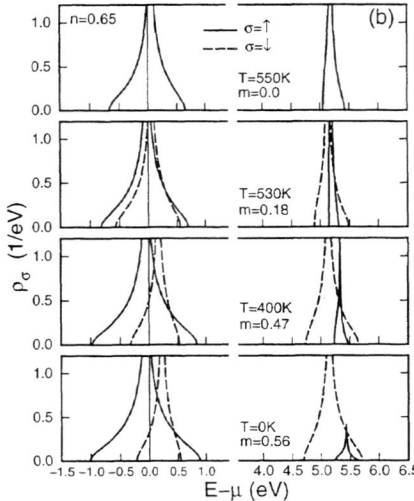

FIGURE 3.3. Left: Quasiparticle density of states as function of the energy for various band occupancies (bcc lattice, $U = 5\mathrm{eV}$, $W = 2\mathrm{eV}$, $T = 0$). Solid lines for $\sigma = \uparrow$, broken lines for $\sigma = \downarrow$. Vertical lines indicate the chemical potential. Right: The same as left but now for different temperatures and a fixed band occupation $n = 0.65$

which are not preoccupied by an electron with opposite spin. A particle which hops predominantly over preoccupied sites, however, propagates in a quasiparticle state of the upper subband. The area under the subband-DOS roughly scales with the probability that the propagating electron encounters the one or the other situation, approximately given by $(1 - n_{-\sigma})$ (lower part) and $n_{-\sigma}$ (upper part). For $n > 0.58$ an additional spin splitting of each quasiparticle subband appears (*exchange splitting*). The lower down-spin band is getting narrower being simultaneously shifted to higher energies. The electron system has a finite spontaneous magnetic moment. The moment reaches saturation $m = n$ for $n \geq 0.78$. We have to distinguish obviously two different correlation - caused bandsplittings, the *quasiparticle splitting* of order U (*Hubbard bands*), which happens in the strong coupling regime ($U \gg W$) for practically all parameter constellations, and the *exchange splitting* of the subbands, that indicates the spontaneous magnetic moment of the ferromagnetic phase.

The temperature dependence is exhibited in the right column of Fig. 3.3 for a fixed band occupation $n = 0.65$ of a less than half-filled band. With increasing temperature the minority-spin band shifts to lower energies and is getting broader therewith steadily diminishing the spontaneous magnetic moment.

It is an interesting point to look at the influence of the non-locality (k-dependence) of the electronic selfenergy (2.65) given by the wave-vector dependent part $F_{\mathbf{k}-\sigma}$ (3.24) of the higher correlation function (2.39). Fig. 3.4 shows for an sc lattice the inverse paramagnetic static susceptibility χ^{-1} as a function of the band occupation n for different coupling strengths U and $T = 0$. The zeros of χ^{-1} indicate instabilities of the paramagnetic state towards ferromagnetism. It is a special feature of the SDA [25,36], possibly even for the Hubbard model itself, that there are two ferromagnetic solutions, i. e. two zeros of χ^{-1}. The first solution sets in at $n_c = 0.34$ $(U \to \infty)$ where the actual value only slightly depends on U. This solution runs into saturation for $n \geq 0.68$, in exact agreement with the results of ref. [37]. The second solution appears for higher bandoccupations, but does never reach saturation and is always less stable than the other solution. Fig. 3.4 shows that a ra   ttice

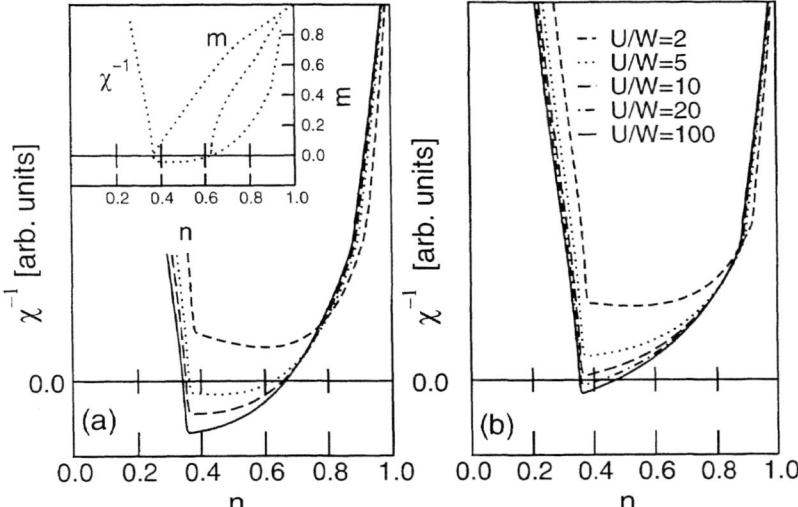

FIGURE 3.4. Inverse paramagnetic static susceptibility χ^{-1} for the sc lattice as a function of the band occupation n for various values of U. (a) System with the full k-dependent selfenergy. (b) System with a local selfenergy ($F_{\mathbf{k}-\sigma} \equiv 0$), (3.24)

a ferromagnetic solution. Part (b) demonstrates the importance of the bandwidth correlation $F_{\mathbf{k}-\sigma}$. Switching off this terms leads to a dramatic further increase of the critical $\frac{U}{W}$ from 4 to 14. More detailed studies, however, show that the influence of $F_{\mathbf{k}-\sigma}$ on magnetic stability strongly depends on the lattice type. Increasing coordination number (sc→bcc→fcc) let the importance of the k-dependence of the selfenergy decrease [36].

B Quasiparticle Damping

A severe shortcoming of the SDA is the neglect of quasiparticle damping (3.23) because of a real selfenergy (2.65).

Does quasiparticle damping influence magnetic stability?

Let us try to improve the SDA by inclusion of damping effects. To prepare this we start with a well-known theory of the Hubbard model which results in a complex selfenergy.

1 Hubbard-Alloy Analogy (AA)

The main idea of an alloy-analogy solution for the Hubbard model [38] is to consider the propagation of a σ electron with the $-\sigma$ electrons being *frozen* at their lattice sites and randomly distributed over the crystal. For the σ electron there are two possibilities when it enters a lattice site. It can meet a $-\sigma$ electron or not. That can be understood as a hopping through a fictitious binary alloy, where the two constituents appear with concentrations $x_{1\sigma}$ and $x_{2\sigma}$. To each alloy atom belongs one atomic level $E_{1\sigma}$, $E_{2\sigma}$. The *coherent potential approximation* CPA is a standard method to perform the configurational average over the positions of the *frozen* $-\sigma$ electrons [39]. The σ-selfenergy can be derived from the following implicit equation [21].

$$0 = \sum_{p=1}^{2} x_{p\sigma} \frac{E_{p\sigma} - \Sigma_\sigma(E) - T_0}{1 - \frac{1}{\hbar} G_\sigma(E)\left(E - \Sigma_\sigma(E) - T_0\right)} \tag{3.33}$$

$$G_\sigma(E) = \frac{1}{N} \sum_{\mathbf{k}} G_{\mathbf{k}\sigma}(E) \tag{3.34}$$

CPA is a single-site approximation, the resulting selfenergy therefore wave-vector independent. To proceed one has to specify the two-component alloy. Hubbard was the first [38] to propose the zero-bandwidth results (2.30),(2.31):

$$E_{1\sigma} = T_0 \quad ; E_{2\sigma} = T_0 + U \tag{3.35}$$
$$x_{1\sigma} = 1 - n_{-\sigma}; x_{2\sigma} = n_{-\sigma}$$

The idea behind this alloy analogy (AA) is that for diverging lattice constant ($a \to \infty$) the quasiparticle band energies should degenerate to $E_{1\sigma}$ and $E_{2\sigma}$, respectively.

With (3.35) in (3.33) the selfenergy $\Sigma_\sigma^{AA}(E)$ can be derived, in general turning out to be a complex function. Σ_σ^{AA} therefore includes quasiparticle damping. It has been proven [40], [41], however, that the AA does exclude any spontaneous ferromagnetism. This is in striking contrast to the SDA results (Sect. III A 2). *Does the quasiparticle damping really kill any spontaneous moment order?*

If we check the trustworthiness of the AA approach we find remarkable support. It has been shown [42] that for infinite lattice dimensions the CPA is an exact (!) treatment of the alloy problem. Furthermore, the so-called *Falicov-Kimball model* (FKM), defined as the special case of the Hubbard model, when the hopping of $-\sigma$ electrons is switched off,

$$\text{FKM}: \quad T_{ij} \to T_{ij\sigma} \quad ; \; T_{ij-\sigma} = T_0 \delta_{ij} \qquad (3.36)$$

is exactly solved by (3.33), (3.35) [43]. On the other hand the AA selfenergy disagrees with the strong as well as weak coupling behaviour (Sects. II C, II G) of the Hubbard model. This discrepancy can only be explained by the conclusion that (3.35) is the *wrong* alloy analogy. In particular the assumption of *frozen* $-\sigma$ electrons is surely unacceptable. The fictitious binary alloy is indeed by no means predetermined!

2 Modified Alloy Analogy (MAA)

We use the exact high-energy expansions in Sect. II E to derive an optimal alloy analogy. First we consider $E_{1,2\sigma}, x_{1,2\sigma}$ as free parameters. Inserting the expansions 2.55 and (2.56) into the CPA-equation (3.33) and sorting the different terms in powers of $\frac{1}{E}$ yields the following set of equations:

$$\sum_{p=1}^{2} x_{p\sigma} = 1 \qquad (3.37)$$

$$\sum_{p=1}^{2} x_{p\sigma} (E_{p\sigma} - T_0) = U n_{-\sigma} \qquad (3.38)$$

$$\sum_{p=1}^{2} x_{p\sigma} (E_{p\sigma} - T_0)^2 = U^2 n_{-\sigma} \qquad (3.39)$$

$$\sum_{p=1}^{2} x_{p\sigma} (E_{p\sigma} - T_0)^3 = U^3 n_{-\sigma} + U^2 (B_{-\sigma} - T_0) n_{-\sigma} (1 - n_{-\sigma}) \qquad (3.40)$$

Deriving from these equations the levels $E_{1,2\sigma}$ and the concentrations $x_{1,2\sigma}$ automatically guarantees the correctness of the first four spectral moments and therewith the correct strong-coupling behaviour. One finds ($p = 1, 2$):

$$E_{p\sigma}^{\text{MAA}} = \frac{1}{2}\Big\{ T_0 + U + B_{-\sigma} \\ + (-1)^p \sqrt{(U + B_{-\sigma} - T_0)^2 + 4U n_{-\sigma} (T_0 - B_{-\sigma})} \Big\} \qquad (3.41)$$

$$x_{1\sigma}^{\text{MAA}} = \frac{B_{-\sigma} + U(1 - n_{-\sigma}) - E_{1\sigma}^{\text{MAA}}}{E_{2\sigma}^{\text{MAA}} - E_{1\sigma}^{\text{MAA}}} = 1 - x_{2\sigma}^{\text{MAA}} \qquad (3.42)$$

Surprisingly, the energies and weights coincide exactly with the corresponding SDA-entities, if $\epsilon(\mathbf{k})$ is simply replaced by T_0. Because of the single-site aspect of the CPA the correlation $B_{\mathbf{k}-\sigma}$ (2.39) is restricted here to its local part $B_{-\sigma}$ (2.41), the decisive band shift. We conclude:

optimal alloy analogy:

$$E_{p\sigma}^{\text{MAA}} = \left(E_{p\sigma}^{\text{SDA}}(\mathbf{k})\right)_{\epsilon(\mathbf{k})=T_0} = f_p(T_0, U, n_{-\sigma}, B_{-\sigma}) \quad (3.43)$$

$$E_{p\sigma}^{\text{MAA}} = \left(z_{p\sigma}^{\text{SDA}}(\mathbf{k})\right)_{\epsilon(\mathbf{k})=T_0} = g_p(T_0, U, n_{-\sigma}, B_{-\sigma}) \quad (3.44)$$

Inserting this into the CPA-equation (3.33) yields the MAA selfenergy. Before presenting some special results let us list up some general aspects of the MAA solution [44]:

1. As a CPA result the MAA includes quasiparticle damping, and that without giving up the advantages of the SDA procedure (Sect. III A 2)

2. The expectation values $n_{-\sigma}, B_{-\sigma}$ can be determined selfconsistently leading to in principle spin-dependent atomic levels. Furthermore, the appearance of $B_{-\sigma}$ takes into account in a certain sense the itineracy of the $-\sigma$ electrons (*correlated hopping* of the $-\sigma$ electrons), completely neglected in the AA [44].

3. By construction the first four spectral moments $M_\sigma^{(0,1,2,3)}$ are fulfilled guaranteeing the correct (local) strong-coupling behaviour.

4. It is interesting that the strong-coupling behaviour can also be checked by an exact statement of the CPA theory to the so-called split band regime $U \gg W$ [39]. The wave-vector dependent spectral density of a binary alloy consists in the split band regime of two separated peaks with centres of gravity at

$$T_{p\sigma}^{\text{CPA}} = E_{p\sigma} + x_{p\sigma}\left(\epsilon(\mathbf{k}) - T_0\right) \quad : p = 1, 2 \quad (3.45)$$

Inserting (3.41 and 3.42) for $U \gg W$ yields exactly the Harris-Lange results (2.36), (2.37), a further strong confirmation for the modified alloy analogy [44].

5. Applying the MAA to the Falicov-Kimball model (3.36) results because of $B_{-\sigma} = T_0$ in the exact $d = \infty$ solution.

6. Most important in our context here is the fact that contrary to the *conventional* AA the MAA allows for *spontaneous bandferromagnetism*

Let us inspect a typical example: Fig.3.5 shows the spectral density $S_{\mathbf{k}\sigma}$ of a strongly correlated ($\frac{U}{W} = 5$) electron system of an fcc lattice. As Bloch density of states a tight-binding version [45] has been chosen. For less than half-filled bands ($n < 1$) the system is paramagnetic, no spontaneous spin order can be observed. The band occupation $n = 1.6$ used in Fig. 3.5, however, allows bandferromagnetism provided the Coulomb interaction U exceeds a critical value. As for the SDA (Fig. 3.3) two types of splitting occur. At first the spectral density consists for each k-vector of a high-energy and a low-energy peak which are separated by an energy amount of order U. The finite widths of the peaks are due to quasiparticle damping. The

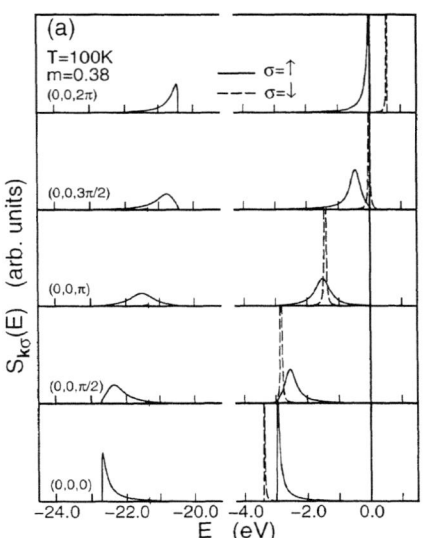
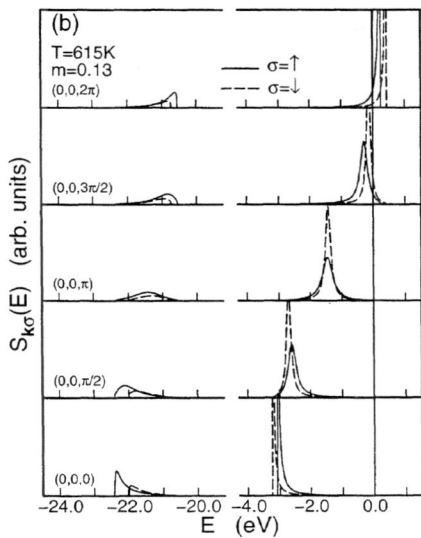

FIGURE 3.5. Spectral density as a function of energy for an fcc lattice calculated within the MAA, (a) $T = 100K$, (b) $T = 615K$. Different **k**-vectors equidistant along the (110)-direction of the 1. Brillouin zone. Further parameters: $n = 1.6$, $U = 20\text{eV}$, $W = 4\text{eV}$. The vertical line indicates the position of the chemical potential.

weight (area) of the low-energy peak corresponds to the probability that the propagating $(\mathbf{k}, \sigma)$ electron in the more than half-filled band enters an empty lattice site, while the weight of the upper peak scales with the probability that it meets any $-\sigma$ electron. This splitting is not at all bound to ferromagnetism, it appears also in the paramagnetic phase. It demonstrates the correct strong-coupling behaviour (2.36), (2.37). Ferromagnetism arises when the two spectral density peaks show an additional spin splitting. At low temperatures ($T = 100K$ in Fig. 3.5a) the system is very close to its saturation ($m = 2 - n$), i. e. , the up-spin states are almost fully occupied. A down-spin electron cannot avoid to meet an up-spin electron at every lattice site and has to perform a Coulomb interaction. Consequently the low-energy peak of $S_{\mathbf{k}\downarrow}(E)$ disappears. At higher temperatures (part (b) of Fig. 3.5) the peak reappears because of a partial demagnetization of the up-spin states. At low temperatures the high-energy peaks of $S_{\mathbf{k}\downarrow}(E)$ are very sharp, indicating long-living quasiparticles. An interesting **k**-dependence of the peak position (quasiparticle energy) is observed in the upper part. At the top of that branch a *normal* exchange splitting appears, i. e., the down-spin peak is above the up-spin peak. At the bottom, below the chemical potential, however, the ↑-quasiparticle energy is higher than that of a ↓- quasiparticle (*inverse exchange splitting*). The quasiparticle dispersions of the two spin parts are crossing as functions of the wave-vector **k**. Two competing correlation effects are responsible for this behaviour. The one is a spin-dependent exchange shift of the centres of gravity of the quasiparticle subbands, the other a spin-dependent band narrowing. In any case the low-temperature exchange splitting exhibits a strong wave-vector dependence, even with a sign change.

A wave-vector summation of the spectral density yields the quasiparticle density of states. The result for $\frac{U}{W} = 5, n = 1.6$ and four different temperatures is plotted in part (b) of Fig. 3.6. In order to elaborate the influence of quasiparticle damping the SDA-results for the same $\frac{U}{W}, n$, and lattice structure are given in part (a) of Fig. 3.6. At first glance the curves appear fairly similar. Most obvious are the weaker structures of the MAA curves due to the non-zero imaginary part of the electronic selfenergy. However, the basic mechanism that leads to spontaneous ferromagnetism of the itinerant band electrons seems to be the same. In both methods the two above-described band splittings in consequence of strong electron correlations determine the magnetic behaviour of the partially filled energy band. For low temperatures SDA as well as MAA predict almost saturated ferromagnetism. As explained above, the lower down-spin subband then disappears while the upper subband has exactly the shape of the *free* fcc-Bloch-DOS. The quasiparticle damping takes care for the fact that for higher temperatures the Q-DOS of the MAA strongly deviates from the B-DOS $\rho_0(E)$, while the SDA-subband densities of states always retain a certain similarity to the original B-DOS.

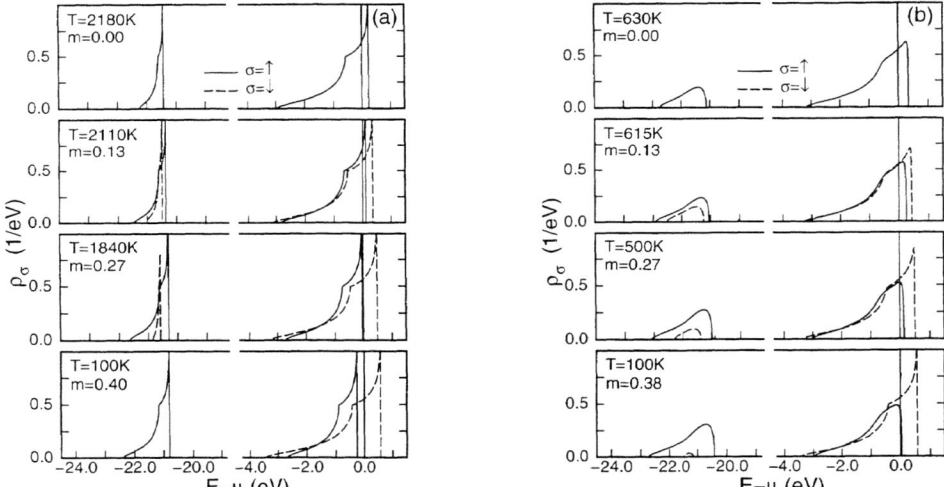

FIGURE 3.6. Quasiparticle density of states as a function of energy and different temperatures T up to T_C, (a) calculated by SDA, (b) by MAA. Parameters as in Fig. 3.5. The vertical line stands for the chemical potential μ.

The most drastic influence of the quasiparticle damping attacks the Curie temperature T_C, the most important key-quantity of bandferromagnetism. T_C sensitively depends on the model parameters n and U. The band occupation dependence is demonstrated in Fig. 3.7. To compare the results with the numerically essentially exact quantum-Monte Carlo (QMC) calculations of Ulmke [46] we have used as B-DOS a generalization of the $d = 3$ fcc lattice, as in Figs. 3.5, 3.6, to infinite dimensions (Sect. III) [47]

$$\rho_0(E) = \frac{\exp\left(-\frac{1}{2}\left(1 + \frac{\sqrt{2}E}{t^*}\right)\right)}{t^*\sqrt{\pi\left(1 + \frac{\sqrt{2}E}{t^*}\right)}} \quad (3.46)$$

Energy units are chosen such that

$$t^* = t\sqrt{2d(d-1)} \stackrel{!}{=} 1 \quad (3.47)$$

It resembles the $d = 3$-fcc B-DOS for holes (!) being sharply peaked at the lower edge. The symbol n in Fig. 3.7 is therefore to be interpreted as hole density ($2 - n$: electron density). The high density of states at the lower band edge is sufficient to produce stable ferromagnetism even within the Hubbard-I solution (3.14). The ferromagnetic region in the $n - T$ plane, however, is confined to very low densities, only. As discussed in Sect. III A 1 this is typical for the early Hubbard approach [18]. The main shortcoming is the lack of a spin-dependent centre of gravity shift, which is, in contrast to the Hubbard approach, present in the SDA. As can be seen in Fig. 3.7 that leads to a considerable increase of the Curie temperature. Ferromagnetic solutions are found for all fillings up to half-filling $n = 1$ (n: hole-density). This effect is mainly due to the spin-dependent band shift (2.41) being absent in the Hubbard-I approach. At $n = 0.58$ the $T = 0$-system is fully polarized ($m = n$). The spin splitting of the centre of gravity of the respective Hubbard band amounts to $n_\uparrow B_\uparrow - n_\downarrow B_\downarrow = 0.73$. This has to be removed as T approaches T_C while the Hubbard gap $\approx U$ still exists even above T_C. The spin splitting is extraordinarily smaller than the exchange splitting $Um = 2.32$ within the Stoner approach (3.7) where correlation effects are totally neglected. Consequently, the SDA yields a Curie temperature smaller by more than a factor three at $n = 0.58$ and $U = 4$. The difference becomes larger and larger with increasing U since as a function of U the Curie temperature is unbounded in the Stoner theory while it reaches saturation within the SDA [25]. The unrealistically high Stoner-Curie temperature results from the necessity to bridge an exchange splitting of order U by the thermal energy $k_B T_C$.

A spin-dependent band shift is also realized in the MAA. Stable ferromagnetism is found for fillings $0 < n < 1$ (n: hole concentration $\hookrightarrow$ more than half-filled bands). Compared to the SDA, the Curie temperature is significantly lower (Fig. 3.7) for all fillings. While the mechanism leading to ferromagnetism is the same in the SDA as in the MAA, quasiparticle damping is destabilizing to some extent the ferromagnetic order [44]. The broadening of the respective spin-dependent bands enhances their overlap and leads to a (self-consistent) depression of the magnetization and therewith of T_C. Again we recognize the decisive influence of the band shift $B_{-\sigma}$. If we replace $B_{-\sigma}$ by T_0, i. e., giving up the validity of the $m = 3$-spectral

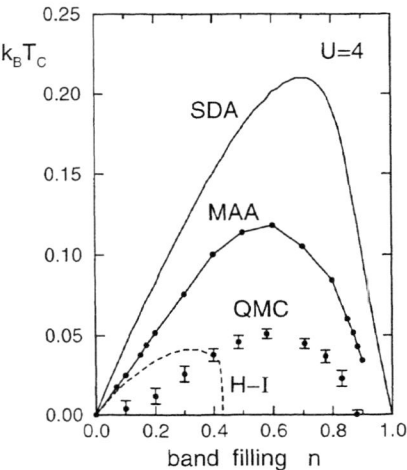

FIGURE 3.7. Filling dependence of the Curie temperature T_C for SDA, MAA, and the Hubbard-I approach in comparison with the QMC results (error bars) of ref. [46] for the $d = \infty$ fcc-type lattice (3.44). [48]

moment, then we come from the SDA to the Hubbard-I solution and from the MAA to the conventional AA. There is no ferromagnetism at all in the AA as has been proven in [40] and also tested numerically for the fcc-type lattice considered here. We conclude, that ignoring the possible spin-dependent band shift $n_{-\sigma}B_{-\sigma}$ misses the decisive route to ferromagnetic order. The methods we are now going to inspect will strongly support this interpretation.

C Interpolation Schemes

A severe draw back of the alloy analogies (AA, MAA) discussed so far lies in the fact that they violate the weak-coupling behaviour (Sects. II G, II H). The question is whether or not this fact affects the strong-coupling phenomenon ferromagnetism. To check this the CPA-equation can be considered as a mere starting point for a reasonable interpolation formula being demanded to be exact for small U as well as for strong couplings. Standard perturbation theory in the interaction U provides us with the exact result for the selfenergy in the weak-coupling regime (see Sect. II G). Let us look for interpolation schemes which guarantee the correctness of the solution in the weak-coupling regime. The main question remains if this correctness is important with respect to ferromagnetism.

1 Edwards-Hertz-Approach (EHA)

An interpolation scheme that is based on the alloy analogy idea but correctly accounts for the weak-coupling limit has been suggested by Edwards and Hertz [49]. Within the Edwards-Hertz Approach (EHA) the CPA equation (3.32) is manipulated in the following way. First the *conventional* alloy analogy (3.34) is used, and furthermore, the propagator $G_\sigma(E)$ in the denominator is replaced, more or less *ad hoc*, by

$$\hat{G}_\sigma(E) = \frac{\hbar}{U^2 n_{-\sigma}(1 - n_{-\sigma})} \Sigma_\sigma^{SOC}\left(E - \Sigma_\sigma^{EHA}(E) + E_\sigma\right) \tag{3.48}$$

Σ_σ^{SOC} is the local ($i = j$) second order perturbation theory contribution to the selfenergy, defined in eq. (2.88). A simple calculation shows that the EHA reproduces the zero-bandwidth limit (2.29) for all band occupations n provided that the quantity E_σ in (3.48) is chosen just so that

$$n_\sigma \stackrel{!}{=} n_\sigma^{(1)} \tag{3.49}$$

$n_\sigma^{(1)}$ is the spin-dependent occupation number for the *free* system being shifted in energy by E_σ. Eq. (3.49) is therefore equivalent to

$$0 \stackrel{!}{=} \int_{-\infty}^{+\infty} dE\, f(E)\left(S_{-\sigma}(E) - S_\sigma^{(0)}(E - E_\sigma)\right) \tag{3.50}$$

The introduction of E_σ is an extension of the original approach [49] proposed in ref. [50].

Expanding the EHA self-energy in powers of U up to order U^2 one indeed gets the exact weak-coupling result (2.87); no wonder because the fairly arbitrary replacement (3.48) has been motivated just so. The correctness up to U^2 also implies that the expansion in U satisfies the Luttinger theorem (Sect. II H). However, Fermi-liquid behaviour is recovered only up to $U \approx W$ where the series diverges [49]. Since the EHA is also exact in the zero-bandwidth limit it should be regarded as an interpolation scheme. The solution for the $d = \infty$-FKM, too, is correctly reproduced.

However, the EHA fails with respect to the strong-coupling behaviour (Sect. II C). In the selfenergy expansion (2.56) the first two coefficients $C_\sigma^{(0)}$ and $C_\sigma^{(1)}$ are exact while in the third,

$$C_\sigma^{(2,EHA)} = C_\sigma^{(2)}\Big|_{B_{-\sigma} \to B_{-\sigma}^{(1)}}, \tag{3.51}$$

the bandshift $B_{-\sigma}$ appears only in its Hartee-Fock version [51]

$$B_{-\sigma}^{(1)} = T_0 + \frac{2n_\sigma^{(1)} - 1}{n_{-\sigma}^{(1)}\left(1 - n_{-\sigma}^{(1)}\right)} \sum_{i,j}^{i \neq j} T_{ij}\langle c_{i-\sigma}^+ c_{j-\sigma}\rangle^{(1)} \tag{3.52}$$

Here we have defined

$$\langle c^+_{i-\sigma} c_{j-\sigma} \rangle^{(1)} = \frac{1}{\hbar} \int_{-\infty}^{+\infty} dE f(E) S^{(0)}_{ji-\sigma}(E - E_\sigma) \tag{3.53}$$

Only for the zero-bandwidth limit, the FKM (3.35), and the symmetric case $n = 1$ we have $B_{-\sigma} = B^{(1)}_{-\sigma} = T_0$. It therefore remains to improve the strong-coupling behaviour.

2 Interpolating Alloy Analogy (IAA)

The goal is to achieve correctness for a maximum number of limiting cases [51]:

1. $\frac{U}{W} \gg 1$; 2. $\frac{U}{W} \ll 1$; 3. FKM; 4. $W \to 0$

The procedure follows that of the EHA, except for the *conventional* alloy analogy. The latter is improved in the same manner as in the AA $\to$ MAA transition (Sect. III B 2). Starting point is the CPA-equation (3.32). In this equation $G_\sigma(E)$ is substituted by the modified propagator (3.48). Finally the high-energy expansion is performed as for the MAA (see eqs. (3.36) to (3.39)) thereby regarding that the moments $M^{(m)}_\sigma$ in the expansion (2.55) are now to be interpreted as the moments of the modified propagator (3.48). The improved alloy analogy then reads:

$$E^{IAA}_{p\sigma} = \left(E^{SDA}_{p\sigma}(\mathbf{k}) \right)_{\substack{\epsilon(\mathbf{k})=T_0 \\ B_{-\sigma} \to B_{-\sigma} - B^{(1)}_{-\sigma} + T_0}} \tag{3.54}$$

$$x^{IAA}_{p\sigma} = \left(z^{SDA}_{p\sigma}(\mathbf{k}) \right)_{\substack{\epsilon(\mathbf{k})=T_0 \\ B_{-\sigma} \to B_{-\sigma} - B^{(1)}_{-\sigma} + T_0}} \tag{3.55}$$

For this fictitious binary alloy the CPA-equation (3.32) is solved to yield Σ^{IAA}_σ. By construction the first four spectral moments are fulfilled by the IAA therewith guaranteeing the correct strong coupling behaviour. Also by construction, the weak-coupling behaviour is reproduced to the same extent as in the EHA. Up to order U^2 the IAA is exact. Furthermore, the zero-bandwidth limit is fulfilled as well as the FKM special case (3.35). The IAA interpolates between EHA (Sect. III C 1) and MAA (Sect. III B 2). The MAA emerges from the IAA by the replacement $B^{(1)}_{-\sigma} \to T_0$.

The IAA allows for spontaneous bandferromagnetism. Since quasiparticle damping becomes stronger when turning from MAA to IAA a considerable decrease of the Curie temperature is to recognize. Fig. 3.8 shows that the maximum T_C in the IAA is less than half of the maximum T_C predicted by the MAA. In addition the ferromagnetic region shrinks to $0 < n < 0.61$. Comparing with EHA one notices again the influence of the band shift $B_{-\sigma}$ with respect to ferromagnetic stability. However, since $B_{-\sigma}$ is replaced by $B^{(1)}_{-\sigma}$ (3.52), EHA partially accounts for the band shift. It can be shown [51] that in the ferromagnetic phase the selfconsistently

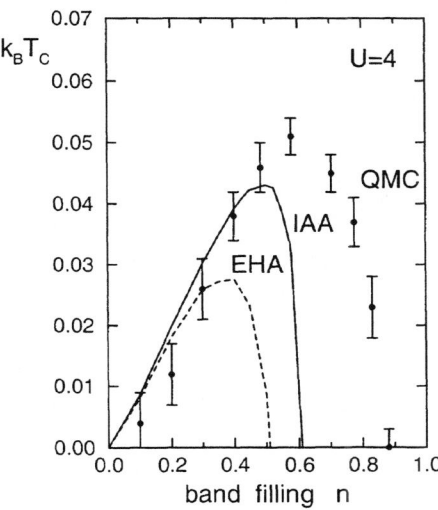

FIGURE 3.8. Filling dependence of the Curie temperature T_C for the IAA (solid line) and the EHA (dashed line) compared with QMC results (error bars) of ref. [46] for the fcc-type lattice (3.44). [48]

calculated $B_{-\sigma}$ is larger than $B_{-\sigma}^{(1)}$. Therefore IAA favours ferromagnetism stronger than EHA, but the difference is not so drastic as for SDA↔HI and in particular in the case of MAA↔AA.

A severe drawback of the alloy-analogy based theories EHA and IAA is the fact, that the correct weak-coupling behaviour has to be enforced artificially. The use of (3.46) instead of $G_\sigma(E)$ in (3.32) is by no means justifiable.

3 Dynamical Mean Field Theory (DMFT)

Basic for the following considerations are the findings of Sect. II J. We assume that the locality (2.112) of the selfenergy, exact for infinite lattice dimension d, represents a reasonable approach for finite dimensions, too. It has been demonstrated in Sect. II J that in such a case the selfenergy in the Hubbard model can be expressed via (2.118) by the diagrammatic functional C_σ of the impurity scattering (2.117). Let us exploit this fact by the inspection of special well-known impurity problem [52,53].

4 Mapping on the Single-Impurity Anderson model (SIAM)

The model is defined by the Hamiltonian

$$H = \sum_\sigma (\epsilon_d - \mu) n_{d\sigma} + \sum_{\mathbf{k}\sigma} (\epsilon(\mathbf{k}) - \mu) n_{\mathbf{k}\sigma} +$$
$$+ \sum_{\mathbf{k}\sigma} V_{\mathbf{k}d} \left(c^+_{d\sigma} c_{\mathbf{k}\sigma} + c^+_{\mathbf{k}\sigma} c_{d\sigma} \right) + \frac{1}{2} U \sum_\sigma n_{d\sigma} n_{d-\sigma} \qquad (3.56)$$

$c^+_{d\sigma}$ ($c_{d\sigma}$) is the creation (annihilation) operator for an electron in the single localized d-level (*impurity*) while $c^+_{\mathbf{k}\sigma}$ ($c_{\mathbf{k}\sigma}$) is that for the band electrons. $n_{d\sigma} = c^+_{d\sigma} c_{d\sigma}$ is the occupation number operator for the impurity level, $n_{\mathbf{k}\sigma} = c^+_{\mathbf{k}\sigma} c_{\mathbf{k}\sigma}$ has already been used. $V_{\mathbf{k}d}$ is the hybridization matrix element, and U is the on-site Coulomb repulsion for the impurity.

The equation of motion of the d-impurity Green function

$$G_{d\sigma}(E) = \langle\langle c_{d\sigma}; c^+_{d\sigma} \rangle\rangle \qquad (3.57)$$

is formally solved by introducing the respective selfenergy $\Sigma_{d\sigma}(E)$ via

$$\langle\langle \left[c_{d\sigma}, \frac{1}{2} U \sum_{\sigma'} n_{d\sigma'} n_{d-\sigma'} \right]_{-} ; c^+_{d\sigma} \rangle\rangle \equiv \Sigma_{d\sigma}(E) G_{d\sigma}(E) \qquad (3.58)$$

and defining a *hybridization function*

$$\Delta(E) = \sum_\mathbf{k} \frac{|V_{\mathbf{k}d}|^2}{E + \mu - \epsilon(\mathbf{k})} \qquad (3.59)$$

which absorbs the band energies $\epsilon(\mathbf{k})$ and the hybridization matrix element $V_{\mathbf{k}d}$:

$$G_{d\sigma}(E) = \frac{\hbar}{E + \mu - \epsilon_d - \Delta(E) - \Sigma_{d\sigma}(E)} \qquad (3.60)$$

Defining the *free* counterpart by

$$G^{(0)}_{d\sigma}(E) = \frac{\hbar}{E + \mu - \epsilon_d - \Delta(E)} \qquad (3.61)$$

allows to write eq. (3.60) as Dyson equation:

$$G_{d\sigma}(E) = G^{(0)}_{d\sigma}(E) + G^{(0)}_{d\sigma}(E) \frac{1}{\hbar} \Sigma_{d\sigma}(E) G_{d\sigma}(E) \qquad (3.62)$$

The structure is the same as the version (2.115) for the Hubbard model. The corresponding selfenergy diagrams belong to the class C_σ of impurity diagrams (2.117):

$$\Sigma_{d\sigma}(E) = C_\sigma\left[\{G_{d\sigma}^{(0)}(E)\}, U\right] \tag{3.63}$$

provided U is here the same as in (2.117)

Let us now choose $\Delta(E)$ in such a way that

$$G_{d\sigma}^{(0)}(E) \stackrel{!}{=} F_{ii\sigma}(E) \tag{3.64}$$

where $F_{ii\sigma}(E)$ is defined in (2.116). It can be shown that (3.64) can be realized only if the chemical potential μ in both systems are the same and also holds:

$$\epsilon_d = T_0 \tag{3.65}$$

Comparison of (2.118) with (3.63) and (3.64) yields

$$\Sigma_\sigma(E) = \Sigma_{d\sigma}(E) \tag{3.66}$$

and therewith because of (2.115) and (3.62)

$$G_{ii\sigma}(E) = G_{d\sigma}(E) \tag{3.67}$$

The Hubbard model is now traced back to the *simpler* SIAM. Let us gather the above results. Eq. (2.115) can be written as

$$F_{ii\sigma}^{-1}(E) = G_{ii\sigma}^{-1}(E) + \frac{1}{\hbar}\Sigma_\sigma(E)$$

This we use in (3.64),

$$\left(G_{d\sigma}^{(0)}(E)\right)^{-1} = \frac{1}{\hbar}(E + \mu - \epsilon_d - \Delta(E))$$
$$= F_{ii\sigma}^{-1}(E) = G_{ii\sigma}^{-1}(E) + \frac{1}{\hbar}\Sigma_\sigma(E)$$

to arrive with (3.65) at the following selfconsistency-equation:

$$\Delta(E) = E + \mu - T_0 - \Sigma_\sigma(E) - \hbar G_{ii\sigma}^{-1}(E) \tag{3.68}$$

Because of the k-independence of the selfenergy the propagator $G_{ii\sigma}(E)$ can be expressed by the *free* Bloch-DOS:

$$G_{ii\sigma}(E) = \hbar\int_{-\infty}^{+\infty} dx \frac{\rho_0(x)}{E + \mu - x - \Sigma_\sigma(E)} \tag{3.69}$$

The DMFT now solves the many-body problem of the Hubbard model by the following selfconsistency-circle:

1. Choose an initial value for $\Sigma_\sigma(E)$, e. g. $\Sigma_\sigma(E) \equiv 0$!
2. Calculate with (3.69) $G_{ii\sigma}(E)$!
3. Determine $\Delta(E)$ via (3.68)
4. Solve the SIAM (3.60) with the given $\Delta(E)$, i. e. find $\Sigma_{d\sigma}(E)$!
5. Use (3.66) to get a new initial value for $\Sigma_\sigma(E)$.

The remaining task is to solve in step 4) the SIAM, a non-trivial problem, but very much better established than the original Hubbard problem. The QMC results of ref. [46] shown in Figs. 3.7 and 3.8 are found by use of the above described DMFT and a finite temperature Quantum Monte Carlo method [54] to solve the SIAM.

The methods we are going to discuss in the next sections all use DMFT, but with different approaches to step 4).

5 Iterative Perturbation Theory (IPT)

In the alloy-analogy based theories of Sect. III C the correct weak-coupling behaviour has been introduced in an *ad hoc*-manner without deeper justification. It can be automatically accounted for by perturbational methods. It remains a challenging problem to find out how important the weak-coupling behaviour is for a correct description of the strong-coupling phenomena *ferromagnetism*.

To solve the SIAM within the DMFT-scheme a second order perturbation theory around the Hartree-Fock solution has been proposed [52]:

$$\Sigma_{d\sigma}^{SOPT}(E) = U n_{d-\sigma} + \Sigma_{d\sigma}^{SOC}(E) \qquad (3.70)$$

The procedure is the same as that for the Hubbard model described in Sect. II G, [27,28]:

$$\Sigma_{d\sigma}^{SOC} : \qquad (3.71)$$

$$(3.72)$$

The diagrams have the same formal meaning as in Sect. II G. Here they stand for the impurity-Green functions and selfenergies. The ansatz (3.70) for the SIAM-selfenergy used in the DMFT concept is called *iterative perturbation theory* (IPT). Eq. (3.70) has been proven to be a well-behaved approximation of the *symmetric* SIAM $n = 1$ [55]. That transfers to the IPT of the Hubbard problem:

$n = 1$: $M_\sigma^{(m)}$; $m = 0, 1, 2, 3$ correct

                 atomic limit $(\Delta = 0)$ exact

                 Luttinger theorem $(T = 0,\ \rho_\sigma(\mu) = \rho_0(\mu_0))$ fulfilled

                 Mott-Hubbard transition well described

The comparison to numerically exact QMC calculations is indeed very convincing. But all these advantages are lost for $n \neq 1$. In particular, the strong coupling behaviour, decisive for ferromagnetism, is violated.

6   Kajueter-Kotliar Approach (KK)

To correct the shortcomings of the IPT for $n \neq 1$ an interpolating ansatz for the SIAM selfenergy has been proposed [56]

$$\Sigma_{d\sigma}(E) = U n_{d-\sigma} + \frac{a_\sigma \Sigma_{d\sigma}^{\text{SOC}}(E)}{1 - b_\sigma \Sigma_{d\sigma}^{\text{SOC}}(E)} \tag{3.73}$$

For the symmetric case $(n = 1)$ this ansatz reduces to the IPT. The second order perturbational contribution $\Sigma_{d\sigma}^{\text{SOC}}(E)$ (3.71) contains the Hartree-Fock spectral density:

$$S_{d\sigma}^{(1)}(E) = -\frac{1}{\pi} \text{Im} \frac{\hbar}{E + \eta_\sigma - \epsilon_d - \Delta_\sigma(E) - U n_{d-\sigma}} \tag{3.74}$$

To describe ferromagnetism the *hybridization function* must get a formal spin-dependence $(\Delta \to \Delta_\sigma(E))$. An additional parameter η_σ has been introduced (cf. E_σ in the EHA (3.48)) to fulfill the Luttinger sum rule for $d = \infty$ [32]

$$\eta_\sigma : \quad \mu = \mu|_{U=0} + \Sigma_\sigma(0) \tag{3.75}$$

a_σ and b_σ in (3.73) are fixed to fulfill the spectral moment $M_{d\sigma}^{(2)}$ and the exact $\Delta_\sigma = 0$-limit ($\widehat{=}$ zero bandwidth limit) for all particle densities n. Note that $M_{d\sigma}^{(0)}$ and $M_{d\sigma}^{(1)}$ are automatically fulfilled by the ansatz (3.73) for arbitrary a_σ, b_σ. The evaluation yields:

$$a_\sigma = \frac{n_{d-\sigma}(1 - n_{d-\sigma})}{n_{d-\sigma}^{(1)}\left(1 - n_{d-\sigma}^{(1)}\right)} \tag{3.76}$$

$$b_\sigma = \frac{U(1 - 2n_{d-\sigma}) + \eta_\sigma - \mu}{U^2 n_{d-\sigma}^{(1)}\left(1 - n_{d-\sigma}^{(1)}\right)} \tag{3.77}$$

$n_{d-\sigma}^{(1)}$ is calculated with (3.74) and the spectral theorem (2.20).

The disadvantage of this approach, especially with respect to ferromagnetism is the incorrect result for the important fourth ($m=3$) spectral moment. The true SIAM-band shift $B_{d\sigma}$,

$$n_{d-\sigma}(1-n_{d-\sigma})(B_{d-\sigma}-\epsilon_d) = \sum_{\mathbf{k}} V_{\mathbf{k}d} \langle c^+_{\mathbf{k}-\sigma} c_{d-\sigma}(2n_{d\sigma}-1)\rangle \tag{3.78}$$

is replaced by its Hartree-Fock version (cf. (3.52) for EHA, IAA):

$$n_{d-\sigma}^{(1)}\left(1-n_{d-\sigma}^{(1)}\right)\left(B_{d-\sigma}^{(1)}-\epsilon_d\right) = \left(2n_{d\sigma}^{(1)}-1\right)\sum_{\mathbf{k}} V_{\mathbf{k}d} \langle c^+_{\mathbf{k}-\sigma} c_{d-\sigma}\rangle^{(1)} \tag{3.79}$$

These expressions are related to the corresponding definitions (2.41) and (3.52), respectively, by means of the selfconsistent mapping (Sect. III C 4). Similarly to (2.45) $B_{d-\sigma}$ can exactly be expressed by the impurity terms $\Sigma_{d\sigma}(E)$ and $G_{d\sigma}(E)$ [57]:

$$n_{d-\sigma}(1-n_{d-\sigma})(B_{d-\sigma}-\epsilon_d) =$$
$$-\frac{1}{\pi\hbar}\text{Im}\int dE f(E)\Delta_{-\sigma}(E)\left(\frac{2}{U}\Sigma_{d-\sigma}(E)-1\right)G_{d-\sigma}(E) \tag{3.80}$$

Because of

$$B_{d-\sigma}^{(1)} \not\equiv B_{d-\sigma} \tag{3.81}$$

the spectral moment $M_{d\sigma}^{(3)}$ is violated.

7 Modified Perturbation Theory (MPT)

To remove the mentioned shortcomings of the KK-approach it is proposed in ref. [57] to use the moments $M_{d\sigma}^{(2)}$ and $M_{d\sigma}^{(3)}$ to fix the coefficients a_σ, b_σ in the selfenergy ansatz . This yields the same a_σ as in (3.76), while b_σ now explicitly contains the spin-dependent band shift:

$$b_\sigma = \frac{B_{d-\sigma}-\mu-B_{d-\sigma}^{(1)}+\eta_\sigma+U(1-2n_{d-\sigma})}{U^2 n_{d-\sigma}^{(1)}\left(1-n_{d-\sigma}^{(1)}\right)} \tag{3.82}$$

Unlike the condition (3.75) of the KK-approach η_σ is here fixed to get:

$$n_{d-\sigma}^{(1)} \stackrel{!}{=} n_{d-\sigma} \tag{3.83}$$

(cf. (3.49) for the interpolation methods EHA, IAA) The Luttinger theorem used in (3.75) only holds for $T=0$. For finite temperature the KK-requirement seems to be inappropriate because of prefixing a pure $T = 0$ theory. It can be shown [58] that for $T = 0$ the Luttinger sum rule is fulfilled even with the boundary condition (3.83) as long as the band filling is not too far away from $n = 1$.

There are a lot of aspects which confirm the concept of the *modified perturbation theory* (MPT):

1. weak-coupling behaviour correct up to U^2.

2. Fermi liquid behaviour for all n.

3. Luttinger theorem $(T = 0)$ fulfilled for n not too far away from 1.

4. $n = 1$: MPT $\equiv$ KK $\equiv$ IPT.

5. Kondo-resonance for low temperatures.

6. $W \to 0$ limit exact for all n.

7. Strong-coupling behaviour correct.

8. Spontaneous bandferromagnetism possible!

Two inadequacies should be mentioned. The first is that the exact solution of the $d = \infty$ Falicov-Kimball model (3.35) cannot be reproduced by the MPT. The second concerns the ansatz (3.73) being more or less *ad hoc*.

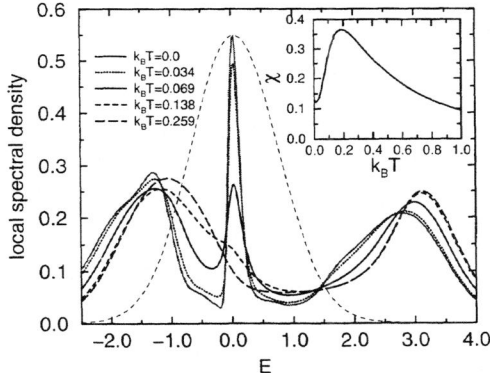

FIGURE 3.9. Spectral density $\frac{1}{\hbar}S_{ii\sigma}(E) = \rho_\sigma(E+\mu)$ for hypercubic lattice (2.108), $n = 0.94$ and $U = 4$ (energy unit $t^* = 1/\sqrt{2}$ (2.109)) as function of the energy and for various temperatures T. Results are derived within the MPT [58]. Thin dashed line: $\rho_0(E+\mu)$, $U = 0$. Inset: static susceptibility χ as function of the temperature.

Fig. 3.9 shows as a typical example of the MPT the local spectral density $\frac{1}{\hbar}S_{ii\sigma}(E) = \rho_\sigma(E+\mu)$ [58] for $n = 0.94$ and various temperatures for an hypercubic lattice described by the density of states (2.108). The energy units are chosen so that $t^* = \frac{1}{\sqrt{2}}$ (2.109). The two charge excitation peaks (*Hubbard bands*) are clearly recognizable, the dominant feature is, however, the Q-DOS resonance at the chemical potential μ. It is interesting to observe that with decreasing temperature the resonance starts growing as soon as the static susceptibility

$$\chi = \frac{\partial}{\partial B}(n_\uparrow - n_\downarrow)|_{B \to 0} \qquad (3.84)$$

significantly differs from a Curie-law behaviour. This indicates a quenching of the local magnetic moment by itinerant electrons at the Fermi edge. The resonance can therefore be interpreted as Kondo-type resonance. For a temperature T, for which χ exhibits the $\frac{1}{T}$-behaviour (e. g. $k_B T = 0.259$ in Fig. 3.9), no Kondo resonance appears at the Fermi energy. On the other hand, the Kondo resonance increases with decreasing temperature reaching at $T = 0$ the value of the *free* density of states. This is in accordance with the Luttinger sum rule (3.75), which can be rewritten to

$$(d = \infty) \qquad S_{ii\sigma}(0) \stackrel{!}{=} S_{ii\sigma}^{(0)}(0) \qquad (3.85)$$

Let us finally come back to the question of ferromagnetism in the Hubbard model. For all approaches discussed in this section ferromagnetism is impossible for the hc-lattice used in Fig. 3.9. So we take, as in Figs. 3.7 and 3.8, the fcc-type $d = \infty$-density of states (3.46) with the energy scaling (3.47). In Fig. 3.10 MPT-results for the band-filling dependence of the Curie temperature T_C are compared to those of the KK-approach. As in the MAA/AA and SDA/HI cases of Fig. 3.7 and for IAA/EHA in Fig. 3.8 the correct incorporation of the band shift $B_{-\sigma}$ by fulfilling the $m = 3$ spectral moment obviously leads to a stronger ferromagnetic stability. Since the KK approach does not fully neglect the band shift, but incorporating it in the Hartree-Fock version (3.79), the differences between MPT and KK are not so drastic as between SDA and HI, and in particular between MAA and AA (Fig. 3.7). They are comparable to those between IAA and EHA (Fig. 3.8).

We have used in Figs. 3.7, 3.8, and 3.10 the filling-dependence of the Curie temperatures to compare the predictions of several analytical approaches for the Hubbard model with the quasi-exact QMC result [46]. It is known from the Stoner solution (3.6) that it overestimates T_C by one or even more orders of magnitude. The simplest method, which recovers the correct strong-coupling behaviour in the sense of Harris and Lange [22], the SDA (Sect. III A 2), substantially improves upon the Stoner result. Its shortcoming is the neglect of damping effects. These are taken into account by the MAA (Sect. III B 2) that represents an improvement with respect to the QMC results. As will be demonstrated in the concluding section damping effects are still underestimated in the MAA, $T_C(n)$ is systematically too high. Obviously the influence of quasiparticle damping is much more realistically taken into account by IAA and MPT. The latter yields the best analytical approach to the QMC results.

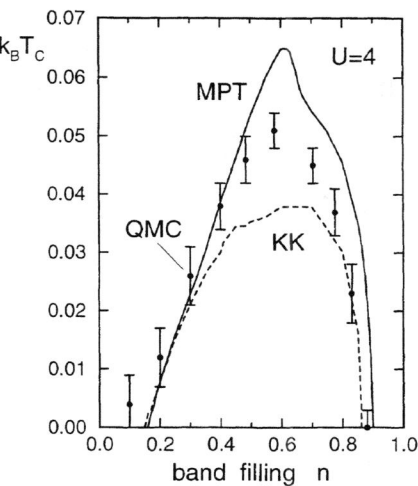

FIGURE 3.10. $T_C(n)$ calculated within the MPT (solid line) and the KK approach (broken line) compared with the QMC results of ref. [46] for fcc-type lattice (3.46). n refers to the concentration of holes in the energy band

To each of the just mentioned theories there exists a counterpart (SDA ↔ HI; MAA ↔ AA; IAA ↔ EHA; MPT ↔ KK) which differs by a violation of the fourth spectral moment (2.52), while all other exactly solvable limiting cases are commonly fulfilled. The comparison with the QMC results stresses the importance of the spin-dependent band shift $B_{\mathbf{k}\sigma}$ (2.39) for a correct description of bandferromagnetism in the Hubbard model.

We close our considerations of ferromagnetism in itinerant moment systems with some comparing remarks to the four presented methods SDA, MAA, IAA, MPT which altogether reproduce the strong-coupling behaviour incorporating the spin-dependent band shift. The different quality of their results has mainly to be ascribed to a different quality of treating the quasiparticle damping.

D   Concluding Remarks

We finally inspect the two decisive components for bandferromagnetism in the Hubbard model

a) quasiparticle damping

b) spin-dependent band shift

Damping effects are due to the imaginary part of the selfenergy. Fig. 3.11 shows a typical example calculated for the hypercubic density of states (2.108) (paramagnetism). In any case the total weight is given by

$$\int_{-\infty}^{+\infty} \mathrm{Im}\Sigma_\sigma(E) dE = -\pi U^2 n_{-\sigma}(1 - n_{-\sigma}) \qquad (3.86)$$

This can be derived from the high-energy expansion (2.56). The pole of the SDA selfenergy (2.65) at

$$E = B_{-\sigma} - U(1 - n_{-\sigma}) - \mu \qquad (3.87)$$

gives rise to a single δ-peak in $\mathrm{Im}\Sigma_\sigma(E)$. It has, however, no influence since it falls into the Hubbard gap The quasiparticles are stable. The main motivation to construct the MAA was to introduce quasiparticle damping. A careful inspection of $\mathrm{Im}\Sigma_\sigma(E)$ shows that it is unequal zero just in those regions with a finite Q-DOS. However, there is an additional δ-peak in the gap which takes away a great part of the weight (3.86). The δ-peak does not contribute to the quasiparticle damping which is therefore still underestimated in the MAA, probably responsible for the too high T_C-values (Fig. 3.7). The dampings in the IAA and MPT are of the same order of magnitude. As mentioned the Fermi-liquid behaviour is reproduced by the IAA only in a restricted region of the $U - n$ plane. The parameters in Fig. 3.11 are so that $\mathrm{Im}\Sigma_\sigma(E = 0) \neq 0$ at $T = 0$, violating therewith (2.93).

Contrary, the MPT selfenergy vanishes quadratically at $E = 0$ reproducing therewith the correct Fermi-liquid behaviour at $T = 0$. The dip in $\mathrm{Im}\Sigma_\sigma(E = 0)$ at $E \approx 0$ in Fig. 3.11 is reminiscent to the correct $T = 0$-behaviour (2.93).

The spin-dependent band shift $B_{-\sigma}$ shows a strong U-dependence (Fig. 3.12), running, however, into a saturation for large U. The qualitative behaviour is again for all the presented analytical methods almost the same. The quantitative differences refer directly to the predicted T_C-values in Figs. 3.7, 3.8, 3.10, confirming once more that the ferromagnetic stability is decisively influenced by the band shift.

Fig. 3.12 contains also the Hartree-Fock versions $B_{-\sigma}^{\mathrm{HF}}$ of the band shift for IAA and MPT. Eq. (3.52) fixes $B_{-\sigma}^{(1)}$ for the IAA. Because of the boundary condition (3.49) $n_{-\sigma}^{(1)}$ as well as $\langle c_{i-\sigma}^+ c_{j-\sigma} \rangle^{(1)}$ are independent of U. For given β, n $B_{-\sigma}^{(1)}$ is therefore a constant in IAA. Because of (3.83) there is no U-dependence of $n_{d-\sigma}^{(1)}$ in MPT, either. The hybridization function $\Delta_{-\sigma}$ in (3.80), however, is U-dependent

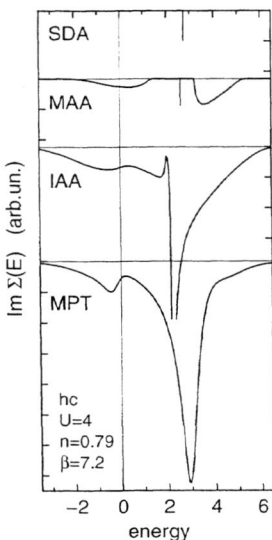

FIGURE 3.11. Imaginary part of the selfenergy as function of energy for the hc lattice (2.108) at $U = 4$, $n = 0.79$, $\beta = 7.2$. Energy unit $t^* = 1/\sqrt{2}$ (2.109). Horizontal thin lines indicate $\text{Im}\Sigma_\sigma = 0$. Vertical bars (SDA, MAA) for δ-peaks. [48]

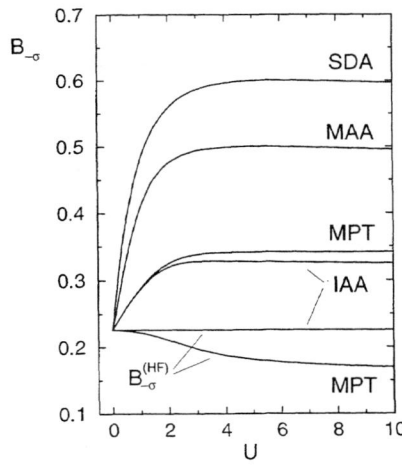

FIGURE 3.12. U dependence of the bandshift $B_{-\sigma}$ for the hc lattice (β, n and t^* as in Fig. 3.11). For IAA and MPT the Hartree-Fock version $B_{-\sigma}^{\text{HF}}$ ($= B_{-\sigma}^{(1)}$ in (3.52) and $= B_{d-\sigma}^{(1)}$ in (3.79)) is also plotted. [48]

due to the selfconsistency process in the DMFT. This transfers to the second factor in (3.79).

The correct taking into account of the quasiparticle damping by the various methods manifests itself in the quasiparticle density of states (Fig. 3.13). The qualitative features (*Hubbard splitting, exchange splitting*) are accounted for by all presented methods. The missing damping is responsible for the fact that in the SDA the quasiparticle subbands densities of states are still very similar to the *free* DOS ((3.46) in Fig. 3.13), while the MPT is closest to the exact QMC results.

Let us finally look at the temperature dependencies of the magnetization and the static susceptibility (Fig. 3.14). The QMC data can be fitted to an $S = \frac{1}{2}$ Brillouin function. Extrapolation to $T = 0$ yields saturation $m = n$. Saturation is also found by SDA and MAA, MPT predicts a slightly lower magnetization. For the chosen hole-concentration ($n = 0.58$) SDA, MAA as well as MPT predict second-order phase transitions at $T = T_C$. However, first-order transition, are also observed. For $n > n_{\max}$, where $n_{\max}$ is the density for which T_C achieves its maximum (Fig. 3.7), discontinuous transitions appear in SDA and MAA. The critical density in the MPT is given by the kink in the $T_C(n)$-curve (Fig. 3.10). The IAA situation in Fig. 3.14 is somewhat special because $n = 0.58$ is close to the quantum-critical

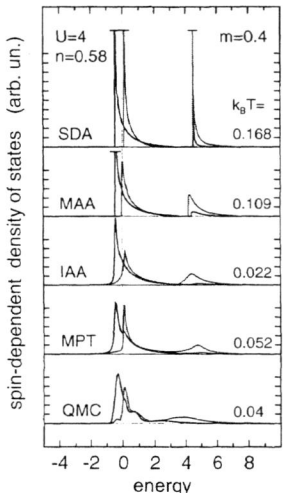
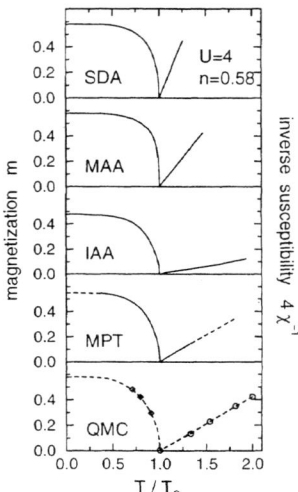

FIGURE 3.13. Q-DOS $\rho_\sigma(E)$ for the fcc-type lattice (3.46) with $U = 4$ and $n = 0.58$ (energy unit (3.47)). Solid line: majority spin σ. Dotted lines: minority spin. QMC result from [46]. Temperatures T chosen so that for all presented methods a magnetisation of 0.4 is found.

FIGURE 3.14. Magnetisation m as function of the reduced temperature $\frac{T}{T_C}$ for the fcc-type lattice (3.46) with U, n and t^* as in 3.13. Bottom figure: QMC results [46], dashed line: $S = \frac{1}{2}$-Brillouin function. For $\frac{T}{T_C} > 1$ the inverse static susceptibility χ^{-1} is shown as function of $\frac{T}{T_C}$.

point $n_c^{IAA} = 0.61$ (Fig. 3.8). So there is a certain deviation from the Brillouin function behaviour. Phase transitions in IAA are always continuous. Whether or not the first order transitions observed for SDA, MAA and MPT are artefacts of the approximations or inherent properties of the Hubbard model remains an open question.

The static susceptibility χ (3.84) obeys the Curie-law for high temperatures:

$$\chi \approx \frac{C}{k_B (T - \Theta)} \tag{3.88}$$

Except for IAA the paramagnetic Curie temperature Θ coincides with T_C. The QMC value for the Curie constant $C = 0.47$ has to be compared to 0.42 (SDA), 0.52 (MAA), 0.50 (IAA), and 0.57 (MPT). Note that the different shapes in Fig. 3.14 are simply due to the different Curie temperatures in the $m - \frac{T}{T_C}$ curves!

We have demonstrated in this section that spontaneous ferromagnetism in the Hubbard model does exist depending on

lattice structure
band occupation
Coulomb coupling

temperature.

We found that two aspects are decisive:

> spin-dependent band
> shift quasiparticle damping

A series of analytical approaches (SDA $\rightarrow$ MAA$\rightarrow$ IAA, MPT) was motivated by an increasing accuracy with respect to these two aspects.

IV FERROMAGNETISM IN LOCAL MOMENT SYSTEMS

There are remarkable differences in the theoretical description of bandferromagnetism, which we discussed up to now, and the so-called *local-moment ferromagnetism* which is prototypically realized in the metallic rare earths Gd, Tb and Dy [59] and the insulating EuO and EuS [60]. As we have already discussed in Sect. I D the essential features of *local-moment systems* can be traced back to an intimate correlation between strictly localized *magnetic* (f) electrons and *quasi-free* band (s) electrons. It is commonly accepted that the so-called s-f model [9], which we introduced by eq. (1.56), represents a good starting point for an at least qualitative understanding of the physical properties of these materials. The scientific interest in the s-f model has recently got a striking upsurge with the discovery of the *colossal magnetoresistance* of the manganite perovskites $T_{1-x}\, D_x\, MnO_3$ where T is a trivalent lanthanide cation (e.g. La) and D is a divalent alkaline-earth (e.g. Ca, Sr, Ba) cation [61]. In this connection the s-f model is frequently called *ferromagnetic Kondo-lattice model* or *double-exchange model*. The additive *ferromagnetic* points to a positive exchange coupling constant J in the interaction H_{sf} (1.63), which favours the parallel (*ferromagnetic*) alignment of the conduction electron spin and the localized spin. The *normal* Kondo-lattice model is nothing than the s-f model with $J < 0$. It is considered as candidate for the extraordinary physics of *heavy fermion systems* ($CeCu_{6-x}Au_x$, $CeSi_x$) [62]. Here we are exclusively interested in the above-mentioned *local moment* semiconductors and metals, which are characterized by a ferromagnetic s-f exchange $J > 0$.

As for the Hubbard model in Sect. III almost all information we need can be extracted from the single-electron Green function (2.4). Its equation of motion reads

$$\sum_m (E\delta_{im} - T_{im})G_{mj\sigma}(E) = \hbar\delta_{ij} - \frac{1}{2}J\hbar\left(z_\sigma I_{ii,j\sigma}(E) + F_{ii,j\sigma}(E)\right) \quad (4.1)$$

The two types of interaction terms in (1.63) let appear in (4.1) the *spinflip function* $F_{im,j\sigma}(E)$ and the *Ising function* $I_{im,j\sigma}(E)$:

$$F_{im,j\sigma}(E) = \langle\langle S_i^{-\sigma} c_{m-\sigma}\, ;\, c_{j\sigma}^+ \rangle\rangle_E \quad (4.2)$$

$$I_{im,j\sigma}(E) = \langle\langle S_i^z c_{m\sigma}\, ;\, c_{j\sigma}^+ \rangle\rangle_E \quad (4.3)$$

For brevity we have written (cf. 2.77):

$$S_i^\uparrow \equiv S_i^+ \;;\; S_i^\downarrow \equiv S_i^- \;;\; z_\uparrow = +1 \;;\; z_\downarrow = -1 \quad (4.4)$$

The *higher* Green functions on the right-hand side of (4.1) prevent a direct solution of the equation of motion. However, there exists a very illustrative and rather realistic limiting case which can be treated rigorously [63,64]. It concerns the situation of a ferromagnetically saturated semiconductor, i. e. a single electron in an otherwise empty conduction band exchange coupled to a fully parallel aligned f spin system (Sect. IV A 2). Furthermore, the zero-bandwidth limit is rigorously accessible. We start the discussion with this limiting case.

A Exactly Solvable Limiting Cases

1 Zero-Bandwidth Limit

As in Sect. II B we assume that the conduction band is shrinked to an N-fold degenerate level T_0:

$$T_{ij} \to T_0 \delta_{ij} \quad ; \quad \epsilon(\mathbf{k}) \to T_0 \; \forall \, \mathbf{k}$$

Nevertheless, we consider the f-spin system as collectively ordered for $T < T_c$ by any direct or indirect exchange interaction. The s-f model Hamiltonian reads in this limit as follows:

$$\begin{aligned} H_{W=0} = &\, (T_0 - \mu) \sum_{i,\sigma} n_{i\sigma} + \frac{1}{2} U \sum_{i,\sigma} n_{i\sigma} n_{i-\sigma} \\ &- \frac{1}{2} J\hbar \sum_{i,\sigma} \left(z_\sigma S_i^z n_{i\sigma} + S_i^\sigma c_{i-\sigma}^+ c_{i\sigma} \right) \end{aligned} \quad (4.5)$$

For purposes which become clear at a later stage we have additionally introduced here a correlation among the conduction electrons in form of the Hubbard interaction (1.85). The full equation of motion reads (cf. (4.1)):

$$\begin{aligned} (E - T_0) G_{ii\sigma}(E) = &\, \hbar - \frac{1}{2} J\hbar \left(z_\sigma I_{ii,j\sigma}(E) + F_{ii,i\sigma}(E) \right) \\ &+ U\Gamma_{iii,i\sigma}(E) \end{aligned} \quad (4.6)$$

The *higher* Green function $\Gamma_{imn,j\sigma}(E)$ is already known to us from the discussion of the Hubbard model, being defined in (2.26). In the next step we have to construct the equations of motion of the *higher* functions I, F and Γ. After a troublesome, but straightforward calculation the hierarchy of equations of motion decouples exactly and can rigorously be solved [65]. The resulting energy levels are given in 4.1. The original Bloch-level T_0 is split by the s-f interaction into four quasiparticle levels. The lower two belong to the situation where our *test electron* enters an empty site, the upper two for the case where it jumps to a place which is already preoccupied by another electron of opposite spin. That is the reason why we switched on the Hubbard interaction because it signals the appearance of double occupancies. Note, however, that even for $U = 0$ the s-f-interaction produces a splitting into four not coinciding quasiparticle levels.

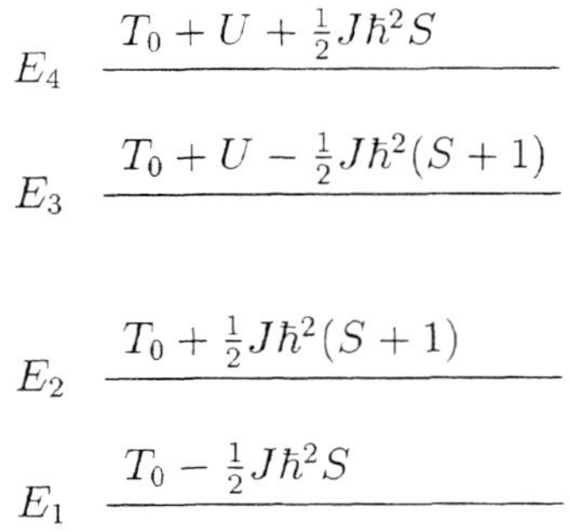

E_4 $\quad \dfrac{T_0 + U + \tfrac{1}{2}J\hbar^2 S}{}$

E_3 $\quad \dfrac{T_0 + U - \tfrac{1}{2}J\hbar^2(S+1)}{}$

E_2 $\quad \dfrac{T_0 + \tfrac{1}{2}J\hbar^2(S+1)}{}$

E_1 $\quad \dfrac{T_0 - \tfrac{1}{2}J\hbar^2 S}{}$

FIGURE 4.1. Energy levels of the zero-bandwidth s-f model with Hubbard interaction

Each of the four energies E_i is connected with a spectral weight $z_{i\sigma}$, which turns out to be strongly particle concentration- and temperature dependent.

$$z_{1\sigma} = \frac{1}{2S+1} \{S + 1 + z_\sigma \langle S^z \rangle + \Delta_{-\sigma} - (S+1)n_{-\sigma}\} \tag{4.7}$$

$$z_{2\sigma} = \frac{1}{2S+1} \{S - z_\sigma \langle S^z \rangle - \Delta_{-\sigma} - Sn_{-\sigma}\} \tag{4.8}$$

$$z_{3\sigma} = \frac{1}{2S+1} \{Sn_{-\sigma} - \Delta_{-\sigma}\} \tag{4.9}$$

$$z_{4\sigma} = \frac{1}{2S+1} \{\Delta_{-\sigma} + (S+1)n_{-\sigma}\} \tag{4.10}$$

For a complete solution one needs the average occupation number $n_{-\sigma}$ and the mixed correlation function $\Delta_{-\sigma}$:

$$\Delta_\sigma = \langle S^\sigma c^+_{-\sigma} c_\sigma \rangle + z_\sigma \langle S^z n_\sigma \rangle \tag{4.11}$$

This can selfconsistently be done with the spectral theorem for the Green functions $G_{ii\sigma}(E)$, $I_{ii,i\sigma}(E)$ and $F_{ii,i\sigma}(E)$. The resulting spectral weights are plotted in Fig. 4.2 as functions of temperature T and band occupation n. Only the up-spin weights are given, the down-spin weights follow by particle-hole symmetry:

$$z_{1\sigma}(T,n) = z_{4-\sigma}(T, 2-n)$$
$$z_{2\sigma}(T,n) = z_{3-\sigma}(T, 2-n)$$

The temperature-dependence can be traced back to that of the 4f magnetization $\langle S^z \rangle$. In Fig. 4.2 we have used an $S = \tfrac{7}{2}$-Brillouin function.

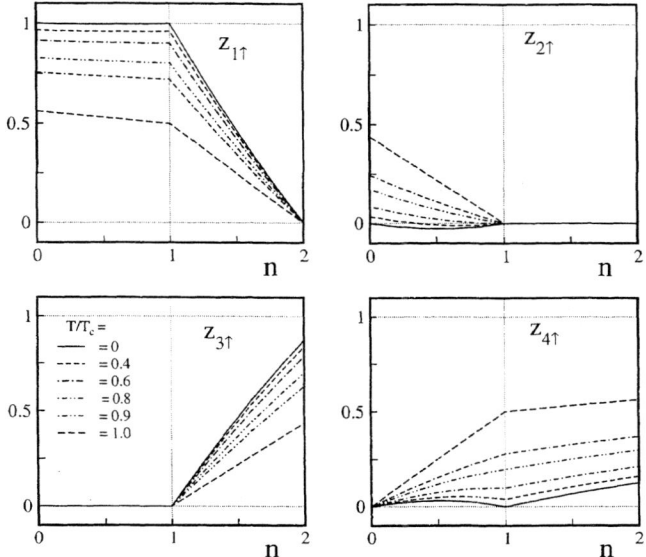

FIGURE 4.2. Spectral weights $z_{i\uparrow}$ ($i = 1, \ldots, 4$) of the quasiparticle levels E_i of the zero-bandwidth s-f model for various temperatures T as function of band occupation n. ($S = \frac{7}{2}$)

For all parameter constellations there exist at most three levels with non-vanishing spectral weights, at least one of the E_i has a vanishing weight. There are a lot of special cases which cause further zero-weights, as, e.g., $T = 0$ and/or $n = 0, 1, 2$.

An interesting special feature should be mentioned. In Fig. 4.2 one recognizes that the spectral weight $z_{2\uparrow}$ becomes slightly negative for sufficiently low temperatures, although spectral weights by definition have to be positive. The reason for this discrepancy is the following: For the discussion of the zero-band width limit we assumed a ferromagnetically ordered f-spin system. The 4f-magnetization $\langle S^z \rangle$, however, has not selfconsistently be calculated, but rather represented by an $S = \frac{7}{2}$-Brillouin function. The spin exchange with the conduction electrons, however, prevents $\langle S^z \rangle$ from reaching its maximum value $\hbar S$. The assumption $\langle S^z \rangle \to \hbar S$ therefore must lead to a contradiction resulting in a negative spectral weight $z_{2\uparrow}$. For $n < 1$ one finds the following upper bound [65]:

$$\langle S^z \rangle \leq \hbar \frac{S(S+1-n)}{S+1} \qquad (4.12)$$

As for the Hubbard model we can conclude from the exact zero-bandwidth limit that at least in the strong-coupling region ($J \gg W$) a three peak structure of the spectral density has to be expected. In the special case of an empty *band* ($n = 0$: semiconductor, insulator) only the lower two levels ($E_{1,2}$) survive.

2 Ferromagnetically Saturated Semiconductor

There is another very instructive limiting case that can be treated rigorously:

$$n = 0 \; ; \; T = 0 \iff \langle S^z \rangle = \hbar S \tag{4.13}$$

By the single-electron Green function we observe a single electron in an empty energy band interacting with a totally aligned localized spin system. For this special case the Ising function (4.3) simplifies to

$$I_{im,j\sigma}(E) \to \hbar S \cdot G_{mj\sigma}(E) \tag{4.14}$$

For solving the equation of motion (4.1) we can therefore concentrate ourselves on the spinflip function (4.2), the structure of which differs strongly for $\sigma = \uparrow$ and $\sigma = \downarrow$.

(a) ↑-spectrum

One easily realizes that the spinflip function becomes trivial for a ↑-electron in the limit (4.13):

$$F_{ii,j\uparrow}^{(0,0)}(E) = \langle\langle S_i^- c_{i\downarrow}; c_{j\uparrow}^+ \rangle\rangle = 0 \tag{4.15}$$

The physical reason is clear. A ↑-electron has no chance to exchange its spin with the parallel aligned local moment system. The ↑-spectrum is therefore very simple, the selfenergy is a (**k**,E)-independent constant:

$$\Sigma_{\mathbf{k}\uparrow}^{(T=0,n=0)}(E) \equiv -\frac{1}{2}J\hbar^2 S \tag{4.16}$$

The quasiparticle density of states is only rigidly shifted compared to the *free* Bloch-density of states:

$$\rho_\uparrow^{(0,0)}(E) = \rho_0(E + \frac{1}{2}J\hbar^2 S) \tag{4.17}$$

The quasiparticle dispersion is undeformed with respect to the Bloch energies:

$$E_\uparrow(\mathbf{k}) = \epsilon(\mathbf{k}) - \frac{1}{2}J\hbar^2 S \tag{4.18}$$

(b) ↓-spectrum

More complicated and more interesting is the $\downarrow$-spectrum, because a $\downarrow$-electron has several possibilities to exchange its spin with the antiparallel f spins. The spinflip function $F_{ii,j\downarrow}^{(0,0)}(E)$ is therefore not at all trivial. The spinflip terms in the s-f exchange will drastically modify the quasiparticle spectrum.

To proceed one can write down the equation of motion of $F_{im,j\downarrow}^{(0,0)}$. There appear as usual a lot of *higher* Green functions. However, all of them can be traced back to *lower* functions, so that the hierarchy of equations of motions decouples exactly. In particular the following Green functions come into play:

$$\langle\langle S_i^+ n_{m\downarrow} c_{m\uparrow} ; c_{j\downarrow}^+ \rangle\rangle \xrightarrow[n=0]{} 0$$

$$\langle\langle S_i^+ S_m^z c_{m\uparrow} ; c_{j\downarrow}^+ \rangle\rangle \xrightarrow[n=0,T=0]{} \hbar(S - \delta_{im}) F_{im,j\downarrow}^{(0,0)}$$

$$\langle\langle S_i^+ S_m^- c_{m\downarrow} ; c_{j\downarrow}^+ \rangle\rangle \xrightarrow[n=0,T=0]{} 2\hbar^2 S \delta_{im} G_{ij\downarrow}^{(0,0)}$$

$$\langle\langle (n_{i\uparrow} - n_{i\downarrow}) S_i^+ c_{m\downarrow} ; c_{j\downarrow}^+ \rangle\rangle \xrightarrow[n=0]{} 0$$

$$\langle\langle S_i^z c_{i\uparrow}^+ c_{i\downarrow} c_{m\uparrow} ; c_{j\downarrow}^+ \rangle\rangle \xrightarrow[n=0]{} 0$$

It remains as equation of motion:

$$\left(E + \frac{1}{2} J\hbar^2 (S - \delta_{im}) \right) F_{im,j\downarrow}^{(0,0)}(E) = \sum_k T_{mk} F_{ik,j\downarrow}^{(0,0)}(E)$$
$$- J\hbar^3 S \delta_{im} G_{ij\downarrow}^{(0,0)}(E) \qquad (4.19)$$

Together with (4.1) and (4.14) we have a closed system of equations which can be solved by Fourier-transformation for $G_{\mathbf{k}\downarrow}^{(0,0)}(E)$ [21]. It results in the following selfenergy:

$$\Sigma_\downarrow^{(0,0)}(E) = \frac{1}{2} J\hbar^2 S \left(1 + \frac{J\hbar G_\uparrow(E)}{1 - \frac{1}{2} J\hbar G_\uparrow(E)} \right) \qquad (4.20)$$

$$G_\uparrow(E) = \frac{\hbar}{N} \sum_\mathbf{k} (E - \epsilon(\mathbf{k}) + \frac{1}{2} J\hbar^2 S)^{-1} \qquad (4.21)$$

The selfenergy turns out to be wave-vector independent because we have neglected any direct exchange and therefore suppressed spin wave dispersions for the local moment system ($\hbar\omega(\mathbf{q}) \equiv 0$). Spin exchange processes between s- and f-subsystem are naturally accompanied by magnon-emission or -absorption what normally would give rise to a wave-vector dependence of the electronic selfenergy.

For very weak couplings the first term in (4.20) dominates and the quasiparticle dispersion is only rigidly shifted compared to the free dispersion

$$S^{(0,0)}_{\mathbf{k}\downarrow}(E) \approx \hbar\delta(E - \epsilon(\mathbf{k}) - \frac{1}{2}J\hbar^2 S) \qquad (4.22)$$

This is just the $T = 0$-mean field result (1.64). However, already for moderate coupling strengths the second term in (4.20) strongly contributes, and that as the consequence of two different types of spin exchange processes. First, the original $\downarrow$-electron can emit a magnon thereby reversing its spin. This leads to a rather broad structure of *scattering states*. Magnon emission is only possible if the $\downarrow$-electron finds a $\uparrow$-state after spinflip. It is therefore to expect that *scattering states* occupy the same energy region as the $\uparrow$-QDOS (4.17). As a second spin exchange pro-

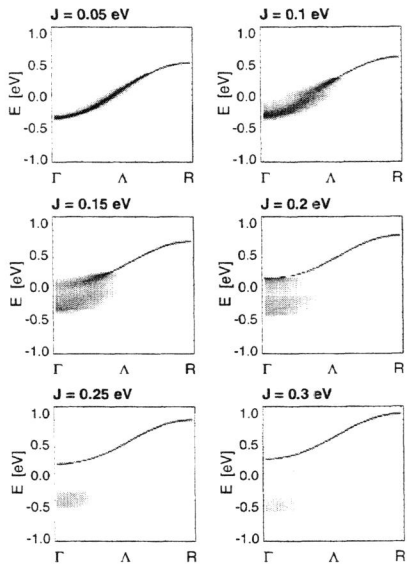

FIGURE 4.3. Quasiparticle bandstructure as density plot of the respective $\downarrow$-spectral density for a ferromagnetically saturated semiconductor (T=0, n=0) and for various exchange couplings J ($\hbar = 1$). Parameters: sc-tight-binding B-DOS, W=1eV, $S = \frac{7}{2}$

cess the $\downarrow$-electron can polarize its immediate f-spin neighbourhood by repeatedly emitting and reabsorbing a magnon. The propagating electron may be considered as being *dressed* by a cloud of virtual magnons. The corresponding quasiparticle is called the *magnetic polaron*. It manifests itself as a pole in the $\downarrow$-Green function leading to a high-energy peak in the spectral density. It can be shown that as long as the polaron peak falls outside the scattering region the magnetic polaron has an infinite lifetime. When it dips into the scattering part the polaron may decay after a certain time into a $\uparrow$-electron and a magnon. Fig. 4.3 presents a density plot of the $\downarrow$-spectral density for various exchange couplings. The degree of blackening scales with the magnitude of the peaks. The plot therefore exhibits a quasiparticle dispersion including damping effects. One recognizes already for a very moderate coupling of $\hbar^2 J/W = 0.1$ the sharp polaron peak at high energies and the somewhat washed out scattering region. The polaron with **k** near the R-point is stable while that near the Γ-point has only a finite lifetime. For $\hbar^2 J/W \geq 0.2$ polaron part and scattering part are separated. Evidently the distribution of spectral weight

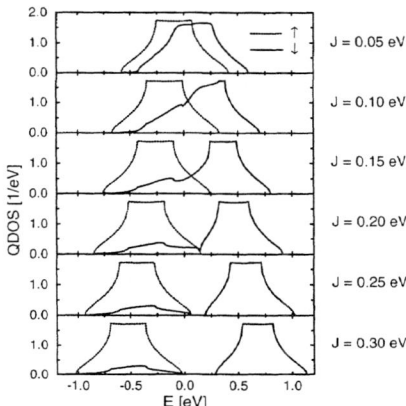

FIGURE 4.4. Quasiparticle density of states as function of energy for various exchange couplings J. Parameters for the ferromagnetically saturated semiconductor as in Fig. 4.3

between polaron and scattering states is strongly wave-vector dependent.

Fig. 4.4 shows the quasiparticle density of states for various exchange couplings J. One recognizes the rigid shift of $\rho_\uparrow^{(0,0)}(E)$ proportional to JS (4.17). Correlation effects appear only in the down-spin Q-DOS. For weak couplings only slight deformations appear besides a *Stoner-like* shift. The first term in (4.20) still dominates. However, already for a moderate coupling of $\hbar^2 J/W \approx 0.2$ a quasiparticle band splitting sets in. The lower subband consists of scattering states, consequently coinciding with the up-spin band ($\hbar\omega(\mathbf{q}) \equiv 0$). The upper subband consists of pure and stable polaron states. With increasing J the gap steadily widens. This band splitting looks like the *Hubbard splitting* discussed in Sect. III (Fig. 3.3), is, however, based on a completely different interaction mechanism. Since we have discussed an exactly solvable limiting case, free of any uncontrollable approximation, the features of Fig. 4.3 and Fig. 4.4 must be regarded as true consequences of the s-f exchange interaction.

However, for finite temperatures and arbitrary band occupations the equation of motion (4.1) is not exactly solvable. Approximations are unavoidable. The required approach has to reveal the mutual influence of localized f-spin states and extended band electron states. We start with a discussion of the moment-subsystem. Note that the model-Hamiltonian (1.56) does not contain a direct exchange interaction between the f-spins. The coupling must therefore be of indirect nature via the s-f interaction H_{sf} (1.65). In a simple version it is the RKKY-interaction discussed in Sect. I D 2.

B Localized Moment System

The main goal is to express the f-magnetization $\langle S_z \rangle$ and pure f-spin correlations such as $\langle S_i^{\pm} S_i^{\mp} \rangle$, $\langle (S_i^z)^2 \rangle$, ... in terms of the conduction electron selfenergy $\Sigma_{\mathbf{k}\sigma}(E)$.

1 Modified RKKY-interaction

The idea is to map the s-f exchange H_{sf} on an effective spin-operator of Heisenberg-type:

$$H_f = -\sum_{i,j} \hat{J}_{ij} \mathbf{S}_i \cdot \mathbf{S}_j \tag{4.23}$$

For this purpose we use the following form of the s-f interaction, completely equivalent to (1.59), (1.63) or (1.65):

$$H_{sf} = -J \frac{\hbar}{N} \sum_{i\sigma\sigma'} \sum_{\mathbf{k},\mathbf{q}} e^{i\mathbf{q}\mathbf{R}_i} (\mathbf{S}_i \cdot \hat{\sigma})_{\sigma\sigma'} c^+_{\mathbf{k}+\mathbf{q}\sigma} c_{\mathbf{k}\sigma'} \tag{4.24}$$

The components of the band electron spin operator $\hat{\sigma}$ are the Pauli spin matrices. We transform H_{sf} into an *effective* spin Hamiltonian by averaging out the conduction electron degrees of freedom, symbolized by $\langle \ldots \rangle^{(s)}$:

$$H_{sf} \to \langle H_{sf} \rangle^{(s)} \equiv H_f \tag{4.25}$$

Averaging only in the conduction electron subspace means that $\langle H_{sf} \rangle^{(s)}$ retains operator character in the f-spin subspace. According to (4.24) we have to calculate the expectation value

$$\langle c^+_{\mathbf{k}+\mathbf{q}\sigma} c_{\mathbf{k}\sigma'} \rangle^{(s)}$$

which does not necessarily vanish for $\mathbf{q} \neq 0$ and for $\sigma \neq \sigma'$ as does the corresponding expectation value in the full Hilbert space. For the determination of the above term we introduce the *restricted* Green function

$$\hat{G}^{\sigma\sigma'}_{\mathbf{k},\mathbf{k}+\mathbf{q}}(E) \equiv \langle\langle c_{\mathbf{k}\sigma} ; c^+_{\mathbf{k}+\mathbf{q}\sigma'} \rangle\rangle^{(s)}_E \tag{4.26}$$

which is formally defined as in (2.4), (2.5), only the averaging has to be done in the s-subspace, while the time-dependence of the Heisenberg operators in (2.4) is due to the full Hamiltonian. The equation of motion reads:

$$(E + \mu - \epsilon(\mathbf{k})) \, \hat{G}^{\sigma\sigma'}_{\mathbf{k},\mathbf{k}+\mathbf{q}}(E) = \hbar \, \delta_{\mathbf{q},0} \, \delta_{\sigma\sigma'} - \qquad (4.27)$$
$$- \hbar J \frac{1}{N} \sum_{i\mathbf{k}'\sigma''} e^{-i(\mathbf{k}-\mathbf{k}')\mathbf{R}_i} \, (\mathbf{S}_i \cdot \hat{\sigma})_{\sigma'\sigma''} \, \hat{G}^{\sigma''\sigma}_{\mathbf{k}',\mathbf{k}+\mathbf{q}}(E)$$

Obviously, this equation can be iterated up to any desired accuracy producing spin products of the type

$$(\mathbf{S}_i \cdot \hat{\sigma})_{\sigma'\sigma''} (\mathbf{S}_j \cdot \hat{\sigma})_{\sigma''\sigma'''} (\mathbf{S}_k \cdot \hat{\sigma})_{\sigma'''\sigma''''}$$

By use of the *free* Green function

$$G^{(0)}_{\mathbf{k}}(E) = \frac{\hbar}{E + \mu - \epsilon(\mathbf{k})} \qquad (4.28)$$

eq. (4.27) can be written as:

$$\hat{G}^{\sigma\sigma'}_{\mathbf{k},\mathbf{k}+\mathbf{q}}(E) = \delta_{\sigma\sigma'} \, \delta_{\mathbf{q},0} \, G^{(0)}_{\mathbf{k}}(E) - \qquad (4.29)$$
$$- J \frac{1}{N} \sum_{i\mathbf{k}'\sigma''} e^{-i(\mathbf{k}-\mathbf{k}')\mathbf{R}_i} \, G^{(0)}_{\mathbf{k}}(E) \, (\mathbf{S}_i \cdot \hat{\sigma})_{\sigma'\sigma''} \, \hat{G}^{\sigma''\sigma}_{\mathbf{k}',\mathbf{k}+\mathbf{q}}(E)$$

This equation is still exact. The first order approximation on the r.h.s.,

$$\hat{G}^{\sigma''\sigma}_{\mathbf{k}',\mathbf{k}+\mathbf{q}}(E) \to \delta_{\sigma''\sigma} \, \delta_{\mathbf{k}',\mathbf{k}+\mathbf{q}} \, G^{(0)}_{\mathbf{k}+\mathbf{q}}(E) \qquad (4.30)$$

leads to the following expression:

$$\left[\hat{G}^{\sigma'\sigma}_{\mathbf{k},\mathbf{k}+\mathbf{q}}(E)\right]^{(1)} = \delta_{\sigma\sigma'} \, \delta_{\mathbf{q},0} \, G^{(0)}_{\mathbf{k}}(E) - \qquad (4.31)$$
$$- J \frac{1}{N} \sum_{i} e^{i\mathbf{q}\mathbf{R}_i} \, G^{(0)}_{\mathbf{k}}(E) \, (\mathbf{S}_i \cdot \hat{\sigma})_{\sigma'\sigma} \, G^{(0)}_{\mathbf{k}+\mathbf{q}}(E)$$

Assuming that the spectral theorem [21] holds in the *restricted* Hilbert space, too, we get for the required expectation value:

$$\frac{1}{N} \sum_{\mathbf{k}} \langle c^+_{\mathbf{k}+\mathbf{q}\sigma} c_{\mathbf{k}\sigma'} \rangle^{(s)} \approx \frac{1}{2} \delta_{\sigma\sigma'} \, \delta_{\mathbf{q},0} \cdot n - \qquad (4.32)$$
$$- J \frac{1}{N} \sum_{i} e^{i\mathbf{q}\mathbf{R}_i} \, D^{(1)}_{\mathbf{q}} \, (\mathbf{S}_i \cdot \hat{\sigma})_{\sigma'\sigma}$$

Here we have defined:

$$D^{(1)}_{\mathbf{q}} = -\frac{1}{\pi} \operatorname{Im} \int_{-\infty}^{+\infty} dE \, f_-(E) \frac{1}{N\hbar} \sum_{\mathbf{k}} G^{(0)}_{\mathbf{k}}(E - \mu) \, G^{(0)}_{\mathbf{k}+\mathbf{q}}(E - \mu) \qquad (4.33)$$

$f_-(E)$ is defined in (2.21). By inserting (4.32) into (4.25) we indeed achieve the operator form (4.23) with the following effective exchange integrals:

$$\hat{J}_{ij}^{(1)} = \frac{1}{N} \sum_{\mathbf{q}} \hat{J}^{(1)}(\mathbf{q}) \, e^{-i\mathbf{q}(\mathbf{R}_i - \mathbf{R}_j)}$$

$$\hat{J}^{(1)}(\mathbf{q}) = -\frac{1}{2} J^2 D_{\mathbf{q}}^{(1)} \qquad (4.34)$$

This is nothing else than the *conventional* RKKY result (1.73) for finite temperatures:

$$\hat{J}^{(1)}(\mathbf{q}) = -\frac{1}{2N} J^2 \hbar^2 \sum_{\mathbf{k}} \frac{f_-(\epsilon(\mathbf{k}+\mathbf{q})) - f_-(\epsilon(\mathbf{k}))}{\epsilon(\mathbf{k}+\mathbf{q}) - \epsilon(\mathbf{k})} \qquad (4.35)$$

The temperature-dependence is of course only very weak. We have seen in Sect. I D 2 that (4.35) is the result of a second order perturbation theory, which starts from an unpolarized conduction electron gas. In order to incorporate to higher order the exchange-induced polarization of the itinerant conduction electrons, one might think to modify the approximation (4.30) by replacing on the right-hand side the free by the full, all polarization processes containing Green function:

$$\hat{G}_{\mathbf{k}',\mathbf{k}+\mathbf{q}}^{\sigma''\sigma}(E) \to \delta_{\sigma''\sigma} \, \delta_{\mathbf{k}',\mathbf{k}+\mathbf{q}} \, G_{\mathbf{k}+\mathbf{q}\sigma}(E) \qquad (4.36)$$

The correct weak-coupling behaviour up to J^2-terms is exactly retained by this ansatz. Instead of (4.34) we have now:

$$\hat{J}(\mathbf{q}) = -\frac{1}{4} J^2 \sum_{\sigma} D_{\mathbf{q}\sigma} \qquad (4.37)$$

$$D_{\mathbf{q}\sigma} = -\frac{1}{\pi} \operatorname{Im} \int_{-\infty}^{+\infty} dE \, f_-(E) \frac{1}{N\hbar} \sum_{\mathbf{k}} G_{\mathbf{k}}^{(0)}(E - \mu) \, G_{\mathbf{k}+\mathbf{q}\sigma}(E - \mu) \qquad (4.38)$$

After the mapping of the s-f interaction on the effective Heisenberg-Hamiltonian (4.23) the exchange integrals $\hat{J}(\mathbf{q})$ are now via $G_{\mathbf{k}\sigma}$ functionals of the conduction electron selfenergy therewith eventually getting a distinct (n,T)-dependence. A proposal how to determine approximatively the selfenergy is offered in Sect. IV C. The selfenergy itself will depend on f-spin correlation functions which we have to derive by use of the effective Hamiltonian (4.23). The effective exchange integrals $\hat{J}(\mathbf{q})$ must therefore be calculated selfconsistently.

2 Magnon Energies and Spin Correlations

To get the magnetic properties of the local-moment system, mediated by the s-f exchange coupling to the band electrons, we use the following spin-Green function first proposed by Callen [66]

$$P_{ij}^{(a)}(E) = \langle\langle S_i^+ \,;\, e^{aS_j^z} S_j^- \rangle\rangle_E \tag{4.39}$$

a is any real parameter. The equation of motion of this function reads:

$$E\, P_{ij}^{(a)}(E) = 2\hbar^2 \delta_{ij}\langle A_a\rangle - 2\hbar \sum_m \hat{J}_{im} \star \tag{4.40}$$

$$\star \langle\langle (S_m^+ S_i^z - S_i^+ S_m^z) \,;\, e^{aS_j^z} S_j^- \rangle\rangle_E$$

For brevity we have written

$$\langle A_a \rangle = \langle\, [\, S_i^+ \,,\, e^{aS_i^z} S_i^- \,]_- \,\rangle \tag{4.41}$$

It is well-known [4] that a simple RPA-decoupling yields in the low- as well as in the high-temperature region surprisingly reasonable results. Doing so we get straightforwardly after Fourier-transformation:

$$P_{\mathbf{q}}^{(a)}(E) = \frac{2\hbar^2 \langle A_a\rangle}{E - E(\mathbf{q}) + i0^+} \tag{4.42}$$

$E(\mathbf{q})$ are *dressed* magnon energies:

$$E(\mathbf{q}) = 2\hbar \langle S^z\rangle (\hat{J}_0 - \hat{J}(\mathbf{q})) \tag{4.43}$$

$$\hat{J}_0 = \sum_i \hat{J}_{im} = \sum_m \hat{J}_{im} = \hat{J}(\mathbf{q}=0) \tag{4.44}$$

In determining the f-magnetization for arbitrary spin S we follow the method of ref. [66] that results in

$$\langle S^z \rangle = \hbar \frac{(1+S+\varphi)\varphi^{2S+1} + (S-\varphi)(1+\varphi)^{2S+1}}{(1+\varphi)^{2S+1} - \varphi^{2S+1}} \tag{4.45}$$

φ can be interpreted as average magnon number:

$$\varphi(S) = \frac{1}{N}\sum_{\mathbf{q}} \frac{1}{\exp(\beta E(\mathbf{q})) - 1} \tag{4.46}$$

Because of (4.43) φ depends on the spin value S. It determines a lot of other local spin correlations,

$$\langle S_i^- S_i^+\rangle = 2\hbar \langle S^z\rangle \varphi(S) \tag{4.47}$$
$$\langle (S_i^z)^2\rangle = \hbar^2 S(S+1) - \hbar\langle S^z\rangle(1+2\varphi(S)) \tag{4.48}$$
$$\langle (S_i^z)^3\rangle = \hbar^3 S(S+1)\varphi(S) + \hbar^2 \langle S^z\rangle(S(S+1) + \varphi(S))$$
$$\qquad - \hbar\langle (S^z)^2\rangle(1+3\varphi(S))\,, \tag{4.49}$$

to mention only some typical examples.

An important key-quantity of magnetism is the Curie-temperature T_c. For

$$T \to T_c^{(-)} \; ; \; \langle S^z \rangle \to 0^+$$

it can be read off from (4.45):

$$k_B T_c = \frac{2}{3}\hbar^2 S(S+1)\left[\frac{1}{N}\sum_{\mathbf{q}}(\hat{J}_0 - \hat{J}(\mathbf{q}))_{T_c}^{-1}\right]^{-1} \quad (4.50)$$

The effective exchange integrals are temperature-dependent and have to be used in (4.50) for $T \to T_c$. As we have seen in Sect. IV B 1 they are mainly influenced by the electronic selfenergy we are going to discuss in the next section.

C Electronic Selfenergy

In Sect. IV A 2 we have presented an exactly solvable but nevertheless non-trivial limiting case of the s-f model, the results of which provide already a good insight into what is to be expected for finite temperatures and arbitrary band occupations. However, for the general case the equation of motion (4.1) is no longer exactly solvable. Approximations concern the equations of motion of the two *higher* Green functions (4.2) and (4.3). More precisely, they have to simplify Green-functions following from

$$\langle\langle S_i^{-\sigma}[c_{p-\sigma}, H_{sf}]_- \; ; \; c_{j\sigma}^+ \rangle\rangle_E$$
$$\langle\langle [S_i^{-\sigma}, H_{sf}]_- c_{p-\sigma} \; ; \; c_{j\sigma}^+ \rangle\rangle_E$$
$$\langle\langle S_i^z[c_{p\sigma}, H_{sf}]_- \; ; \; c_{j\sigma}^+ \rangle\rangle_E$$
$$\langle\langle [S_i^z, H_{sf}]_- c_{p\sigma} \; ; \; c_{j\sigma}^+ \rangle\rangle_E \quad (4.51)$$

Let us exemplify the approach restricting the presentation to the main steps, which are vital for the understanding and the judgement of the subsequent applications. The Green functions are approximated in two steps [67]. The strong intraatomic exchange correlations require a special treatment of diagonal ($i = p$) terms.

(a) *Diagonal ($i = p$) terms*

We first evaluate explicitly the commutators in (4.51). A typical example of the resulting Green functions is

$$X_{ij\sigma}(E) = \langle\langle S_i^{-\sigma} S_i^\sigma c_{i\sigma} \; ; \; c_{j\sigma}^+ \rangle\rangle_E \quad (4.52)$$

which for some relevant special cases can rigorously be expressed by the *lower* functions $G_{ij\sigma}(E)$ and $I_{ii,j\sigma}(E)$ (4.3). So it holds for $S = \frac{1}{2}$ for all temperatures and arbitrary band fillings:

$$X_{ij\sigma}(E) = \frac{\hbar}{2}G_{ij\sigma}(E) - z_\sigma I_{ii,j\sigma}(E) \quad (4.53)$$

The same function $X_{ij\sigma}(E)$ for arbitrary spin S, but for ferromagnetic saturation $\langle S^z \rangle = \hbar S$ reads:

$$X_{ij\sigma}(E) = \hbar S G_{ij\sigma}(E) - z_\sigma I_{ii,j\sigma}(E) \tag{4.54}$$

By these exact limiting cases the following ansatz for the very general situation appears to be plausible:

$$X_{ij\sigma}(E) = \alpha_\sigma G_{ij\sigma}(E) + \beta_\sigma I_{ii,j\sigma}(E) \tag{4.55}$$

All the three Green functions in this ansatz are of type $\langle\langle A; B \rangle\rangle_E$. The respective spectral moments (cf. (2.47))

$$M_{AB}^{(n)} = \langle ((i\hbar \frac{\partial}{\partial t})^n [A(t), B(t')]_+ \rangle_{t=t'} \tag{4.56}$$

can be calculated independently of $\langle\langle A; B \rangle\rangle_E$. The equivalent relationship

$$M_{AB}^{(n)} = -\frac{1}{\pi\hbar} \int_{-\infty}^{+\infty} dE\, E^n \, \text{Im}\langle\langle A; B \rangle\rangle_{E+i0^+} \tag{4.57}$$

allows to fix the parameters in (4.55). In the same manner all the other *diagonal* Green functions coming out of (4.51) are to be elaborated.

(b) *Off-diagonal ($i \neq p$) terms*

Starting point is the definition-equation of the selfenergy (cf. (2.9)) in site-representation:

$$\langle\langle [c_{p\sigma}, H_{sf}]_- ; c_{j\sigma}^+ \rangle\rangle_E = \sum_r \Sigma_{pr\sigma}(E) \langle\langle c_{r\sigma} ; c_{j\sigma}^+ \rangle\rangle_E \tag{4.58}$$

This is formally equivalent to the replacement

$$[c_{p\sigma}, H_{sf}]_- \to \sum_r \Sigma_{pr\sigma}(E)\, c_{r\sigma} \tag{4.59}$$

The two Green functions in (4.58) have the same single-particle poles differing only by the spectral weights of these poles, and the equality in (4.58) is ensured by the selfenergy components $\Sigma_{pr\sigma}(E)$. If we apply the same considerations to the two functions

$$\langle\langle S_i^{-\sigma}[c_{p-\sigma}, H_{sf}]_- ; c_{j\sigma}^+ \rangle\rangle_E \; ; \; \langle\langle S_i^{-\sigma} c_{r-\sigma} ; c_{j\sigma}^+ \rangle\rangle_E$$

then we recognize again that they must have the same pole structure. Because of the additional spinflip operator $S_i^{-\sigma}$ only those single-electron excitations appear as poles of the Green functions which are accompanied by a spin exchange. However, since $S_i^{-\sigma}$ enters both functions, being otherwise the only difference, we propose to use the formal identity (4.59) for the following ansatz:

$$\langle\langle S_i^{-\sigma}[c_{p-\sigma}, H_{sf}]_-\, ;\, c_{j\sigma}^+ \rangle\rangle_E \stackrel{i \neq p}{=} \sum_r \Sigma_{pr-\sigma}(E) \langle\langle S_i^{-\sigma} c_{r-\sigma}\, ;\, c_{j\sigma}^+ \rangle\rangle_E \qquad (4.60)$$

On the r.h.s. appears the spinflip function $F_{ir,j\sigma}(E)$ (4.2), while the selfenergy components $\Sigma_{pr-\sigma}(E)$ have to be determined self-consistently.

An analogous ansatz is used for the third function in (4.51):

$$\langle\langle S_i^z[c_{p\sigma}, H_{sf}]_-\, ;\, c_{j\sigma}^+ \rangle\rangle_E \stackrel{i \neq p}{=} \sum_r \Sigma_{pr\sigma}(E) I_{ir,j\sigma}(E) \qquad (4.61)$$

This *higher* Green function is now expressed by the *Ising function* (4.3).

For the evaluation of the second and the fourth Green function in (4.51) we use the mapping (4.23) on the effective Heisenberg model:

$$[S_i^\sigma, H_{sf}]_- \to 2z_\sigma \hbar \sum_m \hat{J}_{im}(S_m^z S_i^\sigma - S_i^z S_m^\sigma) \qquad (4.62)$$

$$[S_i^z, H_{sf}]_- \to \hbar \sum_m \hat{J}_{im}(S_i^+ S_m^- - S_i^- S_m^+) \qquad (4.63)$$

Performing the same RPA procedure as in Sect. IV B 2 for the f spin system we get

$$\langle\langle [S_i^{-\sigma}, H_{sf}]_- c_{p-\sigma}\, ;\, c_{j\sigma}^+ \rangle\rangle_E \stackrel{i \neq p}{\approx} 2\hbar z_\sigma \langle S^z \rangle \sum_m \hat{J}_{im}\Big(F_{mp,j\sigma}(E) -$$
$$- F_{ip,j\sigma}(E)\Big) \qquad (4.64)$$

$$\langle\langle [S_i^z, H_{sf}]_- c_{p\sigma}\, ;\, c_{j\sigma}^+ \rangle\rangle_E \stackrel{i \neq p}{\approx} 2\hbar \langle S^z \rangle \hat{J}_{ii} G_{pj\sigma}(E) \qquad (4.65)$$

This completes the procedure of evaluating the *higher order* Green functions in (4.51). We eventually arrive at a closed system of equations ending up in an implicit equation for the selfenergy:

$$\Sigma_{\mathbf{k}\sigma}(E) = -\frac{1}{2} J\hbar z_\sigma \langle S^z \rangle + \frac{1}{4} J^2 K_{\mathbf{k}\sigma}(E) \qquad (4.66)$$

The first term, which is exact in the weak coupling limit, represents an induced *Stoner splitting* of the energy band proportional to the f magnetization $\langle S^z \rangle$ in lowest order responsible for the famous *red shift effect* (Sect. I D 1). The second term is dominated by the consequences of spin exchange processes between itinerant (s-) conduction electrons and localized f-moments. These exchange processes have been elaborated in detail for the exactly solvable ($T=0, n=0$)-special case in Sect. IV A 2. $K_{\mathbf{k}\sigma}(E)$ is a complicated functional of the selfenergy for both spin directions. It must be solved by iteration. The full expression for $K_{\mathbf{k}\sigma}(E)$ is given in ref. [67]. It incorporates several types of spin correlations:

(a) *mixed* spin correlations

$$\Delta_\sigma = \langle S_i^z n_{i\sigma}\rangle = -\frac{1}{\pi\hbar}\int_{-\infty}^{+\infty} dE\, f_-(E)\,\mathrm{Im}\,I_{ii,i\sigma}(E-\mu) \tag{4.67}$$

$$\gamma_\sigma = \langle S_i^{-\sigma} c_{i\sigma}^+ c_{i-\sigma}\rangle = -\frac{1}{\pi\hbar}\int_{-\infty}^{+\infty} dE\, f_-(E)\,\mathrm{Im}\,F_{ii,i\sigma}(E-\mu) \tag{4.68}$$

(b) f-spin correlations

$$\langle S_i^z\rangle,\ \langle S_i^\pm S_i^\mp\rangle,\ \langle (S_i^z)^2\rangle,\ \ldots \tag{4.69}$$

As demonstrated, the *mixed* (s-f) spin correlations can be expressed by the Green functions (4.2) and (4.3). The f-spin correlations are derived in Sect. IV B 2 given by eqs. (4.45) to (4.49) as functionals of the electronic selfenergy. Eventually, we have constructed a closed system of equations that can be solved selfconsistently for all quantities of interest.

D Magnetism and Electronic Structure

Let us inspect some typical model results, which are obtained for a s.c. lattice (tight-binding Bloch density of states, bandwidth $W = 1\mathrm{eV}$) and for an f-spin value $S = \frac{7}{2}$. Only paramagnetism and ferromagnetism are considered so that translational symmetry of the lattice can be exploited.

1 Electronic Structure

For a realistic s-f coupling $J = 0.2\mathrm{eV}$ and an electron density $n = 0.2$ the self-consistent solution of our model predicts ferromagnetism with a Curie temperature $T_c = 238K$. The temperature-dependence of the spectral density (angle- and spin resolved photoemission!) is plotted in Fig. 4.5 for wave-vectors from the ΓR direction. The density representation refers to the quasiparticle bandstructure. In spite of the fact that the itinerant electrons are *a priori* uncorrelated, the induced quasiparticle effects are rather striking. The physical background, however, is exactly the same as discussed with the special case in Sect. IV A 2. For low temperatures with the f magnetization close to saturation the up-spin spectrum appears simple being only rigidly shifted compared to the *free* Bloch dispersion. The finite band occupation obviously does not alter very much the exact result (4.18). The down-spin spectrum consists of the sharp polaron dispersion and a broad low-energy region of scattering states. The distribution of spectral weight between the two excitations is remarkably k-dependent.

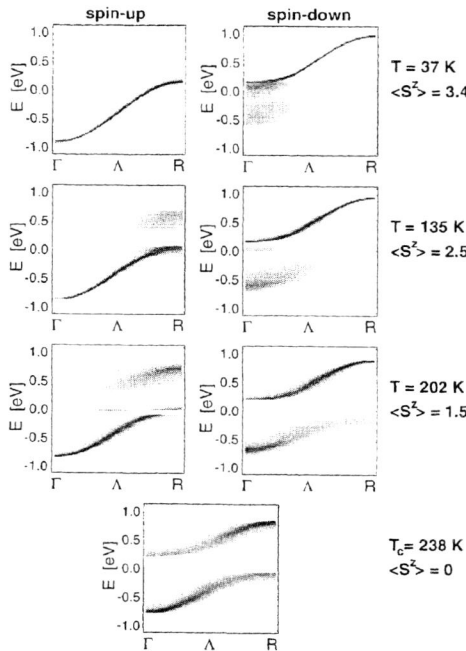

FIGURE 4.5. Spectral density (quasiparticle bandstructure) as a function of wave-vector for four different temperatures. The degree of blackening is a measure of the magnitude of the spectral density. Parameters: $W = 1\text{eV}$, $J = 0.2\text{eV}$, $n = 0.2$, $S = \frac{7}{2}$. Reprinted from Ref. [67], with permission from IOP Publishing Limited.

With increasing temperature the ↑-spectrum becomes more complicated due to induced correlation effects. The reason is that magnon emission by a ↓-electron is completely equivalent to magnon absorption by an ↑-electron. However, the latter has a precondition because magnons have to exist. This is not the case in ferromagnetic saturation ($\langle S^z \rangle = \hbar S$) what makes the ↑-spectrum then relatively simple. With increasing temperature magnons are excited and can be absorbed by the ↑-electron. Therefore scattering and polaron states will appear in the ↑-spectrum, too. For finite temperatures the low energy peak of the spectral density $S_{\mathbf{k}\sigma}$ corresponds to two elementary processes. The excited σ-electron can retain its spin parallel to the f-spin when entering the local frame. The other possibility is that the itinerant electron first flips its spin due to magnon emission or absorption in order to enter the local frame parallel to the f spin. At finite temperature the probability to orientate the spin parallel to the f-spin is finite for both the σ- and the $(-\sigma)$-electron. The high energy branch is in any case due to magnetic polaron formation with or without preceding magnon emission. Because of the normalization of the spectral density each part of the spectrum is connected with a spectral weight which strongly changes with temperature and carries a striking

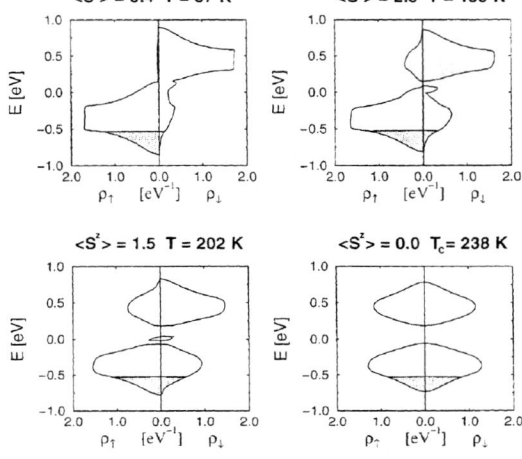

FIGURE 4.6. Quasiparticle density of states of a sc-lattice as a function of energy for four different temperatures. Parameters as in Fig. 4.5. The shaded region gives an impression of the band occupation.
Reprinted from Ref. [67], with permission from IOP Publishing Limited.

wave-vector dependence.

The two elementary processes, magnon emission / absorption and magnetic polaron formation, manifest themselves in the Q-DOS as two quasiparticle subbands. In ferromagnetic saturation, or close to it, the ↑-band is almost identical to the *free* Bloch band only rigidly shifted in agreement with the respective spectral density behaviour. The upper part of the ↓-Q-DOS for low temperatures is built up by stable polaron states, the lower part consists of scattering states. With increasing temperature a correlation-caused gap opens, which even persists above T_c. This striking correlation effect is induced by the interband exchange coupling from highly correlated f electrons into *a priori* uncorrelated s electrons.

2 Magnetic properties

The spin polarization of the conduction electrons induced by the exchange coupling to the local f-moment system leads to an indirect coupling between the localized spins (4.23). In lowest order this is the well-known RKKY-interaction (Sect. I D 2). The resulting magnetic properties are therefore strongly influenced by J and by the band filling n, which enter the effective exchange integrals $\hat{J}_{ij}$ (4.37). The $\hat{J}_{ij}$ are functionals of the s-electron selfenergy getting therewith a distinct temperature dependence which in turn is due to the magnetization state of the f-moment system. The magnetic properties ($\langle S^z \rangle, T_c$) are thus coming out from the same selfconsistent circle as the above-discussed electronic quasiparticle structure.

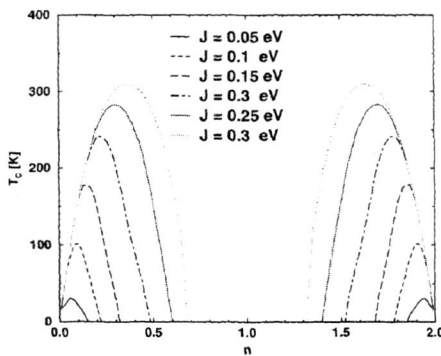

FIGURE 4.7. Curie temperature as a function of the band occupation n ($0 \leq n \leq 2$) for various values of the s-f exchange coupling J. Other parameters: $W = 1\text{eV}, S = \frac{7}{2}$, sc lattice.
Reprinted from Ref. [67], with permission from IOP Publishing Limited.

The Curie-temperature T_C is strongly influenced by the band filling n (Fig. 4.7). No ferromagnetism is possible around the half-filled band $n = 0$, probably because of antiferromagnetic correlations [68]. The Curie-temperature has a maximum at the quarter-filled ($n = 0.5$) and the three-quarter-filled band ($n = 1.5$). The n-interval distinctly increases with increasing J. It cannot be excluded that the RPA-decoupling in the equation of motion of $P_{ij}^{(a)}(E)$ in Sect. IV B 2 overestimates a little bit the stability of ferromagnetism. On the other hand, the calculated T_c-values are obviously in the right order of magnitude.

Since there is no direct exchange, a finite T_c is due to the s-f exchange coupling J. For weak couplings J the conventional RKKY mechanism works predicting for T_c a J^2-dependence. This is correctly reproduced by the present theory as can be seen in Fig. 4.8 (inset).

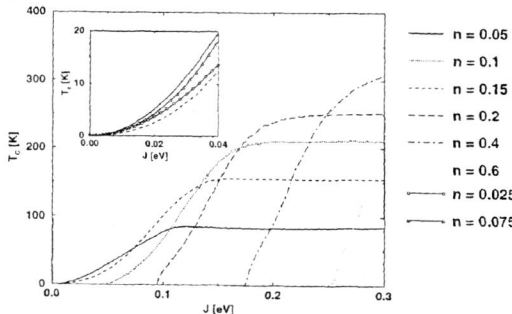

FIGURE 4.8. Curie temperature as a function of the band occupation n ($0 \leq n \leq 2$) for various values of the s-f exchange coupling J. Other parameters: $W = 1\text{eV}, S = \frac{7}{2}$, sc lattice.
Reprinted from Ref. [67], with permission from IOP Publishing Limited.

However, for higher band occupation ($n \geq 0.2$) J must first exceed a critical value (cf. Fig. 4.7) to allow ferromagnetism. Furthermore, with increasing J the critical temperature deviates more and more from the RKKY behaviour, finally even running into saturation. This is surely a physically reasonable result. As soon as the correlation gap is opened (Fig. 4.6) a further increase of J will no longer decisively change the magnetic behaviour of the local moment system, only the gap will be getting larger. The saturation value for T_c is the higher the bigger the particle density. In the case of lower band occupation the saturation is reached for weaker couplings J.

V CONCLUSIONS

The quantum theory of magnetism is covered by a set of rather different model descriptions. A unified theory is not yet available, and the strongly simplifying theoretical models remain mathematically so complicated that a rigorous treatment is in general excluded. Two questions are therefore to be answered when trying to explain the physical properties of a real ferromagnet: Which model is appropriate and how to find a trustworthy approximation to the not exactly solvable many-body problem.

To avoid misunderstandings a strict classification of magnetism can be helpful. Such a classification can exploit the two fundamental preconditions for collective magnetism: We need a solid and the solid must contain permanent magnetic moments. The solids can roughly be divided into insulators (semiconductors) and metals, and the permanent moments into localized and itinerant ones The four possibilities of combination provide a good classification scheme.

The best understood materials are the insulators with localized magnetic moments being excellently represented by the Heisenberg model (Sect. I B). While the operator structure of the Hamiltonian is unique, the interpretation of the exchange integrals is manifold. The direct exchange (Sect. I B 1) is conceptually clear but hard to be realized. More likely are indirect mechanism such as the superexchange (Sect. I B 2).

The striking class of the Mott-Hubbard insulators (NiO, CoO, FeO, ...) can be considered as insulators with itinerant (!) magnetic moments. The partially filled transition metal $3d$ band is split by strong electron correlations into several quasiparticle subbands, each of them either completely filled or empty. The result is an insulator with a permanent magnetic moment originating from a $1-2$ eV broad quasiparticle subband. The appropriate theoretical description is not yet clear. The Hubbard model can provide a qualitative picture of how electron correlations can lead to a splitting of an energy band into quasiparticle subbands (*Mott-Hubbard transition*). It is, however, strictly overcharged with respect to a quantitative fit of the rich variety of experimental data found for these materials.

In so-called *local-moment metals* electrical conductivity and spontaneous magnetization are provoked by two different groups of electrons. Many striking properties of such materials (*4f-systems*) can be traced back to an intimate exchange correlation between the localized *magnetic* electrons and the itinerant conduction electrons. A remarkable temperature dependence is induced into the band system by the coupling to the localized magnetic moment, first observed as "red shift" of the optical absorption edge for the $4f - 5d$ transition in EuO upon cooling below T_C. Furthermore, the mentioned exchange interaction brings about an indirect moment coupling via conduction electron polarization (*RKKY interaction*, Sect. I D 2). As long as such *cross effects* are meant one can believe in the *s-f model* (*Kondo-lattice, double-exchange model*) as a good starting point for an understanding of the respective phenomena.

The classical ferromagnets Fe, Co, Ni belong to the class of *bandferromagnets*, characterized by the existence of a partially filled, narrow energy band, which dominates simultaneously electrical conductivity as well as the spontaneous magnetization. Bandferromagnetism represents a rather involved, up to now not fully understood many-electron problem. The Hubbard model is considered, although surely oversimplified, a good framework. In spite of its simple structure the model Hamiltonian provokes an eigenvalue problem that could not be solved up to now.

The intensive investigation of the Hubbard model, in particular in the last decade, has brought to light a lot of rigorous details which yield already a good insight into the physics of the model. In addition they can be used to test the trustworthiness of unavoidable approximations. The zero-bandwidth limit (Sect. II B) gives a first clue on the quasiparticle structure of the energy excitation spectrum in the strongly coupled Hubbard model. Spontaneous ferromagnetism is, however, excluded for this exotic limit. Of decisive importance for the strong-coupling phenomenon *ferromagnetism* are the results of Harris and Lange [22] concerning the structure of the spectral density in the $U \gg W$-regime. The positions of the centres of gravity and the spectral weights of the two main peaks in the spectral density are known as well as the order of magnitude of the weights of the satellite peaks. The main peaks, located near T_0 and $T_0 + U$, are influenced by a higher correlation function $B_{\mathbf{k}-\sigma}$, which can transfer to the peak positions a direct spin-dependence favouring therewith spontaneous ferromagnetism. $B_{\mathbf{k}-\sigma}$ consist of a local part $B_{-\sigma}$, responsible for the mentioned spin-dependent shifts, and a non-local, wave-vector dependent part $F_{\mathbf{k}-\sigma}$, that can be identified as spin-dependent bandwidth correction. The term $B_{\mathbf{k}-\sigma}$ appears in the $m = 3$-spectral moment. It turns out that the fulfillment of the first four moments ($m = 0, 1, 2, 3$) is necessary for an approximate theory to guarantee the correct strong-coupling behaviour.

Exact statements are also possible for the weak-coupling regime (Sect. II G) by perturbational expansions with respect to $U/W \ll 1$. It is an interesting aspect to check how strongly a correct weak-coupling behaviour influences the possibility and the properties of bandferromagnetism.

Substantial progress in the investigation of the Hubbard model has been made since the discovery that for infinite lattice dimensions ($d \to \infty$) basic many-body functions such as the selfenergy are strictly local terms. This does not at all solve the problem but does open new ways to attack it. So it can be shown that the Hubbard model can be mapped on an effective impurity problem (e. g. *single-impurity Anderson model*). Numerically in principle exact quantum-Monte Carlo techniques for the evaluation of the impurity problem have been used to prove the possibility of bandferromagnetism in the Hubbard model [46].

Analytical approaches to the Hubbard problem are needed to reveal the microscopic mechanism that may lead to bandferromagnetism. The simplest meanfield ansatz, known as *Stoner model*, is in the last analysis based on the concept of an effective magnetic exchange field. Ferromagnetism is dramatically overestimated as a consequence of the total neglect of electron correlations. Curie temperatures come out too high by more than one order of magnitude. The first work which tried to include proper correlations was the pioneering *Hubbard*-I theory. The idea that electron correlations may provoke a splitting of the conventional Bloch-energy band into several quasiparticle subbands was borne (Sect. III A 1). In the meantime one knows that the Hubbard approach [18] violates the correct strong-coupling behaviour providing therewith an inconvenient basis for spontaneous ferromagnetism.

The spectral density in the *Hubbard*-I approach is a two-δ-peak function reproducing exactly the zero-bandwidth limit. One can use such a structure as an ansatz fitting then free parameters by exactly calculated spectral moments. This is the idea of the *spectral density approach (SDA)* [24,35,25] which gives access to a qualitatively convincing description of bandferromagnetism. By definition the fourth ($m = 3$) spectral moment is fulfilled so that the SDA exhibits the important strong-coupling behaviour. The comparison of the Hubbard-I results with those of the SDA impressively underlines the significance of the correlation $B_{\mathbf{k}-\sigma}$, in particular the *band shift* $B_{-\sigma}$, with respect to the possibility and stability of ferromagnetism in the Hubbard model. On the other side, a severe shortcoming of the SDA is the neglect of quasiparticle damping. It must be assumed that damping effects indeed do attack magnetic stability. The CPA-solution of the conventional zero-bandwidth alloy analogy [38], e. g. , yields a non-vanishing imaginary part of the selfenergy and therewith quasiparticle damping, but completely excludes ferromagnetism. It turns out, however, that the strong-coupling behaviour is incorrect, in particular the *band shift* $B_{-\sigma}$ does not appear. The logical extension is therefore the *modified alloy analogy (MAA)* (Sect. III B 2) which starts from the exact high-energy expansion of the Green function and the selfenergy by equating the first four moments. The attractive feature of the MAA is that it retains the advantages of the SDA, but includes in addition quasiparticle damping. Ferromagnetism becomes possible, but much more restricted than in the SDA.

The next question is whether or not the correct weak-coupling behaviour (Fermi liquid, Luttinger theorem, ...) will have some influence on the strong-coupling phenomenon *ferromagnetism*. This question is positively answered by the *interpolating alloy analogy (IAA)*, which uses an *ad hoc*-generalization of the conventional CPA. The latter is based on a proposal of Edwards and Hertz (EHA). IAA and EHA differ again by the appearance of the spin-dependent band shift $B_{-\sigma}$, absent in the EHA. The IAA fulfills a maximum number of exactly calculable limiting cases but suffers from the fact, that the starting-ansatz is by no means physically justifiable (Sect. III C 2). The same objection hits the *modified perturbation theory (MPT)* which, on the other side, is the best to fit the more or less exact *quantum-Monte-Carlo* calculations. It seems to be brought to a common agreement that ferromagnetism really does exist in the properly evaluated Hubbard model. It is therefore suitable at least for a qualitative description of bandferromagnetism. The essentials are already exhibited by the conceptually simple SDA, while the MPT obviously fits best the QMC results. The remaining task has to focus on an extension of the theory to the multiband situation of real bandferromagnets.

Local moment ferromagnets like the metallic rare earth element Gadolinium require another theoretical model than the bandferromagnets. It is commonly accepted that the s-f model can serve as a good frame for the electronic and magnetic properties of such materials. Recently other names for the same model have superfluously been invented, as e. g. *ferromagnetic Kondo-lattice model* or *double exchange model*. As in the Hubbard model a splitting of the conduction band into several quasiparticle subbands happens, but now as a consequence of the interband (s-f) exchange interaction. While the bandsplitting in the Hubbard model sets in when the intraatomic Coulomb interaction U exceeds the bandwidth W, the interband exchange J need to reach only 10 to 20% of W to produce a gap. Free of any uncontrollable approximation this can be demonstrated for a non-trivial limiting case (*ferromagnetically saturated semiconductor*). Furthermore, this special case helps to understand the physical elementary process which lead to the quasiparticle splittings. Typical is the formation of the *magnetic polaron* as a propagating electron dressed by a virtual cloud of magnons. Emitting or absorbing spin waves by the conduction electrons yields somewhat diffuse scattering states.

The s-f exchange interaction is, probably most important, also responsible for an indirect moment coupling. The effective exchange integrals are functionals of the band electron selfenergy, in the weak-coupling regime known as RKKY interaction ($\sim J^2$). For stronger couplings T_C as function of J deviates from the J^2-behaviour, finally running into saturation. Concluding it can be stated that the physics of the *local moment* ferromagnets is better understood than that of the bandferromagnets.

ACKNOWLEDGMENTS

It is a great pleasure to thank all the members of the *Lehrstuhl für Festkörperphysik* of the *Institut für Physik* at the Humboldt-University of Berlin for their substantial contributions to the work. Special thanks are due to Dr. M. Potthoff, who developed numerous ideas, and to C. Santos, P. Sinjukow, F. Lewin and M. Goetsch for carefully preparing the manuscript.

Financial support of the *Deutsche Forschungsgemeinschaft* within the *Sonderforschungsbereich 290 (Metallische dünnen Filmen: Struktur, Magnetismus und elektronische Eigenschaften)* is gratefully acknowledged.

REFERENCES

1. D. C. Mattis, *The Theory of Magnetism* volume 1, (Springer, Berlin) 1988.
2. W. Nolting, *Quantentheorie des Magnetismus*, volume 1, (Teubner, Stuttgart) 1986.
3. W. Heisenberg, Z. Phys. **38**, 441 (1926).
4. W. Nolting, *Quantentheorie des Magnetismus*, volume 2, (Teubner, Stuttgart) 1986.
5. W. Heitler and F. London, Z. Phys. **44**, 455 (1927).
6. P. W. Anderson, Phys. Rev. **79**, 950 (1950).
7. P. W. Anderson, *Magnetism* volume 1, G. T. Rado, H. Suhl (Academic Press) 1963.
8. B. Brandow, Adv. Phys. **26**, 651 (1977).
9. W. Nolting, phys. stat. sol. (b) **96**, 11 (1979).
10. B. Batlogg et al., Phys. Rev. B **12**, 3940 (1975).
11. N. Müller et al., Phys. Rev. B **29**, 1651 (1972).
12. T. Penney et al., Phys. Rev. B **5**, 3669 (1972).
13. C. Godart et al., J. Physique **41**, C5–263 (1980).
14. A. A. Rudermann and C. Kittel, Phys. Rev. **96**, 99 (1954).
15. T. Kasuya, Prog. Theor. Phys. **16**, 45 (1956).
16. K. Yosida, Phys. Rev. **106**, 893 (1957).
17. H. Capellmann, *Metallic Magnetism*, volume 42 of *Springer Topics in Current Physics*, (Springer, Berlin) 1987.
18. J. Hubbard, Proc. R. Soc. London A **276**, 238 (1963).
19. A. A. Abrikosov L. P. Gorkov and I. E. Dzyaloshinsky, *Method of Quantum Field Theory in Statistical Physics*, volume 42, (Prentice-Hall, New Jersey) 1964.
20. G. D. Mahan, *Many Particle Physics*, volume 42, (Plenum Press, New York) 1990.
21. W. Nolting, *Grundkurs Theoretische Physik*, 7 *Viel-Teilchen-Theorie*, (Plenum Press, Braunscweig) 1997.
22. A. B. Harris and R. V. Lange, Phys. Rev. **157**, 295 (1967).
23. Y. M. Vilk and A.-M. S. Tremblay, J. Physique I **7**, 1309 (1997).
24. W. Nolting and W. Borgiel, Phys. Rev. B **39**, 6962 (1989).
25. T. Hermann and W. Nolting, J. Magn. Magn. Mat. **170**, 253 (1997).
26. M. Potthoff, *Habilitation thesis, Humboldt-Universität zu Berlin*, (2000).
27. G. Bulk and J. Jelitto, Phys. Rev. B **41**, 413 (1990).
28. M. Potthoff and W. Nolting, Z. Phys. B **104**, 265 (1997).
29. J. M. Luttinger, Phys. Rev. **121**, 942 (1961).
30. J. M. Luttinger and J. C. Ward, Phys. Rev. **118**, 1417 (1960).
31. W. Metzner and D. Vollhardt, Phys. Rev. Lett. **62**, 324 (1989).
32. E. Müller-Hartmann, Z. Phys. B **74**, 507 (1989).
33. V. Janiš, Z. Phys. B **83**, 227 (1991).
34. E. C. Stoner, Proc. R. Soc. London A **154**, 656 (1936).
35. G. Geipel and W. Nolting, Phys. Rev. B **38**, 2608 (1988).
36. T. Herrmann and W. Nolting, Solid State Commun. **103**, 351 (1997).
37. B. S. Shastry H. R. Krishnamurthy and P. W. Anderson, Phys. Rev. B **41**, 2375 (1990).
38. J. Hubbard, Proc. R. Soc. London A **281**, 401 (1964).
39. B. Velický S. Kirkpatrick and H. Ehrenreich, Phys. Rev. **175**, 747 (1968).

40. H. Fukuyama and H. Ehrenreich, Phys. Rev. B **7**, 3266 (1973).
41. J. Schneider and V. Drchal, phys. stat. sol. (b) **68**, 207 (1975).
42. R. Vlamming and D. Vollhardt, Phys. Rev. B **45**, 4637 (1992).
43. U. Brandt and C. Mielsch, Z. Phys. B **82**, 37 (1991).
44. T. Herrmann and W. Nolting, Phys. Rev. B **53**, 10579 (1996).
45. R. J. Jelitto, J. Phys. Chem. Solids **30**, 609 (1969).
46. M. Ulmke, Europ. Phys. J. B **1**, 301 (1998).
47. E. Müller-Hartmann, Proc. 5th Symp. Phys. Metals ed. E. Talk and J. Szade. Ustron-Jaszowiec (Poland) 1991.
48. M. Potthoff T. Herrmann T. Wegner and W. Nolting, phys. stat. sol. (b) **210**, 199 (1998).
49. D. M. Edwards and J. A. Hertz, Physica **163**, 527 (1990).
50. S. Wermbter and G. Czycholl, Journal of Physics C **7**, 7335 (1995).
51. M. Potthoff T. Herrmann and W. Nolting, Europ. Phys. J. B **4**, 485 (1998).
52. A. Georges and G. Kotliar, Phys. Rev. B **45**, 6479 (1992).
53. M. Jarrell, Phys. Rev. Lett. **69**, 168 (1992).
54. J. E. Hirsch and R. M. Fye, Phys. Rev. Lett. **56**, 2521 (1986).
55. V. Zlatic B. Horvatic and D. Sokvecic, Z. Phys. B **59**, 151 (1985).
56. H. Kajueter and G. Kotliar, Phys. Rev. Lett. **77**, 131 (1996).
57. M. Potthoff T. Wegner and W. Nolting, Phys. Rev. B **55**, 16132 (1997).
58. T. Wegner M. Potthoff and W. Nolting, Phys. Rev. B **57**, 6211 (1998).
59. S. Legfold, *Ferromagnetic Materials* volume 1 chapter 3, E. P. Wohlfarth Amsterdam, North Holland 1980.
60. P. Wachter, *Handbook of Physics and Chemistry of Rare Earths* volume 1 chapter 19, K. A. Gschneider and L. Eyring Amsterdam, North Holland (1979).
61. A. P. Ramirez, Journal of Physics C **9**, 8171 (1997).
62. N. Grewe F. Steglich, *Handbook of Physics and Chemistry of Rare Earth* volume 14, K. A. Gschneider and L. Eyring Elsevier Amsterdam (1991).
63. W. Nolting S. Mathi Jaya and S. Rex, Phys. Rev. B **54**, 14455 (1996).
64. B. S. Shastry and D. C. Mattis, Phys. Rev. B **24**, 5340 (1981).
65. W. Nolting and Matlak, phys. stat. sol. (b) **123**, 155 (1984).
66. H. B. Callen, Phys. Rev. **130**, 890 (1963).
67. W. Nolting S. Rex and S.Mathi Jaya, Journal of Physics C **9**, 1301 (1997).
68. C. Lacroix, Solid State Commun. **54**, 991 (1985).

Magnetic and orbital ordering in cuprates and manganites

Andrzej M. Oleś

Institute of Physics, Jagellonian University, Reymonta 4, PL-30059 Kraków, Poland

Mario Cuoco and Natalia B. Perkins

Dipartimento di Scienze Fisiche "E.R. Caianiello", Universita di Salerno, Via S. Allende, I-84081 Baronissi, Italy

Abstract. We address the role played by orbital degeneracy in strongly correlated transition metal compounds. The mechanisms of magnetic and orbital interactions due to double exchange (DE) and superexchange (SE) are presented. Specifically, we study the effective spin-orbital models derived for the d^9 ions as in $KCuF_3$, and for the d^4 ions as in $LaMnO_3$, for spins $S = 1/2$ and $S = 2$, respectively. The magnetic and orbital ordering in the undoped compounds is determined by the SE interactions that are inherently frustrated, carrying both antiferromagnetic (AF) and ferromagnetic (FM) channels due to low-spin and high-spin excited states, respectively. As a result, the classical phase diagrams consist of several magnetic phases which all have different orbital ordering: either the same orbitals ($x^2 - y^2$ or $3z^2 - r^2$) are occupied, or two different linear combinations of e_g orbitals stagger, leading either to G-AF or to A-AF order. These phases become unstable near orbital degeneracy, leading to a new mechanism of spin liquid. The model for d^4 Mn^{3-} ions in collosal magnetoresistance compounds provides an explanation of the observed A-AF phase, with the orbital order stabilized additionally by the Jahn-Teller effect. Possible extensions of the model to the doped compounds are discussed both for the insulating polaronic regime and for the metallic phase. It is shown that the spin waves are well described by SE in the insulating regime, while they are explained by DE for degenerate e_g orbitals in the metallic FM regime. Orbital excitations contribute to the hole dynamics in FM planes of $LaMnO_3$, characterized by new quasiparticles reminiscent of the t-J model, and a large redistribution of spectral weight with respect to mean-field treatments. Finally, we point out some open problems in the present understanding of doped manganites.

I CORRELATED TRANSITION-METAL OXIDES WITH ORBITAL DEGENERACY

Theory of strongly correlated electrons is one of the most challenging and fascinating fields of modern condensed matter. The correlated electrons are responsible for such phenomena as magnetic ordering in transition metals, heavy-fermion behavior, mixed valence, and metal-insulator transitions [1–3]. They play also a prominent role in transition metal oxides, where they trigger such phenomena as superconductivity with high transition temperatures in cuprates and colossal magnetoresistance (CMR) in manganites. At present, most of the current studies of strongly correlated electrons deal with models of nondegenerate orbitals, such as the Hubbard model, Kondo lattice model, Anderson model, and the like. Strong electron correlations lead in such situations to new effective models which act only in a part of the Hilbert space and describe the low-energy excitations. A classical example is the t-J model which follows from the Hubbard model [4,5], and describes a competition between the magnetic superexchange and kinetic energy of holes doped into an antiferromagnetic (AF) Mott insulator.

The realistic models of correlated electrons are, however, more complex than the Hubbard or Kondo lattice model. Transition metal oxides crystallize in a three-dimensional (3D) perovskite structure, where the oxygen ions occupy bridge positions between transition metal ions, as in $LaMnO_3$, or in similar structures with two-dimesional (2D) planes built by transition metal and oxygen ions, as in CuO_2 planes of high temperature superconductors. The oxygen ligand $2p$ orbitals play thereby a fundamental role in these systems, and determine both the electronic structure and actual interactions between the electrons (holes) which occupy correlated $3d$ orbitals of transition metal ions. The bands in transition metal oxides are built either by p_σ or by p_π oxygen orbitals which hybridize with the respective $3d$ orbitals of either e_g or t_{2g} symmetry. Taking an example shown in Fig. 1, it is clear that the overlap between the p_σ orbitals and $d_{x^2-y^2}$ orbitals is larger than that between the p_π orbitals and the corresponding d orbitals of t_{2g} symmetry. Therefore, the t_{2g} and p_π states are filled in the cuprates, and the relevant model Hamiltonians known as *charge transfer* models include frequently only the e_g orbitals of transition metal ions and the p_σ oxygen orbitals between them.

There are two crucial parameters which decide about the physical properties of a transition metal oxide, provided the $d-p$ hybridization elements are much smaller than the value of the on-site Coulomb interaction U. The latter parameter has to be compared with the splitting between the $3d$ and $2p$ orbitals, given by the so-called charge-transfer energy, $\Delta = |\varepsilon_p - \varepsilon_d|$, where ε_d and ε_p are the energies of an electron (hole) in these states, respectively. These systems are called Mott-Hubbard insulators (MHI) when $U < \Delta$, and it is in this limit that the Hubbard model would apply directly for the description of a metal-insulator transition. In

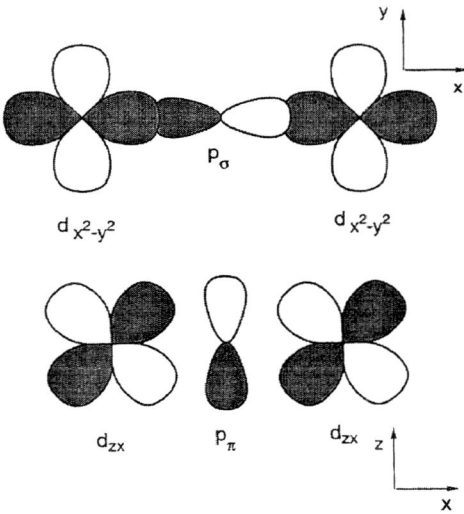

FIGURE 1. Examples of configurations for transition-metal $3d$ orbitals which are bridged by ligand $2p$ orbitals in transition metal oxides (after Ref. [1]).

the opposite case, one deals instead with charge-transfer insulators, as introduced by Zaanen, Sawatzky and Allen fifteen years ago [6,7]. Both classes of correlated (in contrast to band) insulators have quite different spectral properties, but in the strongly correlated regime the charge-transfer insulators resemble MHI, with a charge-transfer energy Δ playing a role of the effective U [8].

In reality, however, many oxides are found close to the above qualitative boarder line between Mott-Hubbard and charge-transfer systems (Fig. 2), and one might expect that the only relevant description has to be based on the charge-transfer models which include explicitly both d and p orbitals. Nevertheless, a reduction of such models to the effective simpler Hamiltonians dealing only with correlated d-like orbitals is possible, and examples of such mapping procedure have been discussed in the literature [9–12]. Unfortunately, there is no general method which works in every case, but the principle of the mapping procedure is clear, at least in perturbation theory. We will follow this idea in the present paper and concentrate ourselves on such simpler models which describe interactions between $3d$ electrons, determined by the *effective* hopping between transition metal ions which follows from intermediate processes involving charge-transfer excitations at the $2p$ oxygen orbitals [13]. It will be clear from what follows that while this simplification is allowed, there is in general no way to reduce these models any further to those of nondegenerate d orbitals, at least not for the oxides with a single electron or hole occupying (almost) degenerate e_g orbitals.

We concentrate ourselves on a class of insulating strongly correlated transition metal compounds, where the crystal field leaves the 3d orbitals of e_g symmetry explicitly degenerate and thus the type of occupied orbitals is not known *a priori*, while the effective magnetic interactions between the spins of neighboring transition metal ions are determined by orbitals which are occupied in the ground state [14–17]. The most interesting situation occurs when e_g orbitals are partly occupied, which results in rather strong magnetic interactions, accompanied by strong Jahn-Teller (JT) effect. Typical examples of such ions are: Cu^{2+} (d^9 configuration, one hole in e_g-orbitals) [18], low-spin Ni^{3+} (d^7 configuration, one electron in e_g-orbitals) [19–21], as well as Mn^{3+} [22] and Cr^{2+} ions (high-spin d^4 configuration with one e_g electron). The situation encountered for d^9 (or d^7) transition metal ions is simpler, as the t_{2g} orbitals are filled. The effective interactions may then be derived by considering only e_g orbital degrees of freedom and spins $s = 1/2$ at every site, and were first considered by Kugel and Khomskii more than two decades ago [18]. In the case of d^4 configuration one needs instead to consider larger spins $S = 2$ which interact with each other, due to virtual excitation processes which involve either e_g or t_{2g} electrons [23]. Finally, the early transition-metal compounds with d^1 or d^2 ions give also some interesting examples of degenerate t_{2g} orbitals [24–29]. In general, the magnetic superexchange and the coupling to the lattice are weaker in such cases due to a weaker hybridization between 3d and 2p orbitals (Fig. 1). Moreover, this problem is somewhat different due to the symmetry of the orbitals involved, and we will not discuss it here. The collective behavior of e_g electrons follows from their interactions. The models of interacting electrons in degenerate 3d states are usually limited to the leading on-site part of electron-electron interaction given by the Coulomb and exchange elements, U and J_H, respectively. The model Hamiltonian which includes these interactions is of the form,

$$H_{int} = (U + 2J_H) \sum_{i\alpha} n_{i\alpha\uparrow} n_{i\alpha\downarrow} + (U - J_H) \sum_{i,\alpha<\beta,\sigma} n_{i\alpha\sigma} n_{i\beta\sigma} + U \sum_{i,\alpha<\beta,\sigma} n_{i\alpha\sigma} n_{i\beta\bar\sigma}$$
$$- J_H \sum_{i,\alpha<\beta,\sigma} d^\dagger_{i\alpha\sigma} d_{i\alpha\bar\sigma} d^\dagger_{i\beta\bar\sigma} d_{i\beta\sigma} + J_H \sum_{i,\alpha<\beta} (d^\dagger_{i\alpha\uparrow} d^\dagger_{i\alpha\downarrow} d_{i\beta\downarrow} d_{i\beta\uparrow} + d^\dagger_{i\beta\uparrow} d^\dagger_{i\beta\downarrow} d_{i\alpha\downarrow} d_{i\alpha\uparrow}), \quad (1)$$

where summations over $\alpha < \beta$ guarantee that every pair of different states interacts only once, and we neglected the anisotropy of the interorbital interactions. It is important to realize that precisely for this reason the multiplet structure of transition metal ions [31] cannot be characterized by two quantities such as U and J_H, but one needs instead three independent parameters, usually chosen as Racah parameters A, B, and C. The commonly used relation between these parameters and the Slater parameters F_0, F_2, and F_4 are given in Table II of Ref. [1]. We also emphasize that the last term which describes the hopping of double occupancies between different orbitals has the same amplitude as the spin exchange and is $\propto J_H$. Such terms are frequently neglected in the Hubbard-like models which thus cannot reproduce the correct multiplet structure and give uncontrolled errors when superexchange is derived from them.

Early applications of the model Hamiltonian (1) were devoted to the understanding of magnetic states of transition metals [2,3,32–34]. More recently, the Hamilto-

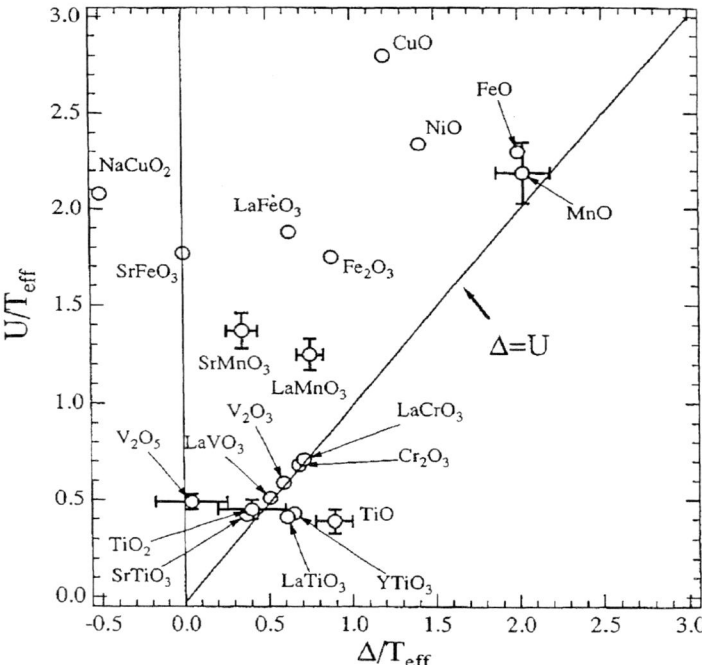

FIGURE 2. Zaanen-Sawatzky-Allen phase diagram for 3d transition metal oxides (after Ref. [30]).

nian (1) has been used to improve the local density approximation (LDA) scheme for determining the electronic structure of correlated transition metal oxides by including the electron-electron interactions in the Hartree-Fock (HF) approximation which gives the so-called LDA+U method [35]. If the electron-electron interactions U and J_H are treated in the HF approximation, they generate local potentials which act on different local states $|i\alpha\sigma\rangle$, and allow thus for the ground states with anisotropic distributions of charge and magnetization over five d orbitals. Such corrections improve the gap values in Mott-Hubbard and charge-transfer insulators, and become particularly important in the cuprates and manganites with partial filling of e_g orbitals.

The consequences of local potentials which follow from the Coulomb and exchange terms in Eq. (1) are well seen on the example of $KCuF_3$, one of the compounds which exhibits the degeneracy of e_g orbitals. We start out with the observation that according to LDA $KCuF_3$ would be an undistorted perovskite, as

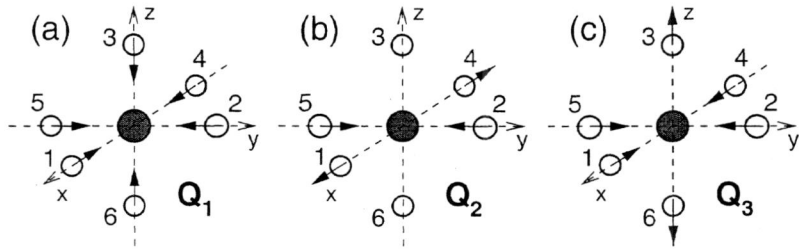

FIGURE 3. Different local modes for an MnO_6 octahedron: (a) breathing mode Q_1, (b) JT mode active in (a,b) planes, and (c) JT tetragonal distortion Q_3. Filled and empty circles show Mn and O ions, respectively.

the energy increases if the lattice distortion is made (see Sec. III.C). The reason is that the band structure of $KCuF_3$ determined by LDA would give a band metal with a Fermi-surface which is not susceptible to a band JT instability. LDA+U yields instead a drastically different picture: it allows both the orbitals and the spins to polarize which results in an energy gain of order of the band gap, i.e., of the order of 1 eV and reproduces the observed orbital ordering [36]. The orbital- and spin polarization is nearly complete and the situation is close to the strong-coupling limit underlying the spin-orbital model of Sec. III.A.

Observed orbital ordering could also be obtained in manganites using the LDA+U approach [37]. As a remarkable success of this method, the orbital ordering which corresponds to the so-called CE phase with the orbital ordering accompanied by the charge ordering was obtained for $Pr_{1/2}Ca_{1/2}MnO_3$ [37]. In the undoped $PrMnO_3$ one finds that orbitals alternate between two sublattices in (a,b) planes, as also expected following more qualitative arguments [17], allowing thus for the ferromagnetic (FM) coupling between spins within the (a,b) planes. In contrast, the orbitals almost repeat themselves along the c-axis, suggesting that the effective magnetic interactions should be AF. However, when the superexchange constants are determined using the band structure calculations [38], they do not agree with the experimental data [39,40]. Not only the FM exchange constants are larger by a factor close to four, but even the sign of the AF superexchange along the c-axis *cannot be reproduced*. Contrary to the suggestions made [38], this result cannot be corrected by effective interactions between further neighbors, as the crystal structure and the momentum dependence of the spin-waves in $LaMnO_3$ indicate that only nearest neighbor interactions should contribute in the effective spin model [39,40], and represents one of the spectacular examples how the electronic structure calculations fail in strongly correlated systems. Therefore, it is necessary to study the effective models which describe the low-energy sector of excited states and treat more accurately the strong electron correlations, as presented in this article. It is impossible to discuss the magnetic and orbital states of cuprates and manganites without paying attention to the lattice distortions. When the cubic crystal distorts, the energies of e_g orbitals change due to the coupling to the lattice. Depending on the type of distortion, the energy of one or the other e_g orbital will be lower. Therefore, particular lattice distortions alone might stabilize orbital ordering. In spite of some other views presented in the literature, we shall argue below that this is not the case for the cuprates, and the orbital ordering observed in KCuF$_3$ follows from the electronic interactions between strongly correlated electrons. We shall discuss this problem in particular for manganites (in Sec. V), where we argue that the electronic interactions alone determine the observed type of magnetic ordering, which is however additionally stabilized by the orbital interactions which follow from the JT effect [23].

Let us recall first the single-ion (noncooperative) JT effect. It was realized long ago by Kanamori that local JT effect leads to the symmetry lowering for Cu^{2+} and Mn^{3+} ions with a single hole (electron) at octahedral sites [41]. He considered a single-site problem of an ion surrounded by six neighbors which may be distorted from their initial symmetric positions which satisfy the octahedral symmetry (see Fig. 3). The normal modes may be written as follows:

$$Q_1 = \frac{1}{\sqrt{3}}(x_1 - x_4 + y_2 - y_5 + z_3 - z_6), \tag{2}$$

$$Q_2 = \frac{1}{\sqrt{2}}(x_1 - x_4 - y_2 + y_5), \tag{3}$$

$$Q_3 = \frac{1}{\sqrt{6}}(2z_3 - 2z_6 - x_1 + x_4 - y_2 + y_5), \tag{4}$$

where x_i, y_i and z_i are the coordinates of atom i. In contrast to the breathing mode Q_1, where all the neighbors move towards/away from the central site and the e_g orbitals do not split, the other two normal modes (Q_2 and Q_3) remove orbital degeneracy and favor the occupancy of either $|x\rangle \equiv |x^2 - y^2\rangle$ or $|z\rangle \equiv |3z^2 - r^2\rangle$ orbital.

Following Kanamori [41], we write the effective Hamiltonian in the form,

$$H_{JT} = H_0 + H_1 + H_2, \tag{5}$$

$$H_0 = \beta\lambda \sum_i Q_{1i}, \tag{6}$$

$$H_1 = \lambda\sqrt{C} \sum_{i\alpha\beta\sigma\sigma'} c_{i\alpha\sigma}^\dagger (Q_{2i}\sigma_i^x + Q_{3i}\sigma_i^z)_{\alpha\beta} c_{i\beta\sigma'}, \tag{7}$$

$$H_2 = \tfrac{1}{2}C \sum_i (Q_{2i}^2 + Q_{3i}^2), \tag{8}$$

where $c_{i\alpha\sigma}^\dagger$ is a creation operator for an e_g electron in orbital α with spin σ, σ_i^x and σ_i^z are the Pauli matrices, and λ and β are the parameters which depend on the system. The ions at different sites are independent and one may solve just a single-site problem, assuming an ansatz for the orbital state,

$$|i\theta\sigma\rangle = \cos\theta |iz\sigma\rangle + \sin\theta |ix\sigma\rangle. \tag{9}$$

Using the uniform angle $\theta_i = \theta$, the classical distortions and the coordinates and the orbital state are given as follows:

$$Q_{2i}^0 = \frac{\lambda}{\sqrt{C}} \sin 2\theta, \qquad Q_{3i}^0 = \frac{\lambda}{\sqrt{C}} \cos 2\theta, \qquad (10)$$

$$\langle \sigma_i^z \rangle = \cos 2\theta, \qquad \langle \sigma_i^x \rangle = \sin 2\theta. \qquad (11)$$

As easily recognized from Eqs. (10) and (11), the orbital state follows the lattice distortions and one finds the energy minimum given by $-\lambda^2/2$, showing that the lowest state cannot be determined uniquely. The situation changes when anharmonic terms are included which lead to the energy contribution of the form,

$$H_A = -A \sum_i \cos 6\theta_i, \qquad (12)$$

with $A > 0$. This term favors directional orbitals, and tetragonal distortions with the elongated tetragonal axis is the most stable structure. One finds identical energy for three different distortions, corresponding to $3z^2 - r^2$ orbital at $\theta = 0$, and to $3x^2 - r^2$ ($3y^2 - r^2$) orbital at $\theta = \pm \pi/3$, respectively.

The above energy contributions occur for the sites occupied by a single e_g electron, as for instance at Mn^{3+} ions. Important deformations of the lattice occur as well around Mn^{4+} ions, when the e_g electron is absent. In this case the breathing mode Q_1 becomes active, and the respective energy contribution takes the form,

$$H_{hole} = \beta \lambda \sum_i (1 - n_i) Q_{1i}, \qquad (13)$$

and typically $\beta \gg 1$.

Although the tendency towards directional orbitals might be considered to be generic for the present systems, such states cannot occur independently of each other in a crystal, as the lattice distortions are correlated. Therefore, a more realistic description requires a coupling between the oxygen distortions realized around different manganese sites. We discuss this problem on the example of $LaMnO_3$, which has been studied in more detail only recently [42,43]. If oxygens around a given Mn^{3+} ion are distorted, there are also distortions of common oxygen atoms around the neighboring Mn^{3+} ions, and in this way the orbital angles are coupled to each other. This is called cooperative JT effect, in contrast to the noncooperative one [41] which concerns single sites. The total energy which follows from the coupling to the lattice was derived by Millis [42]:

$$H_{lat} = -|E_0| \sum_i [n_i^2 + \beta^2 (1 - n_i)^2 + A \cos 6\theta_i]$$

$$+ \kappa \sum_{\langle ij \rangle} n_i n_j \cos 2(\theta_i + \psi_\alpha) \cos 2(\theta_j + \psi_\alpha)$$

$$+ 2\beta \kappa \sum_{\langle ij \rangle} (1 - n_i) n_j \cos 2(\theta_j + \psi_\alpha) + \beta^2 \kappa \sum_{\langle ij \rangle} (1 - n_i)(1 - n_j). \qquad (14)$$

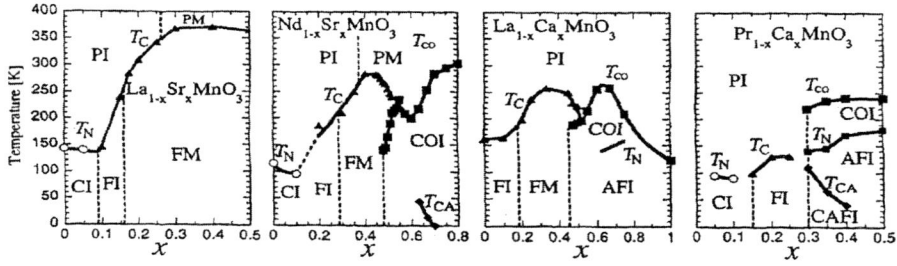

FIGURE 4. Phase diagrams of $A_{1-x}B_xMnO_3$ compounds (after Ref. [1]). Various ordered phases are labeled as follows: FI (PI) – ferromagnetic (paramagnetic) insulator, FM (PM) – ferromagnetic (paramagnetic) metal, CI – spin-canted insulating, COI – charge-ordered insulating, AFI – antiferromagnetic insulating, CAFI – canted antiferromagnetic insulating.

It includes the on-site terms, and the intersite couplings along the bonds, making the JT effect cooperative. A particular tendency for occupying the orbitals of a given type is expressed by the terms $\cos 2(\theta_i + \psi_\alpha)$, with the angle ψ_α depending on the bond direction as follows: $\psi_c = 0$, $\psi_a = -\pi/3$, and $\psi_b = \pi/3$, for the bonds $\langle ij \rangle$ along the c, a, and b-axis, respectively. The coupling constants E_0 and κ depend on the coefficient λ introduced in Eq. (7) and are functions of the respective force constants which describe the coupling between the manganese and oxygens ions, and between the pairs of oxygen ions, respectively. For the purpose of these lectures we will treat them as phenomenological parameters, but an interested reader may find explicit expressions and more technical details in Ref. [42].

Recent extensive research on the CMR manganites has generated a huge number of papers in the scientific literature. The interest is motivated by very spectacular experimental properties, with typically several magnetic phases stable in different doping regimes, most of them insulating, but one metallic FM phase [1,22]. With increasing temperature a transition from the FM metallic phase occurs either to a paramagnetic metal, or at lower doping to a paramagnetic insulator. In the latter case a large change of the resistivity accompanies the phase transition, and the transition temperature is strongly modified by the external magnetic field, giving rise to the phenomenon of CMR [44]. Examples of magnetic phase diagrams for representative distorted perovskites $R_{1-x}A_xMnO_3$ are shown in Fig. 4. The FM metallic phase is found in first three compounds, while in $Pr_{1-x}Ca_xMnO_3$ all magnetic phases are insulating. As the average ionic radius of the perovskite A site increases from (La,Sr) to (Pr,Ca) through (Nd,Sr) or (La,Ca), orthorhombic distortion of the $GdFeO_3$ type [1] increases, resulting in the decrease of the decrease of the one-electron bandwidth W. When the bandwidth gets reduced, the balance between the double-exchange (DE) [45,46] and other interactions changes and such instabilities as JT type distortions, charge and/or orbital ordering may occur.

Moreover, the AF superexchange may play an important role and stabilize the AF order in a broader doping regime.

In order to understand the phase diagrams of manganites, one needs to consider four different kinds of degrees of freedom: charge, spin, orbital, and lattice. Therefore, the models which treat doped manganites in a realistic way are rather sophisticated. A somewhat simpler situation occurs in the undoped $LaMnO_3$ as the charge fluctuations are suppressed by large on-site Coulomb interactions and one may study effective magnetic and orbital interactions, as we present in Sec. V.A. This problem may be also approached in a phenomenological way by postulating model Hamiltonians which contain such essential terms as the Hund's rule exchange interaction between e_g and t_{2g} electrons, the AF interactions between the t_{2g} core spins, and the coupling to the lattice. As an example, we present the Hamiltonian of the degenerate Kondo lattice with the coupling to local distortions of MnO_6 octahedra [47],

$$H = \sum_{ij\alpha\beta\sigma} t_{\alpha\beta} c^\dagger_{i\alpha\sigma} c_{j\beta\sigma} - J_H \sum_{i\alpha\sigma\sigma'} \vec{S}_i \cdot c^\dagger_{i\alpha\sigma} \vec{\sigma}_{\sigma\sigma'} c_{i\beta\sigma} + J' \sum_{\langle ij \rangle} \vec{S}_i \cdot \vec{S}_j$$
$$+ \lambda \sum_{i\alpha\beta\sigma} c^\dagger_{i\alpha\sigma} (Q_{1i}\sigma_i^0 + Q_{2i}\sigma_i^x + Q_{3i}\sigma_i^z)_{\alpha\beta} c_{i\beta\sigma} + \frac{1}{2} \sum_i (\beta Q_{1i}^2 + Q_{2i}^2 + Q_{3i}^2), \quad (15)$$

where $c^\dagger_{i\alpha\sigma}$ is the creation operator of an e_g electron with spin σ in the $d_{x^2-y^2}$ ($d_{3z^2-r^2}$) orbital at site i. The hopping elements $t_{ij}^{\alpha\beta}$ between nearest neighbors follow from the Slater-Koster rules [48]. $\vec{S}_i$ is the localized spin of t_{2g} electrons, and σ_i^x and σ_i^z are Pauli matrices. λ stands for the dimensionless electron-phonon coupling constant. Different distortions Q_{1i}, Q_{2i}, and Q_{3i} are the breathing mode and two JT modes shown in Fig. 3. Hotta et al. [47] took into account the cooperative nature of the JT phonons by introducing the coupling between the neighboring Mn ions in the normal coordinates for distortions of MnO_6 octahedra [42,49]. The important parameter is the ratio of the vibrational energies for manganite breathing (ω_b) and JT (ω_{JT}) modes, $\beta = (\omega_b/\omega_{JT})^2$. The calculations performed by the relaxation technique and by Monte-Carlo [50] for finite 3D clusters are summarized in Fig. 5. Depending on the parameters, four different magnetic phases are found [Fig. 5(a)]: FM phase (Ferro), so-called A-AF phase with staggered FM planes, C-AF phase with staggered FM chains, and, finally, the G-AF phase which is the 3D Néel order. They are identified by investigating the magnetic structure factor $S(\vec{q}) = \frac{1}{N} \sum_{ij} \exp[i\vec{q}(\vec{R}_i - \vec{R}_j)] \langle \vec{S}_i \vec{S}_j \rangle$ [Fig. 5(c)]. It is straightforward to understand that the ground state is FM at $J' = 0$. In this case the lowest energy gain may be obtained from the combination of the kinetic energy of e_g electrons with the local Hund's rule $\propto J_H$. Increasing J' increases the tendency towards the AF order and leads finally to the G-AF phase in the range of large $J'/t > 0.15$ [Fig. 5(b)].

Hotta et al. [47] found that various magnetic orderings are accompanied by the orbital orderings shown in Fig. 5(d)]. The shape of the occupied orbital arrangment is not easy to understand, however. It follows from the cooperative JT effect

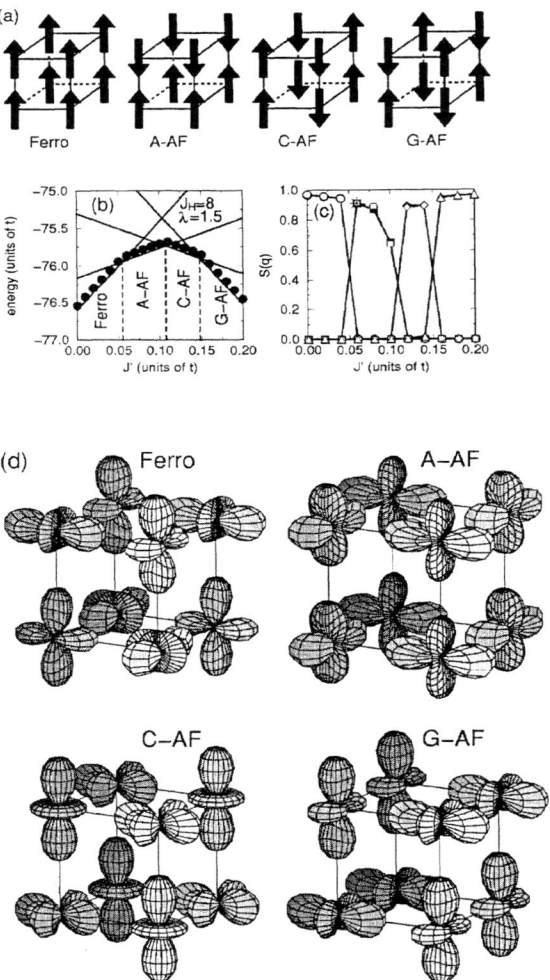

FIGURE 5. Types of magnetic order (a) and orbital order (d) in perovskite structures: Ferro, A-AF, C-AF, and G-AF, as obtained in MC simulations by Hotta et al. [47] using a model of e_g electrons coupled to the core t_{2g} spins and to JT lattice distortions (15). Quantitative results obtained as functions of the AF coupling J' between t_{2g} spins for $J_H = 8$ and $\lambda = 1.5$ are shown by two middle panels: (b) the total energy for a $2 \times 2 \times 2$ lattice; (c) the spin-spin structure factor $S(\vec{q})$ obtained for $4 \times 4 \times 2$ (solid symbols) and $4 \times 4 \times 4$ (open symbols) clusters. The circles, squares, diamonds, and triangles indicate $S(\vec{q})$ for $\vec{q} = (0,0,0)$, $\vec{q} = (\pi,0,0)$, $\vec{q} = (\pi,\pi,0)$, and $\vec{q} = (\pi,\pi,\pi)$, respectively.

and expresses a compromise between the orbital and magnetic energies. The overall picture might seem appealing, but it is questionable whether the JT effect is the dominating mechanism that determines the magnetic and orbital ordering in manganites and related compounds which are known to be primarily MHI [1], i.e., the on-site Coulomb interaction U is the largest parameter, typically $5 < U < 10$ eV, which dominates the hybridization. Although it has been argued that the large Coulomb interaction will not change the main results shown in Fig. 5 [47], we do not think this conclusion is allowed. In fact, in the absence of large Coulomb interaction U the magnetic interactions are dominated by DE [46] and the system is FM at $J' = 0$ as new effective interactions arise in the presence of large U, and they easily might change the delicate balance between different magnetic and orbital ordered phases. We shall discuss the problem of magnetic and orbital ordering in detail in Secs. III and V and show that the electronic interactions alone give a dominating contribution to magnetic interactions.

Magnetism in transition metals and in their compounds is known to be due to intraatomic Coulomb interaction [51]. The simplest model which takes into account the Coulomb interaction is that due to Hubbard (see Sec. II.A). It describes electrons in a narrow and nondegenerate tight-binding band and allows for repulsion U between electrons only when they are at the same site. This model has been studied intensively since Hubbard proposed it, especially in connection to the occurrence of magnetism [3]. Anderson has shown in 1959 [52] that the Hubbard Hamiltonian is equivalent to a Heisenberg Hamiltonian with an AF superexchange interaction given in terms of the hopping amplitude and the Coulomb interaction, if U is large. Indeed, for two neighboring ions an extra delocalization process $d_i^1 d_j^1 \to d_i^0 d_j^2$ is only possible for antiparallel arrangment of neighboring spins, decreasing the energy and favoring this configuration (Fig. 6). Therefore, if each ion has only one nondegenerate orbital, the superexchange is AF, as explained in Sec. II.A.

The question of whether the AF correlations might evolve by changing the electron concentration into ferromagnetism is still controversial. A rigorous proof of Nagaoka [53] of the existence of ferromagnetism applies only in a very special case – in the limit of infinite Coulomb repulsion U when one hole or one extra electron is added to the half-filled band ($n = 1$), in a lattice of particular symmetry. However, the occurrence of ferromagnetism comes in a more natural way if one takes into account the orbital degeneracy as Van Vleck has emphasized [54]. In the case of two-fold orbital degeneracy, applying similar arguments as those used by Anderson [52], one ends up with a richer structure of effective interactions when the processes $d_i^1 d_j^1 \to d_i^0 d_j^2$ are analyzed. For the occupancy of $n = 1$ electron per atom one finds four possible situations as depicted in Fig. 6: (a) same orbital – same spin, (b) same orbital – different spin, (c) different orbital – same spin, (d) different orbital – different spin. This problem has been studied already in the seventies, but mostly starting from simplified model Hamiltonians [55–58]. In order to study the qualitative effects, the simplest case with only diagonal hopping and equal intra- and interorbital Coulomb elements U in Eq. (1) has been usually assumed.

As a qualitative new effect due to the Hund's rule exchange $J_H > 0$, *ferromag-*

netic superexchange becomes possible, if the excitation involves the high-spin state with two parallel electrons [Fig. 6(c)]. Although the processes which contribute to the superexchange for nondegenerate orbitals [Fig. 6(b)] are also present, and there are more AF terms, the FM term has the largest coefficient due to the structure of Coulomb interactions (1). Therefore, one might expect that under certain conditions such terms could promote ferromagnetism. While this is not easy and happens only for rather extreme parameters in the doubly degenerate Hubbard model with isotropic but diagonal hopping elements [55–57], it has been recognized in these early works that the orbital ordering may accompany the magnetic ordering, and orbital superlattice favors the appearance of magnetism at zero temperature. Indeed, the onset of magnetic long-range order (LRO) is obtained for such values of parameters that the usual Stoner criterion is not yet fulfilled. Furthermore, the studies at finite temperature revealed that the orbital order is more stable than the magnetic one. Therefore, two phase transitions are expected in general: at the lower temperature the ferromagnetism disappears and, at the higher one, the orbital order [55–57].

In order to understand the behavior of CMR manganites, it is necessary to include the orbital degrees of freedom for partly occupied e_g orbitals. The motivation comes both from theory and experiment. For quite long time it was believed that the FM state in manganites can be understood by the DE model [45,46]. In fact, it provides not more than a qualitative explanation why the doped manganites should have a regime of FM metallic state (Fig. 4). However, if one calculates the Curie temperature T_C using the DE model the values are overestimated by a factor of the order of five [59]. Also the experimental dependence of the resistivity in the metallic phase [60] cannot be reproduced within the DE model [59]. Finally, in a FM metal one expects a large Drude peak and no incoherent part in the optical conductivity. The experimental result is quite different – most of the intensity is incoherent at low temperatures, and only a small Drude peak appears which absorbs not more than 20 % of the total spectral weight [61,62]. All these results demonstrate the importance of orbital degrees of freedom which should be treated on equal footing as the spins of e_g electrons. The orbital degeneracy leads therefore to a new type of models in the theory of magnetism: *spin-orbital models*. They act in the extended space and describe the (super)exchange interactions between spins, between orbitals, and simultaneous spin-and-orbital couplings. In order to address realistic situations encountered in cuprates and manganites, such models cannot rely on the degenerate Hubbard model [55–57], but have to include the anisotropy in the hopping elements [48], nonconservation of the orbital quantum number, and realistic energetic structure of the excited states [31]. Once such models are derived, as we present in Secs. III and V, their phase diagrams may be studied using the mean-field (MF) approximation [23,63,64]. It turns out that their phase diagrams show an unusual competitions between classical (magnetic and orbital) ordering of different type, in particular close to the degeneracy of e_g orbitals. Therefore, two interesting questions occur for such orbitally degenerate MHI: (i) Which *classical* states with magnetic LRO do exist in the neighborhood

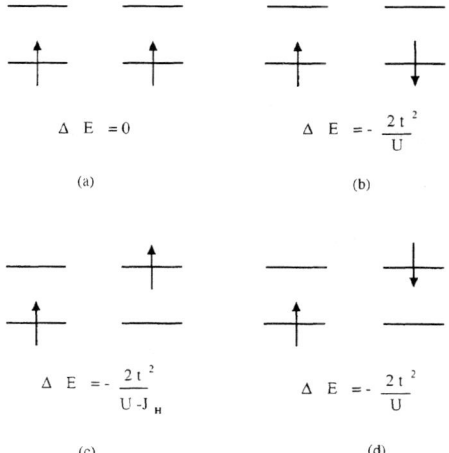

FIGURE 6. The various configurations of nearest neighbors for doubly-degenerate orbitals. The kinetic exchange is always AF in a nondegenerate model (b), while the processes which involve differently occupied orbitals on both sites may be either FM (c) or AF (d). U and J_H are the on-site Coulomb and Hund's rule exchange interactions.
Reprinted from Ref. [17], Copyright 1997, with permission from Elsevier Science.

of orbital degeneracy? (ii) Are those forms of classical order always stable against *quantum* (Gaussian) fluctuations? We will show that the orbitally degenerate MHI represent a class of systems in which spin disorder occurs due to frustration of *spin and orbital* superexchange couplings. This frustration mechanism is different from that which operates in quantum antiferromagnets, and suppresses the magnetic LRO in the ground state *even in three dimensions*.

We organized the remaining chapters of this article as follows. In order to clarify the basic magnetic interactions in strongly correlated oxides, we start with the superexchange and DE in Sec. II. Next we introduce and analyze on the classical level the simplest spin-orbital model for spins $S = 1/2$ (Sec. III) which applies to cuprates, and present its collective modes. The model exhibits an interesting frustration of magnetic interactions, and the classical phases are destabilized by quantum fluctuations (Sec. IV). Therefore, we discuss the problem of quantum disorder in low dimensional spin models and in an idealized spin-orbital model with SU(4) symmetry in one dimension. The physics of manganites is richer than that of cuprates close to the degeneracy of e_g orbitals, and additional interactions between orbital variables occur due to the coupling to the lattice (Sec. V). The understanding of various phase transitions shown in Fig. 4 remains an outstanding problem which requires to study simultaneously the coupling to the lattice responsible for the insulating behavior of doped systems, and the DE model at orbital degeneracy. This latter problem is very actual and was addressed for the first time only last year [65] (Sec. VI). The spin and orbital interactions lead in general to new type of effective t-J models [16], and we analyze spin-waves in FM phase and give an example of the hole which dresses by orbital excitations and compare this situation with the hole motion in the t-J model (Sec. VII). Among many open questions for doped manganites, we selected a few such as the CE-phase, stripes, orbital ordering and phase separation (Sec. VIII). In our opinion, they are crucial to understand the complexity of the experimental phase diagrams. We give a brief summary and our conclusions in Sec. IX.

II MAGNETIC INTERACTIONS FOR NONDEGENERATE ORBITALS

A Superexchange and t-J model

Before discussing the consequences of e_g orbital degeneracy in cuprates and manganites, we review shortly the basic magnetic interactions in the models of nondegenerate orbitals – the superexchange and the DE. The main idea to derive the superexchange is the notion of the Mott-Hubbard insulator at $n = 1$ in which the charge fluctuations are suppressed and the electrons localize, occupying the states of the lower Hubbard band. Therefore, part of degrees of freedom is integrated out and one may study an effective model which captures the essential features of the low-energy excitations.

Before deriving the effective superexchange model for degenerate orbitals, we analyze shortly the nondegenerate orbitals filled by one electron per atom $(n=1)$. This situation plays a fundamental role in strongly correlated systems and elucidates the general principle of introducing magnetic (and orbital) interactions in a Mott-Hubbard insulator by integrating out charge fluctuations and reducing the problem. The starting point is the Hubbard Hamiltonian,

$$H = H_t + H_U, \tag{16}$$

$$H_t = -t \sum_{\langle ij \rangle \sigma} \left(c_{i\sigma}^\dagger c_{j\sigma} + \text{H.c.} \right), \tag{17}$$

$$H_U = U \sum_i n_{i\uparrow} n_{i\downarrow}, \tag{18}$$

where U and t are standing for the on-site screened Coulomb repulsion, and for the hopping amplitude between nearest neighbors, respectively, and the summation runs over the bonds $\langle ij \rangle$ between nearest neighbors. Furthermore, the operator $c_{i\sigma}^\dagger$ creates an electron with spin σ at site i, and $n_{i\sigma} = c_{i\sigma}^\dagger c_{i\sigma}$ is the electron density operator. Recall that the interaction term in the Hubbard model can be reexpressed in the following way:

$$H_U = -\frac{2}{3} U \sum_i \left(\vec{S}_i \right)^2, \tag{19}$$

so that it forces the spin $\vec{S}_i$ to be maximal if U becomes infinitely large, i.e., doubly occupied sites are forbidden. Only $|\uparrow\rangle$ and $|\downarrow\rangle$ states are kept in this large U limit at half-filling. Taking the atomic limit ($t = 0$), the interaction part of the Hamiltonian (16) has infinitely many (2^N, where N is the number of sites) degenerate eigenstates, given by different spin configurations. In order to lift this large degeneracy we will keep the effects of fluctuations induced by the kinetic energy term to leading order in an expansion in (t/U). As usually, this problem has to be solved in degenerate perturbation theory.

Suppose we begin with an arbitrary configuration which can be labeled by the local zth components of the spins S_i^z. In the expansion in powers of (t/U), one includes contributions from intermediate states in which one site will become doubly occupied and, at the same time, the other site becomes empty [4]. The energy of the excited state is U above that of the degenerate ground state manifold. The squared transition matrix element is t^2 and the combinatorial factor of two has to be included since this process can occur in two different ways. Hence we expect that the relevant parameter of the effective spin Hamiltonian should be $2t^2/U$. Also, the final state after the double occupancy dissociates has to be either the same as the initial state, or it may differ at most by a spin exchange. The candidate for the effective Hamiltonian is, of course, the quantum Heisenberg antiferromagnet [4,52,58], since we know that the spin-spin interaction follows from a possibility of permuting the electrons on a lattice.

The formal derivation of the effective Heisenberg model can be performed in a few different equivalent ways: (i) by means of a canonical transformation [4], (ii) with Schrieffer-Wolff procedure [66], and (iii) with Brillouin-Wigner perturbation approach [67]. The first method is the most transparent to use away from the half-filling, where it leads to the t-J model, known as a minimal model to describe the electronic states in high temperature superconductors [68]. We will not repeat here the details of the derivation of the t-J model [4] as it belongs already to the textbook material [2,3]. The common result of all these procedures at $n = 1$ is the removal of degeneracy within the second order perturbation, and the effective Hamiltonian, given by the following expression:

$$\langle \phi | H_{eff} | \psi \rangle = -\langle \phi | H_t \frac{1 - P_0}{H_U} H_t | \psi \rangle = - \sum_{n \neq \{0\}} \langle \phi | H_t | n \rangle \frac{1}{U} \langle n | H_t | \psi \rangle, \tag{20}$$

where $|\phi\rangle, |\psi\rangle$ denote states in the subspace without double occupancies, with a projection operator P_0. The states $|n\rangle$ are configurations with one doubly occupied site, and each term in the sum can be represented by a retraceable exchange path. Thereby we assume that $n \leq 1$; the case of $n > 1$ may be treated by the same method after performing a particle-hole transformation. Since the total spin per two sites is conserved in the excitation process $|\uparrow\rangle_i |\downarrow\rangle_j \to |0\rangle_i |\uparrow\downarrow\rangle_j$, we can express the operators which connect the initial and final states of this transition by means of the projection operators for the singlet and for the triplet state on the bond $\langle ij \rangle$, respectively:

$$Q_S(i,j) = \left(-\vec{S}_i \cdot \vec{S}_j + \frac{1}{4}\right), \qquad Q_T(i,j) = \left(\vec{S}_i \cdot \vec{S}_j + \frac{3}{4}\right). \tag{21}$$

The excitation energy associated to the process $|\uparrow\rangle_i |\downarrow\rangle_j \to |0\rangle_i |\uparrow\downarrow\rangle_j$ which creates a singlet at site j is $\varepsilon_S = U$, (if we start from a triplet configuration, virtual processes are blocked due to the Pauli principle). Taking into account that the double occupancy may be created either at site i or at site j, the effective Hamiltonian can be expressed in the following way,

$$H_{eff} = -\frac{4t^2}{U} \sum_{\langle ij \rangle} Q_S(i,j) = J \sum_{\langle ij \rangle} \left(\vec{S}_i \cdot \vec{S}_j - \frac{1}{4}\right). \tag{22}$$

Thus, one finds the AF Heisenberg model with the superexchange constant $J = 4t^2/U$.

B Kondo lattice model – double exchange

The early theoretical studies of manganites were concentrated on the models introduced in order to understand the FM phase which occurs in the doped materials. The basic understanding of the tendency towards FM order follows from the so-called *double exchange* model [45,46] – it explains that electrons in a partially filled band maximize their kinetic energy when their spins are aligned with the localized spins which order ferromagnetically. In fact, this phenomenon is quite reminiscent of the Nagaoka state in the Hubbard model [53]. However, in spite of this qualitative explanation of the existence of ferromagnetism, several features of the experimental phase diagrams of manganites remain unclear especially at low temperature, where one has to go beyond the DE model. In order to understand the reasons of its shortcomings, let us present briefly the main consequences of the DE model.

The Kondo lattice Hamiltonian with *ferromagnetic* spin-fermion coupling can be defined as follows,

$$H = -t \sum_{\langle ij \rangle \sigma} (c_{i\sigma}^\dagger c_{j\sigma} + H.c.) - J_H \sum_{i\alpha\beta} \vec{S}_i \cdot c_{i\alpha}^\dagger \vec{\sigma}_{\alpha\beta} c_{i\beta}, \tag{23}$$

where $c_{i\sigma}^\dagger$ is a creation operator for an electron at site i with spin σ, and $\vec{S}_i$ is the total spin of the t_{2g} electrons $S = 3/2$, assumed to be localized. The first term describes the kinetic energy of electrons in a nondegenerate band due to the electron transfer between nearest-neighbor Mn-ions, $J_H > 0$ is the FM Hund's coupling between the itinerant electron and the t_{2g} core spin ($S = 3/2$). The average electronic density of e_g electrons, denoted by $n = \langle n_i \rangle$, is adjusted using a chemical potential μ.

Let us consider a bond in the perovskite structure formed by two Mn atoms with an oxygen atom in between. In the ionic configuration the $2p$ shell of the O^{2-} ion is completely filled. In order to treat the problem semiclassically [46], we assume that the Mn ions have rather large spins S_1 and S_2, so that one could assign to them definite directions in space, and a definite angle relative to each other. If an itinerant e_g electron is on the site $i = 1$, it has two states, of energies $E_1 = \pm JS$, if the electron spin is parallel and antiparallel to the spin $\vec{S}_1$, respectively. On atom $i = 2$ it also has similar two states, but defined with respect to the direction of spin $\vec{S}_2$. As the electron spin direction is conserved in the hopping process, the final state has to be projected on the new local axis. This is equivalent to rotating the transfer matrix between these two sites in such a way that its elements refer correctly to the projected spin components in the rotated basis.

Let us label the two electronic spin functions referring to the direction of $\vec{S}_1$ by α and β, and those referring to $\vec{S}_2$ by α' and β'. The energies of the eigenstates on atom $i = 1$ are:

$$E(d_1\alpha) = -J_H S, \qquad E(d_1\beta) = J_H(S+1), \tag{24}$$

for FM coupling $J_H > 0$. The energies of the eigenstates on atom $i = 2$ are given by similar expressions:

$$E(d_2\alpha') = -J_H S, \qquad E(d_2\beta') = J_H(S+1). \tag{25}$$

Here d_1 and d_2 describe the spin function of an electron localized at atom 1 and atom 2, respectively.

The transformation which expresses α and β in terms of α' and β' is of the form,

$$\alpha = \cos(\theta/2)\alpha' + \sin(\theta/2)\beta', \tag{26}$$
$$\beta = -\sin(\theta/2)\alpha' + \cos(\theta/2)\beta', \tag{27}$$

where θ is the angle between the spins $\vec{S}_1$ and $\vec{S}_2$. By considering the Hamiltonian (23) for two sites, one can write the secular equation which has the following four solutions [46]:

$$E = 1/2J \pm \left([J(S+1/2) \pm t\cos(\theta/2)]^2 + t^2\sin^2(\theta/2)\right)^{1/2}. \tag{28}$$

The energies depend on the angle θ between both spins, and in the semiclassical case

$$\cos\left(\frac{\theta}{2}\right) = \frac{|\vec{S}_1 + \vec{S}_2|}{2S}. \tag{29}$$

In the absence of any other interaction, the lowest energy is obtained for the aligned spins, at $\theta = 0$.

The existence of phase separation and ferromagnetism in the ground state of the FM Kondo model can also be studied in the limit of $d = \infty$. The dynamical mean field theory (DMFT) [69] leads to a self-consistent equation which can be

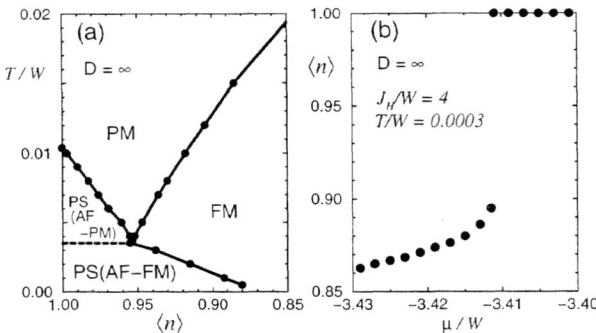

FIGURE 7. (a) Phase diagram in the $d = \infty$ limit working at $J_H/W = 4.0$. The "PS(AF-PM)" region denotes phase separation (PS) between a hole-poor antiferromagnetic (AF) region, and a hole-rich paramagnetic (PM) region. The rest of the notation is standard; (b) Density $\langle n \rangle$ vs μ/W obtained in the $d = \infty$ limit, $J_H/W = 4.0$, and $T/W = 0.0003$. The discontinuity in the density is clear (after Ref. [70]).

solved iteratively starting from a random spin configuration, and as a function of temperature and electron density three solutions have been found with AF, FM, and paramagnetic character. We refer an interested reader to Ref. [70] for more technical details.

The presence of ferromagnetism at finite doping and antiferromagnetism at half-filling are quite clear from Fig. 7. Close to half-filling and at low temperature, the density $n = \langle n_i \rangle$ was found to be discontinuous as a function of μ, in excellent agreement with the results obtained by other numerical calculations. The phase separation observed in Fig. 7(a) occurs between AF and FM regions. However, we note that at higher temperature the phase separation occurs between hole-poor AF and hole-rich paramagnetic regions [Fig. 7(b)].

In order to illustrate the consequences of the DMFT treatment of the Kondo lattice (DE) model (23) we reproduce in Fig. 8 the density of states $A(\omega)$ obtained for the AF and FM phases at low temperature (for details of the calculation see Ref. [70]). The critical value of the chemical potential where the AF and FM phases coexist is $\mu_c \simeq -1.40W$. In the both cases the density of states splits into upper and lower bands due to the large Hund's coupling J_H. The band splitting in the FM phase is due to the half-metalicity of the system. The width of the upper and lower bands is wider for the FM phase, which causes a narrower gaped region centered at $\omega \simeq 0$.

Following Furukawa [71], we calculate the spin excitation spectrum of the DE model, and compare the results with recent data of the neutron inelastic scattering experiments. We use the spin-wave approximation in the ground state, which has

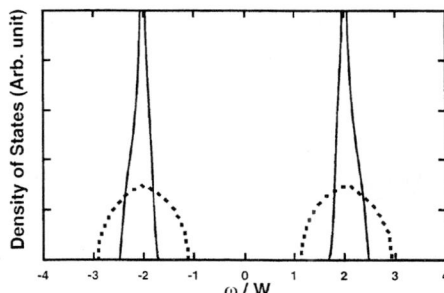

FIGURE 8. Density of states in the $d = \infty$ limit corresponding to the AF (solid line) and FM (dotted line) solutions, found at $J_H/W = 2.0$ and $T/W = 0.005$ (after Ref. [70]).

been introduced by Kubo and Ohata [72]. Expanding the spin operators in terms of boson operators in the FM state,

$$S_i^+ \simeq \sqrt{2S}a_i, \qquad S_i^- \simeq \sqrt{2S}a_i^\dagger, \qquad S_i^z = S - a_i^+ a_i, \qquad (30)$$

the lowest-order effective Hamiltonian can be written as follows:

$$H = \sum_{\vec{k}} \left[(\varepsilon_{\vec{k}} - J_H) c_{\vec{k}\uparrow}^\dagger c_{\vec{k}\uparrow} + (\varepsilon_{\vec{k}} + J_H) c_{\vec{k}\downarrow}^\dagger c_{\vec{k}\downarrow} \right]$$

$$+ J_H \sqrt{\frac{2}{SN}} \sum_{\vec{k}\vec{q}} \left(a_{\vec{q}}^\dagger c_{\vec{k}\uparrow}^\dagger c_{\vec{k}+\vec{q},\downarrow} + a_{\vec{q}} c_{\vec{k}+\vec{q},\downarrow}^\dagger c_{\vec{k}\uparrow} \right) + \frac{J_H}{SN} \sum_{\vec{k}\vec{q}_1\vec{q}_2,\sigma} \sigma a_{\vec{q}_1}^\dagger a_{\vec{q}_2} c_{\vec{k}-\vec{q}_1\sigma}^\dagger c_{\vec{k}-\vec{q}_2\sigma}. \qquad (31)$$

The first line in Eq. (31) describes the electron band split by the exchange interaction with the core spins, while the second line stands for the coupling between the electrons and spin excitations (electron-magnon interaction).

Let us consider the lowest order terms of the $1/S$ expansion at $T = 0$, assuming that J_H is finite but sufficiently large to polarize completely the electronic band, i.e., $n_\uparrow = n$ and $n_\downarrow = 0$, at $T = 0$. The electron concentration is given by $n = 1 - x$. For a simple cubic 3D lattice with nearest-neighbor hopping t one finds,

$$\varepsilon_{\vec{k}} = -2t \left(\cos k_x + \cos k_y + \cos k_z \right) = -6t\gamma_{\vec{k}}, \qquad (32)$$

where $\gamma_{\vec{k}} = \frac{1}{z} \sum_{\vec{\delta}} \exp(i\vec{k} \cdot \vec{\delta})$. For perovskite manganites, estimates of the electron bandwidth and the on-site Hund's coupling being a few eV has been made by the first-principle calculations. However, one might consider that t and J_H in the DE model are effective parameters which could be strongly renormalized from the bare values due to other interactions present in the real systems. Such effective parameters could be determined from a comparison with experiments.

The spin-wave self-energy in the lowest order of $1/S$ expansion is given by

$$\Pi(\vec{q},\omega) = \frac{1}{SN} \sum_{\vec{k}} \left(f_{\vec{k}\uparrow} - f_{\vec{k}+\vec{q}\downarrow} \right) \times \left(J_H + \frac{2J_H^2}{\omega + \varepsilon_{\vec{k}} - \varepsilon_{\vec{k}+\vec{q}} - 2J_H} \right), \qquad (33)$$

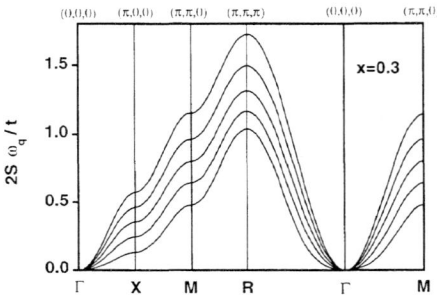

FIGURE 9. Spin wave dispersion in the metallic FM phase calculated using the DE model by Furukawa [71]. Different curves are obtained for $J_H/t = \infty$, 24, 12 and 6 from top to bottom.

where $f_{\vec{k}\sigma}$ is the Fermi distribution function. We have $f_{\vec{k}\downarrow} = 0$, if the system is fully polarized. The spin-wave dispersion relation ω_q is now obtained self-consistently as a solution of the equation $\omega_{\vec{q}} = \Pi(\vec{q}, \omega_{\vec{q}})$. Since $\Pi(\vec{q}, \omega_{\vec{q}}) \propto 1/S$, the lowest order of $1/S$ expansion gives $\omega_{\vec{q}} = \Pi(\vec{q}, 0)$. Therefore, the spin-wave dispersion is described as

$$\omega_{\vec{q}} = \frac{1}{2S}\frac{1}{N}\sum_{\vec{k}} f_{\vec{k}\uparrow} \frac{J_H(\varepsilon_{\vec{k}+\vec{q}} - \varepsilon_{\vec{k}})}{J_H + (\varepsilon_{\vec{k}+\vec{q}} - \varepsilon_{\vec{k}})/2}. \quad (34)$$

In Fig. 9, we show the spin-wave dispersion relation at $x = 0.3$ for various values of J_H/t. As the value of J_H becomes comparable with the electron bandwidth, the softening of the spin-wave dispersion is observed since the effective coupling between spins becomes weak. At $J_H \to \infty$, we have

$$\omega_{\vec{q}} \simeq \frac{1}{2SN}\sum_{\vec{k}} (\varepsilon_{\vec{k}+\vec{q}} - \varepsilon_{\vec{k}}) f_{\vec{k}\uparrow} = \frac{1}{2}W_{sw}(1 - \gamma_{\vec{q}}), \quad (35)$$

where W_{sw} is the spin-wave bandwidth given by the kinetic energy of electrons moving in a polarized band,

$$W_{sw} = \frac{6t}{SN}\sum_{\vec{k}} f_{\vec{k}\uparrow} \cos k_x. \quad (36)$$

The dispersion relation (35) is identical with that given by a FM Heisenberg model with nearest-neighbor spin exchange $J_{eff} = W_{sw}/12$. The above correspondence can be understood as follows. We consider a perfectly polarized FM state at $T = 0$ and then flip a spin at site i_0. In the case of the strong coupling limit $J_H \gg t$, where electrons with spins antiparallel to the localized spin on the same site are disfavored, the electron at site i_0 is localized because it has different spin orientation from that of the localized spins at neighboring sites. Therefore, in this limit the effective spin-spin interaction is short-ranged and the DE model in the strong-coupling limit is mapped onto the Heisenberg model with short-range interactions.

III SPIN-ORBITAL MODEL IN CUPRATES

A Superexchange for degenerate e_g orbitals

Our aim is to construct the effective low-energy Hamiltonian for a 3D perovskite-like lattice, assuming the situation as in the cuprates, i.e., d^9 configuration with single occupancy of one hole in e_g orbitals. This situation was considered already by Kugel and Khomskii [18]; here we present a more recent derivation which uses a correct multiplet structure of the excited d^8 states [64]. From a general point of view, one should approach the problem starting from the charge-transfer multiband model which contains the hybridization between the d orbitals of transition metal ions and the $2p$ orbitals of oxygen ions. Yet, if the Coulomb interaction at the d orbital and the energies required for the electron transfer from the $3d$ to the $2p$ orbital levels are large compared to the other parameters involved, then it is possible to integrate out the oxygen degrees of freedom and to deal instead with a simpler model which describes electrons (holes) in a d band.

We derive the superexchange in a similar fashion as in Sec. II.A for the case of degenerate orbitals. Having in mind the strongly correlated late transition metal oxides, we consider specifically the case of the e_g orbitals, defined by the local basis: $x^2 - y^2 \equiv |x\rangle$ and $(3z^2 - r^2)/\sqrt{3} \equiv |z\rangle$. Although we focus here on the case of the d^9 configuration, though the presented analysis can be easily generalized to the low-spin d^7 configuration with a single electron; in the case of the early transition metal oxides the d^1 case would involve the t_{2g} orbitals occupied by a single electron instead [24–29].

We take as a starting point the following Hamiltonian which describes d-holes on transition metal ions,

$$H_{e_g} = H_t + H_{int} + H_z, \qquad (37)$$

and includes the kinetic energy H_{kin}, and the electron-electron interactions H_{int}, restricted now to the subspace of the e_g orbitals (the t_{2g} orbitals are filled by electrons, do not couple to e_g orbitals by the hoppings via oxygens, and hence can be neglected). The last term H_z describes the crystal-field splitting of the e_g orbitals.

Due to the shape of the two e_g orbitals $|x\rangle$ and $|z\rangle$, their hybridization with oxygen orbitals is unequal in the three cubic directions [16], so that the effective hopping elements are direction dependent and different for $|x\rangle$ and $|z\rangle$. The only nonvanishing hopping in the c-direction connects two $|z\rangle$ orbitals, while the elements in the (a,b) planes fulfill the Slater-Koster relations [48], as presented in Ref. [16]. Taking the hopping t along the c-axis as a unit, the kinetic energy is given by,

$$H_t = \tfrac{1}{4}t \sum_{\langle ij\rangle\|,\sigma} \left[3d^\dagger_{ix\sigma}d_{jx\sigma} + \pm\sqrt{3}(d^\dagger_{iz\sigma}d_{jx\sigma} + d^\dagger_{ix\sigma}d_{jz\sigma}) + d^\dagger_{iz\sigma}d_{jz\sigma} + H.c. \right]$$

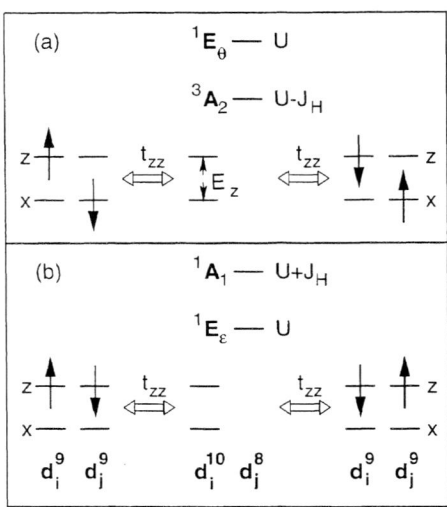

FIGURE 10. Virtual transitions $d_i^9 d_j^9 \to d_i^{10} d_j^8$ which lead to a spin-flip and generate effective interactions for a bond $\langle ij \rangle \parallel$ c-axis, with the excitation energies at $E_z = 0$. For two holes in different orbitals (a), either the triplet $^3A_2$ or the interorbital singlet $^1E_\theta$ occurs as an intermediate d^8 configuration, while if both holes are in $|z\rangle$ orbitals (b), two other singlets, $^1E_\epsilon$ and $^1A_1$, with double occupancy of $|z\rangle$ orbital, contribute. The latter processes are possible either from i to j or from j to i (after Ref. [64]).

$$+t\sqrt{\beta} \sum_{\langle ij \rangle \perp, \sigma} \left(d_{iz\sigma}^\dagger d_{jz\sigma} + \text{H.c.} \right), \tag{38}$$

where $d_{ix\sigma}^\dagger$ ($d_{iz\sigma}^\dagger$) creates a hole in $|x\rangle$ ($|z\rangle$) orbital with spin σ. The sums run over the bonds between nearest neighbors oriented along the cubic axes: $\langle ij \rangle \parallel$ within the (a, b)-planes, and $\langle ij \rangle \perp$ along the c-axis [perpendicular to (a, b)-planes], respectively, and $\beta = 1$ in a cubic system. The $x - z$ hopping in the (a, b) planes depends on the phases of the $x^2 - y^2$ orbitals along a- and b-axis, respectively, included in the factors $\pm\sqrt{3}$ in Eq. (38). The electron-electron interactions are described by the on-site terms, which we write in the following form,

$$H_{int} = (U + \tfrac{1}{2}J_H) \sum_{i\alpha} n_{i\alpha\uparrow} n_{i\alpha\downarrow} + (U - J_H) \sum_{i\sigma} n_{ix\sigma} n_{iz\sigma} + (U - \tfrac{1}{2}J_H) \sum_{i\sigma} n_{ix\sigma} n_{iz\bar\sigma}$$
$$- \tfrac{1}{2}J_H \sum_{i\sigma} d_{ix\sigma}^\dagger d_{ix\bar\sigma} d_{iz\bar\sigma}^\dagger d_{iz\sigma} + \tfrac{1}{2}J_H \sum_i (d_{ix\uparrow}^\dagger d_{ix\downarrow}^\dagger d_{iz\downarrow} d_{iz\uparrow} + d_{iz\uparrow}^\dagger d_{iz\downarrow}^\dagger d_{ix\downarrow} d_{ix\uparrow}), \tag{39}$$

with U and J_H standing for the Coulomb and Hund's rule exchange interaction, respectively, and $\alpha = x, z$. Moreover, we have used the simplified notation $\bar\sigma = -\sigma$. For convenience, U has been defined as the average $d_i^9 d_j^9 \to d_i^{10} d_j^8$ excitation energy of the d^8 configuration, which coincides with the energy of the central $|^1E\rangle$ doublet.

Therefore, U is here not the interorbital Coulomb element. The interaction element J_H stands for the singlet-triplet splitting in the d^8 spectrum (Fig. 10) and is just twice as big as the exchange element K_{xz} usually used in quantum chemistry [31,73]. With the present definition of J_H, the interorbital interaction between holes of opposite (equal) spins is $U - J_H/2$ ($U - J_H$), respectively. This Hamiltonian (39) describes correctly the multiplet structure of d^8 (and d^2) ions [31], and is rotationally invariant in the orbital space [32]. The wave functions have been assumed to be real which gives the same element $J_H/2$ for the exchange interaction and for the pair hopping term between the e_g orbitals, $|x\rangle$ and $|z\rangle$.

The last term in Eq. (37) stands for the crystal field which lifts the degeneracy of the two e_g orbitals and breaks the symmetry in the orbital space,

$$H_z = \sum_{i\sigma}(\varepsilon_x n_{ix\sigma} + \varepsilon_z n_{iz\sigma}) = -\tfrac{1}{2}E_z\sum_{i\sigma}(n_{ix\sigma} - n_{iz\sigma}), \tag{40}$$

if $\varepsilon_x \neq \varepsilon_z$ (and neglecting a constant term $\propto \varepsilon_x + \varepsilon_z$). Here ε_x and ε_z are the energies of a hole at $|x\rangle$ and $|z\rangle$ orbitals, respectively, and

$$E_z = \varepsilon_z - \varepsilon_x. \tag{41}$$

Its effect is like that of a magnetic field in the orbital space, and together with the parameter β in H_{kin} (38) quantifies the deviation in the electronic structure from the ideal cubic local point group.

In the atomic limit, i.e., at $t = 0$, one finds at $E_z = 0$ a highly degenerate problem, with orbital degeneracy next to spin degeneracy. All four basis states per site, with a hole occupying either orbital, $|x\rangle$ or $|z\rangle$, and either spin state, $\sigma = \uparrow$ or $\sigma = \downarrow$, have the same energy. Therefore, the system of N d^9 ions has a degeneracy 4^N, which is, however, removed by the effective interactions between each pair of nearest neighbor ions $\{i,j\}$ that originate from virtual transitions to the excited states, $d_i^9 d_j^9 \rightleftharpoons d_i^{10} d_j^8$, due to hole hopping $t \neq 0$. Hence, we derive the effective spin-orbital model following Kugel and Khomskii [18], starting from the Hamiltonian in the atomic limit, $H_{at} = H_{int} + H_z$, and treating H_{kin} as a perturbation. However, we report here the study which includes the *full multiplet structure* of the excited states within the d^8 configuration which gives corrections of the order of J_H compared with the earlier results of Refs. [18] and [15].

Knowing the multiplet structure of the d^8 intermediate states, the derivation of the effective Hamiltonian can be done in various ways. The most straightforward but lengthy procedure is a generalization of the canonical transformation method used earlier for the Hubbard [4] and the three-band [13] model. A significantly shorter derivation is possible, however, using the cubic symmetry and starting with the interactions along the c-axis. Here the derivation simplifies tremendously as one finds only effective interactions which result from the hopping of holes between the directional $|z\rangle$ orbitals, as shown in Fig. 10. Next the interactions in the remaining directions can be generated by the appropriate rotations to the other cubic axes a and b, and by applying the symmetry rules for the hopping elements between the e_g orbitals [48].

Following the above argument, the derivation of the effective interactions between two d^9 ions at sites i and j takes the simplest form for a bond $\langle ij \rangle$ oriented along the c-axis. In that case due to the vanishing hopping from/to $|x\rangle$ orbital, the orbital occupancies in the initial and final $d_i^9 d_j^9$ states have to be identical (apart from a possible simultaneous and opposite spin flip at both sites), i.e., the zth component of the pseudospin T^z is conserved. The possible initial states are described by a direct product of the total spin state, either a triplet ($S = 1$) or a singlet ($S = 0$), and the orbital configuration, given by one of four possibilities: $|x_i x_j\rangle$, $|x_i z_j\rangle$, $|z_i x_j\rangle$, or $|z_i z_j\rangle$. Moreover, the effective interaction vanishes if the holes occupy the $|x_i x_j\rangle$ configuration. The total spin per two sites is conserved in the $d_i^9 d_j^9 \to d_i^{10} d_j^8$ excitation process, and therefore the spin dependence of the resulting second order Hamiltonian can be expressed in terms of the projection operators on the total spin states, defined for a given bond $\langle ij \rangle$ by Eq. (21).

Depending on whether the initial state is $|z_i x_j\rangle$ or $|z_i z_j\rangle$, the intermediate $d_i^{10} d_j^8$ configuration resulting from the hole-hop $|z_i\rangle \to |z_j\rangle$, involves different d^8 excited states: either the interorbital states, the triplet $^3A_2$ and the singlet $^1E_\theta$ (for $|z_i x_j\rangle$), or the two singlets built from the states with doubly occupied orbitals, $^1E_\epsilon$ and $^1A_1$ (for $|z_i z_j\rangle$). Of course, since the wave function has to be antisymmetric, the spins have to be opposite in the latter case, while in the former case also parallel spin configurations contribute in the triplet channel. The eigenstates within the e_g subspace are:

(i) triplet: $|^3A_2\rangle = \{|z\uparrow\rangle|x\uparrow\rangle, \frac{1}{\sqrt{2}}(|z\uparrow\rangle|x\downarrow\rangle + |z\downarrow\rangle|x\uparrow\rangle), |z\downarrow\rangle|x\downarrow\rangle\}$,

(ii) interorbital singlet $|^1E_\epsilon\rangle = \frac{1}{\sqrt{2}}(|z\uparrow\rangle|x\downarrow\rangle - |z\downarrow\rangle|x\uparrow\rangle)$,

(iii) bonding $|^1E_\theta\rangle$ and antibonding $|^1A_1\rangle$ singlets:
$|^1E_\theta\rangle = \frac{1}{\sqrt{2}}(|z\uparrow\rangle|z\downarrow\rangle + |x\uparrow\rangle|x\downarrow\rangle)$, and
$|^1A_1\rangle = \frac{1}{\sqrt{2}}(|z\uparrow\rangle|z\downarrow\rangle - |x\uparrow\rangle|x\downarrow\rangle)$, with double occupancies of both orbitals.

The energies of the states $|^3A_2\rangle$ and $|^1E_\epsilon\rangle$ are straightforwardly obtained using $\vec{S}_{ix} \cdot \vec{S}_{iz} = +1/4$ and $\vec{S}_{ix} \cdot \vec{S}_{iz} = -3/4$, for $S = 1$ and $S = 0$ states, respectively. The remaining two singlet energies are found by diagonalizing a 2×2 problem in the subspace of doubly occupied states. Hence, the resulting excitation energies which correspond to the *local excitations* $d_i^9 d_j^9 \to d_i^{10} d_j^8$ on a given bond $\langle ij \rangle$ are,

$$\varepsilon(^3A_2) = U - J_H, \tag{42}$$

$$\varepsilon(^1E_\epsilon) = U, \tag{43}$$

$$\varepsilon(^1E_\theta) = U + \frac{1}{2}J_H - \frac{1}{2}J_H \left[1 + (E_z/J_H)^2\right]^{1/2}, \tag{44}$$

$$\varepsilon(^1A_1) = U + \frac{1}{2}J_H + \frac{1}{2}J_H \left[1 + (E_z/J_H)^2\right]^{1/2}, \tag{45}$$

At $E_z = 0$ it consists of equidistant states, with a distance of J_H between the triplet $|^3A_2\rangle$ and the degenerate singlets $|^1E_\theta\rangle$ and $|^1E_\epsilon\rangle$ (which form of course an orbital doublet), as well as between the above singlets and the highest energy singlet $|^1A_1\rangle$. Note that when the pair hopping term $\propto J_H$ is neglected in Hamiltonian (39), the spectrum is incorrect, with $\varepsilon(^1E_\theta) = \varepsilon(^1A_1) = U + J_H/2$.

At this point we have all the elements for deriving the effective spin-orbital model. Hence, its general form is given by the formula which includes all possible virtual transitions to the excited $d_i^8 d_j^{10}$ configurations,

$$H_{\langle ij\rangle} = -\sum_{n,\alpha\beta} \frac{t^2}{\varepsilon_n} Q_S(i,j) P_{i\alpha} P_{j\beta}, \qquad (46)$$

where t stands for the $z-z$ hopping along the c-axis, $Q_S(i,j)$ is one of the projection operators on the total spin state (21), either $S = 0$ or $S = 1$, and $P_{i\alpha}$ is the projection operator on the orbital state α at site i, while ε_n stands for the excitation energies given by Eqs. (42)–(45). The orbital projection operators on $|x\rangle$ and $|z\rangle$ orbital in the initial and final state of the d^9 configuration at site i are, respectively,

$$P_{ix} = |ix\rangle\langle ix| = \tfrac{1}{2} + \tau_i^c, \qquad (47)$$
$$P_{iz} = |iz\rangle\langle iz| = \tfrac{1}{2} - \tau_i^c, \qquad (48)$$

where τ_i^c stands for the zth component of pseudospin and is given by

$$\tau_i^c = \tfrac{1}{2}\sigma_i^z. \qquad (49)$$

The interaction terms along the bonds $\langle ij\rangle \parallel (a,b)$ are represented by the projection operators similar to P_{ix} and P_{iz}, with τ_i^c replaced by the orbital operators τ_i^a and τ_i^b which are expressed in terms of the Pauli matrices as follows:

$$\tau_i^a = -\tfrac{1}{4}(\sigma_i^z - \sqrt{3}\sigma_i^x), \qquad \tau_i^b = -\tfrac{1}{4}(\sigma_i^z + \sqrt{3}\sigma_i^x). \qquad (50)$$

Here, the σ_i^α are Pauli matrices acting on the orbital pseudospins:

$$|x\rangle = \begin{pmatrix} 1 \\ 0 \end{pmatrix}, \qquad |z\rangle = \begin{pmatrix} 0 \\ 1 \end{pmatrix}. \qquad (51)$$

Expanding Eq. (46) for a bond $\langle ij\rangle$ along the c-direction, one finds

$$H_{\langle ij\rangle} = -\frac{t^2}{\varepsilon(^3A_2)}\left(\vec{S}_i \cdot \vec{S}_j + \frac{3}{4}\right)(P_{ix}P_{jz} + P_{iz}P_{jx}) + \frac{t^2}{\varepsilon(^1E_\epsilon)}\left(\vec{S}_i \cdot \vec{S}_j - \frac{1}{4}\right) \times$$
$$(P_{ix}P_{jz} + P_{iz}P_{jx}) + \left[\frac{t^2}{\varepsilon(^1E_\theta)} + \frac{t^2}{\varepsilon(^1A_1)}\right]\left(\vec{S}_i \cdot \vec{S}_j - \frac{1}{4}\right) 2P_{iz}P_{jz}. \qquad (52)$$

As one can see, the magnetic interactions in the first two terms in Eq. (52) cancel each other in the limit of $\eta \to 0$, while the last term favors AF spin orientation independently of η. We recognize that Hamiltonian (52) describes the superexchange along the bond $\langle ij\rangle \parallel c$, with the superexchange constant of $4t^2/U$ [52,4]. However, the hopping in the other directions $\langle ij \rangle \parallel (a,b)$ is reduced and thus we define for convenience $J = t^2/U$ as the energy unit. For simplifying the form (52) we use an expansion of the excitation energies ε_n in the denominators for small J_H, and introduce

$$\eta = J_H/U \qquad (53)$$

as a parameter which quantifies the Hund's rule exchange. Using the explicit form of the orbital projection operators $P_{i\alpha}$ (47) this results in the following form of the effective Hamiltonian for the bond $\langle ij \rangle \parallel c$,

$$H_{\langle ij \rangle} = J \left[(1+\eta) \left(\vec{S}_i \cdot \vec{S}_j + \frac{3}{4} \right) - \left(\vec{S}_i \cdot \vec{S}_j - \frac{1}{4} \right) \right] \times (P_{ix}P_{jz} + P_{iz}P_{jx})$$
$$+ 4J \left(1 - \frac{1}{2}\eta \right) \left(\vec{S}_i \cdot \vec{S}_j - \frac{1}{4} \right) P_{iz}P_{jz}, \qquad (54)$$

which may be represented explicitly by the orbital operators τ_i^c and τ_j^c in the following way,

$$H_{\langle ij \rangle} = J \left(4\vec{S}_i \cdot \vec{S}_j + 1 \right) \left(\tau_i^c - \frac{1}{2} \right) \left(\tau_j^c - \frac{1}{2} \right) + \tau_i^c + \tau_j^c - 1$$
$$+ J\eta \left(\vec{S}_i \cdot \vec{S}_j \right) (\tau_i^c + \tau_j^c - 1) + \frac{1}{2} J\eta \left[\left(\tau_i^c - \frac{1}{2} \right) \left(\tau_j^c - \frac{1}{2} \right) + 3 \left(\tau_i^c \tau_j^c - \frac{1}{4} \right) \right]. \qquad (55)$$

The first line represents the AF superexchange interactions $\propto J$, while the second line describes the weaker FM interactions $\propto J\eta$, which originate from the multiplet splittings of the d^8 excited states.

It is straightforward to verify that the above form of the effective Hamiltonian simplifies in the limit of occupied $|z\rangle$ orbitals to

$$H_{\langle ij \rangle} = 4J \left(1 - \frac{1}{2}\eta \right) \left(\vec{S}_i \cdot \vec{S}_j - \frac{1}{4} \right), \qquad (56)$$

and one recognizes the same constant $-\frac{1}{4}$, and the same superexchange interaction $4J = 4t^2/U$ as in the $t-J$ model at half-filling [4]. However, the effective superexchange is somewhat reduced by the factor $(1 - \frac{1}{2}\eta)$ in the presence of the Hund's rule interaction, which increases the excitation energy $\varepsilon(^1A_1)$. The effective interactions along the bonds within the (a,b) planes may be now obtained by rotating Eq. (52) with the projection operators P_{ix} and P_{iz} [or its simplified version (55) with the orbital operators τ_i^c] by $\pi/2$ to the cubic axes a and b, which generates the orbital operators τ_i^a and τ_i^b (50), respectively [64]. This results in a nontrivial coupling between the orbital and spin degrees of freedom.

Following the above procedure, we have derived the effective Hamiltonian H in spin-orbital space,

$$H(d^9) = H_J + H_\tau, \qquad (57)$$

where the superexchange part H_J can be most generally written as follows (a simplified form was discussed recently in Refs. [63] and [64]),

$$H_J = \sum_{\langle ij \rangle} \left\{ -\frac{t^2}{\varepsilon(^3A_2)} \left(\vec{S}_i \cdot \vec{S}_j + \frac{3}{4} \right) \mathcal{P}_{\langle ij \rangle}^{\zeta\xi} + \frac{t^2}{\varepsilon(^1E_\epsilon)} \left(\vec{S}_i \cdot \vec{S}_j - \frac{1}{4} \right) \mathcal{P}_{\langle ij \rangle}^{\zeta\xi} \right.$$
$$\left. + \left[\frac{t^2}{\varepsilon(^1E_\theta)} + \frac{t^2}{\varepsilon(^1A_1)} \right] \left(\vec{S}_i \cdot \vec{S}_j - \frac{1}{4} \right) \mathcal{P}_{\langle ij \rangle}^{\zeta\zeta} \right\}, \quad (58)$$

and the crystal-field term (59) we rewrite now in the form,

$$H_\tau = -E_z \sum_i \tau_i^c. \quad (59)$$

In general, the energies of the two orbital states, $|x\rangle$ and $|z\rangle$, are different, and thus the complete effective Hamiltonian of the d^9 model (57) includes as well the crystal-field term. It acts as a "magnetic field" for the orbital pseudospins, and is loosely associated with an uniaxial pressure along the c-axis.

The operators $\vec{S}_i$ in Eq. (58) refer to a spin $S = 1/2$ at site i, while $P_{\langle ij \rangle}^{\alpha\beta}$ are projection operators on the orbital states for each bond,

$$\mathcal{P}_{\langle ij \rangle}^{\zeta\xi} = (\frac{1}{2} + \tau_i^\alpha)(\frac{1}{2} - \tau_j^\alpha) + (\frac{1}{2} - \tau_i^\alpha)(\frac{1}{2} + \tau_j^\alpha), \quad (60)$$

$$\mathcal{P}_{\langle ij \rangle}^{\zeta\zeta} = 2(\frac{1}{2} - \tau_i^\alpha)(\frac{1}{2} - \tau_j^\alpha), \quad (61)$$

where $\alpha = a, b, c$ refers to the cubic axes, respectively. The individual projection operators on the orbital state which is parallel (perpendicular) to the bond direction are:

$$P_{i\zeta} = \tfrac{1}{2} - \tau_i^\alpha, \qquad P_{i\xi} = \tfrac{1}{2} + \tau_i^\alpha, \quad (62)$$

and are constructed with the orbital operators (49) and (50) associated with the three cubic axes. The global operators (60) and (61) select orbitals that are either parallel ($P_{i\zeta}$) to the direction of the bond $\langle ij \rangle$ on site i, and perpendicular ($P_{j\xi}$) on the other site j, as in $\mathcal{P}_{\langle ij \rangle}^{\zeta\xi}$, or parallel on both sites, as in $\mathcal{P}_{\langle ij \rangle}^{\zeta\zeta}$, respectively. Hence, we find a Heisenberg Hamiltonian for the spins, coupled into an orbital problem. While the spin problem is described by the continuous symmetry group $SU(2)$, the orbital problem is clock-model like, i.e., there are three directional orbitals: $3x^2 - r^2$, $3y^2 - r^2$, and $3z^2 - r^2$, but they are not independent, and transform into each other by appropriate cubic rotations. In general, the occupied orbital state at a given site i may be expressed by the following transformation of bond basis $\{|z\rangle, |x\rangle\}$ with an assigned angle θ (9). In order to give an idea of the possible orbital configuration one can get by changing θ, we have summarized the results obtained for a few representative angles in Table 1.

TABLE 1. Orbital configuration for a few representative values of the orbital rotation angle θ [see Eq. (9)] for the site $|i\rangle$.

θ	$|i\rangle$
0	$\frac{1}{\sqrt{3}}(3z^2 - r^2) \equiv |z\rangle$
$\frac{\pi}{6}$	$z^2 - y^2$
$\frac{\pi}{4}$	$\frac{1}{\sqrt{6}}\left[2z^2 + (\sqrt{3}-1)x^2 - (\sqrt{3}+1)y^2\right]$
$\frac{\pi}{3}$	$-\frac{1}{\sqrt{3}}(3y^2 - r^2)$
$\frac{\pi}{2}$	$x^2 - y^2 \equiv |x\rangle$
$\frac{3\pi}{4}$	$\frac{1}{\sqrt{6}}\left[2z^2 - (\sqrt{3}+1)x^2 + (\sqrt{3}-1)y^2\right]$

The d^9 spin-orbital model (57)-(59) depends thus on two parameters: (i) the crystal field splitting E_z (41), and (ii) the Hund's rule exchange J_H (53). While the first two terms in (58) cancel for the magnetic interactions in the limit of $\eta \to 0$, the last term favors AF spin orientation. Using again η (53) as an expansion parameter which quantifies the Hund's rule exchange, one finds the following form of the effective exchange Hamiltonian in the d^9 model (57) [63],

$$H_J \simeq J \sum_{\langle ij \rangle} \left[2\left(\vec{S}_i \cdot \vec{S}_j - \frac{1}{4}\right) P^{\zeta\zeta}_{\langle ij \rangle} - P^{\zeta\xi}_{\langle ij \rangle} \right]$$
$$- J\eta \sum_{\langle ij \rangle} \left[\vec{S}_i \cdot \vec{S}_j \left(P^{\zeta\zeta}_{\langle ij \rangle} + P^{\zeta\xi}_{\langle ij \rangle} \right) + \frac{3}{4} P^{\zeta\xi}_{\langle ij \rangle} - \frac{1}{4} P^{\zeta\zeta}_{\langle ij \rangle} \right]. \tag{63}$$

The first term in Eq. (63) describes the AF superexchange $\propto J = t^2/U$ (where t is the hopping between $|\zeta\rangle$ orbitals along the $\langle ij \rangle$ bond), and is the leading interaction term obtained when the splittings between different excited d^8 states $\propto J_H$ are neglected. As we show below, in spite of the AF superexchange $\propto J$, *no LRO can stabilize in a system described by the spin-orbital model (57) in the limit $\eta \to 0$ at orbital degeneracy* ($E_z = 0$) because of the presence of the frustrating orbital interactions $\propto P^{\zeta\zeta}_{\langle ij \rangle}$ which give a highly degenerate classical ground state. We emphasize that even in the limit of $J_H \to 0$ the present Kugel-Khomskii model *does not obey SU(4) symmetry*, essentially because of the directionality of the e_g orbitals. Therefore, such an idealized SU(4)-symmetric model (see Sec. IV.A) does not correspond to the realistic situation of degenerate e_g orbitals and is expected to give different answers concerning the interplay of spin and orbital ordering in cubic crystals.

Taking into account the multiplet splittings, we obtain [see Eq. (63)] again a Heisenberg-like Hamiltonian for the spins coupled into an orbital problem, with a reduced interaction $\propto J\eta$. It is evident that the new terms support FM rather than

AF spin interactions for particular orbital orderings. This net FM superexchange originates from the virtual transitions which involve the triplet state $|^3A_2\rangle$, having the lowest energy and thus providing the strongest effective magnetic coupling.

The important feature of the spin-orbital model (57) is that the *actual magnetic interactions depend on the orbital pattern*. This follows essentially from the hopping matrix elements in H_t (38) being different between a pair of $|x\rangle$ orbitals, between a pair of different orbitals (one $|x\rangle$ and one $|z\rangle$ orbital), and between a pair of $|z\rangle$ orbitals, respectively, and depending on the bond direction either in the (a, b) planes, or along the c-axis [16]. We show below that this leads to a particular competition between magnetic and orbital interactions, and the resulting phase diagram contains a rather large number of classical phases, stabilized for different values of E_z and J_H.

B Classical phases and phase diagrams

The simplest approach to the d^9 spin-orbital model as given by Eqs. (57), and (63) for getting an insight into the competition between spin and orbital interactions is the MF theory which is formally obtained by replacing the scalar products $\vec{S}_i \cdot \vec{S}_j$ by the Ising terms, $S_i^z S_j^z$. We report here the MF study of the phase diagram after Ref. [64] for a distorted system with respect to the cubic perovskite lattice. Therefore, we introduce a parameter β which controls the anisotropy along the c-axis and leads to the different exchange constants in (a, b) planes $(J_a = J_b = J)$, and along c-direction $(J_c = J\beta)$:

$$\mathcal{H}_{\mathrm{MF}} \simeq \sum_{\langle ij \rangle} J_\alpha \left[2 \left(S_i^z S_j^z - \tfrac{1}{4} \right) \mathcal{P}_{\langle ij \rangle}^{\zeta\zeta} - \mathcal{P}_{\langle ij \rangle}^{\zeta\xi} \right]$$
$$- \eta \sum_{\langle ij \rangle} J_\alpha \left[S_i^z S_j^z \left(\mathcal{P}_{\langle ij \rangle}^{\zeta\zeta} + \mathcal{P}_{\langle ij \rangle}^{\zeta\xi} \right) + \tfrac{3}{4} \mathcal{P}_{\langle ij \rangle}^{\zeta\xi} - \tfrac{1}{4} \mathcal{P}_{\langle ij \rangle}^{\zeta\zeta} \right] - E_z \sum_i \tau_i^c. \quad (64)$$

Here $\beta < 1$ ($\beta > 1$) corresponds to the elongation (compression) of the bond $\langle ij \rangle \parallel c$, respectively. The two limiting cases: $\beta = 0$ and $\beta = 1$, stand for the 2D (square) lattice, and the 3D undistorted (perovskite) lattice, respectively. At first sight the MF Hamiltonian (64) contains a dominating AF exchange $\propto J$ which competes with a FM one $\propto \eta J$, and suggests that one should search for a solution with different exchange constants along the three cubic axes. In the following we will consider several magnetic patterns with two- and four-sublattice 3D structures. They include the possibility of having: the G-AF order (AF spin alternating along all three cubic directions), A-AF or 1D-AF phase (FM interaction along two cubic directions and AF along the third axis), and C-AF order (FM exchange along 1D chains, and AF exchange in the directions perpendicular to them).

Moreover, the interaction between orbital variables has also an AF character, $\sim J\tau_i^\alpha \tau_j^\alpha$, suggesting that it might be energetically more favorable to alternate the orbitals in a certain regime of parameters, and pay thereby part of the magnetic energy. This gives the main idea of the complex frustration present in this system. Therefore, to any classical arrangements of spins one has to find the optimal configuration of *occupied orbitals* which minimizes the total energy. Hence, we allow for mixed orbital states of the type as given in Eq. (9),

$$|i\mu\sigma\rangle = \cos\theta_i |iz\sigma\rangle + \sin\theta_i |ix\sigma\rangle, \tag{65}$$

with the set of angles $\{\theta_i\}$ to be found variationally from the minimization of the classical energy. Let us suppose that the orbitals occupied at sites i and j are given by the superposition of the states $\{|iz\sigma\rangle, |ix\sigma\rangle\}$ (65) with an angle θ_i and θ_j, respectively. One finds then the average values of the operator projection operators $\{P_{i\alpha}\}$ for the bonds $\langle ij\rangle \parallel c$:

$$\langle P_{ix}P_{jz} + P_{iz}P_{jx}\rangle = \cos^2\theta_i \sin^2\theta_j + \cos^2\theta_j \sin^2\theta_i, \tag{66}$$

$$\langle 2P_{iz}P_{jz}\rangle = 2\cos^2\theta_i \cos^2\theta_j, \tag{67}$$

while for the bonds $\langle ij\rangle \parallel (a,b)$ they are:

$$\langle P_{i\xi}P_{j\zeta} + P_{i\zeta}P_{j\xi}\rangle = \frac{1}{8}\left[4 - 2\cos 2(\theta_j - \theta_i) + \cos 2(\theta_j + \theta_i) - \sqrt{3}\sin 2(\theta_j + \theta_i)\right], \tag{68}$$

$$\langle 2P_{i\zeta}P_{j\zeta}\rangle = \frac{1}{8}(-2 + \cos 2\theta_i + \sqrt{3}\sin 2\theta_i)(-2 + \cos 2\theta_j + \sqrt{3}\sin 2\theta_j). \tag{69}$$

By means of these expressions one can easily determine the MF energy for any orbital configuration, assuming that the spin structure is assigned. Let us start from the MF solutions with G-AF type of magnetic structure, that is from the 3D Néel state.

It is clear that at large positive E_z, where the crystal field strongly favors $|x\rangle$-occupancy over $|z\rangle$-occupancy, one expects that $\theta_i = \pi/2$ in Eq. (65), and the holes occupy $|x\rangle$ orbitals at every site. In this case the spins do not interact in the c-direction (see Fig. 10), and there is also no orbital energy contribution. Hence, the (a,b) planes will decouple magnetically, while within each plane the superexchange is AF and equal to $9J/4$ along a and b. These interactions stabilize a 2D antiferromagnet, called further AFxx phase. On the contrary, if $E_z < 0$ and $|E_z|$ is large, then the holes occupy $|z\rangle$ orbitals and $\theta_i = 0$ in Eq. (65). By means of the expressions (66) – (69), we find that the spin system has then strongly anisotropic AF superexchange, being $4J$ on the bonds $\langle ij\rangle$ along the c-axis, and $J/4$ on the bonds within the (a,b) planes, respectively. This 3D Néel state with the holes occupying $|z\rangle$ orbitals is called AFzz phase. The spin and orbital order in both AF phases is shown schematically within the (a,b) planes in Fig. 11. In this case the energies normalized per one site are given by:

$$E_{\text{AFxx}} = -3J\left(1 - \frac{\eta}{4}\right) - \frac{1}{2}E_z, \tag{70}$$

$$E_{\text{AFzz}} = -J\left(1 + \frac{\eta}{4}\right) - 2J\beta\left(1 - \frac{\eta}{2}\right) + \frac{1}{2}E_z. \tag{71}$$

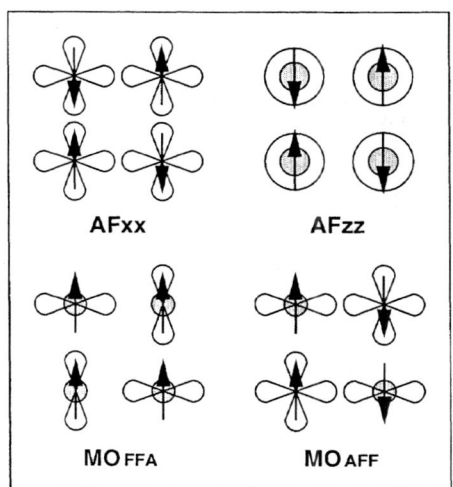

FIGURE 11. Schematic representation of magnetic and orbital long-range orderings in (a,b) planes for the classical phases: AFxx, AFzz, MO$_{\text{FFA}}$ and MO$_{\text{AFF}}$ phases. Grey parts of different e_g orbitals are oriented along the c-axis (after Ref. [64]).

The AFxx and AFzz phases are degenerate in a cubic system ($\beta = 1$) along the line $E_z = 0$, while decreasing β moves the degeneracy point to negative values of E_z, given by $E_z = -2J(1-\beta)(1-\frac{\eta}{2})$.

However, for intermediate values of E_z one may expect to optimize the energy by realizing mixed orbital configurations ($0 < \theta < \pi/2$). In this case, guided by the observation that the orbital interaction is AF-like, we look for solutions with alternating orbitals at two sublattices, A and B. The alternation is chosen in a way to allow the orbitals being parallel (optimizing the magnetic energy) in one direction, and being (almost) orthogonal in the other two (optimizing the orbital energy). Such states are realized by choosing in Eq. (65) the angles alternating between two sublattices in particular planes: $\theta_i = +\theta$ for $i \in A$, and $\theta_j = -\theta$ for $j \in B$, respectively,

$$|i\mu\sigma\rangle = \cos\theta |iz\sigma\rangle + \sin\theta |ix\sigma\rangle,$$
$$|j\mu\sigma\rangle = \cos\theta |jz\sigma\rangle - \sin\theta |jx\sigma\rangle. \qquad (72)$$

Let us assume first the G-AF state. By evaluating the orbital operators following Eqs. (66) – (69) for this case, one finds easily the energy as a function of θ in Eqs. (72),

$$E(\theta) = -\frac{J}{4}(1+\frac{\eta}{2})(7 - 4\cos^2 2\theta) - \frac{J}{4}(1-\frac{\eta}{2})(1 - 2\cos 2\theta)^2$$
$$- \frac{J}{2}\beta(1+\frac{\eta}{2})(1 - \cos^2 2\theta) - \frac{J}{2}\beta(1-\frac{\eta}{2})(1 + \cos 2\theta)^2 + \frac{1}{2}E_z \cos 2\theta. \qquad (73)$$

This expression has a minimum at

$$\cos 2\theta = -\frac{(1-\frac{\eta}{2})(1-\beta) + \frac{1}{2}\varepsilon_z}{(2+\beta)\eta}, \quad (74)$$

where $\varepsilon_z = E_z/J$, if $\eta \neq 0$, and provided that $|\cos 2\theta| \leq 1$ (a similar condition applies to all the other states with MO considered below). So, as long as $2J(\beta - 1) - 3J(\beta+1)\eta \leq E_z \leq 2J(\beta-1) + J(5+\beta)\eta$, there is genuine MO order, while upon reaching the smaller (larger) boundary value for E_z, the orbitals go over smoothly into $|z\rangle$ ($|x\rangle$), i.e., one retrieves the AFzz (AFxx) phase. Taking the magnetic ordering in the three cubic directions $\{a,b,c\}$ as a label to classify the classical phases with MO (72), we call the phase obtained in the regime of genuine MO order MO$_{AAA}$, with classical energy given by

$$E_{\text{MO}_{AAA}} = -\left(2 + \beta + \frac{3}{4}\eta\right)J - J\frac{[(2-\eta)(1-\beta) + \varepsilon_z]^2}{4(2+\beta)\eta}. \quad (75)$$

In a similar fashion we can get the MF solutions for other possible spin configurations of A-AF type. Consider first the MO$_{FFA}$ phase, with FM order within the (a,b) planes, and AF order along the c-axis. The classical energy as a function of θ is given by:

$$E(\theta) = -\frac{J}{4}(1+\eta)(7 - 4\cos^2 2\theta) - \frac{J}{2}\beta\left(1+\frac{\eta}{2}\right)(1-\cos^2 2\theta)$$
$$- \frac{J}{2}\beta\left(1-\frac{\eta}{2}\right)(1+\cos 2\theta)^2 + \frac{1}{2}E_z \cos 2\theta, \quad (76)$$

with a minimum at

$$\cos 2\theta = \frac{\beta(1-\frac{\eta}{2}) - \frac{1}{2}\varepsilon_z}{2 + (2+\beta)\eta}, \quad (77)$$

where again the MO exist as long as $|\cos 2\theta| \leq 1$. Using Eqs. (76) and (77) one finds that the classical energy of the MO$_{FFA}$ phase is given by

$$E_{\text{MO}_{FFA}} = -\frac{J}{4}(11 - 7\eta) - \frac{J}{2}\frac{[\beta(1-\frac{\eta}{2}) - \frac{1}{2}\varepsilon_z]^2}{2 + (2+\beta)\eta}. \quad (78)$$

This solution is stable for $E_z < 0$, while for $E_z > 0$ the other two degenerate phases: the MO$_{FAF}$ and MO$_{AFF}$ phase have a lower energy, as they are characterized by a lower hole density in $|z\rangle$ orbitals which become unfavorable. In this case, due to the breaking of local symmetry of the magnetic interactions within the (a,b) planes, with one direction AF and the other FM, one is forced to look for solutions with different angles on the two sublattices [64].

Finally, one may consider how the degeneracy of the AFxx phase is removed by the interactions along the c-axis. One possibility is the MO$_{AAA}$ phase, with the energy given above by Eq. (75). If the interactions along the c-axis are instead FM, one finds the classical energy of the MO$_{\text{AAF}}$ phase given by

$$E_{\text{MOAAF}} = -\left(2 + \frac{3}{4}\eta\right)J - \frac{1}{2}\beta(1+\eta) - J\frac{(2-\eta+\varepsilon_z)^2}{2[\beta(1+\eta)+2\eta]}, \qquad (79)$$

with the mixing angle

$$\cos 2\theta = -\frac{1 - \frac{\eta}{2} + \frac{1}{2}\varepsilon_z}{\beta(1+\eta) + 2\eta}. \qquad (80)$$

This solution turns out to be stable with respect to the MO$_{\text{AAA}}$ as long as $1 + \cos 2\theta < \eta$. This means that when the hole density in the $|z\rangle$ orbitals $\sim \cos^2\theta$ grows smoothly from zero (at $\theta = \pi/2$) with decreasing E_z, it tends to stabilize first the MO$_{\text{AAF}}$ phase by FM terms $\sim J\eta\cos^2\theta$, while at higher occupancy of $|z\rangle$ orbitals the AF interactions $\sim J\cos^4\theta$ take over.

Thus, one obtains the classical phase diagram of the 3D spin-orbital model (57) by comparing the energies of the six above phases for various values of two parameters, E_z/J and J_H/U: two AF phases with two sublattices and pure orbital character (AFxx and AFzz), three A-AF phases with four sublattices (MO$_{\text{FFA}}$ and two degenerate phases: MO$_{\text{AFF}}$ and MO$_{\text{FAF}}$), one C-AF phase (MO$_{\text{AAF}}$), and one G-AF phase with MO's (MO$_{\text{AAA}}$). By looking at the phase diagram one can see that the generic sequence of classical phases at finite η and decreasing E_z/J is: AFxx, MO$_{\text{AAF}}$, MO$_{\text{AAA}}$, MO$_{\text{AFF}}$, MO$_{\text{FFA}}$, and AFzz, and the magnetic order is tuned together with the gradually increasing $|z\rangle$ character of the occupied orbitals. By making several other choices of orbital mixing and classical magnetic order, it has been verified that no other commensurate ordering with up to four sublattices can be stable in the present situation. Although some other phases have been found, they were degenerate with the above phases only at the $M = (0,0)$ point of the phase diagram, and otherwise had higher energies.

The result for cubic symmetry ($\beta = 1$) is presented in Fig. 12, where one finds all six phases, but the MO$_{\text{AAA}}$ phase does stabilize only in a very restricted range of parameters for $J_H/U < 0.1$, in between AFxx and MO$_{\text{AFF}}$ phases. Only the first of the above transitions is continuous, while the other lines in Fig. 12 are associated with jumps in the magnetic and in orbital patterns. We would like to emphasize that *all the considered phases are degenerate at the $M = (0,0)$ point* [63]. It is a multicritical point, where the orbitals may be rotated freely when the spins are AF, and a few other states with FM planes, and tuned to them orbital order of the MO type gives precisely the same energy.

When $\beta \neq 1$, the phase diagram changes quantitatively but not qualitatively, with either expanded or reduced areas corresponding to the different classical phases [64]. In particular, $\beta > 1$ stabilizes the MO phases [especially the MO$_{\text{AFF}}$(MO$_{\text{FAF}}$) states]. On the contrary, the MO phases are stable in a reduced range of E_z for a fixed value of J_H/U, if $\beta < 1$. It is worth emphasizing that the multicritical point M is a common feature of the classical phase diagram independently of the

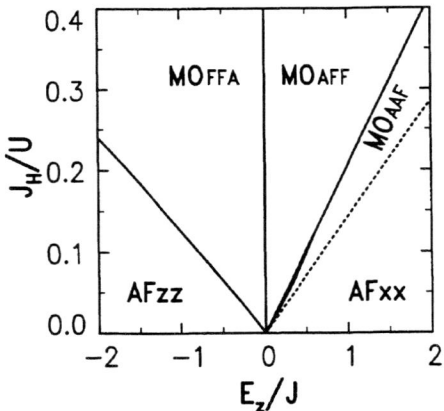

FIGURE 12. Mean-field phase diagram of the 3D spin-orbital model (57) in the (E_z, J_H) plane for $\beta = 1$ (after Ref. [64]). The lines separate the classical states shown in Fig. 11; the transition from AFxx to MOAFF phase is second order (dashed line), while all the other transitions are first order (full lines).

value of β. It follows from the degenerate multiplet structure of d^8 ions, and its coordinate moves along the $\eta = 0$ line, according to the following relation: $E_z = -2J(1-\beta)$. This is a clear demonstration of the frustrated nature of the spin and orbital superexchange in the model, whereas the crystal field term just compensates the enhanced or suppressed magnetic interactions in the (a,b) planes.

A special role plays the case with $\beta = 0$ which corresponds to the 2D spin-orbital model. In this case the MOAFF phase disappears completely while the other two phases with AF order in the (a,b) planes MOAAA and MOAAF collapse into a single MOAA phase. The resulting phase diagram is shown in Fig. 13. The MOFF is still stable in a large region of the parameter space which demonstrates that the strong AF exchange along the c axis in the corresponding 3D MOFFA phase is not instrumental to stabilize this phase, but the orbital energy within the FM planes is a dominating mechanism. It is interesting to compare the results obtained on the classical level with some relevant physical systems. For La_2CuO_4 and Nd_2CuO_4 the crystal field splitting is large, $E_z \simeq 0.64$ eV [73], so that one falls in the region of the 2D AFxx phase observed in neutron scattering. If on the contrary the orbital splitting is small, the orbital ordering sets in and has to couple strongly to the lattice. The net result is a quadrupolar distortion as indicated in Fig. 14. This lattice instability is again related to the question on the origin of the orbital ordering: is it due to JT and/or to electronic mechanism? The deformations found in $KCuF_3$ (or $LaMnO_3$) could in principle be entirely caused by phonon-driven collective JT effects. One might therefore attempt to neglect electron-electron interactions, and focus on the electron-phonon coupling. In case that the ions

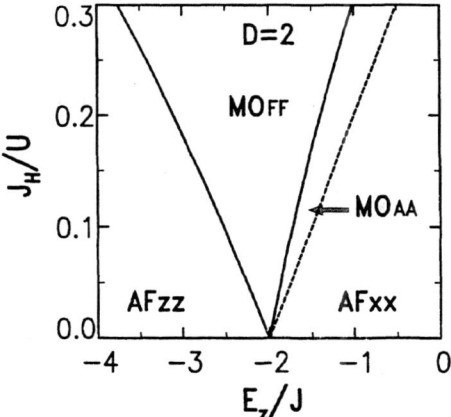

FIGURE 13. Mean-field phase diagram of the spin-orbital model (57) in the (E_z, J_H) plane in two dimensions ($\beta = 0$). Full lines separate the classical states AFxx, AFzz, and MOFF shown in Fig. 11, while the spin order in the MOAA phase is AF, and the orbitals are in between those in AFxx and MOFF phase (after Ref. [64])

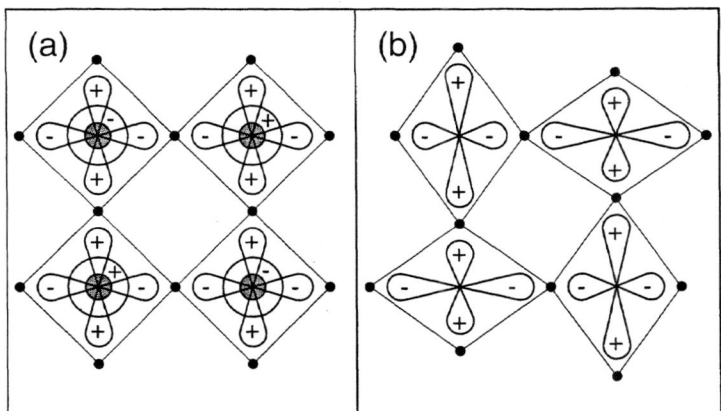

FIGURE 14. Schematic representation of the mixed orbitals in (a,b) planes of the MOFF phase in a 2D model: (a) the orbitals with their phases, and (b) the resulting distortion in the oxygen lattice, stabilized by the orbital ordering (after Ref. [64])

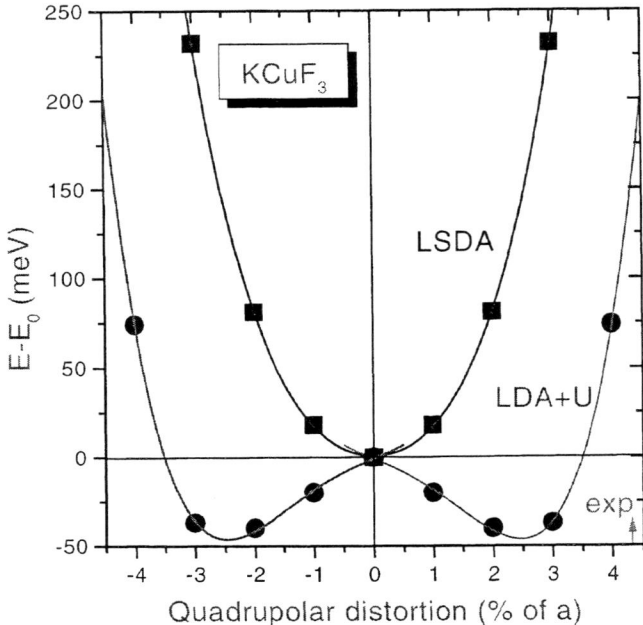

FIGURE 15. The dependence of the total energy of KCuF$_3$ on the quadrupolar lattice distortion according to LSDA and LDA+U band structure calculations (after Ref. [36]).

are characterized by a JT (orbital) degeneracy, one can integrate out the (optical) phonons, and one finds effective Hamiltonians with phonon mediated interactions between the orbitals. In the specific case of e_g degenerate ions in a cubic crystal, these look quite similar to the orbital interactions in the d^9 Hamiltonian, except that the spin dependent term is absent [74]. Any orbital order resulting from this Hamiltonian is now accompanied by a lattice distortion of the same symmetry.

The size of the quadrupolar deformation in the (a,b) plane of KCuF$_3$ is actually as large as 4 % of the lattice constant a_0. It is therefore often argued that the orbital order is clearly phonon-driven, and that the orbital interactions discussed above are less important. Although appealing at first sight, this argument is flawed: large displacements do not necessarily imply that phonons are the driving mechanism. Unfortunately, the deformations of the lattice and the orbital degrees of freedom cannot be disentangled using general principles: they constitute an irreducible subsector of the problem. The issue is therefore a quantitative one, and may be answered by calculating the electronic structure.

We start out with the observation that according to LDA KCuF$_3$ would be an undistorted, cubic system: the energy increases if the distortion is switched on (see Fig. 15). The reason is that KCuF$_3$ would be a band metal according to LDA (the usual Mott-gap problem) with a Fermi-surface which is not susceptible to a band JT instability. Therefore, the effects of strong on-site Coulomb interaction should be included and the LDA+U method [35] is a well designed method to serve this purpose. It is constructed to handle the physics of electronic orbital ordering, keeping the accurate treatment of the electron-lattice interaction of LDA intact. According to LDA+U calculations the total energy gained by the deformation of the lattice is only a small contribution of ~ 50 meV (Fig. 15) to the energies involved in the electronic orbital ordering [36]. Therefore, the coupling to the lattice is here not a driving force for the orbital and magnetic ordering, but *the lattice follows the orbital state.*

Although the energy gained in the deformation of the lattice is rather small, the electron-phonon coupling is quite effective in keeping KCuF$_3$ away from the frustrated interactions associated with the origin of the phase diagram (Fig. 12). Since the FM interactions in the (a,b) plane of KCuF$_3$ are quite small ($J_{(a,b)} = -0.2$ meV, as compared to the '1D' AF exchange $J_c = 17.5$ meV [75–77]), one might argue that the effective Hund's rule coupling $J\eta$ as of relevance to the low energy theory is quite small. Such a strong anisotropy of magnetic interactions $J_{(a,b)}$ and J_c has been reproduced recently within the *ab initio* method, but not in unrestricted HF, demonstrating the importance of electron correlation effects. Although this still needs further study, it might well be that in the absence of the electron-phonon coupling KCuF$_3$ would be close to the origin of Fig. 12. Therefore, although further work is needed to clarify the role played by electron-phonon coupling, it might be that phonons are to a large extent responsible for the stability of KCuF$_3$'s classical ground state. In any case, one cannot rely just on the size of the lattice deformations to resolve this issue.

C   Elementary excitations in the d^9 model

The presence of the orbital degrees of freedom in the Hamiltonian (57) yields excitation spectra that are qualitatively different from those of the quantum antiferromagnet with a single spin-wave mode. In the present case one gets two transverse excitations: *spin waves* and *spin-and-orbital waves* [78]; and also longitudinal excitations – *orbital waves*, thus producing three elementary excitations for the present spin-orbital model (57) [63,64,78,79]. This gives therefore the same number of modes as found in a 1D SU(4) symmetric spin-orbital model (see Sec. IV.A) in the Bethe ansatz method [80,81]. We emphasize that this feature is a consequence of the dimension (equal to 15) of the $so(4)$ Lie algebra of the local operators, as explained below, and is not related to the global symmetry of the Hamiltonian. In this chapter, we report the analysis of the realistic d^9 spin-orbital model for the 3D simple cubic (i.e., perovskite-like) lattice (57), using linear spin-wave (LSW) theory [82,83], generalized in such a way that makes it applicable to the present situation.

Before we introduce the excitation operators, it is convenient to rewrite the spin-orbital model (57) in a different representation which uses a four-dimensional space: $\{|x\uparrow\rangle, |x\downarrow\rangle, |x\downarrow\rangle\}, |z\uparrow\rangle$, instead of a direct product of the spin and orbital subspaces. This will demonstrate explicitly that three different elementary excitations appear in a natural way. Hence, we introduce operators which define purely *spin excitations* in individual orbitals,

$$S_{ixx}^{+} = d_{ix\uparrow}^{\dagger}d_{ix\downarrow}, \qquad S_{izz}^{+} = d_{iz\uparrow}^{\dagger}d_{iz\downarrow}, \qquad (81)$$

and operators for simultaneous spin-flip and transfer between the orbitals, *spin-and-orbital excitations*,

$$K_{ixz}^{+} = d_{ix\uparrow}^{\dagger}d_{iz\downarrow}, \qquad K_{izx}^{+} = d_{iz\uparrow}^{\dagger}d_{ix\downarrow}. \qquad (82)$$

The corresponding operators $S_{i\alpha\alpha}^{z}$ and $K_{i\alpha\beta}^{z}$ are defined as follows,

$$S_{ixx}^{z} = \tfrac{1}{2}(n_{ix\uparrow} - n_{ix\downarrow}), \qquad S_{izz}^{z} = \tfrac{1}{2}(n_{iz\uparrow} - n_{iz\downarrow}), \qquad (83)$$

$$K_{ixz}^{z} = \tfrac{1}{2}(d_{ix\uparrow}^{\dagger}d_{iz\uparrow} - d_{ix\downarrow}^{\dagger}d_{iz\downarrow}), \qquad K_{izx}^{z} = \tfrac{1}{2}(d_{iz\uparrow}^{\dagger}d_{ix\uparrow} - d_{iz\downarrow}^{\dagger}d_{ix\downarrow}). \qquad (84)$$

The Hamiltonian (57) contains also purely orbital interactions which can be expressed using the following *orbital excitation* operators,

$$T_{ixz} = \tfrac{1}{2}(d_{ix\uparrow}^{\dagger}d_{iz\uparrow} + d_{ix\downarrow}^{\dagger}d_{iz\downarrow}), \qquad T_{izx} = \tfrac{1}{2}(d_{iz\uparrow}^{\dagger}d_{ix\uparrow} + d_{iz\downarrow}^{\dagger}d_{ix\downarrow}), \qquad (85)$$

while the anisotropy in the orbital space is expressed by orbital-polarization operators,

$$n_{i-} = \tfrac{1}{2}(d_{ix\uparrow}^{\dagger}d_{ix\uparrow} + d_{ix\downarrow}^{\dagger}d_{ix\downarrow} - d_{iz\uparrow}^{\dagger}d_{iz\uparrow} - d_{iz\downarrow}^{\dagger}d_{iz\downarrow}). \qquad (86)$$

In order to simplify the notation, we also introduce global operators for the spin, spin-and-orbital and orbital excitations,

$$S_{i}^{+} = S_{ixx}^{+} + S_{izz}^{+}, \qquad S_{i}^{z} = S_{ixx}^{z} + S_{izz}^{z}, \qquad (87)$$

$$K_{i}^{+} = K_{ixz}^{+} + K_{izx}^{+}, \qquad K_{i}^{z} = K_{ixz}^{z} + K_{izx}^{z}, \qquad (88)$$

$$T_{i} = T_{ixz} + T_{izx}. \qquad (89)$$

The number of collective modes in a particular phase may be determined as follows. The $so(4)$ Lie algebra consists of three Cartan operators, i.e., operators diagonal on the local basis of the symmetry-broken phase under consideration (e.g. S_{ixx}^{z}, S_{izz}^{z}, and n_{i-} in the AFxx phase), plus twelve nondiagonal operators turning the eigenstates into one another (like S_{ixx}^{+} and S_{izz}^{+} in the AFxx phase). Out of those twelve operators, six connect two excited states (like S_{izz}^{+} in the AFxx phase), and are physically irrelevant (in the lowest order), because they give only rise to the so-called 'ghost' modes, the modes for which the spectral function vanishes identically at $T = 0$. The remaining six operators connect the local ground state with excited states, three of them describing an excitation and three a deexcitation, and only these six operators are physically relevant. Out of the three excitations (deexcitations), two are transverse, i.e., change the spin, and one is longitudinal, i.e., does not affect the spin. For a classical phase with L sublattices one therefore expects $4L$ transverse and $2L$ longitudinal modes. Because of time-reversal invariance they all occur in pairs with opposite frequencies, $\pm\omega_{\vec{k}}^{(n)}$.

Finally, the SU(2) spin invariance of the Hamiltonian guarantees that the transverse operators raising the spin are decoupled from those lowering the spin, and that both sets of operators are described by equivalent equations of motion, so that the transverse modes are pairwise degenerate. Such a simplification does not occur in the longitudinal sector. So, in conclusion, in an L-sublattice phase there are L doubly-degenerate positive-frequency transverse modes and L nondegenerate positive-frequency longitudinal modes, accompanied by the same number of negative-frequency modes. This may be compared with the well-known situation in the quantum antiferromagnet [82], where there is, with only spin excitation operators involved, only one (not two) doubly-degenerate positive-frequency (transverse) mode and the corresponding negative-frequency mode in the two-sublattice Néel state.

For the actual evaluation it is convenient to decompose the superexchange terms (58) in the spin-orbital Hamiltonian (57),

$$\mathcal{H}_J = \mathcal{H}_{\parallel} + \mathcal{H}_{\perp}, \tag{90}$$

into two parts which depend on the bond direction:
(i) for the bonds $\langle ij \rangle \parallel (a, b)$,

$$\mathcal{H}_{\parallel} = \tfrac{1}{4} J \sum_{\langle ij \rangle \parallel} \Big[(1 - \tfrac{1}{2}\eta)(3\vec{S}_{ixx} + \vec{S}_{izz} + \lambda_{ij}\sqrt{3}\vec{K}_i) \cdot (3\vec{S}_{jxx} + \vec{S}_{jzz} + \lambda_{ij}\sqrt{3}\vec{K}_j)$$
$$- 2\eta \vec{S}_i \cdot \vec{S}_j + (1 + 2\eta)(n_{i-} + \pm\sqrt{3}T_i)(n_{j-} + \pm\sqrt{3}T_j) - (3 + \eta)\Big], \tag{91}$$

(ii) for the bonds $\langle ij \rangle \perp (a, b)$, i.e., along the c-axis,

$$\mathcal{H}_{\perp} = J \sum_{\langle ij \rangle \perp} [(4 - 2\eta)\vec{S}_{izz} \cdot \vec{S}_{jzz} - \eta(\vec{S}_{ixx} \cdot \vec{S}_{jzz} + \vec{S}_{izz} \cdot \vec{S}_{jxx})$$
$$+ (1 + 2\eta)n_{i-}n_{j-} - \tfrac{1}{4}(3 + \eta)]. \tag{92}$$

Here and in the following paragraphs we consider a 3D cubic model with $\beta = 1$. We note that the orbital interactions (82) are quite different in $H_{\parallel}$ and $H_{\perp}$; propagating composite spin-and-orbital excitations are possible only within the (a, b) planes, where they are coupled to the spin excitations, while in the c-direction only pure spin excitations and pure spin-and-orbital excitations occur, which are decoupled from one another. This apparent breaking of symmetry between $H_{\parallel}$ and $H_{\perp}$ is a consequence of the choice of basis as $|x\rangle$ and $|z\rangle$ orbitals.

In the following, we report briefly the results obtained for transverse and longitudinal excitations in the various symmetry-broken classical states of the spin-orbital model (57). The transverse excitations, i.e., spin-waves and spin-and-orbital-waves, are calculated using the spin-rising operators which make a transition to a state realized in a classical phase at a given site i. As an example we use the AFxx phase to illustrate the formalism and calculation method with the excitation operators:

$$S^+_{ixx} = d^\dagger_{ix\uparrow} d_{ix\downarrow}, \qquad K^+_{ixz} = d^\dagger_{ix\uparrow} d_{iz\downarrow}. \tag{93}$$

The longitudinal excitations (without spin-flip) are most conveniently obtained starting from spin-dependent orbital excitation operators,

$$T_{ixz\sigma} = d^\dagger_{ix\sigma} d_{iz\sigma}, \qquad T_{izx\sigma} = d^\dagger_{iz\sigma} d_{ix\sigma}, \tag{94}$$

as these excitations conserve the spin component and we ask a question whether such a longitudinal excitation may propagate coherently in a given symmetry-broken classical state.

The nature and dispersion of elementary excitations in the spin-orbital model (57) can be conveniently studied in the leading order of the $1/S$ expansion using the Green function formalism. The starting point are the equations of motion for the Green functions generated by the excitation operators (93) written in the energy representation [84,85],

$$E\langle\langle S^+_{ixx}|...\rangle\rangle = \frac{1}{2\pi}\langle[S^+_{ixx},...]\rangle + \langle\langle[S^+_{ixx},H]|...\rangle\rangle, \tag{95}$$

$$E\langle\langle K^+_{ixz}|...\rangle\rangle = \frac{1}{2\pi}\langle[K^+_{ixz},...]\rangle + \langle\langle[K^+_{ixz},H]|...\rangle\rangle, \tag{96}$$

where the average of the commutator on the right hand side, e.g. $\langle[S^+_{ixx},S^-_{jxx}]\rangle$, is evaluated in the classical ground state. We note, however, that equivalent results for the AFxx and AFzz phases can be obtained using instead an expansion around a classical saddle point with Schwinger bosons [82].

The equations of motion have been derived for the Green functions generated by the set of operators $\{S^+_{ixx}, K^+_{ixz}, S^+_{jxx}, K^+_{jxz}\}$, where $i \in A$ and $j \in B$, and used the random-phase approximation (RPA) for spinlike operators which linearizes the equations of motion by a decoupling procedure [84,85]. Thereby, the operators which have nonzero expectation values in the considered classical state give finite contributions, e.g. for the first spin-flip Green function one uses,

$$\langle\langle S^+_{ixx} S^z_{mxx}|...\rangle\rangle \simeq \langle S^z_{mxx}\rangle\langle\langle S^+_{ixx}|...\rangle\rangle, \tag{97}$$

and a similar formula for the mixed spin-and-orbital excitation described by $\langle\langle K^+_{ixz}|...\rangle\rangle$,

$$\langle\langle K^+_{ixz} S^z_{mxx}|...\rangle\rangle \simeq \langle S^z_{mxx}\rangle\langle\langle K^+_{ixz}|...\rangle\rangle. \tag{98}$$

In the present case of the AFxx phase one uses the respective Néel state average values,

$$\langle S^z_{ixx}\rangle = -\langle S^z_{jxx}\rangle = \tfrac{1}{2}, \tag{99}$$
$$\langle n_{i-}\rangle = \langle n_{j-}\rangle = \tfrac{1}{2}, \tag{100}$$

where $i \in A$ and $j \in B$, and A and B are the two sublattices in a 2D lattice of the AFxx phase, and all the remaining averages vanish. It is crucial that the decoupled operators have different site indices, and thus the decoupling procedure preserves the local commutation rules. Instead, if one uses products of spin and orbital operators, e.g., $K^+_{ixz} = S^+_{ixx}\sigma^+_i$, one is tempted to decouple these operators locally [24,86] which would violate the algebraic structure of the $so(4)$ Lie algebra.

The translational invariance of the Néel state implies that the transformed Green functions are diagonal in the reduced Brillouin zone (BZ). As in the Heisenberg antiferromagnet, the Fourier transformed functions are defined for the Green functions which describe the spin dynamics on a given sublattice, either A or B. For instance, the pure spin-flip Green functions are transformed as follows,

$$\langle\langle S^+_{\vec{k}xx}|\ldots\rangle\rangle_A = \frac{1}{\sqrt{N}}\sum_{i\in A} e^{i\vec{k}\vec{R}_i}\langle\langle S^+_{ixx}|\ldots\rangle\rangle_A,$$
$$\langle\langle S^+_{\vec{k}xx}|\ldots\rangle\rangle_B = \frac{1}{\sqrt{N}}\sum_{j\in B} e^{i\vec{k}\vec{R}_j}\langle\langle S^+_{jxx}|\ldots\rangle\rangle_B, \tag{101}$$

where N is the number of sites in one sublattice. Hence, the problem of finding the elementary excitations of the considered spin-orbital model (57) reduces to the diagonalization of the following 4×4 dynamical matrix at each $\vec{k}$-point:

$$\begin{pmatrix} \lambda_\alpha - \overline{\omega}_{\vec{k}} & 0 & Q_{\alpha\vec{k}} & P_{\alpha\vec{k}} \\ 0 & \tau_\alpha - \overline{\omega}_{\vec{k}} & P_{\alpha\vec{k}} & R_{\vec{k}} \\ -Q_{\alpha\vec{k}} & -P_{\alpha\vec{k}} & -\lambda_\alpha - \overline{\omega}_{\vec{k}} & 0 \\ -P_{\alpha\vec{k}} & -R_{\vec{k}} & 0 & -\tau_\alpha - \overline{\omega}_{\vec{k}} \end{pmatrix} \begin{pmatrix} \langle\langle S^+_{\vec{k}xx}|\cdots\rangle\rangle_A \\ \langle\langle K^+_{\vec{k}xz}|\cdots\rangle\rangle_A \\ \langle\langle S^-_{\vec{k}xx}|\cdots\rangle\rangle_B \\ \langle\langle K^-_{\vec{k}xz}|\cdots\rangle\rangle_B \end{pmatrix} = 0, \tag{102}$$

The symmetric positive and negative eigenvalues $\pm\omega^{(n)}_{\vec{k}}$, with $n=1,2$, solved from the matrix in Eq. (102) may be written in the following form for the AFxx phase,

$$[\omega^{(n)}_{\vec{k}}]^2 = J^2\left(\lambda_x^2 + \tau_x^2 - Q^2_{x\vec{k}} - R^2_{\vec{k}} - 2P^2_{x\vec{k}}\right) \pm J^2\Big[(\lambda_x^2 - \tau_x^2)^2 - 2(\lambda_x^2 - \tau_x^2)(Q^2_{x\vec{k}} - R^2_{\vec{k}})$$
$$- 4(\lambda_x - \tau_x)^2 P^2_{x\vec{k}} + (Q^2_{x\vec{k}} + R^2_{\vec{k}} + 2P^2_{x\vec{k}})^2 - 4(Q_{x\vec{k}}R_{\vec{k}} - P^2_{x\vec{k}})^2\Big]^{1/2}. \tag{103}$$

Here the quantities λ_x and τ_x play the role of local potentials and follow from the model parameters, E_z and J_H,

$$\lambda_x = \tfrac{9}{2} - 3\eta, \tag{104}$$
$$\tau_x = \tfrac{7}{2} - 4\eta - 2 - \eta + \varepsilon_z. \tag{105}$$

The remaining terms are $\vec{k}$-dependent, and depend on

$$\gamma_+(\vec{k}) = \tfrac{1}{2}(\cos k_x + \cos k_y), \tag{106}$$
$$\gamma_-(\vec{k}) = \tfrac{1}{2}(\cos k_x - \cos k_y), \tag{107}$$
$$\gamma_z(\vec{k}) = \cos k_z. \tag{108}$$

The quantities $Q_{x\vec{k}}$ and $P_{x\vec{k}}$ for the AFxx phase take the form,

$$Q_{x\vec{k}} = (\tfrac{9}{2} - 3\eta)\gamma_+(\vec{k}), \tag{109}$$
$$P_{x\vec{k}} = \tfrac{1}{2}\sqrt{3}(3-\eta)\gamma_-(\vec{k}), \tag{110}$$

while the last dispersive term,

$$R_{\vec{k}} = \tfrac{3}{2}\gamma_+(\vec{k}), \tag{111}$$

carries no index and remains identical for both G-AF phases (AFxx and AFzz). We emphasize that the coupling between the spin-wave and spin-and-orbital-wave excitations occurs due to the terms $\propto P_{x\vec{k}}$, as seen from Eq. (102). It vanishes in the planes of $k_x = \pm k_y$, but otherwise plays an important role, as discussed in Sec. III. In the limit of large $E_z \to \infty$, Eq. (103) reproduces the spin-wave excitations for a 2D antiferromagnet with an AF superexchange interaction of $J(\tfrac{9}{4} - \tfrac{3}{2}\eta)$, as given between the occupied $|x\rangle$ orbitals,

$$\omega_{\vec{k}}^{(1)} = J\left(\tfrac{9}{2} - 3\eta\right)[1 - \gamma_+^2(\vec{k})]^{1/2}, \tag{112}$$

while the dispersion of the high-energy spin-and-orbital excitation, $\omega_{\vec{k}}^{(2)} \simeq E_z$, becomes negligible. As explained above, both modes are doubly degenerate.

Consider now the orbital (excitonic) excitations generated by the orbital-flip operators (85). They are found by considering the equations of motion,

$$E\langle\langle T_{i\alpha\beta\sigma}|\ldots\rangle\rangle = \frac{1}{2\pi}\langle[T_{i\alpha\beta\sigma},\ldots]\rangle + \langle\langle[T_{i\alpha\beta\sigma}, H]|\ldots\rangle\rangle, \tag{113}$$

where spin σ corresponds to the occupied state in the symmetry-broken Néel state. By making a Fourier transformations as for the transverse operators (101), one may show that only two operators per sublattice suffice to describe the modes in an antiferromagnet. The structure of the respective RPA dynamical matrix is given by

$$\begin{pmatrix} u_\alpha - \bar{\zeta}_{\vec{k}} & 0 & +\rho_{\alpha\vec{k}} & +\rho_{\alpha\vec{k}} \\ 0 & -u_\alpha - \bar{\zeta}_{\vec{k}} & -\rho_{\alpha\vec{k}} & -\rho_{\alpha\vec{k}} \\ -\rho_{\alpha\vec{k}} & -\rho_{\alpha\vec{k}} & -u_\alpha - \bar{\zeta}_{\vec{k}} & 0 \\ +\rho_{\alpha\vec{k}} & +\rho_{\alpha\vec{k}} & 0 & u_\alpha - \bar{\zeta}_{\vec{k}} \end{pmatrix} \begin{pmatrix} \langle\langle T_{\vec{k}xz\uparrow}|\cdots\rangle\rangle_A \\ \langle\langle T_{\vec{k}zx\uparrow}|\cdots\rangle\rangle_A \\ \langle\langle T_{\vec{k}xz\downarrow}|\cdots\rangle\rangle_B \\ \langle\langle T_{\vec{k}zx\downarrow}|\cdots\rangle\rangle_B \end{pmatrix} = 0, \tag{114}$$

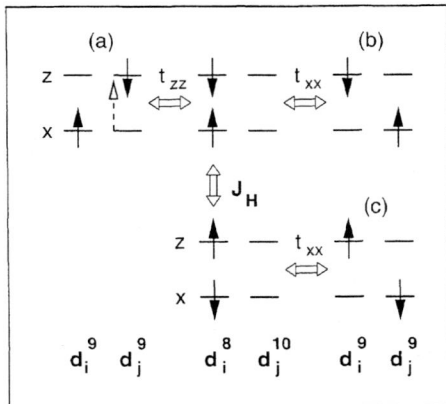

FIGURE 16. Schematic propagation of the orbital (excitonic) excitation (a). If $J_H = 0$, an orbital excitation can propagate only to state (b) and is accompanied by a spin-flip (top), while $J_H > 0$ allows also the spin-flip in the intermediate d_i^8 state, and thus the propagation without spin-flip (c) becomes possible (bottom) (after Ref. [64]).

with

$$u_x = \varepsilon_z - 3\eta, \tag{115}$$

$$\rho_{x\vec{k}} = \tfrac{3}{2}\eta\gamma_+(\vec{k}), \tag{116}$$

and one finds two, in general nondegenerate, positive-frequency modes,

$$\zeta_{\vec{k}} = J\left[u_\alpha(u_\alpha \pm 2\rho_{\alpha\vec{k}})\right]^{1/2}. \tag{117}$$

It is important to realize that the propagation of longitudinal excitations, being equivalent to a finite dispersion of longitudinal modes, becomes possible only at $\eta > 0$. This follows from the multiplet structure of the excited d^8 states, which allows a spin-flip between the orbitals in the $|^1E_\theta\rangle$ and in the $S^z = 0$ component of the $|^3A_2\rangle$-state only if $J_H \neq 0$, as illustrated in Fig. 16. The processes $\sim t_{xz}$ are not shown, as they would also lead to a final state given in Fig. 16(b), i.e., to a propagation of a spin-and-orbital excitation which was already considered above. In contrast, the relevant longitudinal orbital excitation in the symmetry-broken state implies that the exciton has the same spin as imposed by the Néel state of the background; this state is shown in Fig. 16(c). Therefore, in a perfect Néel state without FM interactions due to $\eta \neq 0$, only local orbital excitations are possible. These local excitations cost no energy in the limit of $\varepsilon_z \to 0$ which demonstrates again the frustration of magnetic interactions at the classical degeneracy point, $\varepsilon_z = \eta = 0$. An example of the excitation spectra is presented in Fig. 17 for the main directions in the 2D BZ, with $X = (\pi, 0)$ and $S = (\pi/2, \pi/2)$. Near the

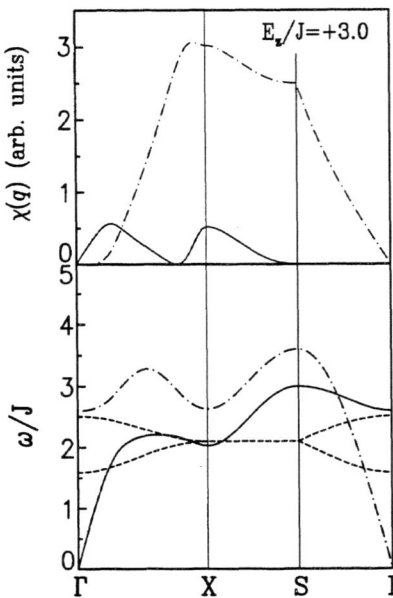

FIGURE 17. Spin-wave and spin-and-orbital-wave transverse excitations (full lines) and longitudinal excitations (dashed lines) in AFxx phase (bottom), and neutron intensities of the transverse excitations (top). Parameters: $E_z/J = 3.0$ and $J_H/U = 0.3$ (after Ref. [64]).

Γ point one finds a (doubly-degenerate) Goldstone mode $\omega_{\vec{k}}^{(1)}$ with dispersion $\sim k$ at $\vec{k} \to 0$, as in the Heisenberg antiferromagnet, and a second (doubly-degenerate) transverse mode at higher energy, $\omega_{\vec{k}}^{(2)} \simeq \omega_0 + ak^2$. Near the Γ point the Goldstone mode is essentially purely spin-wave, the second mode is purely spin-and-orbital wave. With increasing $\vec{k}$ these modes start to mix due to the $P_{x\vec{k}}$ term along the $\Gamma - X$ direction. This is best illustrated by the intensity measured in the neutron scattering experiments, which see only the spin-wave component in each transverse mode. Indeed, the intensity $\chi(\vec{q})$ is transferred from one mode to the other along the $\Gamma - X$ direction in the 2D BZ (Fig. 17), demonstrating that indeed the lowest (highest) mode is predominantly spin-wave-like (spin-and-orbital-wave-like) before the anticrossing point, while this is reversed after the anticrossing of the two modes. Thus, we make here a specific prediction that *two spin-wave-like modes could be measurable in certain parts of the 2D BZ*, in particular in the vicinity of the anticrossing, if only an AFxx phase was realized for parameters not too distant from the classical degeneracy point.

Unfortunately, for the realistic parameters for the cuprates [73], one finds $E_z/J \simeq 10$ which makes the spin-and-orbital excitation and the changes of the spin-wave dispersion hardly visible in neutron spectroscopy. The orbital (longitudinal) excitations are found for the parameters of Fig. 17 at a finite energy, being of the same order of magnitude as the energy of the spin-and-orbital excitation, $\omega_{\vec{k}}^{(2)}$. The weak dispersion of these modes follows from the spin-flip processes in the *excited* states, as explained in Fig. 16 and discussed above. We emphasize that the orbital mode has a gap and *does not couple* to any spin excitation. At the classical degeneracy point M the orbital mode falls to zero energy and is dispersionless, expressing that the orbital can be changed locally without any cost in energy.

The transverse excitations in the AFzz phase are determined by considering the complementary set of Green functions to that given by Eqs. (95) and (96), with the excitation operators S_{izz}^{+} and K_{izx}^{+}. After deriving the RPA equations, one finds the final form of the equations of motion by performing a Fourier transformation and using the following nonvanishing expectation values,

$$\langle S_{izz}^z \rangle = -\langle S_{jzz}^z \rangle = \tfrac{1}{2}, \tag{118}$$
$$\langle n_{i-} \rangle = \langle n_{j-} \rangle = -\tfrac{1}{2}, \tag{119}$$

in the AFzz phase, with $i \in A$ and $j \in B$. This leads again to the general form (102), with the elements λ_x, τ_x, $Q_{x\vec{k}}$, and $P_{x\vec{k}}$ now replaced by,

$$\lambda_z = \tfrac{1}{2} - \eta + 2(2 - \eta), \tag{120}$$
$$\tau_z = -\tfrac{1}{2} - \eta + 2(1 - 2\eta) - \varepsilon_z, \tag{121}$$
$$Q_{z\vec{k}} = (\tfrac{1}{2} - \eta)\gamma_+(\vec{k}) + 2(2 - \eta)\gamma_z(\vec{k}), \tag{122}$$
$$P_{z\vec{k}} = \tfrac{1}{2}\sqrt{3}(1 - \eta)\gamma_-(\vec{k}). \tag{123}$$

Thus, the transverse excitations have the same form (103) as in the AFxx phase, but the above quantities (120)–(123) have to be used.

In the limit of large $E_z \to -\infty$ one finds the spin-wave for a 3D anisotropic antiferromagnet with strong superexchange equal to $2J(2 - \eta)$ along the c-axis, and weak superexchange $\frac{1}{4}J(1 - 2\eta)$ within the (a,b)-planes,

$$\omega_{\vec{k}}^{(1)} = J\left\{\left[(\tfrac{1}{2} - \eta) + 2(2-\eta)\right]^2 - \left[(\tfrac{1}{2}-\eta)\gamma_+(\vec{k}) + 2(2-\eta)\gamma_z\right]^2\right\}^{1/2}, \quad (124)$$

while the spin-and-orbital excitation, $\omega_{\vec{k}}^{(2)} \simeq -E_z$, is dispersionless. Again, both these transverse modes are doubly degenerate.

The representative excitation spectrum for the AFzz phase may be found in Ref. [64]. One finds again a Goldstone mode $\omega_{\vec{k}}^{(1)}$ at the Γ point which is spin-wave-like, accompanied by a finite energy spin-and-orbital mode $\omega_{\vec{k}}^{(2)}$. The first one is linear, while the second changes quadratically with increasing $\vec{k}$. The dispersion in the $\Gamma - X$ direction is, however, only $\sim 0.7J$, while in the AFxx phase a large dispersion of $\sim 2.5J$ was found. This demonstrates the large difference between the superexchange in the (a,b)-planes in these two AF phases. Here one should bear in mind, that in a strongly anisotropic antiferromagnet, such as the AFzz phase, the dispersion of the spin-wave mode in the (k_x, k_y) plane is roughly given by $(zJ_{ab}J_c)^{1/2}S$, so actually enhanced by $(J_c/zJ_{ab})^{1/2}$ compared with the planar exchange constant.

The (longitudinal) orbital excitations in the AFzz phase are found using the equations of motion of the form (113) which lead to Eq. (114) with,

$$u_z = -\varepsilon_z - 3\eta, \quad (125)$$
$$\rho_{z\vec{k}} = -\tfrac{3}{2}\eta\gamma_+(\vec{k}), \quad (126)$$

and we find again zero-energy nondispersive modes at $\varepsilon_z = \eta = 0$. The orbital excitation is found at the $X = (\pi, 0, 0)$ and $L = (\pi/2, \pi/2, \pi/2)$ points at the same energy as that of a *local* excitation from $|z\rangle$ to $|x\rangle$ orbital. It depends only on the energy difference between the orbitals, and has a weak dispersion $\sim J\eta$ due to the same mechanism as described above for the AFxx phase (Fig. 16).

The excitation operators which couple to the local states in the MOFFA phase with mixed orbitals are linear combinations of the operators considered above. It is therefore convenient to make a unitary transformation of the Hamiltonian (57) to new orbitals defined as follows for $i \in A$ or $i \in D$ sublattice,

$$\begin{pmatrix} |i\mu\rangle \\ |i\nu\rangle \end{pmatrix} = \begin{pmatrix} \cos\theta & \sin\theta \\ -\sin\theta & \cos\theta \end{pmatrix} \begin{pmatrix} |iz\rangle \\ |ix\rangle \end{pmatrix}, \quad (127)$$

and for $j \in B$ or $j \in C$ sublattice,

$$\begin{pmatrix} |j\mu\rangle \\ |j\nu\rangle \end{pmatrix} = \begin{pmatrix} \cos\theta & -\sin\theta \\ \sin\theta & \cos\theta \end{pmatrix} \begin{pmatrix} |jz\rangle \\ |jx\rangle \end{pmatrix}. \quad (128)$$

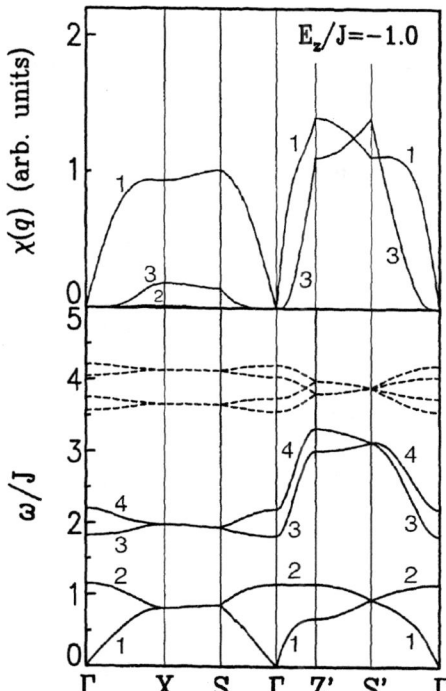

FIGURE 18. The same as in Fig. 17, but for the MOFFA phase, as obtained for $E_z/J = -1.0$ and $J_H/U = 0.3$. Different modes are labelled by the increasing indices $i = 1,\ldots,4$ with increasing energy (after Ref. [64]).

With these definitions and by choosing the angle θ at the value which minimizes the classical energy (77), we guarantee that $|i\mu\rangle$ and $|j\mu\rangle$, respectively, are the orbital states realized in the classical MOFFA phase at each site, which is G-type with respect to the orbital ordering, while $|i\nu\rangle$ and $|j\nu\rangle$ are the excited states, so that one can readily define the excitation operators pertinent to the symmetry-broken ground state of this phase.

Thus the spin $\mathcal{S}^+_{i\mu\mu}$, spin-and-orbital $\mathcal{K}^+_{i\mu\nu}$, and orbital $\mathcal{T}_{i\mu\nu\sigma}$ operators are defined in terms of the new reference orbital states $\{|\mu\rangle,|\nu\rangle\}$, and fulfill the same commutation rules as the non-transformed operators: $S^+_{i\alpha\alpha}$, $K^+_{i\alpha\beta}$, and $T_{i\alpha\beta\sigma}$, respectively. To simplify the notation we also introduce total spin $\mathcal{S}^+_i$ and spin-and-orbital $\mathcal{K}^+_i$ operators, as explained above. The Hamiltonian (57) has to be transformed by the inverse transformations to those given by Eqs. (127) and (128) [64]. Hence, the transverse excitations may be found starting from the relevant raising operators that lead to the local state $|i\mu\uparrow\rangle$ realized in one of the sublattices, analogous to those introduced for the AFxx phase (93), i.e., the set $\{\mathcal{S}^+_{i\mu\mu}, \mathcal{K}^+_{i\mu\nu}, \mathcal{S}^+_{j\mu\mu}, \mathcal{K}^+_{j\mu\nu}, \mathcal{S}^+_{k\mu\mu}, \mathcal{K}^+_{k\mu\nu}, \mathcal{S}^+_{l\mu\mu}, \mathcal{K}^+_{l\mu\nu}\}$, where $i \in A$, $j \in B$, $k \in C$, and

$l \in D$; they lead as usual to the orbitals $\{|i\mu\rangle, |j\mu\rangle\}$ (72) realized in the MOFFA phase. We applied the same RPA procedure as explained above for the AFxx and AFzz phase in order to determine the Green function equations in the $\vec{k}$-space. The longitudinal excitations can be obtained from operators $\mathcal{T}_{i\mu\nu\sigma}$ similar to those used in the AFxx and AFzz phases (94), taking $\sigma = \uparrow$ for the (a, b) planes occupied with $\uparrow$-spins, and $\sigma = \downarrow$ for the (a, b) planes occupied with $\downarrow$-spins. As expected, there are four doubly-degenerate positive-frequency transverse modes, and four non-degenerate positive-frequency longitudinal modes, consistent with the MOFFA phase having four sublattices.

An example of the transverse and longitudinal modes in the MOFFA phase is presented in Fig. 18. The modes are shown in the respective BZ which corresponds to the magnetic unit cell of the MOFFA phase: The 2D part along $\Gamma - X - S - \Gamma$ resembles the modes in the AFxx phase (compare Fig. 17), reflecting the orbital alternation, while the AF coupling along the c-axis results in the folding of the zone along the $\Gamma - Z$ direction, with $Z' = (0, 0, \pi/2)$ and $S' = (\pi/2, \pi/2, \pi/2)$. One finds one Goldstone mode, and three other finite-energy modes at the Γ point. If no AF coupling along the c-axis is present, similar positive-energy modes describe the excitation spectrum in the MOFF phase in the 2D part of the BZ (in the region of stability shown in Fig. 13), and the symmetric negative-frequency modes carry then no weight. In contrast, due to the strong AF interactions in the MOFFA phase, the negative modes give a large energy renormalization due to quantum fluctuations, as discussed in more detail in Sec. IV.B.

The spin-wave and spin-and-orbital-wave excitations are well separated along the $\Gamma - X - S - \Gamma$ path, with a gap of $\sim 0.5J$, as the FM interactions $\propto J\eta$ are considerably weaker than the orbital interactions which are $\propto J$. Therefore, the neutron intensity $\chi(\vec{q})$ is found mainly as originating from the lowest energy mode, $\omega_{\vec{k}}^{(1)}$, with a small admixture of the higher-energy spin-and-orbital excitation, $\omega_{\vec{k}}^{(3)}$. The magnetic interactions are considerably stronger along the c-axis; the modes mix and the higher-energy excitations, $\omega_{\vec{k}}^{(n)}$ with $n = 3, 4$, have a larger dispersion in the remaining directions with $k_z \neq 0$. Strong mixing of the modes in this part of the BZ is also visible in the intensity distribution, with the modes $n = 1$ and $n = 3$ contributing with comparable intensities (Fig. 18). The fact that modes labeled as 2 and 4 have zero intensity is due to the path $\Gamma - Z' - S' - \Gamma$ being in the high-symmetry BZ plane where $k_x = k_y$ so that $\gamma_-(\vec{k}) = 0$. Then modes 2 and 4 have equal amplitude but are exactly out-of-phase between A and B sites as well as between C and D sites, and so their neutron intensities vanish, and only the companion in-phase modes 1 and 3 are observable by neutrons. Unlike in the AF phases, the purely orbital excitation is here energetically separated from the spin wave and spin-and-orbital wave. The dispersion is quite small and decreases with η.

Interestingly, although the order in the (a, b) planes is FM, the energy of the Goldstone mode increases *linearly in all three directions* with increasing $\vec{k}$, and the slopes are proportional to the respective exchange interactions. This behavior is a manifestation of the A-AF spin order; a qualitatively similar spectrum is found experimentally in LaMnO$_3$ [39,40], where, however, the excitation spectra correspond to large spins $S = 2$ of Mn^{3+} ions. The rather small dispersion of the spin-wave part at low energies is due to small values of the exchange constants for the actual optimal orientation of orbitals found at $J_H/U = 0.3$. We note, however, that the AF interactions along the c-axis are much stronger at $J_H \to 0$ than in the present case. The AF structure along the c-axis may be easily recognized from the symmetric spin-wave mode in the $\Gamma - Z$ direction with respect to $Z' = (0, 0, \pi/2)$, while this mode increases all the way from the Γ to the X point. Unfortunately, no experimental verification of these spectra is possible at present, as the spin excitations measured in neutron scattering for KCuF$_3$ are consistent with the Bethe ansatz and thus suggest a spin-liquid ground state with strong 1D AF correlations instead of the A-AF phase with magnetic LRO [75–77].

The elementary excitations in the MO$_{\text{AFF}}$ phase may be obtained using a similar scheme to that described here for the MO$_{\text{FFA}}$ phase. In this case, the transverse excitations have a similar dependence on the $\vec{k}$-vector to those found in the MO$_{\text{FFA}}$ phase, but the value of the crystal-field E_z is effectively smaller by a factor of two in comparison with the MO$_{\text{FFA}}$ phase. This asymmetry is a consequence of the choice of $|x\rangle$ and $|z\rangle$ states as orbital basis to which E_z refers. Most importantly, one finds that the classical phases are all stable on the RPA level in the regions of their stability in the phase diagram of Fig. 12. However, there are characteristic low-frequency modes which follow from the mixing between the spin wave and spin-and-orbital wave modes, and these modes are responsible for enhanced quantum fluctuations (sec. IV.B).

IV   SPIN LIQUID DUE TO ORBITAL FLUCTUATIONS

A   Idealized case: $SU(4)$ model

Motivated by the recent studies of the strongly correlated systems with orbital degeneracy, and by the role played by the orbital degree of freedom in spin systems, the $SU(4)$ symmetric spin-orbital model has attracted a lot of interest [81,87–92]. It describes the localized electrons in the twofold degenerate Hubbard model with the diagonal isotropic hopping t between the same type of orbitals for the filling of one electron per site (the filling by one hole is equivalent by the particle-hole transformation). As the off-diagonal hopping elements are absent, the zth component of pseudospin is conserved, unlike in the d^9 model of Sec. III. If the Coulomb interaction U is large compared with t, $U \gg t$, the effective Hamiltonian may be derived in a similar way as shown in Secs. II.A and III.A, and one finds anisotropic orbital interactions with extra terms $\propto T_i^z T_j^z$, while the spin interactions are as usually $SU(2)$ symmetric [87]. The highly symmetric $SU(4)$ model follows in the limit of vanishing Hund's rule exchange, $J_H \to 0$, when the spectrum of the excited states collapses into a spin triplet multiplied by orbital singlets, and an orbital triplet multiplied by three spin singlets, all at the same energy U. Taking the projection operators on the spin triplet and spin singlet (21), and introducing similar operators for the orbital pseudospin states, one finds the Hamiltonian of the form,

$$H = 2J \sum_{\langle ij \rangle} \left[\left(\vec{S}_i \cdot \vec{S}_j + \frac{3}{4} \right) \left(\vec{T}_i \cdot \vec{T}_j - \frac{1}{4} \right) + \left(\vec{S}_i \cdot \vec{S}_j - \frac{1}{4} \right) \left(\vec{T}_i \cdot \vec{T}_j + \frac{3}{4} \right) \right], \quad (129)$$

where $\vec{S}_i$ and $\vec{T}_i$ are spin $S = 1/2$ and pseudospin $T = 1/2$, corresponding to spin and orbital degrees of freedom, respectively, and we defined the energy unit for the superexchange interaction $J = 2t^2/U$. An interesting observation is that a pure spin $\sim \vec{S}_i \cdot \vec{S}_j$ interaction has here a prefactor $J = 2t^2/U$ which is by a factor of two smaller than in the t-J model (see Sec. II.A). This reduction follows from the competition between the orbital triplet and orbital singlets in the present case. The derived expression (129) explains the physical origin of a fully symmetric Hamiltonian in spin and pseudospin (orbital) space, usually written as,

$$H = J \sum_{\langle ij \rangle} \left(2\vec{S}_i \cdot \vec{S}_j + \frac{1}{2} \right) \left(2\vec{T}_i \cdot \vec{T}_j + \frac{1}{2} \right). \quad (130)$$

It was noticed out only very recently [87–90], that the Hamiltonian (130) has not only the obvious SU(2)×SU(2) symmetry, but the full symmetry of Eq. (130) obeys even the higher symmetry group SU(4). It is worth pointing out that SU(N) symmetric models in one dimension were studied by Affleck, using conformal field theory [93]. He showed that any 1D system of SU(N) symmetry is critical. In this case, the critical exponents and zero temperature correlations at the very low energy scale are equivalent to $N - 1$ free massless bosons. These general results naturally also applies to the case with $N = 4$. We note also that a different SU(4) symmetric model has been introduced by Santoro et al. [94], which has a different low-energy physics from the present Hamiltonian (130). In realistic materials, however, this high SU(4) symmetry is practically always broken by an anisotropic hybridization [16], or by the JT effect [15], as we have emphasized in other Sections.

The advantage of the high SU(4) symmetry is that the rigorous analysis of this model is possible in one dimension. The SU(4) model (130) belongs to a class of models which are exactly soluble in one dimension by the Bethe ansatz [95]. The Bethe ansatz solution obtained by Sutherland gives the exact ground state energy and three branches of low-energy gapless excitations [80], having all a common velocity $v = \pi J/2$. The physical interpretation of these branches is not straightforward though.

The essential complexity comes from the large local degeneracy. For a single bond, the ground state is six-fold degenerate: either spin triplet multiplied by any of the three orbital singlets, or any of spin singlets multiplied by the orbital triplet. It is rotationally invariant not only in $\vec{S}$-space, but also in $\vec{T}$-space. Furthermore, it has an interchange symmetry between spin and orbital operators. In such a case, the standard mean-field approach [24] that leads to FM correlations for one type of variables and AF correlations for the others, is not reliable and more powerful methods have to be applied. The first investigation of the thermodynamic properties of the SU(4) model (130) has been performed by Frischmuth, Mila, and Troyer [81] by means of the continuous time quantum Monte Carlo (QMC) loop algorithm [96,97], adapted to spin-orbital models. The ground state energy for a chain of $L = 100$ with periodic boundary condition amounts to $\epsilon_0(L = 100) = -0.8253(1)J$, and is in perfect agreement with the Bethe Ansatz result for the infinite chain, $-0.8251189\ldots J$ [80]. In contrast, if the MF decoupling is made, one finds the energy of $-0.3863J$ [81] which demonstrates that the MF method cannot be used even for qualitative insight into the nature of the ground state.

The structure of the ground state becomes more transparent when the correlation functions are investigated. The QMC study of Frischmuth et al. [81] gives the zero-temperature correlation function $w_{ij}(T = 0) \equiv \langle S_i^z S_j^z \rangle (T = 0)$ as a function of distance $|i - j|$ along a 1D chain (for $L = 100$). We reproduce their results in Fig. 19. Due to the SU(4) symmetry, all the following correlations are equal:

$$w_{ij} = \langle S_i^\alpha S_j^\alpha \rangle = \langle T_i^\alpha T_j^\alpha \rangle = \langle 4 S_i^\alpha S_j^\alpha T_i^\beta T_j^\beta \rangle, \tag{131}$$

independent of the indices $\alpha, \beta = x, y, z$. This relation is valid for zero as well as for finite temperatures, and is easily violated by any MF decoupling. While the first equality also holds for an arbitrary SU(2)×SU(2) symmetric model with exchange symmetry of the $\vec{S}$ and $\vec{T}$-variable, the second one is a special property of the SU(4) symmetric model. We observe that the correlation function w_{ij} exhibits a clear four-site periodicity (Fig. 19). Its sign is positive if $|i - j| = 4N$ with N integer, and negative otherwise. The reason for this behavior is the tendency for every four neighboring sites to form an SU(4) singlet [98]. Looking at the results for w_{ij}, it can be concluded that they correspond to a disordered state and the two dominant modes are those with $k = \pi/2$ (positive prefactor) and $k = 0$ (negative prefactor) [81]. This is also reflected in the Fourier transform $\mathcal{S}^z(k)$ of the correlation function w_{ij}, having a characteristic cusp structure at $k = 0, \pi/2$ and π (see Fig. 19). While the cusps at $k = 0$ and $\pi/2$ are quite sharp, the one at $k = \pi$, however, is not so pronounced, indicating that the $k = \pi$ mode is the least dominant mode in the correlation function of all the three modes.

The large degeneracy of the SU(4) invariant model (130) becomes transparent in the entropy $s(T)$ per site determined by the QMC loop algorithm [81]; its T-dependence is shown in Fig. 20. With increasing T, the entropy s increases monotonically from zero towards the high temperature value $k_B \ln 4$. This shows that the short-range correlations are gradually lost, and the spins and orbitals are fluctuating individually at every site at high temperatures, in some similarity with the spin fluctuations in the spin model which gives the high temperature value of entropy $k_B \ln 2$. At low temperatures the entropy varies linearly in both models. However, the slope of $s(T)$ in the spin-orbital model (130) is about a factor *three* bigger than that in the Heisenberg antiferromagnet, as shown in the inset of Fig. 20. This is consistent with the statement of Affleck [93] that the AF Heisenberg

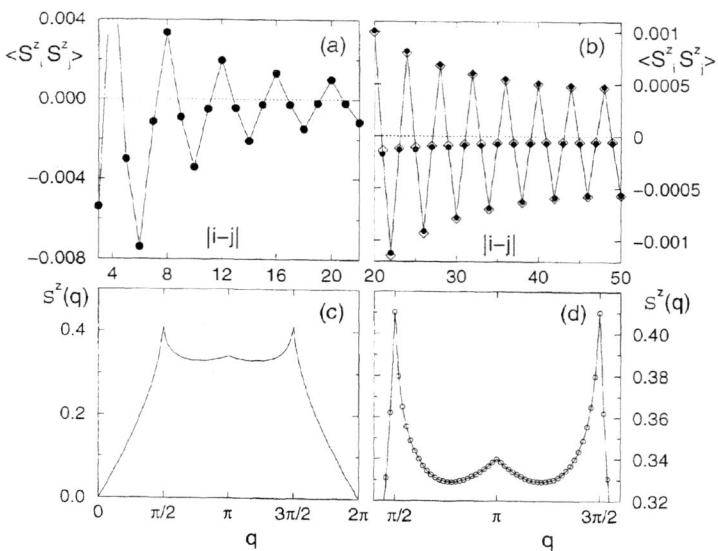

FIGURE 19. (a) QMC results for the correlation function $w_{ij} \equiv \langle S_i^z S_j^z \rangle$ (131) (solid points) as a function of $|i-j|$ for a SU(4) chain of length $L=100$ with PBC which is predominantly in the ground state. The correlations for $|i-j|=1, 2$ and 4 (which are out of the plot range) are -0.07168(1), -0.04011(1) and 0.008261(4), respectively. Part. (b) shows the correlations w_{ij} at large distances $|i-j|$ and the fit to the QMC data (open diamonds). The statistical error bars of the QMC calculations are much smaller than the symbols. Parts (c) and (d) show the Fourier transform $S^z(k)$ of w_{ij} on two different scales (after Ref. [81]).

model is equivalent to *one* free massless boson, while the SU(4) invariant spin-orbital model is equivalent to *three* massless bosons. The velocity of these bosons are all equal to $\pi J/2$ [80]. Therefore we expect the low energy density of states (and hence the entropy) of these two models to differ just by a factor of three. This analysis, however, does not give a clear indication on the nature of the ground state and of the corresponding excitations, and this problem is still open. Modifications of the SU(4) Hamiltonian in one dimension give also interesting physics. If the anisotropy in the orbital sector is introduced, either gapless or gaped phases are found depending on the balance of excitations; in the gaped phase an alternation of spin and orbital singlets gives the lowest energy [99]. The ground state is known exactly and can be written as a product of the fully polarized FM ground state for the spins and the Jordan-Wigner Fermi sea for the local states [100]. On addition to this phase, one finds two different dimerized phases, where the spin and orbital variables are dimerized in a correlated pattern. In one phase spin singlets and orbital singlets arise on alternating rungs, whereas in the other phase spin singlet and orbital triplet share the same rungs. In spite of AF interactions, no gapless AF spin

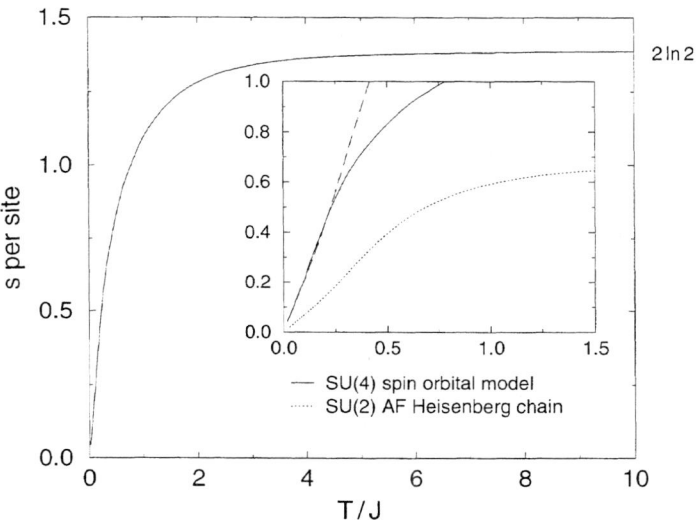

FIGURE 20. Temperature dependence of the entropy s per site for the 1D SU(4) spin-orbital model (130) (solid line). In the inset the entropy per site is shown on larger temperature scale together with the entropy s_{HB} per site of a SU(2) spin-1/2 AF Heisenberg chain (dotted line). For comparison also $3s_{\mathrm{HB}}$ is shown (dashed line) (after Ref. [81]).

phase is found [100]. These results show that spin-gap phases of this type survive anisotropies in the orbital space and are quite generic in spin-orbital models.

Furthermore, the 1D model has also been studied in two limiting cases: the pure XY model both in spin and in orbital space [92], and the dimerized XXZ model [92]. In the pure XY case, a phase separation takes place between two phases with free – fermion like, gapless excitations, while in the dimerized case the low-energy effective Hamiltonian reduces to the 1D Ising model with gaped excitations. In both cases, all the elementary excitations involve simultaneous flips of the spin and orbital degrees of freedom, which gives a clear indication of the breakdown of the traditional MF theory.

The SU(4) symmetric model represents an idealized case for investigating a spin and orbital disordered ground state in higher dimensions. Unfortunately, the QMC method of Frischmuth *et al.* cannot be used [81], as it suffers from a severe minus sign problem in 2D lattices. Hence, unbiased results may be obtained only by an exact diagonalization technique. As pointed out by Li *et al.* [88–90], the SU(4) symmetric model (130) is characterized by a strong tendency towards a liquid ground state with resonating plaquette singlets. This follows from a particular stability of a (spin and orbital) singlet on a square, with the energy of $-4J$, while the first excited state of energy $-2J$ has a degeneracy of fifty [98]. The diagonalization of larger (8-site and 16-site) clusters provides a good evidence that the spin-liquid ground state is realized in fully symmetric SU(4) model in two dimensions. The nature of the disordered ground state is not completely understood, but the results obtained so far indicate that a singlet-multiplet gap survives in the thermodynamic limit, and low-lying SU(4) singlets might exist within this gap [98]. In the view of these results it is interesting to search for a spin-liquid ground state as well in a more realistic situation described by the d^9 model. As a matter of fact, the spin-liquid ground state as resulting from the orbital quantum fluctuations has been predicted first in the d^9 model [63].

B   Quantum corrections in the d^9 model

The size of quantum fluctuation corrections to the classical order parameter determines the stability of the classical phases. The frustration of magnetic interactions might lead in spin models to divergent quantum corrections within the LSW theory [101–109]. Before calculating these corrections in the present situation, a generalization of the usual RPA procedure to a system with several excitations is necessary. Here we present only the relations needed to calculate the quantum corrections to the LRO parameter and to the ground state energy [64]. For that purpose, let us introduce here the local operators constituting the $so(4)$ Lie algebra at site i as Hubbard operators, $X_i^{\alpha\beta} = |i\alpha\rangle\langle i\beta|$. Using the unity operator, $\sum_\beta X_i^{\beta\beta} = \mathcal{I}$, the diagonal operator $X_i^{\alpha\beta}$ that refers to the state $|i\alpha\rangle$ *realized at site i in the classical ground state* of the phase under consideration may be expanded in terms of the excitation operators,

$$X_i^{\alpha\alpha} = \mathcal{I} - \sum_{\beta \neq \alpha} X_i^{\beta\alpha} X_i^{\alpha\beta}, \qquad (132)$$

while the diagonal operators referring to *excited* states $|i\beta\rangle$ are expressed as

$$X_i^{\beta\beta} = X_i^{\beta\alpha} X_i^{\alpha\beta}. \qquad (133)$$

Applying these equations to the z-th spin component $S_i^z = S_{ixx}^z + S_{izz}^z$ of the total spin at site i in one of the G-AF phases with pure orbital character (say AFxx for definiteness), one finds, for i in the spin-up sublattice [64],

$$S_i^z = \frac{1}{2}(X_i^{x\uparrow,x\uparrow} - X_i^{x\downarrow,x\downarrow} + X_i^{z\uparrow,z\uparrow} - X_i^{z\downarrow,z\downarrow})$$

$$= \frac{1}{2}\mathcal{I} - X_i^{x\downarrow,x\uparrow} X_i^{x\uparrow,x\downarrow} - X_i^{z\downarrow,x\uparrow} X_i^{x\uparrow,z\downarrow} = \frac{1}{2}\mathcal{I} - S_{ixx}^- S_{ixx}^+ - K_{izz}^- K_{izz}^+. \qquad (134)$$

Taking the average of Eq. (134) one obtains, with the MF value $\langle S_i^z \rangle_{\rm MF} = \frac{1}{2}$,

$$\langle S_i^z \rangle_{\rm RPA} = \frac{1}{2} - \langle S_i^- S_i^+ \rangle - \langle K_i^- K_i^+ \rangle = \langle S_i^z \rangle - \delta\langle S_i^z \rangle, \qquad (135)$$

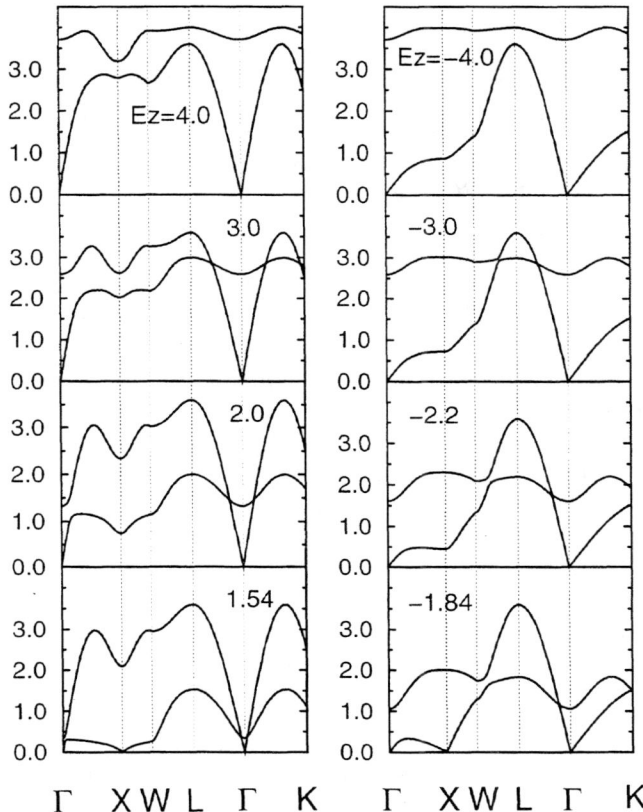

FIGURE 21. Spin-wave and spin-and-orbital-wave excitations in the G-AF phases: AFxx (left) and AFzz (right), in the main directions of the 3D BZ for a few values of E_z (in the units of J), and for $J_H/U = 0.3$. The lower-energy mode becomes soft for $E_z/J < 1.54$ ($E_z/J > -1.84$) in the AFxx (AFzz) phase. To allow a direct comparison, both phases are shown using the 3D BZ for a *bcc* lattice with the standard notation of high symmetry points: $\Gamma = (0,0,0)$, $X = (\pi,0,0)$, $W = (\pi,\pi/2,0)$, $L = (\pi/2,\pi/2,\pi/2)$ and $K = (3\pi/4,3\pi/4,0)$ (after Ref. [64]).

where the averages like $\langle S^-_{ixx}S^+_{izz}\rangle$ are zero since they involve 'ghost' modes, so that one may formally replace S^+_{ixx} by $S^+_{ixx} + S^+_{izz} = S^+_i$, etcetera. The first contribution $\propto \langle S^-_i S^+_i\rangle$ is the usual renormalization due to spin waves, while the second term $\propto \langle K^-_i K^+_i\rangle$ stands for the reduction of $\langle S^z_i\rangle_{\rm RPA}$ due to spin-and-orbital excitations. Both terms involve a local excitation preceded by a deexcitation which reproduces the initial local state. As expected, only transverse excitations contribute to the spin renormalization. Note that, since Eq. (134) is an *exact operator relation*, the present procedure guarantees that Eq. (135) is a *conserving approximation* which respects the sum rule for the occupancies of all states, $\sum_\beta \langle X^{\beta\beta}_i\rangle = 1$. The generalization of Eq. (135) to the MO phases using the operators $\mathcal{S}^\pm_i$ and $\mathcal{K}^\pm_i$ in the expansion of $\mathcal{S}^z_i$, or to other order parameters, like the orbital polarization (86), is straightforward.

The local correlation functions which renormalize the order parameter in Eq. (132) are determined in the standard way [85],

$$\langle B^\dagger_i A_i\rangle = \frac{1}{N}\sum_{\vec{k}}\int_{-\infty}^{+\infty} d\omega\, \mathcal{A}_{AB^\dagger}(\vec{k},\omega)\frac{1}{\exp(\beta\omega)-1}, \qquad (136)$$

where $\beta = 1/k_B T$, and

$$\mathcal{A}_{AB^\dagger}(\vec{k},\omega<0) = 2\mathrm{Im}\langle\langle A_{\vec{k}}|B^\dagger_{\vec{k}}\rangle\rangle_{\omega-i\epsilon} = \sum_{\nu<0}\mathcal{A}^{(\nu)}_{AB^\dagger}(\vec{k})\delta(\omega-\omega^{(\nu)}_{\vec{k}}) \qquad (137)$$

is the respective spectral density for the negative frequencies ($\nu<0$), and $\mathcal{A}^{(\nu)}_{AB^\dagger}(\vec{k})$ are the respective spectral weights. Therefore, the correlation functions at $T=0$ are found by summing up the total spectral weight at the negative frequencies,

$$\langle B^\dagger_i A_i\rangle = \frac{1}{N}\sum_{\vec{k}}\sum_{\nu<0}\mathcal{A}^{(\nu)}_{AB^\dagger}(\vec{k}). \qquad (138)$$

Before discussing the renormalization of the order parameter and the corresponding energies in RPA, we concentrate ourselves on the behavior of the transverse excitations when the crossover lines between the classical phases are approached. As already emphasized in Sec. III.C, the spin-wave and spin-and-orbital-wave excitations couple. As a consequence, the modes in all considered phases *soften* when the transition lines between different classical phases, or classical degeneracy point are approached. This softening is shown for a representative value of $J_H/U = 0.3$ in Fig. 21 for the two AF phases with either $|x\rangle$ or $|z\rangle$ orbitals occupied. In the AFxx phase the energy scales of both excitations are separated for $E_z > 4J$, while the spin-and-orbital mode moves towards zero energy with decreasing E_z, and finally becomes soft at the X point, along $\vec{k} = (\pi,0,k_z)$ and along equivalent lines in the BZ for $E_z \simeq 1.54J$. A similar mode softening is found for the AFzz phase at $E_z < 0$, with the soft mode along $\Gamma - X$ and equivalent directions in the BZ at $E_z \simeq -1.84J$. This peculiar softening along lines and not at points in the BZ shows that the modes behave 2D-like instead of 3D-like: constant-frequency surfaces are cylinders contracting towards lines, not spheres contracting towards a point.

By making an expansion of Eq. (103) around the soft-mode lines, one finds that the (positive) excitation energies are characterized by *finite* masses in the perpendicular directions:

$$\omega_{\text{AFxx}}(\vec{k}) \to \Delta_x + B_x \left(\bar{k}_x^4 + 14\bar{k}_x^2 k_y^2 + k_y^4\right)^{1/2}, \quad (139)$$

independently of k_z (here $\bar{k}_x = k_x - \pi$) for the AFxx phase, and

$$\omega_{\text{AFzz}}(\vec{k}) \to \Delta_z + B_z \left(k_y^2 + 4k_z^2\right), \quad (140)$$

independently of k_x, and similarly along the $\Gamma - Y$ direction with k_y replaced by k_x for the AFzz phase. As an example we give explicit expressions for the AFxx phase at $\eta = 0$,

$$\Delta_x = \frac{9}{2}\frac{\varepsilon_z}{\varepsilon_z + 3}, \qquad B_x = \frac{27}{16}\frac{1}{\varepsilon_z + 3}, \quad (141)$$

where one finds that the gap $\Delta_x \to 0$ when $\varepsilon_z \to 0$, i.e., upon approaching the $M = (E_z, J_H) = (0,0)$ point at which the AF order is changed to the AFzz phase. This illustrates a general principle: $\Delta_i \to 0$ when the crossover line to another phase is approached, and $B_i \neq 0$ when the modes (139) and (140) soften, making quantum fluctuation corrections to the order parameter to diverge logarithmically, $\langle \delta S \rangle \sim \int d^3k/\omega(\vec{k}) \sim \int d^2k/(\Delta_i + B_i k^2) \sim \ln \Delta_i$. We emphasize that for the occurrence of this divergence not only the finiteness of the mass, but also the 2D-like nature of the dispersion is essential. It enables a 3D system to destabilize LRO by what are essentially 2D fluctuations. So the divergence of the order parameter near the crossover lines in the phase diagram and the associated instability of the classical phases, may be regarded as another manifestation of the effective reduction of the dimensionality occurring in the spin-orbital model. We do not present explicitly the softening of the longitudinal modes which also happens at the transition lines, but has no direct relation to the stability of classical phases. A seemingly attractive way to simplify the calculation of the transverse excitations would be to make a decoupling of the spin-waves and spin-and-orbital-waves. However, this is equivalent to violating the commutation rules between the spin and spin-and-orbital operators [78], and this changes the physics. It gives the same excitation energies as Eq. (103), but with $P_{a\vec{k}} = 0$; the numerical result is given in Fig. 22. Of course, the spin-wave excitation does not depend then on the orbital splitting E_z, and the spin-and-orbital-wave excitation gradually approaches the line $\omega_{\vec{k}} = 0$ with decreasing $|E_z|$. It has a weak dispersion which depends on J_H and on the value of $|E_z|$, and gives an instability at the Γ point only, not at certain lines in the BZ. The instability occurs well beyond the transition lines in the phase diagram of Fig. 12, i.e., within the MO$_{\text{FFA}}$ and MO$_{\text{AFF}}$ phase for $E_z < 0$ and $E_z > 0$, respectively. Such spin-wave and spin-and-orbital-wave modes give, of course, much smaller quantum

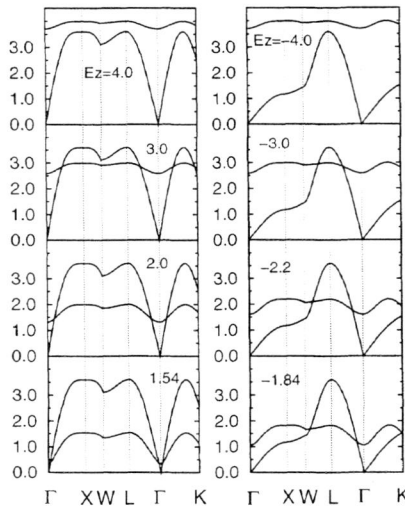

FIGURE 22. The same as in Fig. 21, but for decoupled spin-wave and spin-and-orbital-wave excitations in the G-AF phases (after Ref. [64]).

corrections of the order parameter and energy than the correct RPA spectra of Fig. 21 [78,64].

The spin waves in the MO$_{\text{FFA}}$ phase, stable at $E_z < 0$, soften with decreasing η (53), as shown in Fig. 23. At large η the spin-and-orbital waves at high energies are well separated from the spin-wave modes. The latter have a rather small dispersion at $J_H/U = 0.3$ which follows from relatively weak FM interactions in the (a,b) planes, and AF interactions along the c-axis. The modes start to mix stronger with decreasing η, and finally the gap in the spectrum closes below $\eta = 0.1$. The mode softening occurs again along lines in the BZ, namely along the $\Gamma - X$ direction. Unfortunately, an analogous analytic expansion of the energies near the softening point to those in the AFxx and AFzz phases could not be performed, but the reported numerical results suggest a qualitatively similar behavior to these two phases. The MO$_{\text{AFF}}$ phase gives an analogous instability for $E_z > 0$.

The soft modes in the excitation spectra give a very strong renormalization of the order parameter $\langle S^z \rangle_{\text{RPA}}$ in RPA (135) near the mode softening, as shown in Fig. 24. The quantum corrections *exceed* the MF values of the order parameter in the AFxx and AFzz phases in a region which separates these two types of LRO. Although one might expect that another classical phase with mixed orbitals and FM planes sets in instead, and the actual instabilities where $\delta \langle S_z \rangle \to \infty$ are found indeed beyond the transition lines to another phase, the lines where $\delta \langle S^z \rangle = \langle S^z \rangle$ occur still *before* the phase boundaries in the phase diagram of Fig. 12 (see Fig. 1 of Ref. [63]). This leaves a window where *no classical order is stable* in between

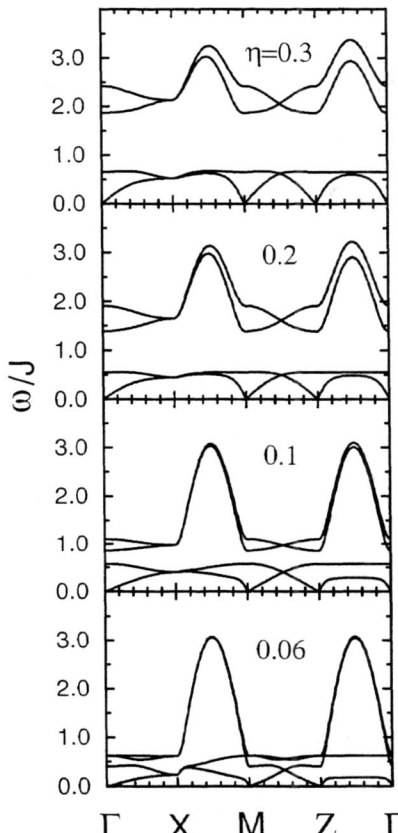

FIGURE 23. Transverse (full lines) and longitudinal (dashed lines) excitations in MOFFA phase in the main directions of the 3D BZ for a few values of J_H/U, and for $E_z/J = -0.5$. The lower-energy mode becomes soft for $J_H/U < 0.06$ (after Ref. [64]).

the G-AF and A-AF spin structures. The origin of such a strong renormalization of $\langle S^z \rangle$ may be better understood by decomposing the quantum corrections into individual contributions as given in Eq. (135). The leading correction comes from the local spin fluctuation expressed by $\langle S_i^- S_i^+ \rangle$ and enhanced with respect to the pure spin model, while the spin-and-orbital fluctuation, $\langle K_i^- K_i^+ \rangle$, increases rapidly when the instability lines $\langle S^z \rangle_{\text{RPA}} = 0$ are approached. Interestingly, the latter fluctuation is stronger in the AFxx than in the AFzz phase for the same values of J_H and $|E_z|$ which demonstrates that the AFzz phase is more robust due to the directionality of the $|z\rangle$ orbitals and the strong AF bonds along the c-axis. This asymmetry is also visible in Fig. 24, where $\langle S^z \rangle_{\text{RPA}}$ decreases somewhat faster towards zero for $E_z > 0$.

In both G-AF phases (AFxx and AFzz) the leading contribution to the renormalization of $\langle S^z \rangle_{\text{RPA}}$ comes from the lower-energy mode, especially at larger values of J_H. In the case of $J_H = 0$ one finds, however, that the contribution from the lower mode either stays approximately constant (in the AFxx phase), or even decreases (in the AFzz phase) when the line of the collapsing LRO is approached at $|E_z| \to 0$. This latter behavior shows again that the coupling between the spin-wave and spin-and-orbital-wave excitations is of crucial importance [78]. This is further illustrated by Fig. 25, which shows the renormalization of $\langle S_z \rangle$ as obtained when spin waves and spin-and-orbital waves are decoupled in the manner discussed above. One observes that significant reduction of $\langle S_z \rangle$ then sets in only very close to the actual divergence. The reduction of $\langle S^z \rangle_{\text{RPA}}$ in the MO$_{\text{FFA}}$/MO$_{\text{AFF}}$ phases, described by a relation similar to Eq. (135), is in general weaker than that in the G-AF phases. This is understandable, as the quantum fluctuations contribute here only from a single AF direction, while the FM order in the planes does not allow for excitations which involve spin flips and stabilizes the LRO of A-AF type. For fixed J_H one finds increasing quantum corrections $\delta \langle S^z \rangle$ when the lines of phase transitions towards the AF phases are approached. These corrections increase faster with increasing $|E_z|$ in the MO$_{\text{FFA}}$ phase, as the increasing occupancy of the $|z\rangle$-orbital makes the AF interaction stronger there than in the MO$_{\text{AFF}}$ phase, where the occupancy of the $|x\rangle$ orbital increases slower roughly by a factor of two. It may be again concluded [64] that the collapse of the LRO in the A-AF (MO) phases is primarily due to increasing spin fluctuations, $\langle S_i^- S_i^+ \rangle$, while the spin-and-orbital $\langle K_i^- K_i^+ \rangle$ fluctuations become of equal importance only when the multicritical point of the Kugel-Khomskii model $M = (E_z, J_H) = (0,0)$ is approached.

The orbital polarization (86) is also renormalized by the quantum fluctuations, but this is a rather mild effect not showing any instability (Fig. 26). In fact, this renormalization involves only the spin-and-orbital and the orbital excitation but not the spin excitation, which gives the largest weight in the lowest transverse mode that goes soft. The value determined in RPA is calculated from an expression similar to Eq. (135), e.g. in the AFxx phase from

$$\langle n_{ix} \rangle_{\text{RPA}} = 1 - 4 \langle T_{izx} T_{ixz} \rangle - \langle K_i^- K_i^+ \rangle. \tag{142}$$

The density $\langle n_{ix} \rangle_{\text{RPA}}$ decreases gradually with decreasing E_z, with somewhat in-

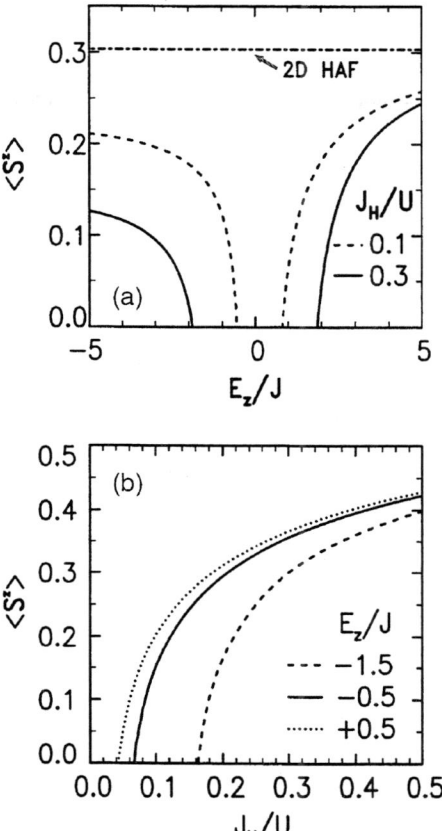

FIGURE 24. Renormalization of the magnetic LRO parameter $\langle S_i^z \rangle$ by quantum fluctuations as obtained in RPA in: (a) AFzz (left) and AFxx (right) phases as functions of E_z/J for $J_H/U = 0.1$ and 0.3; (b) MOFFA phase as function of J_H/U for $E_z/J = 0.5$, -0.5 and -1.5 (after Ref. [64]).

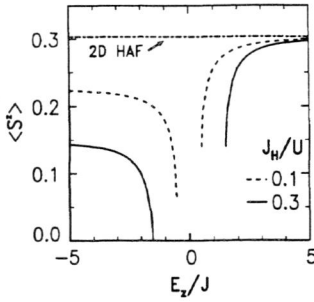

FIGURE 25. Renormalization of the magnetic LRO parameter $\langle S_i^z \rangle$ by quantum fluctuations as obtained in RPA in: (a) AFzz (left) and AFxx (right) phases as functions of E_z/J for $J_H/U = 0.1$ (dashed lines) and 0.3 (solid lines); (b) MOFFA phase as functions of J_H/U for $E_z/J = 0.5$, -0.5 and -1.5 (after Ref. [64]).

creased quantum corrections close to the transition lines between different classical phases. In the MO$_{\text{AFF}}$ phase one finds nonequivalent sublattices, with enhanced/reduced $\langle n_{ix} \rangle_{\text{RPA}}$ from its average value, reflecting the shape of the occupied orbital on a given sublattice. Finally, we note that the dominating contribution to the quantum corrections to the energy comes from the transverse excitations. The longitudinal excitations do not contribute at all at $J_H/U = 0$, where these modes are dispersionless. Otherwise, the orbital excitations have always a significantly smaller dispersion than the value of the orbital gap in the spectrum, and the resulting quantum corrections are therefore almost negligible.

Summarizing, we have reported the case that a generic (Kugel-Khomskii) model for the dynamics of an orbitally degenerate MHI is characterized by a number of peculiar features. Assuming that the ground state exhibits some particular classical spin- and orbital order, the stability of this order can be investigated by considering the Gaussian fluctuations around this state. In this way we find that in various regimes of the zero-temperature phase-diagram, classical order is defeated by the quantum fluctuations, and we expect a qualitative phase diagram with a region of disordered phases [63]. These disordered phases with short-range spin and orbital correlations are discussed in the next Section.

C Spin disorder: VB and RVB states

Generally speaking, the ground state is unknown for most systems described by the Heisenberg model with nonferromagnetic interactions. Even those that are known in analytical form, such as the Bethe solution in one dimension [95], require numerical computations to determine the spin correlations. Under these circumstances, variational wave functions can give good guesses for the ground

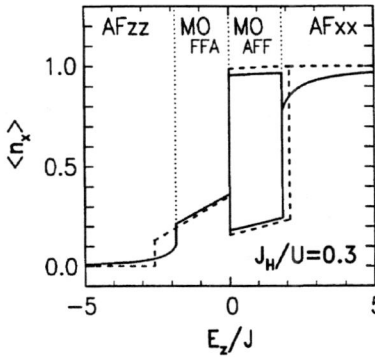

FIGURE 26. Average density of $|x\rangle$-holes $\langle n_x \rangle$ as obtained in RPA for $J_H/U = 0.3$ in MF approximation (dashed lines) and with the quantum corrections calculated in RPA (full lines). The splitting of lines for $E_z/J > 0$ corresponds to the MOAFF phase with two different hole densities $\langle n_x \rangle_A \neq \langle n_x \rangle_B$ on the ions belonging to two sublattices shown in Fig. 11 (after Ref. [64]).

state. In this respect, the valence bond (VB) states and/or plaquettes valence bond (PVB) states are variational wave functions which play a prominent role in approximating the leading spin-spin correlations for the AF Heisenberg model. They have been studied largely in the context of quantum magnetism, especially in recent years when new motivation occurred due to the research aimed at a better understanding of the magnetic states realized in high temperature superconductors.

Their general form can be written in the following way [3,82]:

$$|c_\alpha, S\rangle = \sum_\alpha c_\alpha |\alpha\rangle, \qquad (143)$$

where c_α are variational parameters and

$$|\alpha\rangle = \prod_{\langle ij \rangle \in \Lambda_\alpha} (a_i^\dagger b_j^\dagger - b_i^\dagger a_j^\dagger)|0\rangle, \qquad (144)$$

with a_i and b_i being the Schwinger bosons on site i [82]. Λ_α is a particular configuration of bonds $\langle ij \rangle$ which cover the whole lattice and are occupied by spin singlets. The essence of a VB state is that there are no magnetic correlations between different singlets. The condition on Λ_α is that precisely $2S$ bonds will emanate from each site. In some cases there will be only few configurations to contribute in the large lattice limit. The case were there are macroscopically many configurations and VB states resonate between them has been called resonating valence bond (RVB) states and was introduced by Anderson [110]. Considering the expression given in Eq. (143), it is easy to verify that all bond operators are invariant under SU(2) transformations. Therefore, the VB wave function $|c_\alpha, S\rangle$ is a singlet in the total spin. It provides an example of spin-disordered state, the so-called *spin liquid*.

There exists a class of special Hamiltonians for which the VB states are the *exact* ground states. Majumdar and Ghosh introduced the Hamiltonian for spin $S = 1/2$ [111]:

$$H_{MG} = \frac{4K}{3} \sum_{i=1}^{N} \left(\vec{S}_i \cdot \vec{S}_{i+1} + \frac{1}{2} \vec{S}_i \cdot \vec{S}_{i+2} \right) + \frac{1}{2} N, \quad (145)$$

where i stands for the sites of a 1D chain with an even number of sites N, $K > 0$ and one assumes periodic boundary conditions, with $\vec{S}_{N+1} = \vec{S}_1$ and $\vec{S}_{N+2} = \vec{S}_2$. It is straightforward to show that the dimer state,

$$|d\rangle_{\pm} = \prod_{n=1}^{N/2} \left(|\uparrow_{2n}\rangle |\downarrow_{2n\pm 1}\rangle - |\downarrow_{2n}\rangle |\uparrow_{2n\pm 1}\rangle \right) / \sqrt{2}, \quad (146)$$

has energy zero, that is $H_{MG}|d\rangle_{\pm} = 0$, and all the other eigenenergies are positive. As one can see the Hamiltonian H_{MG} includes AF interactions between next-nearest neighbors which frustrates the nearest neighbor correlations. Thus, one expects the ground state to be more disordered than that in the usual Heisenberg model. Indeed, for the antiferromagnet with only nearest neighbor interactions the spin-spin correlations decay as an inverse of a power of distance, while the wave function $|d\rangle_{\pm}$ has state dimer correlations that is they vanish beyond the nearest neighbors. The proof is interesting since it provides a method for constructing a more general family of Hamiltonians whose ground states are VB states [82].

The basic idea is to express H_{MG} in terms of projection operators. If we consider three spins $S = 1/2$ on an arbitrary triad of sites, the total spin J may be either equal to $3/2$ or to $1/2$. The total spin of three sites $(i-1, i, i+1)$ is given by

$$\vec{J}_i = \vec{S}_{i-1} + \vec{S}_i + \vec{S}_{i+1} \quad (147)$$

and its square is connected to the H_{MG} via the projection operator on a configuration of spin $1/2$ on three sites in the following way:

$$Q_{3/2}(i-1, i, i+1) = \frac{1}{3}\left(J_i^2 - \frac{3}{4} \right) = \frac{1}{2} + \frac{2}{3}\left(\vec{S}_i \cdot \vec{S}_{i-1} + \vec{S}_{i-1} \cdot \vec{S}_{i+1} + \vec{S}_i \cdot \vec{S}_{i+1} \right), \quad (148)$$

so that the Hamiltonian can be rewritten as

$$H_{MG} = K \sum_i Q_{3/2}(i-1, i, i+1). \quad (149)$$

Of course, $Q_{3/2}(i-1, i, i+1)$ annihilates any state with $J = 1/2$ at the three sites $\{i-1, i, i+1\}$. Since the dimer states $|d\rangle_{\pm}$ (146) do not contain states with total $J_z > 1/2$ on any three sites and due to the rotational invariance of the state $|d\rangle_{\pm}$ there cannot be any $J > 1/2$ component in it. Therefore, each operator $Q_{3/2}(i-1, i, i+1)$ annihilates $|d\rangle_{\pm}$. Moreover, since the $Q_{3/2}(i-1, i, i+1)$ have only positive eigenvalues, $|d\rangle_{\pm}$ is the ground state of H_{MG}.

In a similar fashion Affleck, Kennedy, Lieb, and Tasaki [112] have constructed the Hamiltonians for the VB solids as ground states which cover the whole lattice,

$$H_{\text{AKLT}} = \sum_{\langle ij \rangle} \sum_{J=2S-M+1}^{2S} K_J Q_J(ij), \tag{150}$$

where the bond projector $Q_J(ij)$ projects the total bond spin of magnitude J. In this case it can be proven as before, that the ground state is

$$|\Omega_{\text{VBS}}\rangle = \prod_{\langle ij \rangle} \left(a_i^\dagger b_j^\dagger - b_i^\dagger a_j^\dagger \right)^M |0\rangle, \tag{151}$$

where all the nearest neighbor bonds $\langle ij \rangle$ are included in the product, and $M = 2S/z$ is an integer which is related to the spin S and to the coordination number z of the considered lattice. By construction, the smallest S which allows to construct such states is $S = 1$ for a 1D lattice.

Spontaneous spin liquid states occur in 2D lattices due to frustrating spin interactions. The basic model which shows such a behavior is the 2D AF $J_1 - J_2$ Heisenberg model on a square lattice [101–106]. There are two main reasons why this model holds a special place in the physics of spin systems. It is one of the simplest models which exhibits quantum transitions between long-range ordered phases and a quantum disordered phase – a topic of fundamental interest [113]. Moreover, even though the $J_1 - J_2$ model (Fig. 27) does not deal with charge dynamics, it can represent a good starting point for the understanding of how translational symmetry is broken in a purely insulating spin background for approaching this question in spin systems with finite doping. The $J_1 - J_2$ model has been discussed in numerous works over the last ten years and some of the important issues that have been addressed are: (i) how is the Néel order, present for small frustration (J_2), destroyed as frustration increases, and (ii) is a quantum disordered phase present in a finite window of frustration, and what is the structure of this phase? Spin-wave calculations, both at the non-interacting level as well as including interactions perturbatively in powers of $1/S$ (S is the spin value), have found that the magnetization decreases with increasing frustration, ultimately vanishing at a critical value [114]. These calculations however cannot predict the structure of the phase beyond the instability point, or the location of the phase boundary with high accuracy, since as the magnetization decreases more and more powers of $1/S$ have to be included (strong spin-wave interactions). Exact diagonalization (ED) of clusters as large as $N = 36$ [115,116] have found a finite region of quantum disordered (gaped) phase, but have failed to determine accurately the dominant short-range correlations or type of order (e.g. dimer, plaquette, *etcetera*) characterize this phase. The ED calculations also suffer from large finite-size corrections, especially for strong frustration. An insight into the structure of the disordered phase was possible with the help of the large N expansion technique [117,103]. These authors predicted the quantum disordered phase to be spontaneously dimerized in a particular (columnar) configuration (Fig. 27). High order dimer series

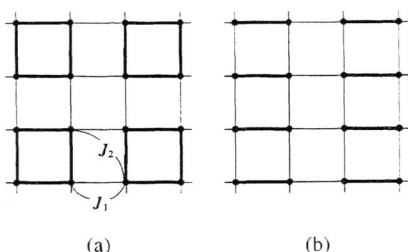

FIGURE 27. Disordered magnetic states for a 2D square lattice with AF interactions J_1 and J_2: (a) plaquette-RVB, and (b) columnar (VB) dimer ground states. The bold lines indicate the bonds occupied by spin singlets.

expansions around this configuration were performed [118,119], all confirming its stability in a window of frustration. Thus the spontaneously dimerized state has emerged as the most probable candidate for a disordered ground state. Let us mention in this connection that in 1D systems the Lieb-Schultz-Mattis (LSM) theorem guarantees that a gaped phase in a quantum spin system breaks translational symmetry [120]. Extension of the LSM theorem to two-dimensions was proposed [121] but not yet proven in the most general case. The large N and dimer series results, however, seem to confirm the validity of the LSM theorem in two dimensions as well, including, in fact, the case of finite doping [122].

The previous case is a classical example that inclusion of additional interactions, such as dimerization and/or frustration, leads to increased quantum fluctuations, and ultimately to vanishing LRO at a critical coupling. In this context, further examples of transitions caused by local alternation of the exchange couplings are the dimerized Heisenberg antiferromagnets [123], the two-layer Heisenberg model [107–109] and the CaV_4O_9 lattice (1/5- depleted square lattice) [124–127]. For these situations, the local dimer or plaquette correlations eventually win over the long-range Néel order, leading to a disordered ground state.

An important issue concerning the quantum transitions mentioned above is their universality class. It is generally accepted that the effective low-energy theory for the 2D Heisenberg systems with a collinear (Néel) order parameter is the O(3) non-linear sigma model (NLSM) in 2+1 dimensions [128]. This field theory contains a single effective coupling constant g and, at $T = 0$, describes the ordered Néel phase for $g < g_c$. For $g > g_c$ the NLSM is in a quantum disordered phase with a finite correlation length. However the determination of g_c and the nature of the disordered phase are beyond the field theory formulation and depend on the specific details of the model. In addition, Berry phases associated with instanton tunneling between topologically different configurations are present in the NLSM [129]. In one dimension the Berry phase effects are known to be important, essentially leading to the difference between the excitations in the integer and half odd-integer spin chains [121] (Haldane conjecture). In two dimensions Berry phases are also present, but their role is less clear. If one neglects these purely quantum effects, the universality class of the quantum transitions in the 2D Heisenberg antiferromagnet should be the same as that of the classical O(3) vector model in three dimensions. QMC simulations performed on the two-layer antiferromagnet [130], and on the CaV_4O_9 lattice [126] confirm with high accuracy that the quantum transitions in the above two models happen in the O(3) universality class.

Furthermore, the $J_1 - J_2$ model, which exhibits a quantum transition due to frustration, could represent a case of difference from the O(3) universality class. Read and Sachdev showed that there are two correlation lengths close to the criticality [103], suggesting to think that there are deviations in the universality class. On the other hand, they showed that only one of the two is relevant implying that even though the Berry phases are relevant in the disordered phase, ultimately, near the critical point, their effect disappears.

As we have discussed, the main characters in this topic are low-dimensional quantum spin systems (spin chains [131] and ladders [132-134]), and it proves difficult to achieve quantum melting of magnetic LRO in empirically relevant systems in higher dimensions. It is therefore worth pointing out that there is a class of systems in which quantum-melting occurs due to a *unique mechanism* which operates in three dimensions: small spin, orbital degenerate MHI, the so-called Kugel-Khomskii systems [15]. There might exist already a physical realization of such a 3D *quantum spin-orbital liquid*: spin disorder in $LiNiO_2$ [135,136].

Here, we report how the orbital degeneracy operates through the same basic mechanisms known from spin systems, to produce quantum melting in the spin-orbital d^9 systems. The novelty is that these systems tend to "self-tune" to (critical) points of high classical degeneracy. There are interactions which may lift the classical degeneracy, but they are usually weak. An interaction of this kind is the electron-phonon coupling – the degeneracy is lifted by a change in crystal structure, the conventional collective JT instability. As we have discussed in Sec. III.C, the lattice has to react to the symmetry lowering in the orbital sector, but it was recently convincingly shown, at least in the archetypical compound $KCuF_3$, that the structural distortion is a side effect [36]. The fundamental question which arises in this context is what happens when the classical order becomes unstable against quantum fluctuations. Although the subject is much more general (singlet-triplet models in rare earth compounds [137], V_2O_3 [24], $LaMnO_3$ [138], heavy fermions [2,3,139]), we focus here on the simplest situation encountered in $KCuF_3$ and related systems, described by the d^9 model [63].

If the classical limit is as sick as explained in Sec. IV.B, what is happening instead? *A priori* it is not easy to give an answer to this question. There are no 'off the shelf' methods to treat quantum spin problems characterized by classical frustration, and the situation is similar to what is found in, e.g. $J_1 - J_2 - J_3$ problems [101-105]. A first possibility is quantum order-out-of-disorder [106]: quantum fluctuations can stabilize a particular classical state over other classically degenerate

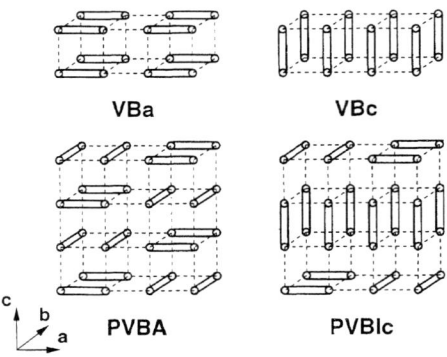

FIGURE 28. A variety of VB (VBa and VBc) and PVB (PVBA and PVBIc) solids discussed in the text for the d^9 model. (After Ref. [140], Copyright 1999, with permission from Elsevier Science.)

states, if this particular state is characterized by softer excitations than any of the other candidates. Khaliullin and Oudovenko [86] have suggested that this mechanism is operative in the present spin-orbital model, where the 3D anisotropic AFzz antiferromagnet is the one becoming stable. Their original argument was flawed because of the decoupling procedure they used which violates the so(4) dynamical algebra constraints [78]. Nevertheless, there is yet another possibility: VB singlet (or spin-Peierls) order, which at the least appears in a more natural way in the present context than is the case of higher dimensional spin-only problems, because it is favored by the directional nature of the orbitals.

In similarity to the purely spin systems, in the presence of orbital interactions either one particular covering of the lattice with these 'spin-dimers' might be favored (VB or spin-Peierls state), or the ground state might become a coherent superposition of many of such coverings (RVB state). On a cubic lattice the difficulty is that although much energy is gained in the formation of the singlet pairs, the bonds between the singlets are treated poorly. Nevertheless, both in 1D spin systems (Majumdar-Ghosh [111], AKLT-systems [112]) and in the large N limit of $SU(N)$ magnets in two dimensions [113], ground states are found characterized by spin-Peierls/VB order [103]. In principle, various topologically different coverings of the 3D lattice may be considered, in analogy with the square lattice [141]. Two obvious choices, defined by the sets Λ_α of parallel singlets in Eq. (144), and suggested by the tendency towards particular directional orbitals for positive (negative) E_z are: (i) singlets along the a-axis with orbitals close to $3x^2 - 1$ (VBa), expected to be favored for $E_z > 0$, and (ii) singlets along the c-axis with orbitals $\sim |z\rangle$ (VBc), preferred if $E_z < 0$ (see Fig. 28). Interestingly, the immanent frustration of the magnetic interactions in the spin-orbital model (57) causes that these phases have lower energies than the classical phases in a broad parameter regime. This result

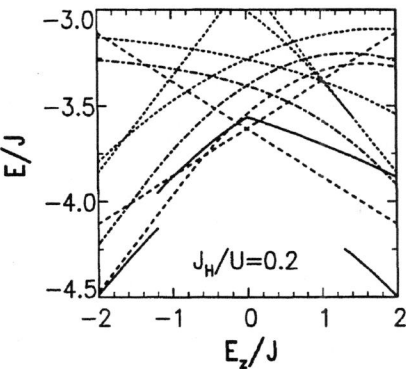

FIGURE 29. Energies of the various ordered and disordered phases for the d^9 model (57)–(59) as functions of E_z/J, for $J_H/U = 0.2$. The energies of classical phases were calculated within MF (short-dashed lines) and RPA approach (full lines), while the energies of VBa and VBc states (see Fig. 28) are given by dashed-dotted lines. The energies of disordered states (RVBc, PVBI, and PVBA) stable close to the orbital degeneracy are given by the long-dashed lines (after Ref. [142]).

is already quite spectacular when one realizes that such VB phases thus appear to be better approximations to the exact ground state than the classical phases with magnetic LRO in *three dimensions*. An example is shown in Fig. 29 for $J_H/U = 0.2$; as expected, VBa (VBc) has a lower energy for $E_z > 0$ ($E_z < 0$), and both phases are degenerate at $E_z = 0$. We have recognized that the exceptional stability of these (nonresonating) VB states is due to a unique mechanism involving the orbital sector. Unlike in the spin system with a simple Heisenberg exchange, the bonds not occupied by the singlets *contribute orbital energy* and this is optimized when singlets in orthogonal directions are connected (which is not the case in the VBa and VBc phases; see further below). Further improvement of energy may be expected by including the leading quantum fluctuations in the VB (VBa and VBc) states. In the case of the VBa phase this leads directly to constructing the PRVB states, with the plaquette wave functions of the type,

$$|\Psi_\Box\rangle = \frac{1}{\sqrt{2|1+S|}} (|\Psi_a\rangle + e^{i\phi}|\Psi_b\rangle), \quad (152)$$

where the two components are composed of the singlet pairs along a and b-axis, $|\Psi_a\rangle$ and $|\Psi_b\rangle$, respectively, and $S = e^{i\phi}\langle\Psi_a|\Psi_b\rangle$ is the overlap. It is straightforward to show that this Ansatz gives an exact energy $E_0 = -2J_0$ for a plaquette occupied by spins $S = 1/2$, assuming the AF interaction J_0. The wave function $|\Psi_\Box\rangle$ turns out to be identical with the exact wave function of an isolated plaquette [143], if we make the choice $\phi = 0$ in Eq. (152). When averaged over the square lattice, this gives an energy of $-J_0/2$ per site, the same energy as that of the classical Néel state. However, in the present case the resonance between the two singlet structures is much weaker, as the optimized orbitals are either $\| a$, or $\| b$ in the two components of the plaquette wave function (152). This results in a much smaller overlap $S = 1/32$ and typically very small offdiagonal energy elements, $\langle\Psi_a|H|\Psi_b\rangle$. As a result, the improvement due to this resonance is marginal, and it turns out to be better to optimize the singlet distribution over the lattice with respect to the energy contributions which originate from the bonds not occupied by the singlets.

There are three different possible energy contributions for these bonds depending on whether they connect: (i) two singlets along a single (a, b, or c) line, (ii) two singlets oriented along two different lines with an angle of $\pi/2$, or (iii) two parallel singlets, with the bond making itself an angle of $\pi/2$ to both of them. As the second type of the non-singlet bonds is energetically the most favorable, more energy than in the PRVB state (152) is gained if the plaquette wave functions $|\Psi_a\rangle$ and $|\Psi_b\rangle$ alternate and form a superlattice (see Fig. 30). This results in a plaquette VB (PVB) state for each (a,b) plane. As also the energy of the vertical ($\| c$) bonds is optimized when these bonds connect one singlet $\| a$ and another $\| b$, the optimal phase with a lower energy that the quantum-corrected MO$_{\text{AFF}}$ phase found for $E_z > 0$ (Fig. 29) is given by a PVB alternating (PVBA) state, with the alternation of two along the c-axis (see Figs. 28 and 30). We note that the energy comparison between the AFxx state and the PVBA state resembles qualitatively the situation in a 2D 1/5-depleted lattice [124,125], but the spin-disordered state is remarkably more stable in our case, and provides an approximate Ansatz for the ground state in *three dimensions*. In contrast, the energy of the VBc phase can be improved by considering the resonance between the singlets along the vertical ($\| c$) direction. The energy of the resulting resonating VBc (RVBc) state could be obtained using the Bethe ansatz result for the 1D Heisenberg antiferromagnet [95], and adding the orbital energy contributions due to the bonds $\perp c$. This qualitatively different behavior has to do with the symmetry breaking in the orbital space between $E_z > 0$ and $E_z < 0$, as the system has to choose between two equally favorable directional orbitals $\zeta\rangle$ ($\| a$ and $\| b$) if $E_z > 0$, which makes the PVBA phase optimal due to the intersinglet contributions, while a single possibility ($|z\rangle$ orbitals) remains if $E_z < 0$. As shown in Fig. 29, the energy of the RVBc phase is lower than the energy of the MO$_{\text{FFA}}$ phase corrected by quantum fluctuations.

In the crossover regime between the RVBc and PVBA phases one may expect that some other arrangment of singlets in the cubic lattice gives still a better energy that the two above magnetically disordered phases. In fact, we have found a PVB interlayered (PVBI) state, composed of single planes of the PVBA phase (Fig. 30) interlayered with two planes of VBc phase along the c-direction (Fig. 28), to be more stable in the crossover regime, as shown in Fig. 29. The energy in the PVBI state is gained both from more $|z\rangle$ orbital character in the singlets $\| c$, and from the orbital energy contributions due to the bonds not occupied by the singlets, connecting a single layer of the PVBA phase with the singlets $\| c$ in the neighboring

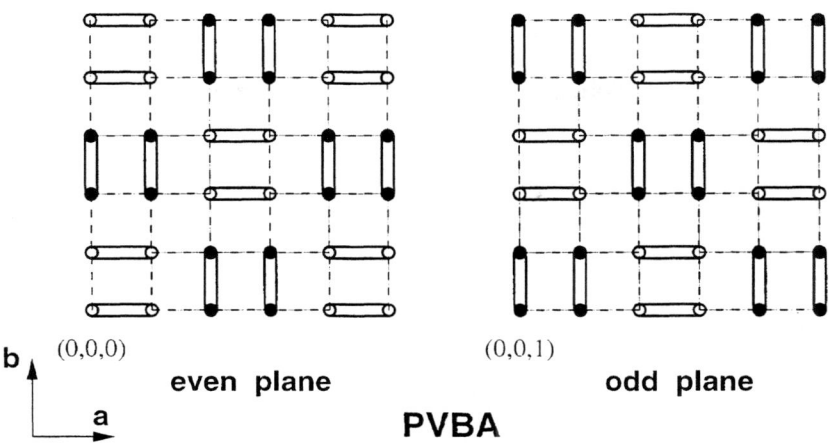

FIGURE 30. Schematic representation of the PVBA state, stable in a range of J_H/U and for $E_z/J > 0$. The open and full circles refer to the directional $|\zeta\rangle$ orbitals parallel to the bonds occupied by the singlets along a and b axis, respectively (after Ref. [142]).

planes. Of course, this phase is destabilized for $E_z > 0$, where the holes occupying $|z\rangle$ orbitals have higher energy.

It is straightforward to understand that the interplay of orbital- and spin degrees of freedom tends to stabilize VB order. Since the orbital sector is governed by a discrete symmetry, the orbitals tend to condense in some classical orbital order. It is precisely for this reason that the best ground state wave functions do not resemble the AKLT models (150) with alternating spin and orbital singlets [100], but form instead composite spin-and-orbital singlets on the same bonds, as for instance in the PVBA phase (Fig. 30). Different from the fully classical phases, one now looks for orbital configurations optimizing the energy of the spin VB configurations. The spin energy is optimized by having directional orbitals $|\zeta\rangle$ parallel to the bond $\langle ij \rangle$ at both sites i and j at which the VB spin-pair also lives. This choice maximizes the overlap between the wave functions, and thereby the binding energy of the singlet. At the same time, this choice of orbitals minimizes the unfavorable overlaps with spin pairs located in directions orthogonal to $|\zeta\rangle$. The net result is that VB states are much better variational solutions for the d^9 model (57), than in the standard spin systems.

Addressing the problem of *spin-liquid* in the d^9 model systematically, it has been found [63] that two families of VB states are most stable: (i) The 'staggered' VB states like the PVBA and PVBIc states of Fig. 28. These states have in common that the overlap between neighboring VB pairs is minimized: the large lobes of the $|\zeta\rangle$ orbitals of different pairs are never pointing to each other. (ii) The 'columnar'

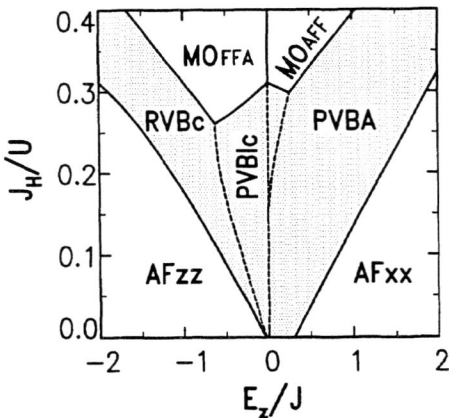

FIGURE 31. The same as in Fig. 12, but including quantum fluctuations as determined in RPA. The spin liquid (RVBc, PVBIc, PVBIa, and PVBA) takes over in the shaded region between G-AF and A-AF phases (after Ref. [63]).

VB states like the VBc (or VBa) state of Fig. 28. In the orbital sector, this is nothing else than the AFzz state of Fig. 11 ($3z^2 - r^2$ orbitals on every site). Different from the AFzz state, the spin system living on this orbital backbone is condensed in a 1D spin-Peierls state along the c-direction which is characterized by strong exchange couplings. The spins in the $a(b)$-directions stay uncorrelated, due to the weakness of the respective exchange couplings as compared to the VB mass gap. These considerations lead to a phase diagram of the d^9 model (57)–(59) shown in Fig. 31. If $\eta < 0.30$, the PVBA state is stable at $E_z > 0$, while a similar PVBIc interlayered state with alternating layers of (a,b)-plane/c-axis bonds (Fig. 28), and the RVBc state are stable at $E_z < 0$. Thus, a spin liquid is stabilized by the *orbital degeneracy* over the ordered MO phases with RPA fluctuations in a broad regime. This resembles the situation in a 2D 1/5-depleted lattice [124,125], but the present instability is much stronger and happens in *three dimensions*.

Summarizing, there is a strong theoretical arguments supporting the conjecture that quantum-melting might occur in orbital degenerate MHI. Why does it not occur always (e.g., in KCuF$_3$)? Next to the Hund's rule coupling, the electron-phonon coupling λ (7) may destroy quantum disorder. The lattice will react to the orbital fluctuations, dressing them up in analogy with polaron physics, and thereby reducing the coupling constant. In order to quantum melt KCuF$_3$-like states, one should therefore look for ways to reduce both the effective J_H and λ. It was first suggested in Ref. [63] that this situation might occur in LiNiO$_2$: although the spin-spin interactions in the (111) planes should be very weakly FM according to the Goodenough-Kanamori rules, magnetic LRO is absent [135] and the system might represent the spin-orbital liquid. More strikingly, LiNiO$_2$ is cubic and should undergo a collective JT transition, which absence is actually an old chemistry mystery! Upon electron-hole transformation, d^7 low-spin Ni^{3+} maps on d^9 Cu^{2+} in $KCuF_3$, but with a difference in chemistry. While the e_g hole in $KCuF_3$ is nearly entirely localized on the Cu, the e_g electron in $LiNiO_2$ is rather strongly delocalized over the Ni and surrounding O ions which reduces both J_H and λ, and explains the absence of classical ordering. A more precise experimental characterization of $LiNiO_2$ is needed to decide whether it provides an example for a quantum disorder of the kind discussed here.

Comparing with spin models, the understanding of the disordered ground states in spin-orbital models is much less developed. We have discussed in this Section only these results which could be obtained so far using simple methods based on variational wave functions of the VB and RVB type. There are a few questions which deserve a future study. For instance, by deriving similar rules to those known for the spin systems [144], for calculating the overlaps between different VB-type configurations on the lattice would simplify the analysis of these situations in which the resonance of different configurations might play a role in spin-orbital systems. Formulating the problem on the abstract level, one might attempt to find parent Hamiltonians similar to the AKLT models (150) to particular dimerized states with alternating spin and orbital singlets. However, the e_g orbitals are more classical than the pseudospins in the symmetric SU(4) model of Sec. IV.A. Further studies should clarify to what extent the e_g orbitals exhibit really quantum behavior in realistic situations, such as described by the d^9 model (57)–(59). However, we believe that even if the orbitals are more classical than the spins, they amplify the quantum effects just by their fluctuations which couple to spin fluctuations.

V SPIN-ORBITAL MODEL IN MANGANITES

A Superexchange and orbital interactions in LaMnO$_3$

As we have shown in Fig. 4, the phase diagrams of manganites are very rich and one would like to understand the microscopic origin of the experimentally observed A-AF order in $LaMnO_3$ [145] in first place, and next why the magnetic interactions change so drastically that the system becomes FM at larger doping. Naively, the FM order follows just from the DE model as presented in Sec. II.B. However, we have presented arguments in Sec. I that this qualitative argument does not explain the physics of manganites and the realistic approach has to include the double degeneracy of e_g orbitals occupied by one or less electron per Mn ion.

As we already know from the earlier studies [15,146] and from Sec. III of the present paper, the magnetic order is coupled to the orbital one. Therefore, one possible explanation of the observed A-AF phase might be that it follows from a cooperative JT effect [42] which stabilizes a particular order of the singly occupied e_g orbitals [18]. Although this might sound quite attractive and attempts were made to understand the superexchange constants observed in the A-AF phase of $LaMnO_3$ in *ab initio* calculations [38,147], these studies did not give the superexchange constants close to their experimental values. A model of degenerate e_g orbitals [148] is more successful, but only when the orbital ordering is not calculated but fitted to the experiment. Here we argue that the leading mechanism comes instead from strong Coulomb interaction U is the dominating energy scale in late transition metal oxides which gives superexchange induced by the hopping of e_g electrons [23]. At the same time the coupling to the lattice is much smaller than the Coulomb interactions which is consistent with the existence of the insulating phase above the JT transition [149]. Of course, the JT effect has to be included in a complete model, as we will also show below, as it leads to particular orbital interactions and stabilizes the orbital ordering well before (when the temperature is decreased) the magnetism sets in. Therefore, we believe that the JT effect is of crucial importance in the doped regime where it drives the transition from the insulating polaronic regime to a metal [150].

If transition metal ions have partly filled e_g orbitals close to orbital degeneracy, as we have discussed for the $KCuF_3$, the strong Coulomb interactions lead to an effective low energy Hamiltonian, where the spin and orbital degrees of freedom are coupled [18,63,64]. As an important difference with respect to the d^9 configuration (Sec. III.A), one has to construct *superexchange between total spins* $S = 2$ of Mn^{3+} ions in $LaMnO_3$ compound [23]. A simpler approach in which the superexchange is considered separately for the e_g electrons with spins $s = 1/2$ and for t_{2g} electrons with spins $S_t = 3/2$ [151,152] is not realistic as the spin dynamics of e_g and t_{2g} spins does not occur independently of each other and the spin waves measured by neutrons correspond to the spin excitations of spins $S = 2$ [39,40]. This simplified approach proposed recently by Ishihara, Inoue and Maekawa [151] emphasizes correctly the role of orbitals, but violates the $SU(2)$ spin symmetry and involves a Kondo coupling K between e_g and t_{2g} spins, which by itself is not a faithful approximation to the multiplet structure, as we explain below. The complete effective model of the undoped $LaMnO_3$ has to include also the t_{2g}-part of superexchange and the orbital interactions induced by the JT effect [23]. We will show that these terms, while unessential qualitatively, are very important for a quantitative understanding.

The superexchange between the d^4 Mn^{3+} ions originates in the large-U regime from virtual (e_g or t_{2g}) excitations, $d_i^4 d_j^4 \rightleftharpoons d_i^3 d_j^5$. The spin-orbital model presented below follows from the full multiplet structure of the manganese ions in octahedral symmetry, both in the d^4 ($t_{2g}^3 e_g$) configuration of the Mn^{3+} ground state, and in the d^3 (t_{2g}^3) and d^5 ($t_{2g}^3 e_g^2$) virtually excited states of Mn^{4+} and Mn^{2+} ions, respectively. Our starting point is that each Mn^{3+} (d^4) ion is in the high-spin ($t_{2g}^3 e_g$) ground state in agreement with large Hund's rule interaction J_H, i.e., the high-spin ($S = 2$) orbital doublet ^5E. First, we analyze the strongest channel of superexchange, which follows from the hopping of e_g electrons between nearest neighbor sites i and j. The derivation is similar to that analyzed in detail for the d^9 configuration in Sec. III.A, but the important difference is found in the spin sector, where the derived effective interactions have to be represented by the superexchange terms between

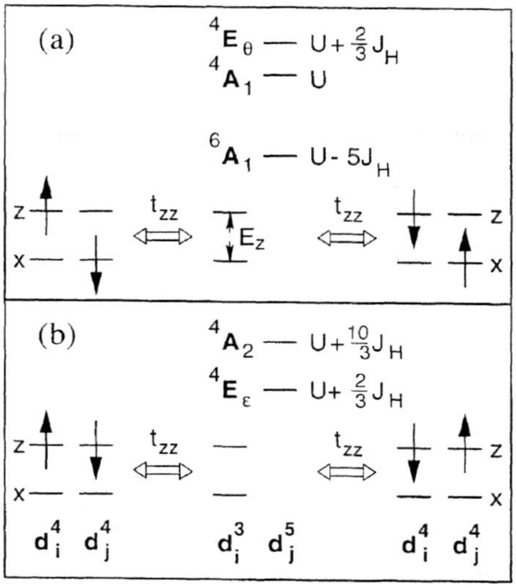

FIGURE 32. Virtual $d_i^4 d_j^4 \to d_i^3 d_j^5$ excitations which generate effective interactions for a bond $(ij) \parallel$ c-axis: (a) for one $|x\rangle$ and one $|z\rangle$ electron, and (b) for two $|z\rangle$ electrons (after Ref. [23]).

total spins $S = 2$ per site [23]. When we consider a bond oriented along the cubic c-axis, only a $3z^2 - r^2$ electron can hop as in the d^9 case (Sec. III.A), and four d^5 states may be reached: the high-spin $^6A_1$ state ($S = 5/2$), and the lower-spin ($S = 3/2$) $^4A_1$, ^4E, and $^4A_2$ states (Fig. 32). The $d_i^4 d_j^4 \rightleftharpoons d_i^3(t_{2g}^3) d_j^5(t_{2g}^3 e_g^2)$ excitation energies require for their description in principle *all three* Racah parameters, A, B and C [31]: $\varepsilon(^6A_1) = A - 8B$, $\varepsilon(^4A_1) = A + 2B + 5C$, $\varepsilon(^4E) \simeq A + 6B + 5C$, $\varepsilon(^4A_2) = A + 14B + 7C$. In view of the realistic values of $B = 0.107$ and $C = 0.477$ eV for Mn^{2+} (d^5) ions [153], one may use an approximate relation $C \simeq 4B$, and write the excitation energies in terms of Coulomb, $U \equiv A + 2B + 5C$, and Hund's exchange, $J_H \equiv 2B + C \simeq 0.69$ eV, parameters:

$$\varepsilon(^6A_1) = U - 5J_H, \tag{153}$$
$$\varepsilon(^4A_1) = U, \tag{154}$$
$$\varepsilon(^4E) = U + \frac{2}{3}J_H, \tag{155}$$
$$\varepsilon(^4A_2) = U + \frac{10}{3}J_H. \tag{156}$$

We emphasize that U is here by definition *not* the Coulomb matrix element, but the reference energy of the $|^4A_1\rangle$ state; this definition is different from some other conventions [153,154]. A value of $U = 7.3$ eV was deduced for LaMnO$_3$ by Feiner and Oleś [23] from the available spectroscopic data [7,153,154].

Using the spin algebra (Clebsch-Gordon coefficients), the reduction of product representations in cubic site symmetry [31] for the intermediate states, and making a rotation of the terms derived for a bond $\langle ij \rangle \parallel c$ to the bonds $\rangle ij \langle \parallel a$ and $\rangle ij \langle \parallel b$ as in Sec. III.A. one finds a compact expression,

$$H_J(e_g) = \frac{1}{16}\sum_{\langle ij \rangle}\left\{-\frac{8}{5}\frac{t^2}{\varepsilon({}^6A_1)}\left(\vec{S}_i\cdot\vec{S}_j + 6\right)\mathcal{P}^{\zeta\xi}_{\langle ij \rangle}\right.$$
$$+\left[\frac{t^2}{\varepsilon({}^4E)} + \frac{3}{5}\frac{t^2}{\varepsilon({}^4A_1)}\right]\left(\vec{S}_i\cdot\vec{S}_j - 4\right)\mathcal{P}^{\zeta\xi}_{\langle ij \rangle}$$
$$\left.+\left[\frac{t^2}{\varepsilon({}^4E)} + \frac{t^2}{\varepsilon({}^4A_2)}\right]\left(\vec{S}_i\cdot\vec{S}_j - 4\right)\mathcal{P}^{\zeta\zeta}_{\langle ij \rangle}\right\}, \quad (157)$$

where t is the hopping element along the c-axis, while $\mathcal{P}^{\zeta\xi}_{\langle ij \rangle}$ and $\mathcal{P}^{\zeta\zeta}_{\langle ij \rangle}$ are the projection operators introduced in Eqs. (60) and (61). A similar expression was derived by Shiina, Nishitani and Shiba using the group theory arguments [155]. However, this method does not allow to determine accurately the coefficients of different terms, and thus the balance between AF and FM interactions is different.

The terms in the first line of Eq. (157) are obtained from the virtual processes which involve different occupancies at the sites i and j – at one site the orbital along the bond $|\zeta\rangle$ is occupied by an e_g electron, while an orthogonal orbital $|\xi\rangle$ is occupied at the other [Fig. 32(a)]. In agreement with the general rule as presented in Secs. I and III.A, the processes which involve the high-spin state $|{}^6A_1\rangle$ lead to FM superexchange, while the low-spin state $|{}^4A_1\rangle$ gives an AF interaction. The terms in the second line arise from the configurations with occupied directional orbitals $|\zeta\rangle$ at both sites which give the excited states with a double occupancy in one of the $|\zeta\rangle$ orbitals [Fig. 32(b)]. After projecting this double occupancy onto the eigenstates $|{}^4E\rangle$ and $|{}^4A_2\rangle$ one finds the AF interactions. Thus, the structure is similar to that of Eq. (58), but the coefficients are different and follow from the multiplet structure of the excited d^5 states and from the spin algebra.

A similar derivation gives the t_{2g} superexchange [23],

$$H_J(t_{2g}) = \frac{1}{4}J_t\sum_{\langle ij \rangle}\left(\vec{S}_i\cdot\vec{S}_j - 4\right), \quad (158)$$

where J_t is an average of the processes which couple different low-spin d^5 ($t^4_{2g}e_g$) and d^3 (t^3_{2g}) excited states ${}^4T_1$ and ${}^4T_2$ of Mn^{2+} and Mn^{4+} ions, respectively, $J_t = (J_{11} + J_{22} + J_{12} + J_{21})/4$. The individual exchange elements, $J_{mn} = t_\pi^2/\varepsilon({}^4T_m, {}^4T_n)$, result from different local $d_i^4 d_j^4 \rightleftharpoons d_i^5(t^4_{2g}e_g)d_j^3(t^2_{2g}e_g)$ excitations within an $\langle ij \rangle$ bond, with energies $\varepsilon({}^4T_1, {}^4T_1) \simeq U + 8J_H/3$, $\varepsilon({}^4T_1, {}^4T_2) \simeq U + 2J_H/3$, $\varepsilon({}^4T_2, {}^4T_1) \simeq U + 4J_H$, $\varepsilon({}^4T_2, {}^4T_2) \simeq U + 2J_H$, expressed by the same elements U and J_H as above, where the states ${}^4T_m$ (${}^4T_n$) label the symmetry of d_i^5 (d_j^3) excited configurations, respectively. The hopping element between the t_{2g} orbitals involves two-step processes via the oxygen orbitals in bridge positions (see Fig. 1), and thus one finds that $t_\pi = t/3$. We neglected the correction terms which describe the anisotropy of t_{2g} superexchange depending on the actual configuration of e_g electrons.

As we already pointed out, the complete spin-orbital model for the undoped manganites [23],

$$\mathcal{H} = H_J(e_g) + H_J(t_{2g}) + H_{\mathrm{JT}} + H_\tau, \tag{159}$$

includes both above superexchange terms due to e_g and t_{2g} excitations [$H_J(e_g)$ and $H_J(t_{2g})$], the effective interactions between orbital degrees of freedom which follow from the JT effect (H_{JT}), and a low-symmetry crystal field (H_τ). The JT term may be derived from the general energy expression (15) for the cooperative JT effect [42]. The operator form of the intersite orbital interaction $\propto \kappa$ (15),

$$H_{\mathrm{JT}} = \kappa \sum_{\langle ij \rangle} \left(\mathcal{P}^{\zeta\zeta}_{\langle ij \rangle} - 2\mathcal{P}^{\zeta\xi}_{\langle ij \rangle} + \mathcal{P}^{\xi\xi}_{\langle ij \rangle} \right) = 2\kappa \sum_{\langle ij \rangle} \left(1 - 2\mathcal{P}^{\zeta\xi}_{\langle ij \rangle} \right), \tag{160}$$

favors orbital alternation, i.e., one $|\zeta\rangle$ and one $|\xi\rangle$ as occupied orbitals at two ends of the same bond $\langle ij \rangle$. In analogy to Eqs. (60) and (61) we define $\mathcal{P}^{\xi\xi}_{\langle ij \rangle} = 2P_i^\xi P_j^\xi$. The second equality in Eq. (160) follows from the completeness relation, $\mathcal{P}^{\zeta\zeta}_{\langle ij \rangle} + 2\mathcal{P}^{\zeta\xi}_{\langle ij \rangle} + \mathcal{P}^{\xi\xi}_{\langle ij \rangle} = 2$. The coefficient κ is a parameter which was estimated in Ref. [23] from the experimental temperature of the structural phase transition T_s. Eq. (160) gives therefore the same energetically favored orbital configurations as those involved in the superexchange terms in Eq. (157) which occur due to the $|^6A_1\rangle$ and $|^4A_1\rangle$ excited states. The last term is the crystal-field splitting of the e_g orbitals (59) which we reproduce here for completeness,

$$H_\tau = -E_z \sum_i \tau_i^c. \tag{161}$$

The strength of e_g and t_g superexchange can be estimated fairly accurately from the basic electronic parameters for the Mn ion as determined from spectroscopy [7,153,154], with an estimated accuracy of $\sim 20\%$. Using $U = 7.3$ eV and $J_H = 0.69$ eV, and taking into account that the Mn-Mn hopping occurs in effective processes via the bridging oxygen, one finds $t = 0.41$ eV which follows from $t = t_{pd}^2/\Delta$, with Mn-O hopping $t_{pd} = 1.5$ eV and charge transfer energy $\Delta = 5.5$ eV. This yields $J = t^2/U = 23$ meV which we take as a measure of the e_g-part of superexchange, and $J_t = 2.1$ meV. The accuracy of these parameters may be appreciated from the resulting prediction for the Néel temperature of $CaMnO_3$, where a similar derivation gives for spin $S = 3/2$,

$$\hat{H}_t = \hat{J}_t \sum_{\langle ij \rangle} \left(\frac{4}{9} \vec{S}_i \cdot \vec{S}_j - 1 \right), \tag{162}$$

where $\hat{J}_t \simeq J_t(1 + J_H/U)$. Using the estimates from spectroscopy, one obtains $\hat{J}_t = 4.6$ meV and thus $T_N = 124$ K [23], in excellent agreement with the experimental value $T_N = 110$ K [145].

Here it is worthwhile to discuss the difference between the present approach and the popular simplified model introduced by Ishihara, Inoue, and Maekawa [151]. These authors also emphasized the role of orbital variables in the e_g-superexchange, but assumed that the total spin is conserved within the e_g-subband, and one can thus use the same derivation as presented in Sec. III.A, with the excitation energies adopted to the new situation when the e_g electrons interact also with the t_{2g} core states by the Hund's rule interaction K between one e_g and one t_{2g} electron defined in the same way as in Eq. (23). Although the original notation was somewhat more involved [151], using our projection operators (60) and (61), and the spin projection operators (21), their result may be rewritten as follows,

$$H_{\text{IIM}} = \sum_{\langle ij \rangle} \left\{ -\frac{t^2}{\bar{U}' - \bar{J}_H} \left(\vec{S}_i \cdot \vec{S}_j + \frac{3}{4} \right) \mathcal{P}_{\langle ij \rangle}^{\zeta\xi} + \frac{t^2}{\bar{U}' + \bar{J}_H + K} \left(\vec{S}_i \cdot \vec{S}_j - \frac{1}{4} \right) \mathcal{P}_{\langle ij \rangle}^{\zeta\xi} \right.$$
$$\left. + \frac{2t^2}{\bar{U} + K} \left(\vec{S}_i \cdot \vec{S}_j - \frac{1}{4} \right) \mathcal{P}_{\langle ij \rangle}^{\zeta\xi} \right\}, \qquad (163)$$

Therefore, the multiplet structure given by Eqs. (153)–(156) is now replaced by:

$$\varepsilon(^6A_1) \simeq \bar{U}' - \bar{J}_H, \qquad (164)$$
$$\varepsilon(^4E) \simeq \bar{U}' + \bar{J}_H + K, \qquad (165)$$
$$\varepsilon(^4A_2) \simeq \bar{U} + K. \qquad (166)$$

It is evident that the Hamiltonian (163) *is not equivalent* to the expression derived using the correct multiplet structure as given by Eq. (157) [23]. First of all, note that $\bar{J}_H$ has here the same meaning as the Hund's rule element used to describe the exchange interactions between a pair of d electrons. Then the excitation spectrum (164)–(166) should be equivalent to the excitation spectrum given by Eqs. (42)–(45) in the absence of t_{2g} spins at $K = 0$. The correspondence between the Coulomb and exchange elements in Eqs. (163) and (1) is given by $\bar{U}' = U - J_H$ and $\bar{J}_H = J_H$, and the first two energies of the $|e_g^2\rangle$ excited state agree, but the highest energy does not. The reason is that the intraorbital Coulomb interaction $\bar{U}$ is not an independent parameter, but $\bar{U} = \bar{U}' + 2\bar{J}_H$ [32]. Second, the state $|^4A_2\rangle$ comes out to be doubly degenerate in Eq. (163) as the processes $\propto \bar{J}_H$ which transfer a pair of electrons with opposite spins between different orbitals in Eq. (1) were not included and thus the doublet occurs as the highest-energy state rather than the state in the middle. Furthermore, for the choice of parameters of Ref. [151] with $\bar{U}' = 5$, $\bar{U} = 7$ and $\bar{J}_H = 2$ eV, not only the structure of levels is incorrect, but even the $|^4E\rangle$ and $|^4A_2\rangle$ states have accidental degeneracy. Third, the parameter K in the case of $d_i^4 d_j^4 \rightleftharpoons d_i^3(t_{2g}^3)d_j^5(t_{2g}^3 e_g^2)$ excitations is not a free parameter, but $K = 3\bar{J}_H$ (the factor of three comes about due to three t_{2g} electrons which form a large spin $S = 3/2$ interacting with an e_g electron by the Kondo term). This last condition is obeyed by the actual choice of parameters in Ref. [151]. Finally, the state $|^4A_1\rangle$ has been completely missed in Eqs. (164)–(166). Therefore, we conclude that the

Hamiltonian (163): (i) violates the SU(2) invariance of superexchange interactions in spin space, and (ii) does not correspond to the correct multiplet structure of d^5 ions in any nontrivial limit. A common feature with Eq. (157) is that FM interactions are enhanced due to the lowest excited $|^6A_1\rangle$ state, but the dependence of the magnetic interactions on $\bar{J}_H$ is quite different, and it gives a different answer concerning the stability of the A-AF phase.

The model of Ishihara et al. [151] contains also a superexchange term between core spins t_{2g} which takes the same form as that given in Eq. (162). In contrast to the present derivation of Ref. [23], no formula for the interaction $\hat{J}_t$ was derived in Ref. [151], but instead the value of $\hat{J}_t$ was fitted in order to reproduce the stable A-AF ordering, as observed in LaMnO$_3$. This results in a value of $\hat{J}_t$ overestimated by a factor close to seven and having no relation to the experimental value of the Néel temperature for CaMnO$_3$.

A comparison between Eqs. (157) and (163) is possible only when the magnetic interactions concern the same spins $S = 2$. Therefore, we have renormalized the terms $\sim t^2/\varepsilon_n$, where ε_n is the excitation energy, and $\hat{J}_t$ in the model of Ishihara, Inoue and Maekawa [151] to act on the total spins $S = 2$. After this modification, the superexchange terms obtained from both models may be written in the same form:

$$\mathcal{H}_{SE} = \sum_{\langle ij \rangle} \left(-J_1 \mathcal{P}^{\zeta\zeta}_{\langle ij \rangle} + J_2 \mathcal{P}^{\zeta\zeta}_{\langle ij \rangle} + J_3 \mathcal{P}^{\zeta\zeta}_{\langle ij \rangle} + J_{AF} \right) \vec{S}_i \cdot \vec{S}_j, \qquad (167)$$

where the first interaction $\propto J_1$ is FM, and the remaining interactions are AF. We present the values of different constants in Table 2. First of all, the superexchange terms are much smaller than the hopping integral t which justifies *a posteriori* the perturbative approach. Second, taking the multiplet structure of Mn^{2+} ions one finds a competition between the FM and AF contributions to the e_g-promoted superexchange (157), with the AF terms being roughly half of the largest FM term [23]. As we show below, this ratio between the FM and AF interactions gives automatically the A-AF order as observed in LaMnO$_3$.

In contrast, the (effective) interactions J_n found from Eq. (163) are not balanced due to overestimated excitation energies $\varepsilon_n \simeq 13.0$ eV to the low-spin states ($n = 2, 3$). On one hand, one finds a dominating FM term $J_1 = 3.48$ meV, not far from $J_1 = 4.40$ meV found from Eq. (157), while on the other hand the AF terms are smaller by a factor close to two. Therefore, the e_g part of the superexchange alone gives a FM state in all three directions, and one is forced to simulate the missing AF interactions by a large superexchange term coming from the hopping of t_{2g} electrons [151] (see Table 2). Another consequence of too low values of J_2 and J_3 is a rather narrow parameter regime for the stable A-AF order, much narrower than in the approach using the correct multiplets, where the A-AF order is generic [23].

The classical phases found in the present d^4 model (Fig. 33) are similar to those shown before in Fig. 11, but at present the spins of e_g and t_{2g} electrons are aligned and form total spins $S = 2$. Due to the identical interactions in the orbital

TABLE 2. Magnetic interactions J_n in Eq. (167), and the excitation energies ε_n used in perturbation theory, as obtained from the d^4 spin-orbital model (157) [23] and from the model of Ishihara, Inoue and Maekawa (163) [151]. The excitation energies ε_2 and ε_3 in the case of Eq. (157) were averaged over two states which contribute to J_n.

superexchange	orbital operator	model (157) [23] ε_n (eV)	J_n (meV)	J_i/t (10^{-2})	model (163) [151] ε_n (eV)	J_n (meV)
J_1	$\mathcal{P}^{\zeta\xi}_{\langle ij\rangle}$	3.80	4.40	1.07	3.0	3.48
J_2	$\mathcal{P}^{\zeta\xi}_{\langle ij\rangle}$	7.53	2.21	0.54	13.0	0.81
J_3	$\mathcal{P}^{\zeta\zeta}_{\langle ij\rangle}$	8.68	2.44	0.60	13.0	1.62
J_{AF}	–	8.91	0.53	0.13	–	3.50

sector, the MF phase diagram of the e_g-part of the manganese d^4 model (159), $H = H_J(e_g) + H_\tau$, at $T = 0$ is similar to that of the cuprate d^9 spin-orbital model (57) [63,64] analyzed in Sec. III.B: at large positive (negative) E_z, one finds AF phases with either $|x\rangle$ (AFxx) or $|z\rangle$ (AFzz) orbitals occupied, while MO phases with occupied orbitals given as linear combinations (65) of the basis states $|x\rangle$ and $|z\rangle$ are favored by increasing J_H. As in the d^9 case, the e_g part of superexchange (157) is frustrated at $J_H = 0$ and $E_z = 0$. However, due to the large value of total spin $S = 2$, the quantum corrections are expected to be much smaller and we may assume that the classical order is robust, at least certainly at finite J_H and in the presence of t_{2g} AF superexchange (158). The A-AF order is found at finite $J_H/U \simeq 0.1$ [Fig. 33(a)]. If $E_z < 0$ the spin order is FM (AF) in the (a,b) planes (along the c-axis) in the MOFFA phase, while at $E_z > 0$ two similar phases, MOAFF and MOFAF, are degenerate. For the parameters appropriate for LaMnO$_3$, with $J_H/U \simeq 0.095$, one finds a MOFFA/MOAFF ground state, i.e., *A-AF magnetic order*, while a FM (MOFFF) phase is stable only at $J_H/U > 0.12$. The result is therefore qualitatively similar to the classical phase diagram of the d^9 model (Fig. 12). We emphasize again that the region of stability of the A-AF phase is somewhat modified by t_{2g}-superexchange [Fig. 33(a)], but this change is small as $J_t \ll J$.

The value of κ in Eq. (160) is the only parameter of the manganite model (159) which had to be determined from experiment [23]. First of all, the e_g part of superexchange alone favors the alternation of orbitals, i.e., maximizes the average $\langle \mathcal{P}^{\zeta\xi}_{\langle ij\rangle}\rangle$ and one can verify that a structural transition follows already from this term. However, the transition temperature obtained in this way amounts to $T^e_s \simeq 440$ K (see Fig. 34) and is far below the experimental value of $T^{exp}_s \simeq 750$ K. Therefore, the structural phase transition is induced to a larger extent by the orbital interactions generated by the JT effect included via the H_{JT} term (160). Their contribution to the mean-field value of the transition temperature is equal $T^\kappa_s = 6\kappa$. Using the typical ratio between the MF value T_s and the experimental transition temperature T^{exp}_s for pseudospins 1/2, $T_s/T^{exp}_s = 1.6$, it follows from $T_s = T^e_s + T^\kappa_s = 1200$ K that $6\kappa \simeq 760$ K. Thus $\kappa \simeq 11$ meV which agrees with the estimate of Millis that $\kappa > 10$

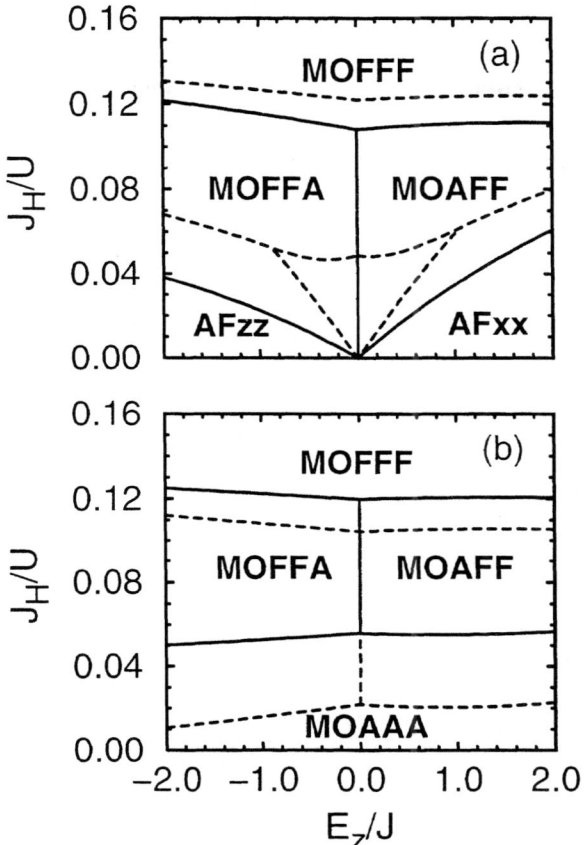

FIGURE 33. Classical phase diagram of the manganite model (159): (a) no JT effect ($\kappa = 0$): $J_t = 0$ (full lines) and $J_t = 0.092J$ (dashed lines), with the AFxx and AFzz phases separated by a MOAAA phase; (b) including JT effect ($\kappa = 0.5J$): $J_t = 0$ (dashed lines) and $J_t = 0.092J$ (full lines) (after Ref. [23]).

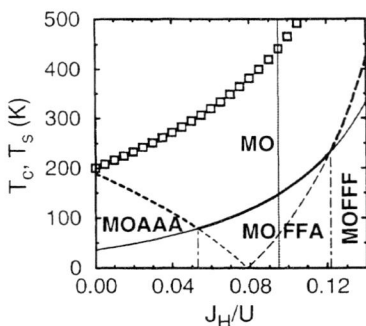

FIGURE 34. Magnetic transition temperatures T_c ($J = 23$ meV, $E_z = 0$, $J_t = 0.092J$, $\kappa = 0.5J$) for: MOAAA (dashed line), MOFFF (long-dashed line), and MOFFA (full line) phases, and T_s for the structural (MO) phase transition at $\kappa = 0$ (squares). The dotted line indicates realistic $J_H/U = 0.095$ for LaMnO$_3$ (after Ref. [23]).

meV (where he did not separate the orbital interactions into those originating from the JT effect and from superexchange). Therefore, the representative value $\kappa/J = 0.5$ was considered in Ref. [23]. Due to the structural transition which occurs in LaMnO$_3$ at $T_s \simeq 750$ K, one has to use a MF theory for the phases with coupled order parameters, assuming that the order parameters corresponding to $\langle \vec{S} \rangle$, $\langle \vec{\tau} \rangle$, and $\langle \vec{S}\vec{\tau} \rangle$ are independent variables [156]. When the orbital interactions induced by the JT effect are included via H$_{\rm JT}$ term (160), one gets with decreasing temperature first a transition due to the largest orbital interactions to the phase with $\langle \vec{\tau} \rangle \neq 0$. This induces significant changes in the phase diagram close to the origin $(E_z, J_H) = (0,0)$ [Fig. 33(b)]. In fact, the JT coupling κ enforces alternating orbitals with $\theta = \pi/4$ in Eqs. (72) which stabilizes G-AF spin order in the MOAAA phase at small J_H/U, while the pure AFxx and AFzz phases are suppressed. However, in the physically interesting regime of larger J_H/U the orbital order is mainly driven by the e_g-superexchange interactions (157). Besides, from Fig. 33 one can see that the A-AF phase is not affected by the JT coupling in the physical regime of parameters for the LaMnO$_3$.

At finite temperature one may study the competition of different magnetic phases. One finds that the same magnetic phases develop in the presence of orbital ordering at finite temperature (Fig. 34) as those found independently at $T = 0$. For the experimental value of $J_H/U \simeq 0.095$ the magnetic transition into the A-AF phase is obtained at $T_N^{\rm MF} \simeq 148$ K which after reduction gives a prediction for the experimental transition temperature of $T_N \simeq 106$ K, in reasonable agreement with the actually observed value of 136 K [157]. As a remarkable success of the model (159), the magnetic interactions obtained for the A-AF phase are close to the experimental values. The FM interactions in the (a,b) planes $J_{(a,b)}$, and the AF interactions along the c-axis J_c are found from Eqs. (157) and (158)

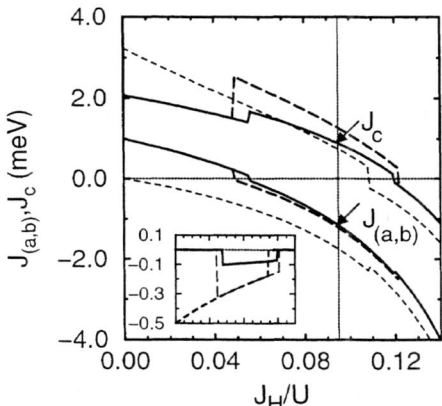

FIGURE 35. Exchange interactions $J_{(a,b)}$ and J_c in the ground state for increasing J_H/U, for $J = 23$ meV and: $J_t = 0$, $\kappa = 0$ (dashed lines), $J_t = 0.092J$, $\kappa = 0$ (long-dashed lines), and $J_t = 0.092J$, $\kappa = 0.5J$ (full lines). The jumps in $J_{(a,b)}$ and J_c correspond to phase transitions of Fig. 33. The inset gives the values of $\cos 2\theta$ defined as in Eqs. (72), in the same range of J_H/U for the optimal orbital state (after Ref. [23]).

by averaging over the orbital operators in the actual classical ground state (Fig. 35). The values obtained at $J_H/U = 0.095$: $J_{(a,b)} = -1.15$ meV and $J_c = 0.88$ meV [23] are somewhat larger than the experimental values of -0.83 and 0.58 meV [39,40], respectively. It is interesting to realize that although the observed A-AF phase would be obtained from the e_g superexchange alone, the ratio of the above exchange constants is then $J_c/|J_{(a,b)}| = 2.25$. Instead, if the t_{2g} superexchange and the JT term are included in the present model, the value $J_c/|J_{(a,b)}| = 0.77$ is found which agrees very well with the experimental ratio of 0.7 [39,40]. The t_{2g} term alone enhances the AF interactions, but does modify the orbital ordering, and gives thus a different ratio $J_c/|J_{(a,b)}| = 1.04$. Therefore, we conclude that all the interaction terms in Eq. (159) are important in order to explain the experimentally observed exchange interactions in LaMnO$_3$. These values could not be reproduced up to now by *ab initio* calculations performed in LSDA which fails even to give the correct sign of J_c [38], while the values somewhat closer to experiment and to those of Ref. [23] were obtained recently in the HF approximation: $J_{(a,b)} = -0.44$ meV and $J_c = 0.11$ meV [158].

B Polaronic regime at low hole doping

The spin-orbital model (159) which includes the complete superexchange and the orbital interactions which follow from the JT effect provides a good starting point to describe the interplay between magnetic and orbital interactions in doped manganites. It reproduces very well the spin and orbital pattern observed in the ground state of LaMnO$_3$ [23] which can be well understood as mainly due to the spin and orbital interactions contained in the e_g superexchange (157), with small corrections given by the JT effect. Thus, the undoped case involves spin, orbital and lattice degrees of freedom. When the holes are doped into this ordered ground state, the charge degrees of freedom occur in addition and modify the ground state properties. Their understanding represents one of the most challenging and complex problems in the physics of CMR compounds [1,22].

We need to generalize the interaction terms which appeared in the undoped case to the present situation, taking into account the constraint which follows from large U that prevents the occupancy of any Mn ion by more that one e_g electron. Therefore, we restrict the space to Mn^{3+} and Mn^{4+} configurations. Furthermore, at low hole concentrations the doped holes are trapped and form polarons, as follows from the *insulating* magnetic phases found at low doping in the phase diagrams of Fig. 4. Therefore, a polaronic model was developed in Ref. [159], with the low-energy effective model Hamiltonian given by:

$$\mathcal{H} = H_J(e_g) + H_J(t_{2g}) + H_{\mathrm{JT}} + H_\tau + H_{pol}. \tag{168}$$

It includes superexchange terms due to e_g and t_{2g} electrons [$H_J(e_g)$ and $H_J(t_{2g})$], orbital interactions (H_{JT}), the crystal-field splitting (H_τ) induced by the JT effect, and the polaronic energy (H_{pol}) which contributes at finite doping with new FM interactions due to the effective processes involving the excitations around a Mn^{4+} ion created by a hole.

The superexchange between total spins $S = 2$ is generalized with respect to the undoped case by including the constraint on the e_g occupation to,

$$H_J(e_g) = \frac{1}{4}J \sum_{\langle ij \rangle} f_i f_i^\dagger \left\{ -\frac{8}{5} \frac{U}{\varepsilon(^6A_1)} \left(\frac{1}{4}\vec{S}_i \cdot \vec{S}_j + \frac{3}{2} \right) \mathcal{P}_{\langle ij \rangle}^{\zeta\xi} \right.$$
$$+ \left[\frac{U}{\varepsilon(^4E)} + \frac{3}{5} \frac{U}{\varepsilon(^4A_1)} \right] \left(\frac{1}{4}\vec{S}_i \cdot \vec{S}_j - 1 \right) \mathcal{P}_{\langle ij \rangle}^{\zeta\zeta}$$
$$\left. + \left[\frac{U}{\varepsilon(^4E)} + \frac{U}{\varepsilon(^4A_2)} \right] \left(\frac{1}{4}\vec{S}_i \cdot \vec{S}_j - 1 \right) \mathcal{P}_{\langle ij \rangle}^{\zeta\zeta} \right\} f_j f_j^\dagger, \tag{169}$$

where $J = t^2/U$, and t is the hopping between $3z^2 - r^2$ orbitals along the c-axis. The excitation energies are given by Eqs. (153)–(156), and the orbital projection operators have been introduced in Sec. III. The hole operators $f_i^\dagger$ and f_i guarantee that this superexchange term contributes only if both sites carry a single e_g electron ($f_i f_i^\dagger = 1$), and the orbital operators $\mathcal{P}_{\langle ij \rangle}^{\zeta\xi}$ and $\mathcal{P}_{\langle ij \rangle}^{\zeta\zeta}$ decide about the strength of a particular superexchange contribution.

The t_{2g}-superexchange is isotropic in leading order,

$$H_J(t_{2g}) = \sum_{\langle ij \rangle} \left\{ J_t \left(\frac{1}{4} \vec{S}_i \cdot \vec{S}_j - 1 \right) f_i f_i^\dagger f_j f_j^\dagger + \hat{J}_t \left(\frac{4}{9} \vec{S}_i \cdot \vec{S}_j - 1 \right) f_i^\dagger f_i f_j^\dagger f_j \right.$$
$$\left. + \bar{J}_t \left(\frac{1}{3} \vec{S}_i \cdot \vec{S}_j - 1 \right) \left(f_i^\dagger f_i f_j f_j^\dagger + f_i f_i^\dagger f_j^\dagger f_j \right) \right\}, \quad (170)$$

with the AF superexchange constants J_t, $\hat{J}_t$, and $\bar{J}_t$ obtained from the hopping of t_{2g} electrons for the pairs of Mn^{3+}–Mn^{3+}, Mn^{4+}–Mn^{4+}, and Mn^{3+}–Mn^{4+} ions, respectively. The spin operators $\vec{S}_i$ correspond to spins $S = 2$ and $S = 3/2$ of Mn^{3+} and Mn^{4+} ions, when the number of holes at site i is either $f_i^\dagger f_i = 0$ or $f_i^\dagger f_i = 1$, respectively. The JT term (160) leads to static distortions [42] which induce intersite orbital interactions ($\propto \kappa$) between the e_g orbitals and is rewritten in a similar way,

$$H_{JT} = \kappa \sum_{\langle ij \rangle} f_i f_i^\dagger \left(\mathcal{P}_{\langle ij \rangle}^{\zeta\zeta} - 2\mathcal{P}_{\langle ij \rangle}^{\zeta\xi} + \mathcal{P}_{\langle ij \rangle}^{\xi\xi} \right) f_j f_j^\dagger. \quad (171)$$

The problem of a Mott insulator doped by a low number of holes is one of the fascinating problems in the field of strongly correlated electrons. In the t-J model the motion of a hole added to an AF background is hindered by the string effect [160], and the free propagation disappears. In manganites the situation is even more complex due to the presence of lattice and orbital degrees of freedom, which are in first place responsible for hole localization at lower doping concentrations. In fact, in an orbitally degenerate Mott-Hubbard system there exists a direct coupling between holes and orbitals due to the polarization of e_g orbitals in the neighborhood of a hole [161]. This coupling might be strong enough to form a bound state of a hole with the surrounding orbitals, leading to a dressing of hole by orbital excitations [162] (Sec. VII.B), and yielding a strong reduction of the bandwidth. Such orbital-hole bound states lead to an exponential suppression of the bandwidth which makes the system unstable towards hole localization. Here we assume that the holes loose most of their kinetic energy due to the lattice distortions of the breathing mode type (Fig. 3), and only the effective processes survive which excite a hole to its nearest neighbors [159]. The cubic symmetry of perovskite manganites is locally broken close to a hole which appears as the removal of the e_g degeneracy on the sites adjacent to a doped hole. The origin of this splitting is in the distortions of the oxygen ions that point towards the empty site, and by the electrostatic potential between positive hole and negative electrons. The occupied e_g orbitals around a hole are likely to be modified, and if any other orbital interactions were absent, an *orbital polaron* bound state would be formed around a hole, with occupied $|\zeta\rangle$ orbitals oriented along the bonds towards the hole site (Fig. 36).

In the insulating regime, the localized holes are randomly distributed and form locally lattice polarons with energy E_p. The new interactions originate from the DE mechanism which induces a FM superexchange for Mn^{4+}–Mn^{3+} pairs $\propto J_p = \bar{J}_p/(1 + E_p/2J_H)$ (with $\bar{J}_p = t^2/2E_p$) [159],

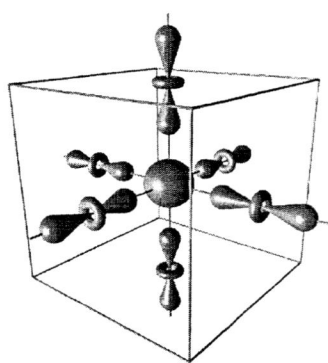

FIGURE 36. Orbital polaron in the strong coupling limit. Six $|\zeta\rangle$ occupied e_g orbitals at Mn^{3+} ions point towards a central hole at a Mn^{4+} site (after Ref. [161]).

$$H_{pol} = -E_p \sum_i f_i^\dagger f_i - \sum_{\langle ij \rangle} f_i^\dagger f_i \left[\bar{J}_p + \frac{1}{8} J_p \left(\vec{S}_i \cdot \vec{S}_j - 3 \right) \right] P_{j\zeta} f_j f_j^\dagger. \qquad (172)$$

Similar to the undoped case, the exchange terms do depend now on the orientation of the occupied e_g orbital on the Mn^{3+} ion by the projection operator $P_{j\zeta}$ (62), with the largest FM contribution when $\langle P_{j\zeta} \rangle = 1$, i.e., for the directional $3z^2 - r^2$-type orbitals $|\zeta\rangle$ oriented along the bonds towards the Mn^{4+} ion (Fig. 36). The classical phase diagram can be now investigated in a similar fashion as in Sec III.B, assuming that the holes are localized and randomly distributed over the lattice. This assumption corresponds to the dilute limit and describes well the features of polaronic phase in the low doping regime $x = 1 - n < 0.16$.

We have seen that for realistic parameters taken from spectroscopy the A-AF observed in $LaMnO_3$ at $x = 0$ is reproduced (Sec. V.A). Instead of studying the phase diagram as a function of parameters, we address here a simpler question: How the anisotropic exchange constants $J_{(a,b)} = -1.15$ and $J_c = 0.88$ meV found in $LaMnO_3$ [23] change as a function of doping. Localized polarons in doped systems stabilize locally FM order around Mn^{4+} defects as J_p is the largest (FM) exchange element, with $J_p \simeq 4J$ taking $E_p \simeq 0.67$ eV [159], which might provide a natural explanation of a gradual magnetic transition *within the insulating phase* by the DE mechanism. Indeed, if one assumes that the Mn^{4+} are randomly distributed and that the orbital ordering on the Mn^{3+}–Mn^{3+} bonds remains unchanged, one finds that the AF coupling J_c is gradually weakened with increasing doping x, while the FM interaction $|J_{(a,b)}|$ increases much slower (Fig. 37), in good agreement with recent experiments [163]. However, a good agreement with the experimental points was obtained assuming that the occupied orbitals around a hole *are not modified* to $|\zeta\rangle$ orbitals shown in Fig. 36. In contrast, if the directional orbitals $|\zeta\rangle$ are assumed, a much faster transition to the FM phase follows (Fig. 37) which does not agree with experiment. This suggests that the structure of polaron is rather

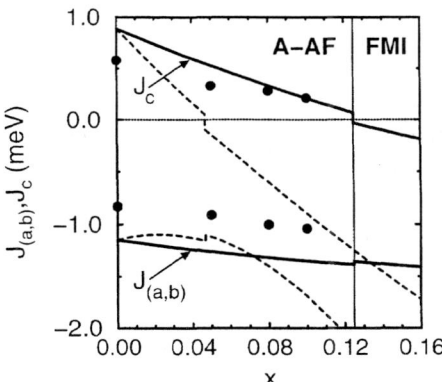

FIGURE 37. Exchange interactions $J_{(a,b)}$ and J_c in the A-AF and FMI ground state of $La_{1-x}Ca_xMnO_3$ ($J_H/U = 0.095$, $\kappa = 0.5J$, $E_p = 0.67$ eV) for rigid orbital order (full lines) and with $|\zeta\rangle$-type occupied e_g orbitals around Mn^{4+} ions as in Fig. 36 (dashed lines). The experimental data points [163] are shown by filled circles. Reprinted from Ref. [159], Copyright 1999, with permission from Elsevier Science.

rigid and determined primarily by lattice distortions and local JT effect rather than by the electronic interactions. It might be that some corrections to the perturbative treatment which leads to Eq. (172) are needed, but the present result suggests that the polarons realized in $La_{1-x}Sr_xMnO_3$ are not orbital polarons of Fig. 36. A better understanding of polarons is very important in order to get more insight into the observed insulator-metal transition from the insulating to metallic FM phase with increasing doping. It is experimentally observed that $La_{1-x}Ca_xMnO_3$ is metallic with FM LRO for hole concentrations in the range of $0.17 < x < 0.5$. This metallic state can be turned into an insulating one by increasing the temperature above the Curie temperature T_C, or by decreasing the hole concentration below $x_{crit} \simeq 0.17$.

The character of the first transition, from a metallic ferromagnet to an insulating paramagnetic state, can be addressed at first sight in the framework of the interplay between the DE mechanism and the lattice-polaron formation [92]. Indeed, in this case the crossover from metallic to insulating state is controlled by the ratio $\lambda = E_b/E_{kin}$ of polaron binding energy E_b to the average kinetic energy E_{kin} of the charge carriers. The formation of a bound state with the lattice distortion is favored only if the loss in kinetic energy is balanced by a gain in the binding energy. This condition can be reached by raising the temperature in the DE system, since the loss of FM correlations is accompanied by a shrinking of the bandwidth, so that it eventually increases the value of λ and induces a carrier localization.

A different situation is represented by the metal-insulator transition induced by the varying hole concentration within the FM phase at low temperature. The total Hamiltonian in the metallic phase has to include explicitly the kinetic energy of e_g electrons determined by the symmetry of allowed hopping processes, as presented in Sec. VI. One arrives at the effective t-J like model which is characterized by two energy scales for orbital and charge excitations: W_{orb} and W_{ch}. Assuming that the orbitals are disordered, i.e., in an orbital liquid state [164], the former quantity describes the orbital fluctuations determined by the value of J, and a fraction of the hopping amplitude proportional to the hole density $\propto xt$ which follows from strong correlations (see Sec. VI.B). The quantity W_{ch} determines the rate of the charge fluctuations and the amount of free kinetic energy, determined the bandwidth of $6t$, which corresponds to the uncorrelated electrons.

If the orbitals order, the charge fluctuations are reduced, i.e., the kinetic energy is lowered below its value in a correlated metal. Since the holes can move coherently even within an antiferro-type orbital arrangement [162], it has been argued [161] that the reduction of the kinetic energy is only of $\sim 5\%$ including the incoherent processes, while it amounts to $\sim 30\%$ without these contributions. This suggests that orbital ordering cannot suffice to explain the observed localization and one might expect that a more plausible mechanism of localization involves either the formation of orbital polarons, as proposed recently by Kilian and Khaliullin [161], or the lattice deformations.

In summary, the crossover from a free-carrier to a small-polaron picture can be triggered in a DE system either by a reduction of the doping concentration, or by an increase in temperature; the former acts via an enhancement of the polaron binding energy, while the latter by constraining the motion of holes via the DE mechanism. An interesting suggestion that orbital polarons play a major role in both transitions was put forward recently [161]. It seems, however, that lattice contribution to localization is more important and the question about the possible role of orbital polarons remains open.

VI ELECTRONIC STRUCTURE AND EXCITATIONS IN DOPED MANGANITES

A Double exchange in uncorrelated e_g orbitals

As we have already shown, the orbital degeneracy gives a rich structure of superexchange which contains competing FM and AF terms, and allows thus for the formation of anisotropic magnetic structures, both in cuprates (Sec. III.A) and in manganites (Sec. V.A). In this Section we will discuss a problem of *double-exchange via degenerate orbitals* in doped manganites. It is easier to consider first a simpler situation with only few electrons in the e_g band that is realistic, for example, for $Ca_{1-y}Sm_yMnO_3$ [165], where $y \ll 1$ ($y = 1 - x$ for conventional notation as used in $La_{1-x}A_xMnO_3$). For $y = 0$ this system is a Mott insulator, with high-spin states $S = 3/2$ per site due to the t_{2g} orbitals filled by three electrons, while e_g states are empty. Upon doping electrons enter into the doubly degenerate e_g orbitals ($E_z = 0$). In $Ca_{1-y}Sm_yMnO_3$ a canted spin structure was observed for $0.07 < x < 0.12$ [165]. Other data show that C-AF order is realized in $Nd_{1-x}Sr_xMnO_3$ for $0.6 < x < 0.8$, while A-AF phase is stable for $0.5 < x < 0.6$ [166]. In $\text{Pr}_{1-x}\text{Sr}_x\text{MnO}_3$ the A-AF order exists for $0.5 < x < 0.7$ [167].

The effective Hamiltonian which describes the low energy properties of electron doped manganites and represents a generalization of the DE model to degenerate orbitals may be written in the form suggested by van den Brink and Khomskii [65],

$$H = -\sum_{\langle ij \rangle \alpha \beta \sigma} t_{ij}^{\alpha\beta}(c^\dagger_{i\alpha\sigma}c_{j\beta\sigma} + \text{H.c.}) - J_H \sum_{i\alpha\sigma\sigma'} \vec{S}_i \cdot c^\dagger_{i\alpha\sigma}\vec{\sigma}_{\sigma\sigma'}c_{i\alpha\sigma'} + J_{AF} \sum_{\langle ij \rangle} \vec{S}_i \cdot \vec{S}_j. \quad (173)$$

The first term describes the kinetic energy of e_g electrons which are labeled by the site index i, spin index σ, and also by the orbital index $\alpha(\beta) = 1, 2$, corresponding to the local basis, e.g., to the $|x\rangle$ and $|z\rangle$ orbitals. As in the DE model, e_g electrons interact by Hund's rule exchange $\propto J_H$ with core t_{2g} spins, and the model is made more realistic by adding the superexchange between t_{2g} spins $\propto J_{AF}$ which might be determined as in Sec. V.A. However, for the present purpose it was treated as a parameter [65]. The presence of the orbital degeneracy, together with the very particular relations between the hopping matrix elements $t_{ij}^{\alpha\beta}$ [48,16], makes this problem and its outcome very different from the usual DE model (Sec. II.B).

The quasiclassical approach to study the model (173) follows the standard route as for the DE mechanism for electrons which move in a nondegenerate band [46]. In the first step we have to determine the spectrum of the e_g-electrons ignoring their interaction with the localized spins. This spectrum is given by the solution of the matrix equation [168],

$$||t_{\mu\nu} - \epsilon\delta_{\mu\nu}|| = 0, \quad (174)$$

where

$$t_{11} = -2t_{ab}(\cos k_x + \cos k_y), \quad (175)$$

$$t_{12} = t_{21} = -\frac{2}{\sqrt{3}}t_{ab}(\cos k_x - \cos k_y), \quad (176)$$

$$t_{22} = -\frac{2}{3}t_{ab}(\cos k_x + \cos k_y) - \frac{8}{3}t_c \cos k_z. \quad (177)$$

Here t_{11} is the dispersion due to an overlap between two $|x\rangle$ orbitals, t_{12} — between a pair of different orbitals, one $|x\rangle$ and one $|z\rangle$ orbital, and t_{22} between two $|z\rangle$ orbitals on neighboring sites. In writing Eqs. (175)–(177) we have taken into account the ratios of different hopping integrals [48,16], which are determined by the symmetry of the e_g wave functions, and introduced the notation t_{ab} and t_c, to be defined further on. The solutions $\epsilon_\pm(\vec{k})$ of Eq. (174) are:

$$\epsilon_\pm(\vec{k}) = -\frac{4t_{ab}}{3}(\cos k_x + \cos k_y) - \frac{4t_c}{3}\cos k_z$$
$$\pm \left\{ \left[\frac{2t_{ab}}{3}(\cos k_x + \cos k_y) - \frac{4t_c}{3}\cos k_z\right]^2 + \frac{4t_{ab}^2}{3}(\cos k_x - \cos k_y)^2 \right\}^{1/2}, \quad (178)$$

These bands were obtained in Refs. [168] and [65].

Following van den Brink and Khomskii [65], we now take into account the interaction of the e_g electrons with the magnetic background. We assume that the underlying magnetic structure is characterized by two sublattices, with a possible canting, so that the angle between neighboring spins in the (a,b)-plane is θ_{ab} and along the c-direction is θ_c. This rather general assumption covers all magnetic phases shown in Fig. 5 –AF phases: G-type (two-sublattice structure, $\theta_{ab} = \theta_c = \pi$), A-type (FM planes coupled antiferromagnetically, $\theta_{ab} = 0$ and $\theta_c = \pi$), and C-type structures (FM chains coupled antiferromagnetically, $\theta_{ab} = \pi$ and $\theta_c = 0$), and the FM phase with $\theta_{ab} = \theta_c = 0$. As assumed in the nondegenerate DE model [46], we then have the effective hopping matrix elements determined by the spin background: $t_{ab} = t\cos(\theta_{ab}/2)$ and $t_c = t\cos(\theta_c/2)$. Note that here t is chosen as the hopping element between two $|x\rangle$ orbitals in (a,b) planes, unlike in the other Sections. We show below that solving this problem in the quasiclassical approximation introduced for the nondegenerate DE model (Sec. II.B), the energy spectrum (178) is renormalized by the magnetic order and, because of the orbital-dependent hopping matrix elements in a degenerate system, this results in an anisotropic magnetic structure.

When we dope the system by adding electrons, these go first into states with minimal energy, in our case into the states close to the Γ-point at $\vec{k} = 0$. Let us first assume that all doped charges go into the state with the lowest energy, which is strictly speaking only the case for very small doping ($y \simeq 0$), and neglect for the moment the effects promoted by a finite filling of the bands. At the Γ-point the energies are $\epsilon_{\pm}(0) = -\frac{4}{3}(2t_{ab} + t_c) \pm \frac{4}{3}|t_{ab} - t_c|$. Using this simplifying assumption which overestimates the kinetic energy, the total energy per site of the system containing y electrons reads:

$$\begin{aligned}E(\theta_{ab}, \theta_c) &= \frac{J}{2}(\cos\theta_c + 2\cos\theta_{ab}) + y\epsilon_-(\vec{k}=0)\\ &= -\frac{3J}{2} + 2J\cos^2(\theta_{ab}/2) + J\cos^2(\theta_c/2)\\ &\quad - \frac{4}{3}yt\left[2\cos(\theta_{ab}/2) + \cos(\theta_c/2) + |\cos(\theta_{ab}/2) - \cos(\theta_c/2)|\right].\end{aligned} \quad (179)$$

Minimizing the energy of the doped state (179) over two angles θ_{ab} and θ_c, one encounters two possible situations [65]. If $\cos(\theta_{ab}/2) > \cos(\theta_c/2)$, then the magnetic structure is A-type-like; in this case the minimization of the energy with respect to the angles θ_{ab} and θ_c gives

$$\cos\left(\frac{\theta_{ab}}{2}\right) = \frac{t}{J}y, \qquad \cos\left(\frac{\theta_c}{2}\right) = 0, \qquad (180)$$

and the energy of the corresponding state amounts to

$$E^{(1)} = -\frac{3}{2}J - \frac{2t^2}{J}y^2. \qquad (181)$$

Physically this state corresponds to an (a,b)-plane with a canted structure, with the spins in neighboring planes being antiparallel. If instead $\cos(\theta_{ab}/2) < \cos(\theta_c/2)$, then the magnetic structure is C-type-like and one finds for θ_{ab} and θ_c:

$$\cos\left(\frac{\theta_{ab}}{2}\right) = \frac{t}{3J}y, \qquad \cos\left(\frac{\theta_c}{2}\right) = \frac{4t}{3J}y. \qquad (182)$$

The energy $E^{(2)}$ of this state is exactly equal to that of the A-type state $E^{(1)}$. In this situation we have a canted structure in all three directions, with the spin correlations in the c-direction being closer to the FM state. Thus, in the lowest-order approximation in J, the two solutions (180) and (182) are degenerate. We note that these phases have in general canted spin structures, and are therefore different from the magnetic structures shown in Fig. 5 in a broad range of parameters. One can easily show that in this approximation we would have degenerate solutions up to a concentration $y_c = 3J/4t$, beyond which the A-type solution becomes energetically favorable. In fact, this solution never evolves into a FM state – for $y > J/t$ the canting angle $\theta_{ab} = 0$ in Eq. (180) and the basal plane becomes FM, while the moments of the neighboring planes are opposite to each other, i.e., one finds pure A-type antiferromagnetism. In contrast to that, the C-type-like solution (182) would give with increasing y first the state with completely polarized FM chains, but with the magnetic moments of neighboring chains pointing in the directions which differ by a certain angle θ_{ab}, and finally, at $y = 3J/t$, also the angle $\theta_{ab} = 0$ and one finds an isotropic FM state. However, the A-type solution has a lower energy in this regime of parameters; thus one never finds a FM state in this approximation, contrary to experiment.

With increasing electron doping the higher lying band states will be filled which will modify the above picture. We report the same calculation as above, but taking into account that the e_g electrons gradually fill the available band states [65]. One finds that at very low doping concentrations the A-type solution given by Eq. (180) and the C-type solution given by Eq. (182) are indeed the magnetic structures of lowest energy, but the degeneracy of these states is lifted and the A-type solution has always a somewhat lower energy than the C-type solution. This has a simple physical reason. In the A-type structure the dispersion of the bands is strictly 2D, so that the density of states (DOS) at the band edge is finite. For the C-type solution of Eq. (182), the bands have a highly anisotropic but already 3D character, leading typically to a vanishing DOS at the band edge. Therefore, the A-type magnetic structure is stabilized as it has a larger DOS at the band edge. At a somewhat higher doping level, however, the quasi-1D peak in the DOS close to the band edge starts to play a role and can cause the transition to a C-type state.

The complete phase diagram of DE model with degenerate e_g orbitals (173) is presented in Fig. 38. As we have discussed above, the sequence of phases follows from the modulation of the DOS by the DE mechanism. The phase diagram has some similarity to that obtained by Maezono, Ishihara and Nagaosa [169], and qualitatively reproduces a transition from the FM to C-AF phase, as obtained in

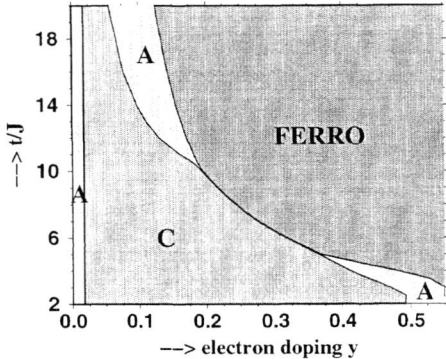

FIGURE 38. Phase diagram of the DE model with degenerate e_g bands. Depending on the electron doping concentration and the ratio of the e_g bandwidth and the t_{2g} superexchange one finds either AF A-type (A), or C-type (C), or FM (FERRO) order (after Ref. [170]).

$Nd_{1-x}Sr_xMnO_3$ for $x > 0.5$ ($y < 0.5$) [166]. Experimentally, the phases found for $y < 0.5$ are insulating (Fig. 4), and the present model gives a C-phase which is susceptible to disorder and likely to become insulating in this range of parameters for $t/J_{AF} \simeq 0.25$. However, for realistic values of parameters $t/J_{AF} \simeq 0.01$ (Table 2), it would give a FM phase except for a narrow range of low doping $y < 0.1$ (Fig. 38). Thus, the model (173) is not complete and we believe that lattice distortions play an important role, changing the balance between the kinetic energy and magnetic interactions and leading to a different phase diagram.

B Spin-waves in a metallic phase

Unfortunately, the phase diagram shown in Fig. 38 does not allow to conclude anything about the favored magnetic phases near the filling of one e_g electron per atom, i.e., in hole doped $La_{1-x}Sr_xMnO_3$ with $x < 0.5$, and other related compounds. In this case the e_g electrons are strongly correlated and one cannot use the DE for degenerate orbitals as discussed in the previous Section. Due to strong on-site Coulomb interactions, the motion of e_g electrons is then allowed only in the restricted space, without creating double occupancies which leads to the superexchange as explained in Sec. V.A. If a half-filled system is doped, the motion of holes in the e_g band becomes possible when it is accompanied by a backflow of electrons which carry spin and orbital index. Thus, one has to consider a hopping problem which resembles the t-J model. In order to make such a model realistic for doped manganites, the degeneracy of e_g orbitals and the previously derived form of superexchange in the doped system (169) and (170) have to be included.

The problem of DE and the resulting phase diagram in the regime of hole doping belongs to the unsolved problems. The approximate solutions were presented by several groups [169,161,152], but either the models were oversimplified, or only certain aspects of the phase diagram were treated. We shall discuss this problem in Sec. VIII; here we analyse only the FM metallic phase and show how the spin-waves follow from DE and superexchange magnetic interactions. Thus we consider an effective t-J model for doped $La_{1-x}A_xMnO_3$ compounds in the FM regime [171],

$$\mathcal{H} = H_t + H_J(e_g) + H_J(t_{2g}), \tag{183}$$

where H_t describes the correlated hopping in the e_g-band (38), with the operators $\hat{c}_{i\alpha\sigma}^\dagger$ which act in the projected space without double occupancies in e_g orbitals, and $H_J(e_g)$ and $H_J(t_{2g})$ stand for the SE terms (157) and (158), respectively.

The spectrum of magnetic excitations in FM phase is very important for a better understanding of the physics of manganites. Experimentally the spin-waves were measured by Perring et al. in $La_{0.7}Pb_{0.3}MnO_3$ [172] along the main high-symmetry directions, and an isotropic dispersion was found. Similar observation were made by Endoh et al. in $La_{1-x}Sr_xMnO_3$ [173]. They found out that unlike in localized FM systems, the energy of the spin-wave near the zone boundary is not approximately equal to $k_B T_C$, where T_C is the Curie temperature, but is significantly larger. This can be interpreted as an experimental proof that the ferromagnetism in doped manganites is itinerant. However, the stiffness constant D at low momenta is almost constant for different compounds at $x \simeq 0.3$ doping [174]. This universality is puzzling and suggests that a common mechanism is responsible in first place for the Goldstone modes and the spin dynamics at low momenta $\vec{q}$, while some other processes might be responsible for a much broader spectrum of the observed Curie temperatures T_C. First we will address the origin of spin-waves and show that the stiffness constant is approximately determined by the DE mechanism in a *correlated* e_g band, $D \simeq J_{DE}S$, where J_{DE} is the effective exchange constant which couples spins $S \simeq 2$. The corrections will come from superexchange which operates as well in doped systems and *counteracts* the DE mechanism.

In order to construct the spin excitations, we separate first the spin dynamics from charge and orbital dynamics. In reality all these processes which involve different degrees of freedom are coupled and are described by fermion operators $\hat{c}_{i\alpha\sigma}^\dagger$ in H_t. We represent such electron operators by fermion operators $\{f_{ix}^\dagger, f_{iz}^\dagger\}$ which carry an orbital index, and by a Schwinger boson operator, either $a_{i\uparrow}^\dagger$ or $a_{i\downarrow}^\dagger$, depending on the spin σ of the moving electron,

$$\hat{c}_{i\alpha\sigma}^\dagger = f_{i\alpha}^\dagger a_{i\sigma}^\dagger. \tag{184}$$

As in Sec. V.A, we use a local basis for the occupied e_g orbitals: $|x\rangle \equiv |x^2-y^2\rangle$ and $|z\rangle \equiv |3z^2-r^2\rangle$. The decomposition (184) allows to simplify the MF analysis in the large-U limit, where the Hilbert space contains only two kind of stated: Mn^{3+} (d^4) ions if a single e_g electron is present, and Mn^{4+} (d^3) ions if an e_g electron was removed. Formally this can be written using the configurations [175]:

$$|i\theta, M\rangle_4 = f_{ix}^\dagger b_{ix}^\dagger |M\rangle, \qquad |i\epsilon, M\rangle_4 = f_{iz}^\dagger b_{iz}^\dagger |M\rangle, \qquad |i, m\rangle_3 = b_{i0}^\dagger |m\rangle, \tag{185}$$

with M being the component of spin $S = 2$ for Mn^{3+} sites, and m being the component of the core t_{2g}-spin $S = 3/2$ for Mn^{4+} sites. The spin part is described by Schwinger bosons in a standard way [82]: $|M\rangle = (a_{i\uparrow}^\dagger)^{S+M}(a_{i\downarrow}^\dagger)^{S-M}|0\rangle$, where $|0\rangle$ is the vacuum state, and similar for $|m\rangle$. The operators $b_{ix}^\dagger$ and $b_{iz}^\dagger$ in Eqs. (185) are slave boson operators which carry the orbital index and are introduced in order to restrict the physical space in analogy to the Kotliar-Ruckenstein bosons in the Hubbard model [176]. The empty boson $b_{i0}^\dagger$ counts Mn^{4+} ions and has a similar meaning. Using these boson operators we implement the local constraints,

$$a_{i\uparrow}^\dagger a_{i\uparrow} + a_{i\downarrow}^\dagger a_{i\downarrow} + b_{i0}^\dagger b_{i0} = 2S, \tag{186}$$

$$b_{ix}^\dagger b_{ix} + b_{iz}^\dagger b_{iz} + b_{i0}^\dagger b_{i0} = 1. \tag{187}$$

They restrict the physical space to contain no double occupancy in the e_g orbitals. In addition, the number of fermions is equal to the number of bosons for each orbital, $f_{i\alpha}^\dagger f_{i\alpha} = b_{i\alpha}^\dagger b_{i\alpha}$, $\alpha = x, z$.

As a next step, we rewrite the hopping Hamiltonian H_t assuming the FM metallic phase and derive the effective fermion problem with the constraints replaced by the operator expressions $z_{i\alpha}$ which contain the bosons used in Eq. (187) that accompany individual hopping processes in Eq. (38) [175],

$$H_t = -\tilde{t} \sum_{\langle ij\rangle\perp,\sigma} a_{i\sigma}^\dagger z_{iz}^\dagger f_{iz}^\dagger f_{jz} z_{jz} a_{j\sigma}$$

$$-\tfrac{1}{4}\tilde{t} \sum_{\langle ij\rangle\|,\sigma} a_{i\sigma}^\dagger \left[3 z_{ix}^\dagger f_{ix}^\dagger f_{jx} z_{jx} + z_{iz}^\dagger f_{iz}^\dagger f_{jz} z_{jz}\right.$$

$$\left.\pm\sqrt{3}\left(z_{ix}^\dagger f_{ix}^\dagger f_{jz} z_{jz} + z_{iz}^\dagger f_{iz}^\dagger f_{jx} z_{jx}\right)\right] a_{j\sigma}, \tag{188}$$

where the renormalized hopping $\tilde{t}$ is introduced in order to compensate for the extra factors resulting from the Schwinger boson operators, $\sim a_{i\sigma}^\dagger a_{j\sigma}$, when the kinetic energy is determined from H_t. The terms $z_{i\alpha}$ contain the orbital bosons $b_{i\alpha}^\dagger$ and an empty boson $b_{i0}^\dagger$, and are similar to the respective factors $z_{i\sigma}$ used in the spin problem [176]. The first sum includes the bonds $\langle ij\rangle \perp$ along the c-axis, while the second sum includes the bonds $\langle ij\rangle \|$ within the (a, b) planes. The MF theory is constructed by averaging over the orbital and empty bosons, and one finds the simpler form of the hopping Hamiltonian,

$$H_t = -\tilde{t} \sum_{\langle ij\rangle\perp,\sigma} a_{i\sigma}^\dagger q_{iz}^* f_{iz}^\dagger f_{jz} q_{jz} a_{j\sigma}$$

$$-\tfrac{1}{4}\tilde{t} \sum_{\langle ij\rangle\|,\sigma} a_{i\sigma}^\dagger \left[3 q_{ix}^* f_{ix}^\dagger f_{jx} q_{jx} + q_{iz}^* f_{iz}^\dagger f_{jz} q_{jz}\right.$$

$$\left.\pm\sqrt{3}\left(q_{ix}^* f_{ix}^\dagger f_{jz} q_{jz} + q_{iz}^* f_{iz}^\dagger f_{jx} q_{jx}\right)\right] a_{j\sigma}, \tag{189}$$

where the renormalization factors at site i ($x = 1 - n$),

$$q_{ix} = \left(\frac{2x}{1+x+(1-x)\cos 2\theta_i}\right)^{1/2}, \quad q_{iz} = \left(\frac{2x}{1+x-(1-x)\cos 2\theta_i}\right)^{1/2}, \quad (190)$$

depend on the angle θ_i which defines the occupied orbital state (65). In the FM state we assume that the orbitals are homogeneous, and the angles correspond to an average orbital state, $\theta_i = \theta$.

Now we consider the Schwinger boson operators in Eq. (189). At low temperatures the magnetic moment of FM metallic manganites is almost fully saturated. It is therefore reasonable to expand Eq. (189) around a FM ground state. Technically this is done by condensing the spin-up Schwinger bosons $a_{i\uparrow} \simeq \sqrt{2\bar{S}}$ (if the magnetic moments point upwards), and by treating the spin-down Schwinger bosons in leading order of $1/\bar{S}$ expansion to describe spin-wave excitations around this ground state. Assuming that the spins are pointing upwards in the MF state, we derive from the constraint (186) the following expansion of the $a^\dagger_{i\uparrow}$-bosons [175],

$$a_{i\uparrow} = \sqrt{2\bar{S} - a^\dagger_{i\downarrow}a_{i\downarrow}} \simeq \sqrt{2\bar{S}}\left(1 - \frac{1}{4\bar{S}}a^\dagger_{i\downarrow}a_{i\downarrow}\right), \quad (191)$$

where the effective spin $2\bar{S} = 2S - x$ is defined using the MF state in Eq. (186), with $\langle b^\dagger_{i0}b_{i0}\rangle = x$. We see that the extra factor generated by Schwinger bosons is equal $2\bar{S}$, and thus one has to use $\tilde{t} = t/2\bar{S}$ in Eq. (188). The final form of the hopping Hamiltonian in the correlated e_g band is obtained after replacing the Schwinger bosons by Holstein-Primakoff boson operators a_i. As the expansion (191) is performed around the FM state, we use therefore $a_{i\downarrow} \equiv a_i$ to study the fluctuations,

$$\sum_\sigma a^\dagger_{i\sigma}a_{j\sigma} \simeq 2\bar{S} - \frac{1}{2}\left(a^\dagger_i a_i + a^\dagger_j a_j - 2a^\dagger_i a_j\right). \quad (192)$$

This expansion leads to the following form of the kinetic energy term,

$$H_t = -\sum_{\langle ij\rangle\alpha\beta} t^{\alpha\beta}_{ij} q_{i\alpha}q_{j\beta} f^\dagger_{i\alpha}f_{j\beta} + \frac{1}{4\bar{S}}\sum_{\langle ij\rangle\alpha\beta} t^{\alpha\beta}_{ij} q_{i\alpha}q_{j\beta} f^\dagger_{i\alpha}f_{j\beta}\left(a^\dagger_i a_i + a^\dagger_j a_j - 2a^\dagger_i a_j\right). \quad (193)$$

Consider first the zeroth-order Hamiltonian,

$$H^{(0)}_t = -\sum_{\langle ij\rangle\alpha\beta} t^{\alpha\beta}_{ij} q_{i\alpha}q_{j\beta} f^\dagger_{i\alpha}f_{j\beta}. \quad (194)$$

It describes the orbital model with correlated fermions; the hopping amplitudes $\propto t^{\alpha\beta}_{ij} q_{i\alpha}q_{j\beta}$ depend on the actual shape of occupied e_g orbitals [177]. This problem was studied qualitatively by van Veenendaal and Fedro [178] who pointed out that the asymmetry in the magnetic phase diagrams of doped manganites between $x > 0.5$ and $x < 0.5$ doping follows from the essentially uncorrelated electrons in the former, and strongly correlated electrons that avoid each other in the latter case.

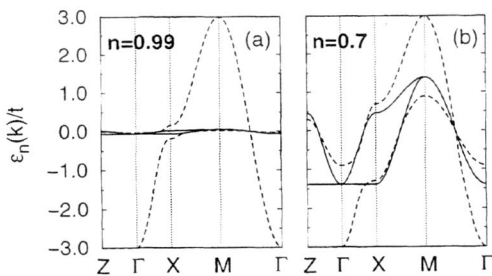

FIGURE 39. Correlated band structure as obtained for two densities of e_g electrons: (a) $n = 0.99$, and (b) $n = 0.70$, in the FM disordered state with the same electron density in both e_g orbitals, $\langle n_x \rangle = \langle n_z \rangle = n/2$ (full lines), and for the fully polarized orbital liquid state [164] with $\langle n_x \rangle = n$ (dashed lines) (after Ref. [177]).

The correlations cause strong band narrowing in the regime of $x \to 0$ due to the reduction of the hopping elements by the q_{ix} and q_{iz} factors in Eq. (190). One finds that the correlated band structure is drastically reduced in the disordered states with $\cos 2\theta = 0$, while smaller reductions are found when the occupied orbitals are close to either $|x\rangle$ or $|z\rangle$ states (Fig. 39). By this mechanism one finds a tendency towards an *orbital liquid* state as observed by Ishihara, Yamanaka and Nagaosa [164]. Such a symmetry breaking in the orbital space happens easily in a 2D case, where the kinetic energy drives the system into the phase with $|x\rangle$ orbitals occupied [179], while there are also other possibilities to stabilize a disordered state in three dimensions [177], and thus the orbital liquid of this type is unlikely.

The orbital model (194) in the limit of $U \to \infty$ is equivalent to the spinless fermion problem. The band structure collapses at $n = 1$ to a line at $\omega = 0$, and the kinetic energy vanishes. In analogy to the Hubbard model in $U \to \infty$ limit, the kinetic energy, $E_{\text{kin}} = \langle H_t^{(0)} \rangle$, is symmetric with respect to the filling of $n = 0.5$, i.e., the system gains kinetic energy at increasing hole concentration in the regime of $0 < x < 0.5$ $(1.0 > n > 0.5)$ (Fig. 40). In contrast, free e_g electrons would give a metallic behavior with the largest kinetic energy near $n = 1$. This demonstrates that the correlations due to the orbital degree of freedom have to be explicitly included in order to reproduce the insulating state in the limit of $x \to 0$. After studying DE in degenerate e_g orbitals (Sec. VI.A), it might be expected that this mechanism produces in first instance anisotropic magnetic phases, such as A-AF and C-AF states. However, these solutions are *de facto* stabilized really by the superexchange interactions in Ref. [65]. However, we have shown in Sec. V.A (see Table 2) that the realistic superexchange constants are much smaller that the hopping element t, and thus the magnetic properties in the highly doped regime are determined primarily by the DE term (194). Consequently, it is allowed to assume that the ground state is FM, and to expand around this state which leads to the DE part of the spin dynamics described by the first order term in Eq. (193) when

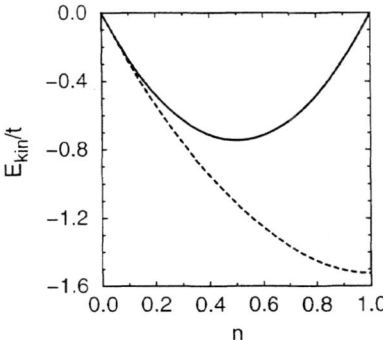

FIGURE 40. Kinetic energy as a function of e_g band filling n for the correlated (194) (solid line) and uncorrelated (173) electrons (dashed line) (after Ref. [177]).

the orbital and charge dynamics are averaged out,

$$H_t^{(1)} \simeq \frac{1}{4\bar{S}} \sum_{\langle ij \rangle \alpha\beta} t_{ij}^{\alpha\beta} q_{i\alpha} q_{j\beta} \langle f_{i\alpha}^\dagger f_{j\beta} \rangle \left(a_i^\dagger a_i + a_j^\dagger a_j - 2a_i^\dagger a_j \right). \quad (195)$$

Using the renormalization factors for the correlated band structure (190), this averaging yields the magnon dispersion due to the DE mechanism [175],

$$\omega_{\vec{q}} = \frac{t}{2\bar{S}} \left(2[1 - \gamma_+(\vec{q})] R_{ab} + [1 - \gamma_z(\vec{q})] R_c \right), \quad (196)$$

where $\gamma_+(\vec{q})$ and $\gamma_z(\vec{q})$ are defined by Eqs. (106) and (108), respectively, and $t = 0.40$ eV is the largest local hopping element between two $3z^2 - r^2$ orbitals along the c-axis [23]. The lattice sums R_{ab} and R_c can be easily derived and depend on the average occupancy of $|x\rangle$ and $|z\rangle$ orbitals, n_x and n_z.

For a disordered state represented by the orbitals with complex coefficients [177], and $\cos 2\theta = n_x - n_z = 0$, the Gutzwiller renormalization factors (190) become equal, $q_{ix} = q_{iz} = q$, and it is convenient to express the result as,

$$\omega_{\vec{q}} = zJ_{\mathrm{DE}} \bar{S}(1 - \gamma_{\vec{q}}), \quad (197)$$

with $\gamma_{\vec{q}}$ defined as in Eq. (32), $z = 6$, and the effective FM exchange constant determined by the kinetic energy of correlated electrons,

$$J_{\mathrm{DE}} = \frac{tq^2}{2\bar{S}^2} \frac{1}{N} \sum_{\vec{k}\nu} \langle f_{\vec{k}\nu}^\dagger f_{\vec{k}\nu} \rangle \cos k_\alpha. \quad (198)$$

The summation in Eq. (198) runs over the occupied band states in each subband ν, and the band structure of degenerate e_g orbitals (178) is renormalized by $q^2 =$

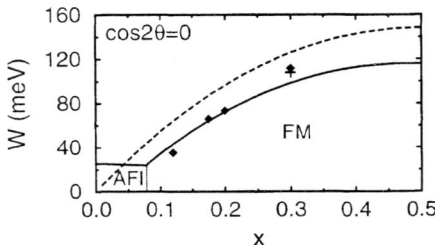

FIGURE 41. Width W of the magnon dispersion in FM manganites $La_{1-x}A_xMnO_3$ as a function of doping x, as obtained including only the DE mechanism (197) (dashed line), and both DE and superexchange contributions (full line). In the AF insulating (AFI) phase at $x < 0.08$ only the anisotropic superexchange interactions contribute. Experimental points correspond to: $La_{1-x}Sr_xMnO_3$ [173] (diamonds) and $La_{0.7}Pb_{0.3}MnO_3$ [172] (cross) (after Ref. [175]).

$2x/(1+x)$. It gives the same result for any cubic direction $\alpha = a, b, c$, so only one component $\propto \cos k_\alpha$ was included. Thus, one finds isotropic spin-waves with the spin-wave stiffness constant $D = J_{DE}\bar{S}$. The qualitative result of the Kondo lattice model with a nondegenerate band [71] is therefore reproduced in the correlated e_g band. The advantage is that the present approach [175] is more straightforward and may be directly used to study the corrections of the DE result due to the superexchange terms (169) and (170). Both superexchange terms may be expanded using Schwinger bosons around the FM state and lead to an isotropic reduction of the effective FM DE interactions, if a disordered orbital state with $\langle \mathcal{P}_{\langle ij \rangle}^{\zeta\xi} \rangle = 1/2$ and $\langle \mathcal{P}_{\langle ij \rangle}^{\zeta\zeta} \rangle = 1/2$ is considered. Under these circumstances, the e_g superexchange (169) is isotropic and weakly AF, taking the parameters $U = 7.3$ eV and $J_H = 0.69$ eV, as given in Sec. V.A. The second AF term comes from the t_{2g} superexchange (170), with the constants $J_t = 2.1$ meV, $\hat{J}_t = 4.6$ meV, and $\bar{J}_t = 5.5$ meV for the pairs of Mn^{3+}–Mn^{3+}, Mn^{4+}–Mn^{4+}, and Mn^{3+}–Mn^{4+} ions, respectively (see Sec. V.B). If hole doping x increases, first the A-AF insulating (AFI) state is modified as reported in Sec. V.B. In this regime the superexchange dominates over the DE term, and a different expansion around the A-AF phase with the polaronic FM superexchange (172) has to be used to derive the spin-waves (Sec. V.B). We present the numerical result for the total magnon width W obtained by these different approaches for the AFI and FM phases in Fig. 41. The theory predicts an observed increase of the magnon width W with increasing doping x [175] due to the DE which dominates in the metallic regime of $x > 0.08$. The DE contribution to W vanishes in the $x \to 0$ limit, in contrast to the unphysical result of band structure calculations that ignore electron correlations, and give the largest FM interactions at $x = 0$ [180], precisely at the point where the A-AF ordering in insulating $LaMnO_3$ is observed.

The magnon dispersion is *isotropic* if the orbitals are disordered. We emphasize that orbital ordering leads instead to anisotropic magnon dispersion, and there-

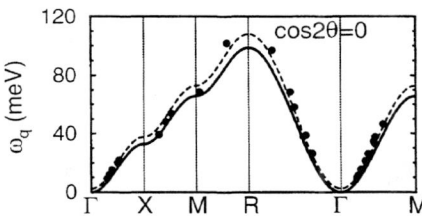

FIGURE 42. Magnon dispersion ω_q as obtained at $x = 0.3$ doping using DE and superexchange contributions (heavy line); parameters as in Ref. [23]. Experimental data for $La_{0.7}Pb_{0.3}MnO_3$ (circles and dashed line) are reproduced from Ref. [172] (after Ref. [175]).

fore such states as obtained recently by Okamoto, Ishihara and Maekawa [152] would lead to anisotropic spin-wave dispersions not only in the A-AF phase, but also in the FM regime. The quality of the model may be best appreciated by an excellent agreement with the experimental results for $x = 0.3$ doped manganite $La_{0.7}Pb_{0.3}MnO_3$ [172]. The exchange interactions found in Ref. [175] at $x = 0.3$: $J_{ab}\bar{S} = J_c\bar{S} = 8.24$ meV are isotropic and reproduce very well the experimental points (Fig. 42). In conclusion, the magnon dispersion derived from DE for degenerate e_g orbitals supplemented by smaller superexchange terms agrees well with the experimental findings in FM metallic manganites [172,173].

C Magnon softening in ferromagnetic manganites

One of the puzzling features is nonuniversality of the magnetic and transport properties of FM manganites at doping $x \simeq 0.30$ [173]. As we have shown in Sec. VI.B, the spin-wave dispersion which follows from the DE model for correlated e_g electrons (197) is isotropic in all three cubic directions and of the nearest-neighbor Heisenberg type. Recently, unexpected deviations from this dispersion with a peculiar softening of the magnon spectrum close to the magnetic zone boundary has experimentally been observed in compounds with low values of T_C. Quite prominent in this respect are measurements of the spin dynamics of the FM manganese oxide $Pr_{0.63}Sr_{0.37}MnO_3$ [181]. While exhibiting conventional Heisenberg behavior at small momenta, the magnetic excitations soften at the boundary of the BZ. Assuming the magnon dispersion to be of Heisenberg type with an exchange constant J_{eff}, the dispersion is quadratic in the regime of $\vec{q} \simeq 0$, $\omega_{\vec{q}} \simeq Dq^2$, with the spin-wave stiffness $D \propto J_{eff}$. Since the latter also controls the Curie temperature $T_C \propto J_{eff}$, the ratio of D and T_C is expected to be a universal constant. Manganites, however, exhibit a pronounced deviation from this behavior: D/T_C increases significantly as one goes from compounds with high to compounds with low values of T_C [174]. This feature together with the above magnon softening at the magnetic zone boundary indicate that the DE model does not suffice and some specific feature of magnetism in manganites has not yet been identified.

Here we report shortly an interesting recent proposal by Khaliullin and Kilian [182] that charge and coupled orbital-lattice fluctuations are responsible for the unusual magnon softening. The starting point is not the t-J model for manganites (183), but a Kondo lattice model with orbital degeneracy which is a generalization of the DE model (23) and takes into account the correlated nature of e_g electrons. We begin by analyzing the DE part,

$$H = -\sum_{\langle ij\rangle \alpha\beta\sigma} t_{ij}^{\alpha\beta}\left(\hat{c}_{i\alpha\sigma}^\dagger \hat{c}_{j\beta\sigma} + H.c.\right) - J_H \sum_{i\alpha\sigma\sigma'} \vec{S}_i \cdot c_{i\alpha\sigma}^\dagger \vec{\sigma}_{\sigma\sigma'} c_{i\alpha\sigma'}, \quad (199)$$

where the constrained operators $\hat{c}_{i\alpha\sigma}^\dagger$ act in the projected space without double occupancies in the e_g orbitals, as in Eq. (183). Due to the strong Hund's coupling $\propto J_H$, core spins $\vec{S}_i$ and itinerant e_g spins $\vec{s}_i$ are not independent of each other; rather a high-spin state with total on-site spin $S = S_t + \frac{1}{2}$, where $S_t = 3/2$, is formed. This unification of band and local spin subspaces suggests to decompose the e_g electron into its spin and orbital/charge components. As we have shown in Sec. VI.B following Ref. [175], the e_g spin can then be absorbed into the total spin, allowing an independent treatment of spin and orbital/charge degrees of freedom. The formal procedure which allows for this separation scheme is realized by introducing Schwinger bosons $d_{i\uparrow}$ and $d_{i\downarrow}$ to describe the e_g spin subspace,

$$s_i^+ = d_{i\uparrow}^\dagger d_{i\downarrow}, \quad s_i^- = d_{i\downarrow}^\dagger d_{i\uparrow}, \quad s_i^z = \tfrac{1}{2}(d_{i\uparrow}^\dagger d_{i\uparrow} - d_{i\downarrow}^\dagger d_{i\downarrow}), \quad (200)$$

as well as Schwinger bosons $a_{i\uparrow}^\dagger$ and $a_{i\downarrow}^\dagger$ to model the *total* on-site spin $S = 2$,

$$S_i^+ = a_{i\uparrow}^\dagger a_{i\downarrow}, \quad S_i^- = a_{i\downarrow}^\dagger a_{i\uparrow}, \quad S_i^z = \tfrac{1}{2}(a_{i\uparrow}^\dagger a_{i\uparrow} - a_{i\downarrow}^\dagger a_{i\downarrow}). \quad (201)$$

These auxiliary particles are subject to the following constraints that depend on the e_g occupation number n_i:

$$d_{i\uparrow}^\dagger d_{i\uparrow} + d_{i\downarrow}^\dagger d_{i\downarrow} = n_i, \quad (202)$$

$$a_{i\uparrow}^\dagger a_{i\uparrow} + a_{i\downarrow}^\dagger a_{i\downarrow} = 2S - 1 + n_i. \quad (203)$$

This is in fact a different way of writing the constraints (186) and (187). By construction, one assumes in both approaches that high-spin states $S = 2$ are realized at the sites occupied by e_g electrons which corresponds to the large J_H limit. The creation and annihilation operators for e_g electrons can then be expressed in terms of spinless fermions $f_{i\alpha}^\dagger$ which carry charge and orbital pseudospin and Schwinger bosons which carry spin in analogy to Eq. (184),

$$\hat{c}_{i\alpha\sigma}^\dagger = \hat{c}_{i\alpha}^\dagger d_{i\sigma}^\dagger. \quad (204)$$

The Bose operators are subject to the constraint (202) that enforces the operators $d_{i\sigma}$ and $d_{i\sigma}^\dagger$ to act only in the projected Hilbert space with one or zero Schwinger bosons, respectively. Our aim is to absorb the e_g spin into the total spin, which requires to map the e_g operators $d_{i\sigma}$ onto the operators $a_{i\sigma}$ for the total spin. It is easy to obtain the following mapping

$$d_{i\sigma} = \frac{1}{\sqrt{2S}} a_{i\sigma}. \tag{205}$$

The kinetic energy in Hamiltonian (199) now describes the simultaneous transfer of pairs of spinless fermions and Schwinger bosons, and can be rewritten in the form analogous to Eq. (189). However, one can still use the fermion operators which carry the orbital index, unlike in Sec. VI.B, where orbital bosons were introduced as well in Eq. (189). One finds,

$$H_t = -\frac{1}{2S} \sum_{\langle ij \rangle \alpha\beta\sigma} t_{ij}^{\alpha\beta} \left(\hat{c}_{i\alpha}^\dagger \hat{c}_{j\beta} a_{i\sigma}^\dagger a_{j\sigma} + H.c. \right), \tag{206}$$

Thus, the above consideration provides a formal proof that the above construction with *two* Schwinger bosons $d_{i\sigma}$ and $a_{i\sigma}$ [182] is completely equivalent to introducing only *one* type of Schwinger bosons referring to the total spins $S = 2$ [64], as shown in Sec. V.B. One finds therefore the same expansion of the kinetic energy as given by Eq. (193), except that the Kotliar-Ruckenstein bosons have not yet been introduced in the present case (206), and one may therefore study the renormalization of magnons by quantum fluctuations beyond the MF theory.

To study the propagation of the magnetic excitations in hole-doped DE systems, we now derive first the correct spin operator taking into account the fact that the total on-site spin depends on whether a hole or an e_g electron is present at that site. The spin quantum number is $S - \frac{1}{2}$ in the former and S in the latter case. In general, a spin excitation is created by the operator S_i^+. Expressing this operator in terms of Schwinger bosons $S_i^+ = a_{i\uparrow}^\dagger a_{i\downarrow}$, and next condensing $a_{i\uparrow}$ and mapping $a_{i\downarrow}$ onto the magnon annihilation operator a_i as in Sec. V.B, the following representation is obtained,

$$S_i^+ = \begin{cases} \sqrt{2S}\, a_i, & \text{for sites with } e_g \text{ electron,} \\ \sqrt{2S-1}\, a_i, & \text{for sites with hole.} \end{cases}$$

Assuming S to be the "natural" spin number of the system, the magnon operator a_i hence has to be rescaled by a factor $[(2S-1)/(2S)]^{1/2}$ when being applied to hole sites,

$$A_i = \begin{cases} a_i, & \text{for sites with } e_g \text{ electron,} \\ \sqrt{(2S-1)/(2S)}\, a_i, & \text{for sites with hole.} \end{cases}$$

The general magnon operator A_i that automatically probes the presence of an e_g electron can finally be written as

$$A_i = a_i \left[n_i + \sqrt{\frac{2S-1}{2S}} (1 - n_i) \right] \approx a_i - \frac{1}{4S}(1 - n_i)\, b_i, \tag{207}$$

where n_i is the number operator of e_g electrons at site i. It turns out that A_i represents the true Goldstone operator of hole-doped DE systems. Its composite character comprises local and itinerant spin features which is a consequence of the fact that static core and mobile e_g electrons together build the total on-site spin. While the itinerant part of A_i is of order $1/S$ only, it nevertheless is of crucial importance to ensure consistency of the spin dynamics with the Goldstone theorem, i.e., to yield an excitation mode whose energy vanishes at zero momentum [182].

Having derived the correct magnon operators A_i for doped DE systems (207) allows to study the energies of magnetic excitations, and their renormalization caused by the coupling of magnons to other quasiparticles present in the correlated e_g band, and the coupling to the lattice. If such processes are neglected, one finds the magnon dispersion determined primarily by the DE in a strongly correlated e_g band, as shown in Sec. VI.B. If, however, the fermion and orbital variables are not averaged out, dynamical processes become possible which dress the magnons and result in finite selfenergy $\Sigma(\omega,\vec{q})$. Therefore, the magnon spectrum in an interacting system $\tilde{\omega}_{\vec{p}}$ contains the many-body correction expressed by the magnon selfenergy,

$$\tilde{\omega}_{\vec{q}} = \omega_{\vec{q}} + \mathrm{Re}[\Sigma(\omega_{\vec{q}},\vec{q})]. \quad (208)$$

The MF magnon dispersion $\omega_{\vec{q}}$ is of conventional nearest-neighbor Heisenberg form (197), and we have seen in Sec. VI.B that it gives the spin-wave stiffness constant $D = J_{\mathrm{DE}} S$.

Apart from electron dynamics in the correlated e_g orbitals which is treated by the projected operators $\hat{c}_{i\alpha}^\dagger$, virtual charge-transfer processes across the Hubbard gap contribute to the superexchange. At low and intermediate doping levels the superexchange due to the hopping of e_g electrons is of importance, and this contributes with a FM component as the high-spin state gives the only nonvanishing magnetic term, if electrons are polarized. These superexchange processes $\propto J_{\mathrm{SE}}$ establish an intersite interaction, which in the limit of a strong Hund's coupling could be written in the form [182],

$$H_J = -J_{\mathrm{SE}} \sum_{\langle ij \rangle} \left(\tfrac{1}{4} - \tau_i^\alpha \tau_j^\alpha\right) \left[\vec{S}_i \vec{S}_j + S(S+1)\right] n_i n_j. \quad (209)$$

Thus one might naively expect that the stiffness constant D is increased by the superexchange J_{SE} which thus amplifies the DE effect $\propto J_{\mathrm{DE}}$ in the correlated e_g band, resulting in $D \propto J_{\mathrm{DE}} + J_{\mathrm{SE}}$ [182]. While the FM term is the only term which contributes in the ground state, the situation is different in the excited states and the above simplified picture is incorrect. It has been shown in Sec. VI.B that the DE effect is *reduced by superexchange interactions*. The reason is twofold: (i) a few AF interactions are generated by the effective processes which involve e_g electrons (157), and they have to be included for the realistic parameters with finite J_H (and not using $J_H = \infty$), and (ii) the processes which involve t_{2g} electrons contribute as well with AF superexchange terms (170). Therefore, the AF superexchange due to both e_g and t_{2g} electrons always dominates over a single FM e_g contribution, and thus the total superexchange decreases the stiffness constant and *de facto counteracts* the DE in the FM manganites.

Fortunately, the above incorrect interpretation of Ref. [182] is of quantitative nature and one may still use the same expansion of the DE processes around the FM state in order to study the consequences of magnon interactions with orbital excitations and with the lattice. Using the slave formalism similar to that introduced in Sec. VI.B, Khaliullin and Kilian derived the effective processes which describe the coupling of magnons to charge and orbital fluctuations separately [182]. These processes involve always scattering of either charge (fermionic) or orbital (bosonic) states on the magnons, and lead in lowest order to the contributions to the magnon selfenergy $\Sigma(\omega, \vec{q})$ shown in Figs. 43(a) and 43(b).

The softening of magnons at the zone boundary was observed in $Pr_{0.63}Sr_{0.37}MnO_3$, the compound which has a lower value of T_C [181], while no softening was found earlier in metallic $La_{0.7}Pb_{0.3}MnO_3$ [172]. The compounds with lower values of T_C are worse metals and become in some cases insulating [174], as for instance $Pr_{1-x}Ca_xMnO_3$. This suggests that the lattice degrees of freedom are likely to play an important role in the magnon softening. The crystal dynamics is given by the Hamiltonian,

$$H_{\rm ph} = \frac{1}{2} K \sum_i \vec{Q}_i^2 + K_1 \sum_{\langle ij \rangle} Q_i^\alpha Q_j^\alpha + \frac{1}{2M} \sum_i \vec{P}_i^2, \qquad (210)$$

where $\vec{Q}_i = (Q_{2i}, Q_{3i})$, $Q_i^{a(b)} = (Q_{3i} \pm \sqrt{3} Q_{2i})/2$, $Q_i^c = Q_{3i}$, and Q_{2i} and Q_{3i} are JT phonons of Fig. 3. The coupling of the spin-waves to phonons depends on the ratio $k_1 = K_1/K$.

One of the central results of Ref. [182] is that the coupling between spins and phonons is indirect – it is mediated via the orbital channel. Orbital fluctuations couple to the spins be the DE term (206), while the coupling of orbitals to phonons (7) admixes low phononic frequencies into orbital fluctuations. The corresponding effective spin-phonon interaction is of the form,

$$H_{\rm s-ph} = \sum_{\vec{p}\vec{q}\lambda} g_{\vec{p}\vec{q}}^\lambda (b_{\vec{q}\lambda}^\dagger + b_{-\vec{q}\lambda}) A_{\vec{p}}^\dagger A_{\vec{p}+\vec{q}}, \qquad (211)$$

where $b_{\vec{q}\lambda}^\dagger$ are the phonon creation operators for the mode $\lambda = 1, 2$ with momentum $\vec{q}$, $A_{\vec{p}}^\dagger$ are the Fourier transforms of the boson operators A_i (207), and $g_{\vec{p}\vec{q}}^\lambda$ are the respective coupling constants which depend on the frequency of the involved phonon mode. The corresponding diagram which contributes to the magnon selfenergy is shown in Fig. 43(c). Using perturbation theory, different processes shown in Fig. 43 lead to the expressions which contain summations over internal momenta. Such summations were performed numerically using a Monte-Carlo algorithm in Ref. [182] and give the result shown by solid lines in Fig. 44. We note that the bare MF dispersion $\omega_{\vec{p}}$ is somewhat overestimated for the chosen parameters, as the contributions coming from the AF superexchange were neglected [182]. Therefore,

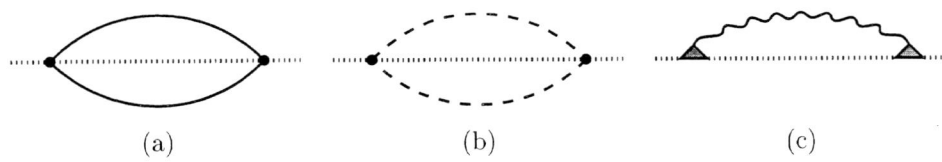

FIGURE 43. Magnon selfenergies describing the effect of magnon scattering on (a) orbital fluctuations, (b) charge fluctuations, and (c) phonons. Solid, dashed, dotted, and wiggled lines denote orbiton, holon, magnon, and phonon propagators, respectively (after Ref. [182]).

the good agreement with experiment claimed for the stiffness constant D using the spectroscopic value of $t = 0.4$ eV [30] is fortuitous; in fact it implies that the value of t is somewhat different in $Pr_{0.63}Sr_{0.37}MnO_3$ from the above value which gives a very good agreement for the stiffness constant in $La_{0.7}Pb_{0.3}MnO_3$ when the superexchange contributions are correctly included [175]. It is clear, however, that the experimental magnon dispersion can be reproduced by Eq. (208) in the parameter space of the model which includes the coupling to the lattice (211). In the actual calculation, with the result shown in Fig. 44 [182], the phonon contribution was determined using: $E_{JT}a_0^2 \equiv (g_2 a_0)^2/2K = 0.004$ eV, $\omega_0 = 0.08$ eV [62], and $\Gamma = 0.04$ eV. More details may be found in Ref. [182].

As the main result, a pronounced softening of magnons at large momenta can be reproduced in the theory which treats the coupling of spin waves to the fluctuations of the orbital and lattice degrees of freedom [182]. Interestingly, charge fluctuations are found to play only a minor role. This follows qualitatively from the energy scales – the spectral density of charge fluctuations $\propto U$ lies well above the magnon band. In contrast, orbital and lattice fluctuations have rather low characteristic frequencies ($\propto xt$ and $\propto \omega_0$, respectively) and hence may couple stronger to spin-waves. The precise mechanism of this coupling is not yet completely understood, however. The numerical study of Ref. [182] suggests that the presence of JT phonons amplifies the magnon softening which follows in first instance from orbital fluctuations. The softening at the zone boundary, which occurs simultaneously with practically unaffected spin dynamics at small momenta that enters the spin-wave stiffness D [174], indicates that the instability towards an orbital-lattice ordered state is responsible for this phenomenon. The unusual magnon dispersion experimentally observed in low-T_C manganites [181] can hence be understood as a precursor effect of orbital-lattice ordering.

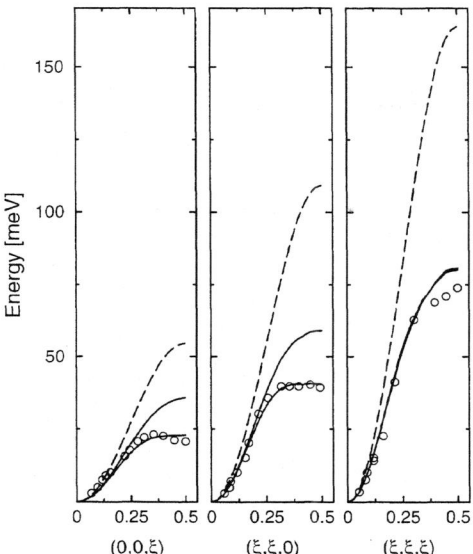

FIGURE 44. Magnon dispersion along $(0,0,\xi)$, $(\xi,\xi,0)$, and (ξ,ξ,ξ) directions, where $\xi = 0.5$ is at the cubic zone boundary. Experimental data from Ref. [181] are indicated by circles, the MF dispersion $\omega_{\vec{q}}$ is marked by dashed lines. Solid lines represent the theoretical result for the dispersion $\tilde{\omega}_{\vec{q}}$ defined by Eq. (208); it includes charge, orbital, and lattice effects. The upper curve is obtained for dispersionless phonons with $k_1 = 0$, the lower one is a fit to the experimental data with $k_1 = -0.33$ corresponding to ferrotype orbital-lattice correlations (after Ref. [182]).

VII ORBITAL DEGREES OF FREEDOM IN A FERROMAGNET

A Orbital excitations

While the electronic interactions in cuprates might stabilize a spin liquid in particular situations (Sec. IV), the spin-orbital model in manganites is in the opposite limit. Large Hund's rule interactions $\propto J_H$ stabilize the orbital ordered states at the filling of one e_g electron per site and at low doping, and FM planes stagger in the A-AF phase. Thus, the spin degree of freedom can be integrated out and one is left with orbital dynamics. In the undoped case one finds an anisotropic pseudospin model with an interesting behavior, as the pseudospin quantum number is not conserved [183].

The proper understanding of pure orbital excitations is very important, both for fundamental reasons, and as a starting point to consider weakly doped FM (a,b) planes of $La_{1-x}A_xMnO_3$. Therefore, we consider an idealized uniform FM phase. In this case the only superexchange channel which contributes is the effective interaction via the high-spin state, and one finds the effective Hamiltonian which describes e_g electrons in a cubic crystal at strong-coupling [63,151,155],

$$\mathcal{H} = H_J(e_g) + H_\tau. \tag{212}$$

It consists of the superexchange part $H_J(e_g)$, and the orbital splitting term due to crystal-field term H_τ (59) which is connected with the uniaxial pressure and was introduced before in Sec. III.A. An example of such a model is the superexchange interaction in the FM state of $LaMnO_3$ which originates from $d_i^4 d_j^4 \rightleftharpoons d_i^3 d_j^5$ excitations into a high-spin d_j^5 state $|^6A_1\rangle$ (with $d_i^4 \equiv t_{2g}^3 e_g$, $d_i^3 \equiv t_{2g}^3$, and $d_i^5 \equiv t_{2g}^3 e_g^2$). It gives the effective Hamiltonian with orbital interactions,

$$H_J(e_g) = -\frac{t^2}{\varepsilon(^6A_1)} \sum_{\langle ij \rangle} \mathcal{P}_{\langle ij \rangle}^{\zeta\xi}, \tag{213}$$

where t is the hopping element between the directional $3z^2 - r^2$ orbitals along the c-axis, and $\varepsilon(^6A_1)$ is the excitation energy [23]. The orbital degrees of freedom are described by the projection operators $\mathcal{P}_{\langle ij \rangle}^{\zeta\xi}$ (61) which select a pair of orbitals $|\zeta\rangle$ and $|\xi\rangle$, being parallel and orthogonal to the directions of the considered bond $\langle ij \rangle$ in a cubic lattice.

The Hamiltonian (213) has cubic symmetry and may be written using any reference basis in the e_g subspace. For the conventional choice of $|x\rangle$ and $|z\rangle$ orbitals, the above projection operators $\mathcal{P}_{\langle ij \rangle}^{\zeta\xi}$ are represented by the orbital operators τ_i^α, with $\alpha = a,b,c$ for three cubic axes, defined by Eqs. (49) and (50). We replace them by *pseudospin operators* $T_i^x = \frac{1}{2}\sigma_i^x$ and $T_i^z = \frac{1}{2}\sigma_i^z$, where σ_i^x and σ_i^z are the Pauli matrices. It is convenient to use the prefactor $J = t^2/\varepsilon(^6A_1)$ in Eq. (213) as the energy unit for the superexchange interaction [note that this definition of J is different from that used in Sec. V.A by a factor of $U/(U - 5J_H) \simeq 2$]. Thus, one finds a pseudospin Hamiltonian,

$$H_J(e_g) = \tfrac{1}{2}J \sum_{\langle ij \rangle \|} \left[T_i^z T_j^z + 3T_i^x T_j^x \mp \sqrt{3}(T_i^x T_j^z + T_i^z T_j^x) \right] + 2J \sum_{\langle ij \rangle \perp} T_i^z T_j^z, \qquad (214)$$

where the prefactor of the mixed term $\propto \sqrt{3}$ is negative in the a-direction and positive in the b-direction, and the meaning of the pseudospin components $|\uparrow\rangle = |x\rangle$ and $|\downarrow\rangle = |z\rangle$ is the same as in Sec. III.A. We choose the same convention as in Eqs. (91) and (92) that the bonds labeled as $\langle ij \rangle \|$ ($\langle ij \rangle \perp$) connect nearest-neighbor sites within (a, b) planes (along the c-axis). By construction, the superexchange interaction occurs only between the pairs of ions with singly occupied *orthogonal* e_g orbitals $|\zeta\rangle$ and $|\xi\rangle$ at two nearest-neighbor sites (two orthogonal e_g orbitals are then singly occupied in the intermediate high-spin excited states). The virtual excitations which lead to the interactions described by Eq. (214) are shown in Fig. 45. Here we neglected a trivial constant term which gives the energy of $-J/2$ per bond, i.e., $-3J/2$ per site in a 3D system. We emphasize that the SU(2) symmetry is explicitly broken in H_J, and the interaction depends only on two pseudospin operators, T_i^x and T_i^z. Moreover, there is an interesting balance of symmetry-breaking $\sim T_i^z T_j^z$ and fluctuating $\sim T_i^x T_j^x$ terms, with Ising-like c-bonds, and more quantum fluctuations on the bonds within the (a, b) planes, assuming the symmetry breaking with $\langle T_i^z \rangle \neq 0$. However, the overall coefficients of both types of terms are equal to $3J$ which shows that the symmetry breaking may happen in any spatial direction and will give *the same classical energy*. However, the models in lower dimension will have different properties and will be more classical. In fact, the 1D model has only Ising interactions, if the orbitals $|\xi\rangle$ and $|\zeta\rangle$ are chosen as a basis. Before analyzing the excitation spectra of the Hamiltonian (214), one has to determine first the classical ground state of the system. The classical configurations which minimize the interaction terms (214) are characterized by the two-sublattice pseudospin order, with two angles describing orientations of pseudospins, one at each sublattice. As usually, the classical ground state is obtained by minimizing the energy with respect to these two rotation angles, i.e., by choosing the optimal orbitals.

Let us consider first the 3D case at orbital degeneracy $E_z = 0$. The superexchange interaction (213) seems to induce the alternation of *orthogonal orbitals* in the ground state in all three directions, which would be equivalent to the *two-sublattice G-AF order in the pseudospin space*. Indeed, this configuration gives the lowest energy on the MF level for individual directions, as the virtual transitions represented in Fig. 45 give the largest contribution, if the hopping involves one occupied and one unoccupied orbital of the same type (e.g., either directional or planar with respect to the direction of the bond $\langle ij \rangle$). However, it is not possible to realize a G-AF state in a 3D lattice, as the orbitals that are orthogonal in (a, b) planes are not orthogonal along the c-direction.

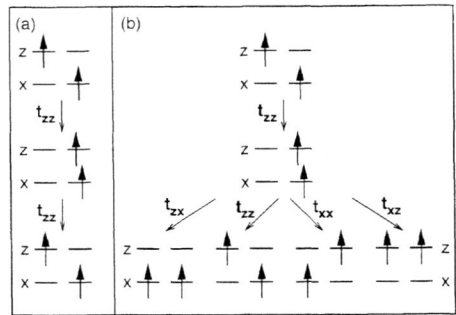

FIGURE 45. Schematic representation of the virtual $d_i^4 d_j^4 \to d_i^3 d_j^5$ excitations in LaMnO$_3$ for the starting FM configuration $d_{iz\uparrow}^\dagger d_{jx\uparrow}^\dagger |0\rangle$ which involve the high-spin $|^6A_1\rangle$ state and generate effective orbital superexchange interactions (214): (a) for a bond along the c-axis, $(ij) \perp$; (b) for a bond within the (a,b) plane, $(ij) \parallel$ (after Ref. [183]).

In order to investigate the ground state on the classical level, we perform a uniform rotation of $\{|z\rangle, |x\rangle\}$ orbitals at each site,

$$\begin{pmatrix} |i\bar{\mu}\rangle \\ |i\bar{\nu}\rangle \end{pmatrix} = \begin{pmatrix} \cos\theta & \sin\theta \\ -\sin\theta & \cos\theta \end{pmatrix} \begin{pmatrix} |iz\rangle \\ |ix\rangle \end{pmatrix}, \qquad (215)$$

and generate the new orthogonal orbitals, $|i\bar{\mu}\rangle$ and $|i\bar{\nu}\rangle$, which are used to determine the energy as a function of θ. The rotation (215) leads to the following transformation of the pseudospin operators,

$$\begin{aligned} T_i^x &\to T_i^x \cos 2\theta - T_i^z \sin 2\theta, \\ T_i^z &\to T_i^x \sin 2\theta + T_i^z \cos 2\theta, \end{aligned} \qquad (216)$$

and the interaction Hamiltonian H_J is then transformed into,

$$\mathcal{H}^\theta = H_\parallel^\theta + H_\perp^\theta, \qquad (217)$$

$$H_\parallel^\theta = \tfrac{1}{2} J \sum_{\langle ij\rangle\parallel} [\,(2+\cos 4\theta \mp \sqrt{3}\sin 4\theta) T_i^x T_j^x + (2-\cos 4\theta \pm \sqrt{3}\sin 4\theta) T_i^z T_j^z \\ -(\sin 4\theta \pm \sqrt{3}\cos 4\theta)(T_i^x T_j^z + T_i^z T_j^x)\,], \qquad (218)$$

$$H_\perp^\theta = J \sum_{\langle ij\rangle\perp} [\,(1-\cos 4\theta)\, T_i^x T_j^x + (1+\cos 4\theta)\, T_i^z T_j^z + \sin 4\theta (T_i^x T_j^z + T_i^z T_j^x)\,]. \qquad (219)$$

The Hamiltonian given by Eqs. (218) and (219) has the symmetry of the cubic lattice, but *surprisingly one finds the full rotational symmetry* of the present interacting problem on the classical level at orbital degeneracy. The occupied orbitals

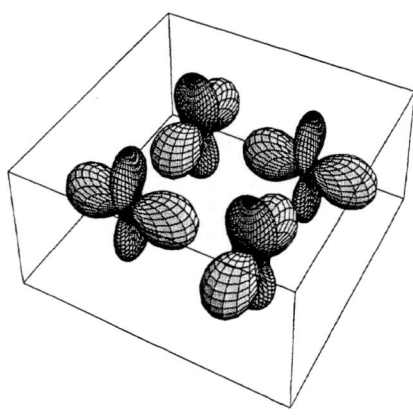

FIGURE 46. Alternating orbital order in FM cubic LaMnO$_3$ and in a 2D model of a FM (a,b) plane in LaMnO$_3$: $(|x\rangle + |z\rangle)/\sqrt{2}$ an $(|x\rangle - |z\rangle)/\sqrt{2}$ as found at $E_z \to 0$ (after Ref. [183]).

are $|i\bar{\mu}\rangle$ and $|i\bar{\nu}\rangle$ on A and B sublattice, respectively, and classically the lowest energy is $E_{\text{MF}} = -3J/4$ per site, independent of the rotation angle θ, as long as the occupied orbitals are staggered. A finite orbital field $E_z \neq 0$ breaks the rotational symmetry on the classical level. It acts along the c-axis, and it is therefore easy to show that the ground state in the limit of $E_z \to 0$ is realized by the alternating occupied orbitals being symmetric/antisymmetric linear combinations of $|z\rangle$ and $|x\rangle$ orbitals, i.e., the occupied states correspond to the rotated orbitals (215) $|i\bar{\mu}\rangle$ and $|i\bar{\nu}\rangle$ on the two sublattices with an angle $\theta = \pi/4$, shown in Fig. 46. In particular, this state is different from the alternating directional orbitals, $3x^2 - r^2$ and $3y^2 - r^2$, which might have been naively expected. It follows in the limit of degenerate orbitals from the 'orbital-flop' phase, in analogy to a spin-flop phase for the Heisenberg antiferromagnet at finite magnetic field. With increasing (decreasing) E_z the orbitals tilt out of the state shown in Fig. 46, and approach $|x\rangle$ ($|z\rangle$) orbitals, respectively, which may be interpreted as an increasing FM component of the orbital polarization in the pseudospin model.

To describe the tilting of pseudospins due to the crystal-field $\propto E_z$ we make two different transformations (215) at both sublattices, rotating the orbitals by an angle $\theta = \frac{\pi}{4} - \phi$ on sublattice A,

$$\begin{pmatrix} |i\mu\rangle \\ |i\nu\rangle \end{pmatrix} = \begin{pmatrix} \cos(\frac{\pi}{4} - \phi) & \sin(\frac{\pi}{4} - \phi) \\ -\sin(\frac{\pi}{4} - \phi) & \cos(\frac{\pi}{4} - \phi) \end{pmatrix} \begin{pmatrix} |iz\rangle \\ |ix\rangle \end{pmatrix}, \qquad (220)$$

and using a similar transformation to Eq. (220) with an angle $\theta = \frac{\pi}{4} + \phi$ on sublattice B, so that the relative angle between the *occupied* orbitals $|i\mu\rangle$ ($i \in A$) and $|j\nu\rangle$ ($j \in B$) is $\frac{\pi}{2} - 2\phi$, and decreases with increasing ϕ. The operators T_i^x and T_i^z may be now transformed as in Eqs. (216) using the actual rotations by $\theta = \frac{\pi}{4} \pm \phi$, as given in Eq. (220). As before, the orbitals $|i\mu\rangle$ and $|j\nu\rangle$ are occupied on two sublattices, $i \in A$ and $j \in B$, respectively, and the orbital order in the classical state is described by the transformed operators $\langle T_i^z \rangle = -1/2$ and $\langle T_j^z \rangle = +1/2$, respectively.

Using the new operators T_i^x and T_i^z, the transformed Hamiltonian (212) takes the form,

$$\mathcal{H}^\phi = H_\parallel^\phi + H_\perp^\phi + H_\tau^\phi, \tag{221}$$

$$H_\parallel^\phi = \frac{1}{2} J \sum_{\langle ij \rangle_\parallel} \left[(2\cos 4\phi - 1) T_i^x T_j^x + (2\cos 4\phi + 1) T_i^z T_j^z \right.$$
$$\left. + 2\sin 4\phi (T_i^x T_j^z - T_i^z T_j^x) \pm \sqrt{3}(T_i^x T_j^z + T_i^z T_j^x) \right], \tag{222}$$

$$H_\perp^\phi = J \sum_{\langle ij \rangle_\perp} \left[(\cos 4\phi + 1) T_i^x T_j^x + (\cos 4\phi - 1) T_i^z T_j^z - \sin 4\phi (T_i^x T_j^z - T_i^z T_j^x) \right], \tag{223}$$

$$H_\tau^\phi = E_z \sum_i (\lambda_i \sin 2\phi T_i^z - \cos 2\phi T_i^x), \tag{224}$$

where $\lambda_i = -1$ for $i \in A$ and $\lambda_i = 1$ for $i \in B$. The energy of the classical ground state is given by,

$$E_{3D}^{MF} = -\frac{3}{4} J \cos 4\phi + \frac{1}{2} E_z \sin 2\phi, \tag{225}$$

and is minimized by,

$$\sin 2\phi = -\frac{E_z}{6J}. \tag{226}$$

The above result (226) is valid for $|E_z| \leq 6J$; otherwise one of the initial orbitals (either $|x\rangle$ or $|z\rangle$) is occupied at each site, and the state is fully polarized ($\sin 2\phi = \pm 1$).

In contrast, in the 2D case the cubic symmetry is explicitly broken, and the classical state is of a spin-flop type (Fig. 46). It corresponds to alternatingly occupied orbitals on the two sublattices in the plane, with the orbitals given by $\theta = \pi/4$ in Eq. (215) at $E_z = 0$. A finite value of E_z tilts the orbitals out of the planar $|x\rangle$ orbitals by an angle ϕ, and the Hamiltonian reduces to

$$\mathcal{H}_{2D}^\phi = H_\parallel^\phi + H_\tau^\phi, \tag{227}$$

as there is no bond in the c-direction. The classical energy is

$$E_{2D}^{MF} = -\frac{1}{4} J (2\cos 4\phi + 1) + \frac{1}{2} E_z \sin 2\phi. \tag{228}$$

Therefore, one finds the same energy of $-3J/4$ as in a 3D model at orbital degeneracy. This demonstrates a particular *frustration of orbital superexchange interactions*, where the orbital energy cannot be gained from the third direction once the orbitals have been optimized with respect to the other two. The bonds along the third direction only allow for restoring the rotational symmetry in the 3D model on the classical level by rotating the orthogonal orbitals in an arbitrary way. The energy (228) is minimized by,

$$\sin 2\phi = -\frac{E_z}{4J}, \qquad (229)$$

if $|E_z| \leq 4J$; otherwise $\sin 2\phi = \pm 1$. Interestingly, the value of the field at which the orbitals are fully polarized is reduced by one third from the value obtained in three dimensions (226). This shows that although the orbital exchange energy can be gained in a 3D model only on the bonds along two directions in the alternating (orbital-flop) phase either at or close to $E_z = 0$, one has to counteract the superexchange on the bonds in all three directions when the field is applied.

The superexchange in the orbital subspace is AF and one may map the orbital terms in the Hamiltonian (221) onto a spin problem in order to treat the elementary excitations within the LSW theory. It is convenient to derive the excitations for the spin-flop phase induced by an orbital-field starting from the rotated Hamiltonian (221). Here we choose the Holstein-Primakoff transformation [3,82] for localized pseudospin operators ($T = 1/2$),

$$T_i^+ = \bar{a}_i^\dagger (1 - \bar{a}_i^\dagger \bar{a}_i)^{1/2}, \qquad T_i^- = (1 - \bar{a}_i^\dagger \bar{a}_i)^{1/2} \bar{a}_i, \qquad T_i^z = \bar{a}_i^\dagger \bar{a}_i - \frac{1}{2}, \qquad (230)$$

for $i \in A$ sublattice and

$$T_j^+ = (1 - \bar{b}_j^\dagger \bar{b}_j)^{1/2} \bar{b}_j, \qquad T_j^- = \bar{b}_j^\dagger (1 - \bar{b}_j^\dagger \bar{b}_j)^{1/2}, \qquad T_j^z = \frac{1}{2} - \bar{b}_j^\dagger \bar{b}_j, \qquad (231)$$

for $j \in B$ sublattice. In the harmonic approximation the terms $\sim T_i^z T_j^x$ do not contribute to the boson Hamiltonian as they give only odd numbers of boson operators. Therefore, the phase dependence in the terms $\propto \pm\sqrt{3}$ is lost in the LSW approximation.

After performing a Fourier transformation to $\{\bar{a}_{\vec{k}}^\dagger, \bar{b}_{\vec{k}}^\dagger\}$ operators, the Hamiltonian may be further simplified by using the symmetry in $\vec{k}$-space and introducing new boson operators,

$$a_{\vec{k}} = \frac{1}{\sqrt{2}}(\bar{a}_{\vec{k}} - \bar{b}_{\vec{k}}), \qquad b_{\vec{k}} = \frac{1}{\sqrt{2}}(\bar{a}_{\vec{k}} + \bar{b}_{\vec{k}}). \qquad (232)$$

which leads to the effective orbital Hamiltonian of the form,

$$H_{\text{LSW}} = J \sum_{\vec{k}} \left[A_{\vec{k}} a_{\vec{k}}^\dagger a_{\vec{k}} + \frac{1}{2} B_{\vec{k}} (a_{\vec{k}}^\dagger a_{-\vec{k}}^\dagger + a_{-\vec{k}} a_{\vec{k}}) \right]$$
$$+ J \sum_{\vec{k}} \left[A_{\vec{k}} b_{\vec{k}}^\dagger b_{\vec{k}} - \frac{1}{2} B_{\vec{k}} (b_{\vec{k}}^\dagger b_{-\vec{k}}^\dagger + b_{-\vec{k}} b_{\vec{k}}) \right], \qquad (233)$$

where the coefficients $A_{\vec{k}}$ and $B_{\vec{k}}$ depend on angle ϕ,

$$A_{\vec{k}} = 3 - B_{\vec{k}}, \tag{234}$$

$$B_{\vec{k}} = \frac{1}{2}\left[(2\cos 4\phi - 1)\gamma_+(\vec{k}) + (\cos 4\phi + 1)\gamma_z(\vec{k})\right], \tag{235}$$

and the $\vec{k}$-dependence is given by $\gamma_-(\vec{k})$, and by $\gamma_z(\vec{k})$, defined by Eqs. (106) and (108), respectively. After a Bogoliubov transformation,

$$a_{\vec{k}} = u_{\vec{k}}\alpha_{\vec{k}} + v_{\vec{k}}\alpha^{\dagger}_{-\vec{k}}, \qquad b_{\vec{k}} = u_{\vec{k}}\beta_{\vec{k}} + v_{\vec{k}}\beta^{\dagger}_{-\vec{k}}, \tag{236}$$

with the coefficients

$$u_{\vec{k}} = \sqrt{\frac{A_{\vec{k}}}{2\zeta_{\vec{k}}} + \frac{1}{2}}, \quad v_{\vec{k}} = -\mathrm{sgn}(B_{\vec{k}})\sqrt{\frac{A_{\vec{k}}}{2\zeta_{\vec{k}}} - \frac{1}{2}}, \tag{237}$$

where $\zeta_{\vec{k}} = \sqrt{A_{\vec{k}}^2 - B_{\vec{k}}^2}$, the Hamiltonian (233) is diagonalized and takes the following form,

$$H_{\mathrm{LSW}} = \sum_{\vec{k}} \left[\omega^-_{\vec{k}}(\phi)\alpha^{\dagger}_{\vec{k}}\alpha_{\vec{k}} + \omega^+_{\vec{k}}(\phi)\beta^{\dagger}_{\vec{k}}\beta_{\vec{k}}\right]. \tag{238}$$

The orbital-wave dispersion is given by

$$\omega^{\pm}_{\vec{k}}(\phi) = 3J\left\{1 \pm \tfrac{1}{3}\left[(2\cos 4\phi - 1)\gamma_+(\vec{k}) + (\cos 4\phi + 1)\gamma_z(\vec{k})\right]\right\}^{1/2}. \tag{239}$$

The orbital excitation spectrum consists of two branches like, for instance, in an anisotropic Heisenberg model [3]. The dependence on the field E_z is implicitly contained in the above relations via the angle ϕ, as determined by Eq. (226) and for the 3D model. The orbital-wave dispersion for a 2D system, can easily be obtained from Eq. (239) by setting $\gamma_z(\vec{k}) = 0$ and selecting ϕ according to Eq. (229). First we discuss the excitation spectra given by Eq. (239) shown in Fig. 47 for different crystal-field splittings E_z for the 3D system, along different high-symmetry directions of the fcc BZ appropriate for the alternating orbital order. Most interestingly, a gapless orbital-wave excitation is found for the 3D system at orbital degeneracy. Obviously, this is due to the fact that the classical ground state energy is independent of the rotation angle θ at $E_z = 0$. At first glance, however, one does not expect such a gapless mode, as the Hamiltonian (214) does not obey a continuous SU(2) symmetry. The cubic symmetry of the model, however, is restored if one includes the quantum fluctuations, as shown in Fig. 48. Note that the quantum corrections found in the 3D orbital model (214) are somewhat smaller than those for the 3D Heisenberg antiferromagnet. They do depend on the rotation angle θ, as the orbital-wave dispersion does.

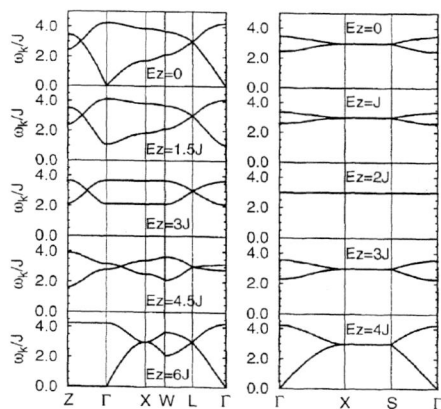

FIGURE 47. Orbital-wave excitations as obtained for different values of the crystal-field splitting E_z for a 3D (left) and 2D (right) orbital superexchange model (212). The result shown for a 3D system at $E_z = 0$ was obtained for the orbitals rotated by $\theta = \pi/4$ in Eqs. (215), and corresponds to the $E_z \to 0$ limit of the orbital-flop phase (after Ref. [183]).

As a special case, the orbital-wave dispersion for a quasi-2D situation in a 3D case can easily be obtained from Eq. (239) by eliminating the term $\propto \gamma_z(\vec{k})$ and assuming $\phi = \pi/4$,

$$\omega_{\vec{k}}^{\pm}(\phi = \pi/4) = 3J\sqrt{1 \pm \gamma_+(\vec{k})}. \qquad (240)$$

In a 3D system it applies to the ground state given by alternating $|x\rangle$ and $|z\rangle$ orbitals on the two sublattices, and one finds the largest quantum corrections, as this dispersion has a line of nodes along the $\Gamma - Z$ direction, i.e., $\omega_{(0,0,q)}^- = 0$ for $0 < q < \pi$. In higher order spin-wave theory, however, it might very well be that a gap opens in the excitation spectrum. We expect, however, that this gap, if it arises, is small, with its size being self-consistently determined by quantum fluctuations.

For the 2D system the situation at orbital degeneracy is quite different (see Fig. 47). The lack of interactions along the c-axis breaks the symmetry of the model already at $E_z = 0$, opens a gap in the excitation spectrum,

$$\omega_{\vec{k}}^{\pm}(\phi = 0) = 3J\sqrt{1 \pm \tfrac{1}{3}\gamma_+(\vec{k})}, \qquad (241)$$

and suppresses quantum fluctuations. Thus, one encounters an interesting example of a more classical behavior in lower dimension. In fact, the 1D model (214) is classical as only Ising interactions are left and the modes are dispersionless (local mode at $\omega_k = 2J$).

At increasing the orbital field $|E_z|$, the 2D and 3D system resemble each other, with a large gap in the excitation spectrum at $|E_z| = 2J$ ($|E_z| = 3J$) in the 2D (3D) system (Fig. 47). At larger $|E_z|$ the gap gradually closes when an orbital field which

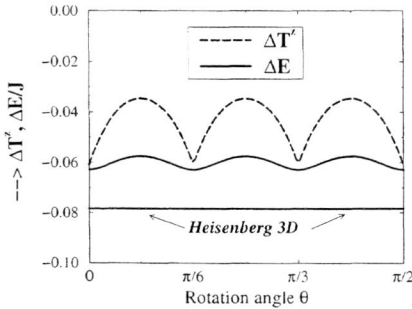

FIGURE 48. Quantum corrections for the 3D orbital model (212) as functions of rotation angle θ (215) for: the order parameter ΔT^z (full lines), and the ground-state energy $\Delta E/J$ (dashed lines) (after Ref. [183]).

compensates the energy loss due to the orbital superexchange between identical (FM) orbitals is approached. At this value of the field ($E_z = 4J$ and $E_z = 6J$ in a 2D and 3D model, respectively), the full dispersion of the orbital waves is recovered, the spectrum is gapless (Fig. 47), and the quantum fluctuations reach a maximal value. We would like to emphasize that this behavior is qualitatively different from the Heisenberg antiferromagnet, both in two and three dimensions, where the anomalous terms $\propto T_i^+ T_j^+$ and $\propto T_i^- T_j^-$ are absent which results in the conserved total spin $T^z = \sum_i T_i^z$, and quantum fluctuations vanish at the crossover from the spin-flop to FM phase. It is instructive to make a comparison between the analytic approximations of the LSW theory and the exact diagonalization (ED) of 2D finite clusters using the finite temperature diagonalization method [184,185]. As in the mean-field approach, one finds also a unique ground state for finite clusters at $E_z = 0$ by ED. Making the rotation of basis (215) is however still useful in the ED method as it gives more physical insight into the obtained correlation functions which become simpler and more transparent when calculated within an optimized basis. Moreover, they offer a simple tool to compare the results obtained by ED with those of the analytic approach. We shall present below the results obtained with 4×4 clusters; similar results were also found for 10-site clusters [186].

Let us look at nearest-neighbor correlation function in the ground state $\langle \tilde{T}_i^z \tilde{T}_{i+R}^z \rangle$, where the operators with a tilde refer to a rotated basis,

$$\begin{aligned}
\tilde{T}_i^z &= \cos 2\phi \, T_i^z + \sin 2\phi \, T_i^x \\
\tilde{T}_{i+R}^z &= \cos 2\psi \, T_{i+R}^z + \sin 2\psi \, T_{i+R}^x,
\end{aligned} \qquad (242)$$

so that the correlation function depends on two angles: ϕ and ψ. In Fig. 49 the intersite orbital correlation in the ground state is shown as a contour-plot. The intensity of the grey scale changes from positive to negative values of the correlation function, $\langle \tilde{T}_i^z \tilde{T}_{i+R}^z \rangle$. One finds that the neighbor correlations have their largest value if the orbitals are rotated by $\phi = \pi/4$ and $\psi = 3\pi/4$ (or $\phi = 3\pi/4$ and $\psi = \pi/4$), i.e., under this rotation of the basis states the system looks like a

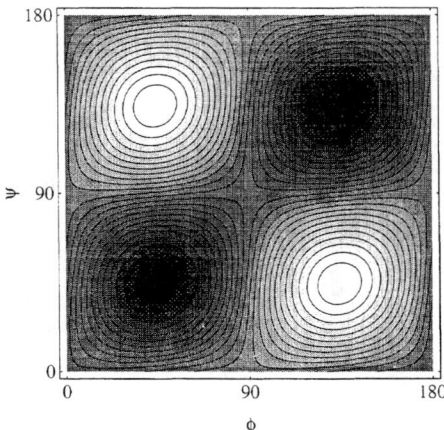

FIGURE 49. Contour plot of the *rotated* nearest-neighbor orbital correlation function $\langle \tilde{T}_i^z \tilde{T}_{i+R}^z \rangle$ as function of the angles ϕ and ψ for a 16-site planar cluster with $E_z = 0$ and $T = 0.1J$. White regions correspond to positive (FM) and black areas to negative (AF) orbital correlations, i.e., $\langle \tilde{T}_i^z \tilde{T}_{i+R}^z \rangle > 0.24$ (< -0.24), respectively. They are separated by 25 contour lines chosen with the step of 0.02 in the interval [-0.24,0.24] (after Ref. [183]).

ferromagnet, indicating that the occupied $(|x\rangle + |z\rangle)/\sqrt{2}$ and $(|x\rangle - |z\rangle)/\sqrt{2}$ orbitals are staggered in a 2D model, as shown in Fig. 46. Note that quantum fluctuations are small as in the ground state and one finds $\langle \tilde{T}_i^z \tilde{T}_{i+R}^z \rangle \simeq 0.246$. One finds that although the symmetry is not globally broken in a finite system, the short-range order resembles that found in the symmetry-broken state with orbital LRO. This demonstrates at the same time the advantage of the rotated basis in the ED study, because in the original unrotated basis one finds instead $\langle T_i^z T_{i+R}^z \rangle \simeq 0$, which might lead in a naive interpretation to a large overestimation of quantum fluctuations.

The ground-state correlations $\langle \tilde{T}_i^z \tilde{T}_{i+R}^z \rangle$ found by ED are in excellent agreement with the results of LSW theory. By comparing the results obtained at $T = 0.1J$, $0.2J$ and $0.5J$, it has been found in Ref. [183] that the calculated low-temperature correlation functions are almost identical in this temperature range, and thus the values shown in Fig. 49 for $T = 0.1J$ are representative for the ground state. They demonstrate an instability of the system towards the symmetry-broken state.

In order to verify the accuracy of the LSW approach for finding the excitation spectrum, we discuss the results for the dynamical orbital response functions in the case of orbital degeneracy. The transverse response function for the orbital excitations evaluated with respect to the *rotated* local quantization axes (242) is defined as follows,

$$\tilde{T}_{\vec{q}}^{+-}(\omega) = \frac{1}{2\pi} \int_{-\infty}^{\infty} dt \, \langle \tilde{T}_{\vec{q}}^+ \tilde{T}_{-\vec{q}}^-(t) \rangle \, \exp(-i\omega t). \tag{243}$$

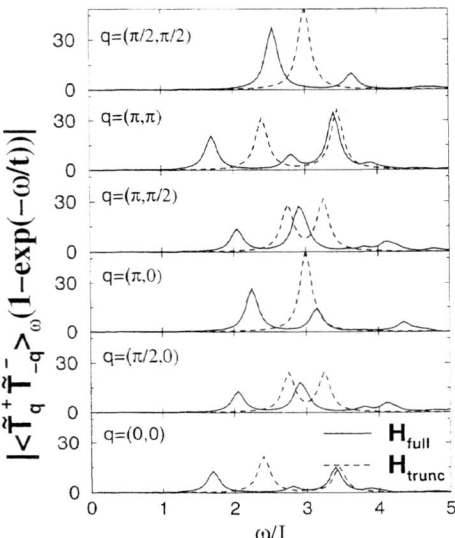

FIGURE 50. Transverse response function $\tilde{T}_q^{+-}(\omega)$ for the rotated orbitals as a function of frequency ω for different momenta at low temperature. Calculations were performed for a 16-site 2D cluster at $E_z = 0$ and $T = 0.1J$ for: (a) the 2D orbital model given by Eq. (214) (H_{full}, full lines), and (b) neglecting the mixed terms $\propto T_i^x T_j^z$ (H_{trunc}, dashed lines). The spectra are broadened by $\Gamma = 0.1J$ (after Ref. [183]).

As we have already mentioned, the LSW approximation does not allow to investigate the consequences of the coupling of single excitonic excitations to the order parameter, represented by the terms $\propto T_i^x T_j^z$. Therefore, strictly speaking the LSW approach corresponds to the truncated Hamiltonian when such terms are not included. The ED gives then a double-peak structure in the response function $\tilde{T}_q^{+-}(\omega)$ which agrees well with the dispersion of two modes found in the LSW approach (Fig. 50). However, if the terms $\propto T_i^x T_j^z$ are included, one finds different structures – the lowest energy excitation moves to lower energies, and satellite structures appear which describe the incoherent processes in orbital dynamics.

In spite of some additional incoherent processes in the spectra, the first moment of $\tilde{T}_q^{+-}(\omega)$ determined by ED agrees very well with the dispersion found within the LSW theory [as determined from Eq. (241)] (Fig. 51). The values of the first moments are only slightly changed when instead the truncated Hamiltonian (without the processes $\propto T_i^x T_j^z$) is used in a numerical approach. This comparison demonstrates that the LSW approach captures the leading term in the orbital dynamics and may be thus used to investigate the consequences of orbital excitations on the hole dynamics, as presented in the next Section.

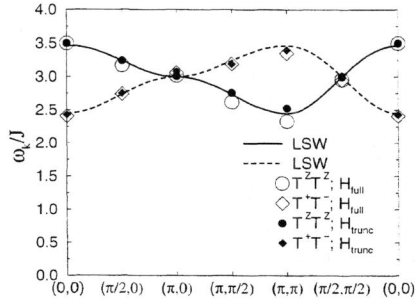

FIGURE 51. Dispersion of orbital waves along the main directions in the 2D BZ. Results for the first moment calculated for a 4×4 cluster with the full (H_{full} and truncated (H_{trunc} Hamiltonian (empty and full symbols) are compared with the dispersions for the two modes found using the LSW theory (solid and dashed line) (after Ref. [183]).

B Hole propagation in an orbital ordered state

As an interesting example of the consequence of orbital degrees of freedom in the hole excitation spectra, the problem of hole propagation in an orbital ordered 2D FM planes, as found in LaMnO$_3$, has been considered recently by van den Brink, Horsch and Oleś [162]. The main idea is that the hole which moves in an orbital ordered plane [Fig. 53(a)] dresses by orbital excitations, and a polaron is formed. Its properties will depend on the actual parameters and it is interesting to investigate: (i) whether a quasiparticle (QP) state will form under these circumstances in analogy with the spin problem – a hole moving in a quantum antiferromagnet [187], and (ii) how much the spectral function changes when the orbital degeneracy is removed.

Let us consider the simplest situation and assume that the e_g orbitals are degenerate, $E_z = 0$. If only few holes are doped to a FM plane as in the A-AF phase of La$_{1-x}$Ca$_x$MnO$_3$ with $x \ll 1$, the ground state is determined in first instance by a 2D version of the orbital model (214),

$$H_J = \tfrac{1}{2}J \sum_{\langle ij \rangle} \left[T_i^z T_j^z + 3T_i^x T_j^x \mp \sqrt{3}(T_i^x T_j^z + T_i^z T_j^x) \right], \tag{244}$$

where the bonds $\langle ij \rangle$ connect nearest neighbors in a single (a,b) plane, and the AF superexchange J implies that the orbitals order. In order to see the consequences of this ordered state for the kinetic energy of doped holes, it is convenient to transform the hopping term (38) along the bonds $\langle ij \rangle$ in the (a,b) FM planes to the basis defined by Eqs. (220) at $\phi = 0$ which corresponds to the symmetry-broken ground state with the orthogonal orbitals $|i\mu\rangle$ and $|j\nu\rangle$ occupied on the two sublattices (Fig. 46). Let us introduce the new fermion (hole) operators which correspond to the occupied orbitals $|i0\rangle$ as $f_{i0}^\dagger$, and the operators which correspond to the excited states $|i1\rangle$ as $f_{i1}^\dagger$. Thus, the operator $f_{i0}^\dagger$ create a hole in the orbital $|i\mu\rangle$ for $i \in A$ and $|i\nu\rangle$ for $i \in B$, respectively. The transformed kinetic energy is [162],

$$H_t = \tfrac{1}{4}t\sum_{\langle ij\rangle}\left[f_{i0}^\dagger f_{j0} + f_{i1}^\dagger f_{j1} + 2(f_{i1}^\dagger f_{j0} + f_{i0}^\dagger f_{j1})\right.$$
$$\left.\pm\sqrt{3}\left(f_{i1}^\dagger f_{j0} - f_{i0}^\dagger f_{j1}\right) + \text{H.c.}\right]. \tag{245}$$

Together with the usual crystal-field splitting term H_τ (59), Eqs. (244) and (245) define the *orbital t-J model*,

$$\mathcal{H} = H_t + H_J + H_\tau. \tag{246}$$

The first interesting observation is that the hole motion is not completely suppressed by the orbital ordering, unlike in the spin t-J model. The processes $\propto f_{i0}^\dagger f_{j0}$ concern the occupied orbitals and thus a hole may always interchange with an electron without disturbing the orbital ordering [Fig. 53(b)]. These processes lead to a band in the limit of $U \to \infty$, where the constraint of no double occupancy is implemented with the slave-boson operators b_{i0} and b_{i1}, standing for the orbital flavors which accompany the hole operators $h_i^\dagger$ according to the prescription:

$$f_{i0}^\dagger = b_{i0}h_i^\dagger, \qquad f_{i1}^\dagger = b_{i1}h_i^\dagger. \tag{247}$$

In the orbital ordered state the b_{i0} bosons are condensed, $b_{i0} = 1$, which leads after a Fourier transformation from the hole operators $h_i^\dagger$ to $h_{\vec{k}}^\dagger$ to a free hole propagation in the lower Hubbard band,

$$H_h = \sum_{\vec{k}} \varepsilon_{\vec{k}}^0(\phi) h_{\vec{k}}^\dagger h_{\vec{k}}, \tag{248}$$

with a dispersion determined by the orbital order via

$$\varepsilon_{\vec{k}}^0(\phi) = (-2\sin 2\phi + 1)t\gamma_+(\vec{k}), \tag{249}$$

where $\gamma_+(\vec{k})$ is defined by Eq. (106). The angle ϕ (220) depends on the optimal orbitals (229): $\phi = 0$ at the orbital degeneracy ($E_z = 0$), while $\phi \neq 0$ if the orbital degeneracy is removed by a finite field $E_z \neq 0$. Note that the largest dispersion is found for $|x\rangle$ orbitals at $\phi = -\pi/4$, while the dispersion vanishes at $\phi = \pi/8$ due to the conflicting phases of $|x\rangle$ and $|z\rangle$ orbitals. If the other processes which couple the moving hole to the orbital excitations could be neglected, the band (248) would give a coherent spectral function of a hole. Furthermore, an electron doped at $n = 1$ would propagate in the upper Hubbard band by a similar dispersion, i.e., due to the term $\propto f_{i1}^\dagger f_{i1}$ in Eq. (245). These subbands are separated by the Coulomb interaction U which acts between two e_g states, in this case between the occupied and unoccupied orbital states,

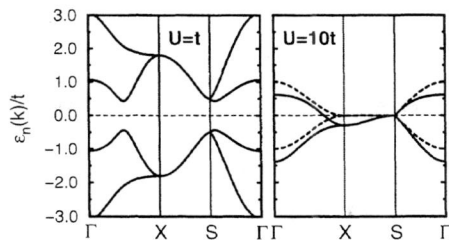

FIGURE 52. Dispersion relations in the reduced BZ [$X = (\pi, 0)$, $S = (\pi/2, \pi/2)$] in the mean-field approximation which simulates the treatment of Coulomb repulsion in LDA+U as obtained for: $U = t$ (left), and in the lower Hubbard band centered at $\omega = 0$ for $U = 10t$ (right, full lines). In the right panel the dispersion obtained in the $U \to \infty$ limit is shown by dashed lines for comparison (after Ref. [162]).

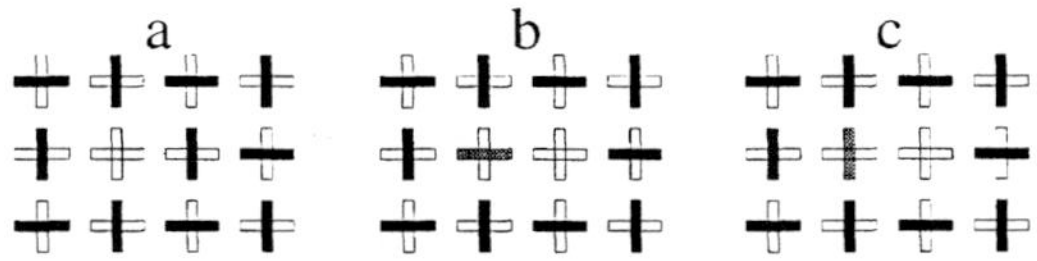

FIGURE 53. A single hole added in an orbital-ordered ground state (a); the occupied (empty) orbitals $|\mu\rangle$ and $|\nu\rangle$ are shown as filled (empty) rectangles. The hole can move either without disturbing the orbital order (b), or by creating orbital excitations (c) (after Ref. [162]).

$$H_U = U \sum_i f_{i0}^\dagger f_{i0} f_{i1}^\dagger f_{i1}. \tag{250}$$

We recall, however, that the interorbital Coulomb interaction is invariant with respect to the choice of orbital basis only if the pair-hopping terms in Eq. (39) are included [32]. In the present case with the occupied $|i0\rangle$ orbitals at $n = 1$, we may take $f_{i0}^\dagger f_{i0} = 1$, and the Coulomb term reduces to a local potential which acts on the *unoccupied* states $|i1\rangle$. When the hopping Hamiltonian (245) is supplemented by this potential $\propto U$, one finds that the bands are separated by a gap which opens at $U = 0$ (Fig. 52). The bands change drastically as a function of U [162]: if U is small, the bands resemble the uncorrelated problem (178) of Sec. VI.A, while the shape of the lower Hubbard band becomes close to the $U \to \infty$ limit (248) already at $U/t \approx 10$ which may be taken a representative value for manganites [23]. These changes of the bands between the small and large U regime simulate the effect of the local potentials which act on the unoccupied states in the LDA+U method [35]. Thus, one finds a free hole band shown in Fig. 52 and the question is how this band changes when the hole starts to dress by orbital excitations of Sec.VII.A [Fig. 53(c)]. It may be expected that the free hole dispersion (248) is drastically modified by the hole-orbiton coupling which allows the hole propagation by frustrating the orbital order, just as the AF order is locally disturbed in the t-J model, and the 'string' of excited bonds is created on the path of the hole [160]. This problem was recently investigated in the orbital model by van den Brink, Horsch, and Oleś [162] who have shown that the interaction between the hole and the orbitals is so strong in the manganites that propagating holes are dressed with many orbital excitations and form polarons that have large mass, small bandwidth and low QP weight.

The orbital background is described by a local constraint for $T = 1/2$ pseudospins,

$$b_{i0}^\dagger b_{i0} + b_{i1}^\dagger b_{i1} = 2T, \tag{251}$$

where $b_{i0}^\dagger$ and $b_{i1}^\dagger$ are boson operators which refer to the occupied and empty state at site i (247). By making the lowest order expansion of the constraint around the ground state with orbital ordering, one recovers the linear orbital-wave (LOW) approximation analyzed in Sec. VII.A. Having chosen the occupied orbitals as $b_{i0}^\dagger b_{i0} \simeq 1$ bosons at every site, the expansion is the same for both sublattices and reads,

$$T_i^x = \tfrac{1}{2}(b_i + b_i^\dagger), \qquad T_i^z = T - b_i^\dagger b_i, \tag{252}$$

with $b_i \equiv b_{i1}$ playing the role of a Holstein-Primakoff boson. It corresponds to the spin problem with the spins rotated by π at one of the sublattices [3,82]. The resulting effective boson Hamiltonian is diagonalized by a Fourier and Bogoliubov transformation (Sec. VII.A). One finds therefore an interacting problem in LOW approximation,

$$\mathcal{H}_{\text{LOW}} = H_h + H_o + H_{ho}, \tag{253}$$

where the orbital waves (orbitons) for a 2D model are given by,

$$H_o = \sum_{\vec{k}} \omega_{\vec{k}}(\phi) \alpha_{\vec{k}}^\dagger \alpha_{\vec{k}}, \tag{254}$$

with

$$\omega_{\vec{k}}(\phi) = 3J \left[1 + \tfrac{1}{3}(2\cos 4\phi - 1)\gamma_+(\vec{k})\right]^{1/2}, \tag{255}$$

standing for the orbiton dispersion. The single mode written now for convenience in the full BZ in Eq. (254) is equivalent to two branches of orbital excitations obtained in the folded zone in Sec. VII.A [183]. The orbital excitations depend sensitively on the orbital splitting E_z. At orbital degeneracy ($E_z = 0$) one finds a maximum of $\omega_{\vec{k}}(\phi)$ at the $\Gamma = (0,0)$ point and a weak dispersion $\sim J$ [see also Eq. (241)]. In contrast, for $E_z = \pm 2J$ orbital excitations are dispersionless, and $\omega_{\vec{k}} = 3J$. The remaining part of Eq. (253) describes the hole-orbiton interaction [Fig. 53(c)],

$$H_{ho} = t \sum_{\vec{k},\vec{q}} h_{\vec{k}-\vec{q}}^\dagger h_{\vec{k}} \left[M_{\vec{k},\vec{q}} \alpha_{\vec{q}}^\dagger + N_{\vec{k},\vec{q}} \alpha_{\vec{q}+\vec{Q}}^\dagger + \text{H.c.}\right], \tag{256}$$

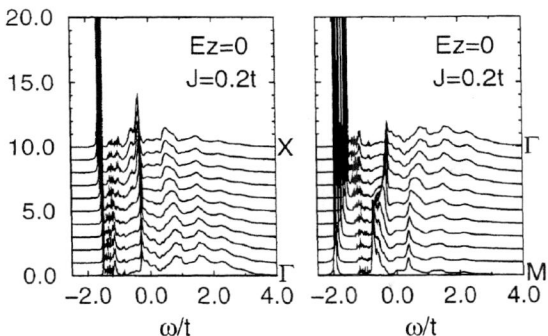

FIGURE 54. Spectral functions for the orbital t-J model as obtained in two high-symmetry directions of the 2D BZ: $\Gamma - X$ (left) and $M - \Gamma$ (right) for $J = 0.2t$ and $E_z = 0$ [$\Gamma = (0,0)$, $X = (\pi, 0)$, $M = (\pi, \pi)$] (after Ref. [162]).

where $\vec{Q} = (\pi, \pi)$, the vertex functions are:

$$M_{\vec{k}, \vec{q}} = 2 \cos 2\phi \left[u_{\vec{q}} \gamma_+(\vec{k} - \vec{q}) + v_{\vec{q}} \gamma_+(\vec{k}) \right], \tag{257}$$

$$N_{\vec{k}, \vec{q}} = -\sqrt{3} \left[u_{\vec{q}} \gamma_-(\vec{k} - \vec{q}) - v_{\vec{q}} \gamma_-(\vec{k}) \right], \tag{258}$$

and $\gamma_-(\vec{k}) = \gamma_+(k_x, k_y + \pi)$ is defined by Eq. (107).

In this way, the orbital t-J model (246) leads to an effective Hamiltonian (253), describing a many-body problem due to the hole-orbiton coupling term H_{ho}. Although the analytic structure and the form of Eq. (253) resembles the usual t-J model written in the slave fermion formalism [187], there are important differences. First of all, a hole may propagate freely in the orbital model by the term (249), as it would be also the case in the quantum antiferromagnet with further-neighbor hopping. Second, the orbital waves do not obey the nesting symmetry, i.e., $\omega_{\vec{k}+\vec{Q}}(\phi) \neq \omega_{\vec{k}}(\phi)$, where $\vec{Q} = (\pi, \pi)$, and are more classical than the spin waves, with a finite gap in the excitation spectrum (255). Furthermore, the hole-orbiton interaction (256) has a richer analytic structure than that of the t-J model, as the scattering processes which conserve the momentum modulo $\vec{Q}$ due to a new vertex $\propto N_{\vec{k}, \vec{q}}$ (258). Finally, an important feature is also that both hole dispersion $\varepsilon_{\vec{k}}^0(\phi)$ and orbiton dispersion $\omega_{\vec{k}}(\phi)$ depend on the crystal-field splitting E_z, and thus the analytic structure is richer. As a special case, at $E_z = -2J$ both modes are dispersionless, and the model orbital becomes equivalent to a hole which moves in a classical antiferromagnet described by the Ising model, with no quantum fluctuations [188]. The many-body problem obtained for a hole propagating in an orbital ordered background (253) may be solved using the self-consistent Born approximation (SCBA) [189,190]. This method gives results of high quality and compares favorably with ED for the hole which moves in a quantum antiferromagnet [187]. Treating $\mathcal{H}_{\text{LOW}}$ in the SCBA, one finds the selfenergy [162],

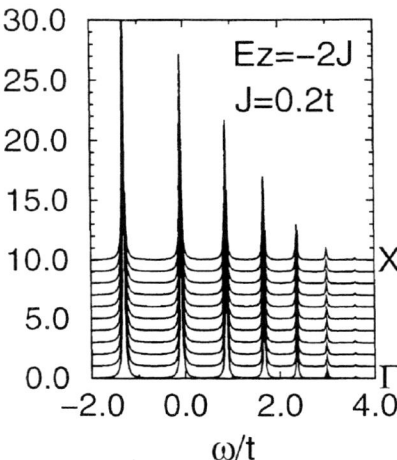

FIGURE 55. Ladder spectrum for the orbital t-J model obtained for $E_z = -2J$ and $J = 0.2t$ (after Ref. [162]).

$$\Sigma(\vec{k}, \omega) = t^2 \sum_{\vec{q}} \left[M^2_{\vec{k}, \vec{k}-\vec{q}} G(\vec{k} - \vec{q}, \omega - \omega_{\vec{q}}) + N^2_{\vec{k}, \vec{k}-\vec{q}} G(\vec{k} - \vec{q}, \omega - \omega_{\vec{q}+\vec{Q}}) \right]. \tag{259}$$

Here $G(\vec{k}, \omega)$ stands for the hole Green function which obeys the Dyson equation,

$$G^{-1}(\vec{k}, \omega) = \omega - \varepsilon^0_{\vec{k}}(\phi) - \Sigma(\vec{k}, \omega). \tag{260}$$

Eqs. (259) and (260) represent a closed set of equations which has to be solved numerically by iteration on a lattice. We discuss below some representative results obtained recently by van den Brink et al. [162]. First of all, the hole spectral function is drastically changed by the coupling processes to the orbital excitations which is particularly strong when the orbital superexchange J is much lower than the hopping t. Examples of rather complex spectra are shown in Figs. 54–56 for $J/t = 0.1$ which corresponds to the realistic parameters for manganites [23]. As in the t-J model, a QP state appears at the threshold, and this results from a strong dressing of a hole by a cloud of orbital excitations. The incoherent part extends over the scale of $\sim 6t$ which corresponds to the full dispersion in the e_g band (178).

The energy and momentum dependence of the incoherent spectra is markedly different from the spin t-J problem. At orbital degeneracy (Fig. 54) particular satellite structures are obtained with a rather weak $\vec{k}$-dependence. Their origin may be understood by looking at the hole spectrum found at $E_z = -2J$ which corresponds to the dispersionless orbiton spectrum (255) and to the vanishing hole dispersion (249) due to the conflicting phases in the e_g electron hopping. In this case a ladder spectrum of the t-J^z model is reproduced [188,187] (Fig. 55), but

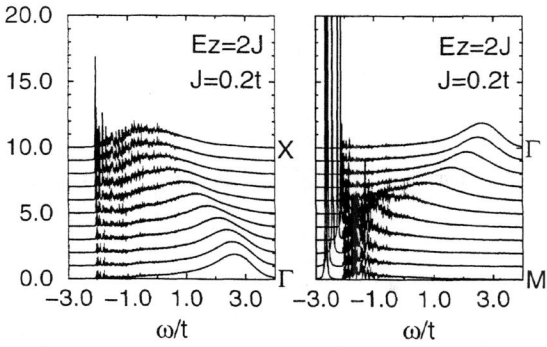

FIGURE 56. The same as in Fig. 54, but for $E_z = +2J$ (after Ref. [162]).

the values of t and J^z have to scaled for the orbital model and the present case corresponds to a larger ratio $J^z/t \simeq 0.3$ in the spin problem [162]. Such ladder features at decreasing distances result in the pronounced satellites observed still in the spectra at orbital degeneracy. In contrast, with increasing value of E_z, the incoherent part changes to a rather smooth curve dominated by a broad maximum that corresponds to the free dispersion which is still visible but strongly damped by the hole scattering on the orbital excitations. An example of such spectra is shown in Fig. 56. The QP states have a narrow dispersion $\propto J$, as in the t-J model (Fig. 57). Note that the QP's are well defined and have a large weight a_M close to the minimum of the QP band at the $M = (\pi, \pi)$ point, while they are weaker at higher energies. As in the t-J model [187], at the maximum of the QP band, found here at the $\Gamma = (0,0)$ point, the QP has the lowest spectral weight a_Γ. Note that the QP minimum is here determined by the minimum of free dispersion (Fig. 52) rather than by the enhanced quantum fluctuations at the boundary of the folded BZ, as encountered in the spin t-J model.

The effective mass m^* which may be defined be the momentum dependence of the QP energy [162], and the QP bandwidth W^* increase first linearly with increasing J in the range of $J/t < 0.3$, while they approach the free values in the weak-coupling regime of $J/t > 1$. However, an additional dependence on the orbital splitting makes the QP dispersion rather narrow close to the orbital degeneracy ($E_z = 0$) and at $E_z < 0$, while it broadens up when $E_z > 0$ and the uniform phase with $|x\rangle$ orbitals occupied is approached. This has interesting consequences for the experimental situation and suggests that the photoemission spectra of manganites should have a strong dependence on the deviations from the cubic symmetry which remove the orbital degeneracy ($E_z \neq 0$). Furthermore, the dependence on the doping might be very interesting in these situations when the orbital ordering changes, as for instance in the layered compounds [179].

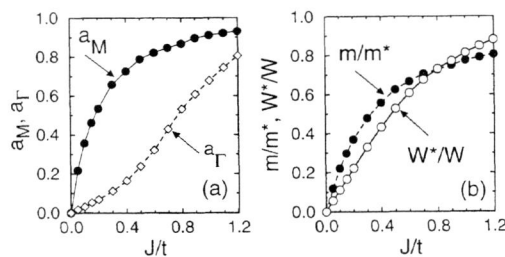

FIGURE 57. Quasiparticle properties at $E_z = 0$ as functions of J/t: (a) weights of the two QP bands at the Γ point: a_M (filled circles), and a_Γ (open squares); (b) inverse effective mass m/m^* (full circles) and the width of QP band W^*/W (empty circles), normalized by the LDA+U values m and W, respectively (after Ref. [162]).

VIII PHASE DIAGRAMS OF MANGANITES – OPEN PROBLEMS

A Magnetic, orbital, and charge ordering in CE-phase

Over the last few years much attention has been focused on the interplay between charge and orbital ordering occurring in half-doped manganites ($x = 0.5$). A direct evidence of the charge ordered (CO) state in half-doped manganite has been provided by the electron diffraction for $La_{0.5}Ca_{0.5}MnO_3$ [191]. Similar observations have also been reported for $Pr_{0.5}Sr_{0.5}MnO_3$ [192] $Nd_{0.5}Sr_{0.5}MnO_3$ [193], and for $Pr_{1-x}Ca_xMnO_3$ with $x = 0.4$ and 0.5 [194]. This CO state is characterized by an alternating arrangement of Mn^{3+} and Mn^{4+} ions in (a, b) planes and the charge stacking in c-direction. In CO state these systems show an insulating behavior with a very peculiar form of AF spin ordering. The observed magnetic structure is the so-called magnetic CE phase and consists of quasi-1D FM zigzag chains coupled antiferromagnetically in both directions. In addition, the occupied orbitals at Mn^{3+} positions show in these systems $d_{3x^2-r^2}/d_{3y^2-r^2}$ orbital ordering, staggered along the FM chains.

The insulating CO state can be transformed into a metallic FM state either by doping, or by applying an external magnetic field [1]. Other interesting observations were done by studying $Pr_{1-x}(Ca_{1-y}Sr_y)_xMnO_3$ crystals with controlled one-electron bandwidth. As already mentioned above, at half-doping $Pr_{0.5}Ca_{0.5}MnO_3$ has a CO CE-type insulating state. However, by substitution Ca with Sr leading to the increase of the carrier bandwidth, one induces the collapse of the CO insulating state, and the A-type metallic state with $d_{x^2-y^2}$ orbital ordering is realized in $Pr_{0.5}Sr_{0.5}MnO_3$ [167]. The coexistence of the A-type spin ordered and CE-type spin/charge ordered states has been detected in the bilayer $LaSr_2Mn_2O_7$ [195] and in 3D $Nd_{0.5}Sr_{0.5}MnO_3$ [193]. These results indicate the competition between the metallic A-type (uniform) $d_{x^2-y^2}$ orbital ordering, and the insulating

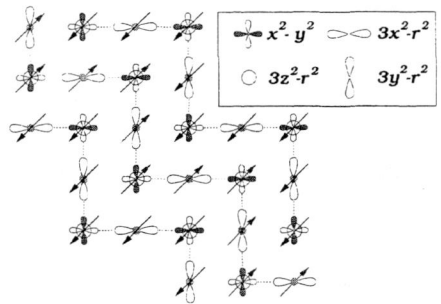

FIGURE 58. View of the CE phase in the (a, b) plane. The $x^2 - y^2$ orbitals at corner sites have positive (white) and negative (grey) lobes, while the phases of the other orbitals are positive. The grey dots at the bridge sites represent a charge surplus (after Ref. [170]).

CE-type $d_{3x^2-r^2}/d_{3y^2-r^2}$ orbital ordering at half-doping and demonstrate the importance of the coupling between magnetic, charge and orbital ordering in these compounds. Being experimentally well established, theoretically, however, the nature of the CO state in half-doped manganites and the origin of the unconventional zigzag magnetic structure remain challenging open problems. As shown in Fig. 58, the CE-phase unit cell contains two geometrically inequivalent sites [170], so-called bridge and corner sites. Note that the specific choice of basis orbitals shown in Fig. 58 is motivated by the convenience of this basis for the calculation of electronic structure. The expectation values of actual observables such as the charge distribution and the type of occupied orbitals are, of course, independent of the choice of basis Wannier orbitals.

It is straightforward to solve the band structure problem at no electron interactions [196]. The hopping elements follow from the Slater-Koster rules [16], and are shown in Fig. 59. The kinetic energy H_t is given by the same expression as in Eq. (38), with the hopping integrals connecting now the orbitals along a single FM chain of the CE phase. We note that the unit cell consists of four atoms, but the electronic problem is simplified by a particular choice of orbital phases [170]. The orbitals perpendicular to the chain direction at the bridge positions $|\xi\rangle$ are decoupled from the directional $|\zeta\rangle$ orbitals and from $|x\rangle$ and $|z\rangle$ orbitals at corner sites – hence the bands are obtained by the solution of a 3×3 matrix. One finds two bands with energies $\epsilon_{\pm}(k) = \pm t\sqrt{2 - \cos 2k}$, where k is the wave vector ($0 < k \leq \pi/2$), and two nondispersive bands at zero energy. In Fig. 59 we reproduce the bands reported in Ref. [170], where a different convention was used and the gap was found at $k = \pi/2$; the conventional definition of momentum k, however, leads with the alternating phases of $x^2 - y^2$ orbitals at corner sites to the gap at $k = 0$.

At $x = 0.5$ the $\epsilon_{-}(k)$ band is fully occupied, and all other bands are empty. The system is insulating as the occupied and empty bands are separated by a gap $\Delta = t$ at $k = 0$. In the C and CE phase the FM chains are decoupled due to the DE mechanism (Sec. II.B), and the kinetic energy is supplemented by the magnetic

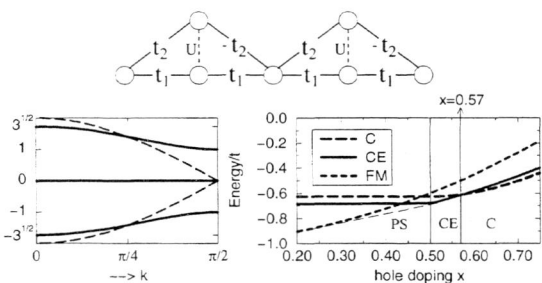

FIGURE 59. Top: topology of the hopping in a zigzag chain, where $t_1 = t/2$, $t_2 = t\sqrt{3}/2$, and U is the Coulomb interaction (261). Bottom left: electron dispersion in the zigzag chain of the CE phase for $U = 0$, and electron dispersion in a C phase (dashed line). Bottom right: total energy per site for the CE, C and FM phase for $t/J = 5$. The Maxwell construction in the phase separated (PS) region is shown by the thin dashed line (after Ref. [170]).

energy, equal for both structures. However, the opening of the gap at the Fermi-energy in the CE phase lowers its kinetic energy if the lowest band is filled, i.e., the system is half-doped. This mechanism is equivalent to the situation known from the lattice-Peierls problem, where the opening of a gap stabilizes the ground state with a lattice deformation. In the half-doped manganites, however, the gap is a direct consequence of the symmetry of the e_g wave functions and is therefore a very robust feature. In the insulating state at $x = 0.5$ the average occupancy of the $|\zeta\rangle$ orbital at the bridge site $(3x^2 - r^2$ or $3y^2 - r^2)$ is $n_b = \langle n_{m\zeta}\rangle = 1/2$, while the orthogonal $|\xi\rangle$ orbitals at bridge positions are empty. The charge is thus uniformly distributed, with the average occupancy at the corner sites equal to $n_c = \langle n_{ix} + n_{iz}\rangle = 1/2$, and the ratio between $|x\rangle$ and $|z\rangle$ occupancy of $n_x : n_z = \sqrt{3} : 1$. This ratio reflects simply the ratio of hopping amplitudes between the bridge $|\zeta\rangle$ orbital and both orbitals at the corner positions. This symmetry in charge distribution is removed when the Coulomb interaction U between e_g electrons occupying different orbitals at the same site is taken into account. Using the above notation we write the Coulomb interaction as follows,

$$H_U = U \left(\sum_{m \in B} n_{m\zeta} n_{m\xi} + \sum_{i \in C} n_{ix} n_{iz} \right), \quad (261)$$

where $n_{i\alpha} = c_{i\alpha}^\dagger c_{i\alpha}$ are the respective electron number operators, and the summations run over bridge $(m \in B)$ and corner $(i \in C)$ positions of zigzag chains. For the e_g electrons in the manganites $U \approx 10t$, so that the system is strongly correlated. We have verified using a HF approximation [197] that the interactions have a very different effect on the corner and bridge-sites. On the bridge positions the $|\xi\rangle$ orbitals are always empty, so that the Coulomb repulsion is ineffective. In contrast, both orbitals ($|x\rangle$ and $|z\rangle$) are partially occupied on the corner sites, and thus the total electron density n_c decreases with increasing U. The same effect was found by ED of finite clusters and with the Gutzwiller projection method by van den Brink, Khaliullin, and Khomskii [170].

The phase diagram at different doping and $U = 0$ in depicted in Fig. 59. We observe that CE-phase is stable only in the nearest vicinity of half-doping. For $x > 1/2$ the holes that are doped into the lower ϵ_--band efficiently suppress the CO state. In this doping range the CE phase becomes unstable with respect to the C phase, and the kinetic energy of the C phase is lower for $x > 0.57$ (Fig. 59). For $x < 1/2$ the energy per site of the CE phase is constant because the extra electrons are doped in the nondispersive bands at zero energy; this is reflected by a kink obtained in the energy as a function of doping at $x = 1/2$. For lower hole-doping (higher electron concentration) the homogeneous FM phase is more stable, as expected.

The interest in the origin of CO is also motivated by recent experimental data. The charge order functions are markedly stronger than the orbital fluctuations, both below and above the magnetic transition [194]. Thus it appears that the transition is driven by CO fluctuations, and the orbital state follows. Recognizing that the on-site Coulomb interaction between the orbitals at corner sites (261) leads to the CO state, we point out that another possibility to stabilize this state follows from the long-range Coulomb repulsion. Thus we consider the two-orbital FM Kondo lattice model (173) filled by $n = 0.5$ electron per site,

$$H = - \sum_{ij\alpha\beta\sigma} t_{ij}^{\alpha\beta} c_{i\alpha\sigma}^{\dagger} c_{j\beta\sigma} - J_H \sum_{i\alpha\sigma\sigma'} \vec{S}_i \cdot c_{i\alpha\sigma}^{\dagger} \vec{\sigma}_{\sigma\sigma'} c_{i\alpha\sigma'} + J_{AF} \sum_{\langle ij \rangle} \vec{S}_i \cdot \vec{S}_j + V \sum_{\langle ij \rangle} n_i n_j, \quad (262)$$

which is extended by the intersite Coulomb repulsion V. On-site Coulomb interaction is not included, however, so the previous mechanism is absent. Studying the model (262) within the MF approximation, a competition between different types of magnetic ordering was established [198]. Let us compare the phase diagrams obtained for a nondegenerate Kondo lattice model (23) extended by a similar intersite Coulomb interaction term $\propto V$, and for the two-orbital model (262). In the case of the one-orbital model the MF theory predicts a 'continuous increase of the CO due to the intersite Coulomb interaction V. Depending on the value of the intersite Coulomb interaction V and on superexchange coupling J_{AF}, different types of spin ordering shown in Fig. 60 (A-AF, C-AF, G-AF, and FM) coexist with the CO in the ground state of the one-orbital model. In contrast, in the two-orbital model the transition to the CO states occurs only at a finite critical value of V; thus magnetic states are obtained either with or without CO, depending on the value of V. The presence of orbital degeneracy with the peculiar anisotropic hopping amplitudes between e_g orbitals introduces a new magnetic state which was absent in the nondegenerate model, CE-type spin ordering (Fig. 60). In contrast to the nondegenerate case, the C-AF ordering is never found within the model (262) due to its instability against the effective "dimerization" and the onset of the zigzag FM order. The alternation of the FM bonds in $a(b)$ directions leads to the alter-

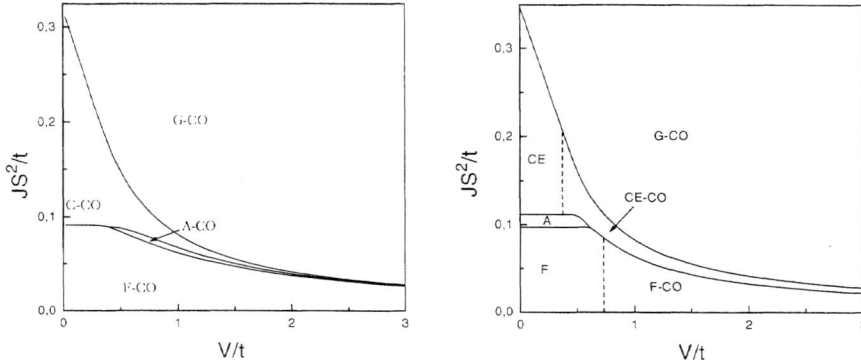

FIGURE 60. Phase diagram of the one-orbital model (left) two-orbital model (262) (right) for different values of JS^2/t ($J = J_{AF}$) and V/t parameters. Here A, C, and G stand for different magnetic ordering: A-AF, C-AF, G-AF and FM, respectively, and CO is charge ordering. The dashed line in the two-orbital case separates the uniform phase from the CO states in corresponding magnetic phases (after Ref. [198]).

nation of the hopping amplitude. As a result the bare band is split into bonding and antibonding states and the "dimerization" gap opens on the Fermi surface at half-doping. The CE-type spin ordered state is accompanied with $d_{3x^2-r^2}/d_{3y^2-r^2}$ orbital ordering that fits naturally to the topology of the zigzag structure as it was discussed above. The CE phase with magnetic and charge ordering as shown in Fig. 58 also wins over the charge disordered A-AF ordering in the regime of intermediate parameters. One may speculate that the competition between A-AF charge disordered and CE CO states indicates that parameters of the system are close to the critical values. Therefore, the small change of the bandwidth or the coupling to the lattice might stabilize one or the other state.

We have to conclude that these answers concerning the stability and the properties of the CE phase have to be treated as rather preliminary. The insulating behavior which follows from a topological phase factor in the hopping is certainly an interesting observation, and the CE phase seems to be indeed a particular type of order driven by the degeneracy of e_g orbitals. The on-site and intersite Coulomb interactions are likely to help each other in stabilizing the CO states [199]. Although the intersite interaction stabilizes the CO state in a 2D lattice, it favors the same charge alternation in the third direction, contrary to the structure of the CE phase. This qualitative trend may by reversed only by other interactions. However, the precise role of different Coulomb interactions and of the JT effect in the stability of the CE phase [200] are not understood at present and have to be established by future studies.

B   Stripes in manganites

As we have discussed above, one of the unique aspects of physics of manganites is the unusually strong interaction between charge carriers and lattice degrees of freedom, due to which markedly distinct types of charge, orbital and magnetic ordering are observed in different doping regimes. The strong electron-phonon coupling, which can be tuned by varying the electronic doping, electronic bandwidth and disorder, gives rise to a complex phenomenology, in which crystallographic structure, magnetic structure and transport properties are intimately interrelated. Below a certain temperature T_{CO}, electronic carriers become localized onto specific sites, which display LRO throughout the crystal structure (CO states). Moreover, the filled e_g ($3z^2 - r^2$-like) orbitals at Mn^{3+} ions and the associated lattice distortions (elongated Mn–O bonds) also develop LRO (orbital ordering). Finally, the magnetic exchange interactions between neighboring Mn ions, mediated by oxygen ions, become strongly anisotropic which gives rise to complex magnetic ordering in the stable structures. Historically, magnetic ordering was the first to be experimentally investigated in manganites [145], where the magnetic structures of a series of manganese perovskites with general formula $La_{1-x}Ca_xMnO_3$ were studied using neutron powder diffraction. It was not until recently, however, that the crystallographic superstructure of this compound was experimentally observed by electron diffraction [201]. The curious static patterns in the spin, charge, and orbital densities observed in manganites are currently attracting much attention [202,203]. It is very interesting, that the charge and orbital ordering (COO) observed experimentally is always concomitant with the stripe-AF (S-AF) phase. At $x = 1/2$, the COO, known as a CE phase (Sec. VII.B), has been confirmed by the synchrotron X-ray diffraction experiment [204]. At $x = 2/3$, however, two different COO patterns have been reported: (i) the *bistripe* (BS) structure [202] (Fig. 61), in which the main building block of the COO pattern at $x = 1/2$ persists even at $x = 2/3$, and (ii) the *Wigner-crystal* (WC) structure [205], in which the COO occurs with the maximized distance between e_g electrons. The appearance of these two different structures suggests that: (i) the corresponding energies are very close to each other and the ground state has (quasi-)degeneracy, and (ii) the conversion between them is prohibited by a large energy barrier. Under these circumstances, it is of limited relevance to investigate which of the two states has a lower energy using some model Hamiltonian. Rather than attempt to investigate which one gives the ground state for particular parameters, we present *in extenso* the recent ideas of Hotta *et al.* [206] on the origin of the near BS-WC degeneracy.

Let us discuss the importance of the role of the 1D conducting zigzag paths in the (a, b) basal plane, and considered parallel-transport of an e_g electron along these paths through the JT centers composed of MnO_6 octahedra. The transport invokes the Berry-phase connection and we can introduce the "winding number" w as a direct consequence of topological invariance which should be conserved irrespective of the details of H. Consider e_g electrons coupled both to localized t_{2g} spins with the Hund's rule coupling J_H, and to JT distortions of the MnO_6 octahedra. Since

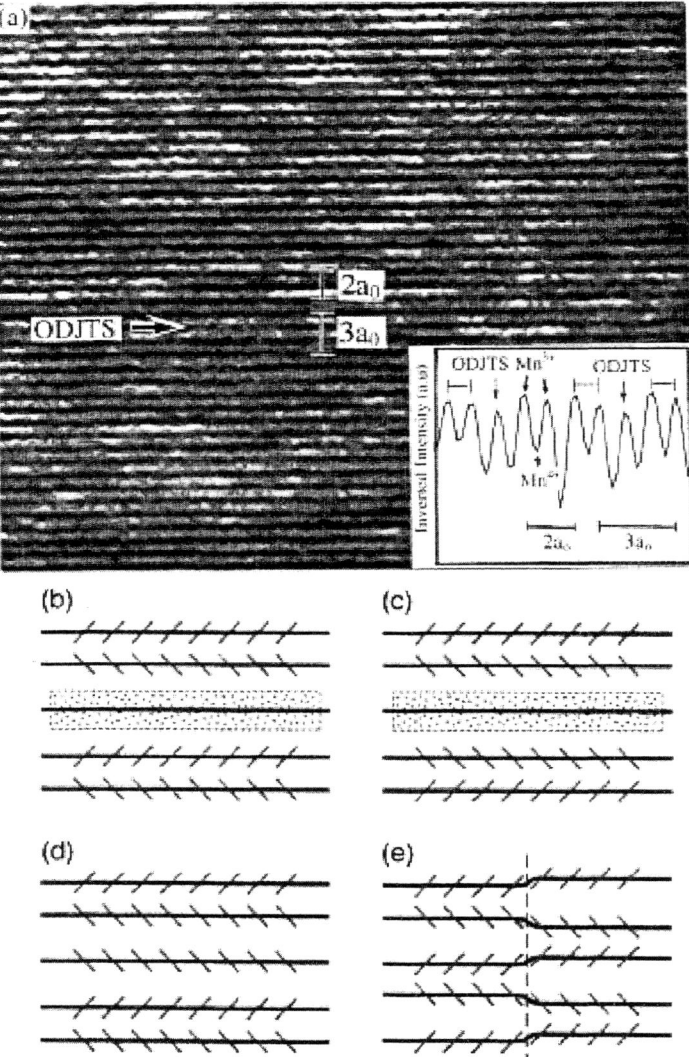

FIGURE 61. (a) High resolution lattice image in the FM phase at 206 K showing a mixture of incommensurate charge-ordered and FM charge disordered microdomains. The arrow stands for an unpaired JT stripes (JTS) between paired ones. An inverted intensity profile of this configuration is plotted in the inset of (a). (b) and (c) show in-phase and out-of-phase configurations, respectively, of the paired JTS surrounding the orbital disordered JTS. The lines represent Mn^{3+} JTS while the slanted slashes (random dotted strip) stand for d_{z^2} orbital ordering (disordering). (b) and (c) go into residual discommensuration (d) and antiphase boundary (e), respectively, in the incommensurate phase when complete orbital order is realized (after Ref. [202]).

J_H is the largest characteristic energy among those considered here, it is taken to be infinite for simplicity. This implies that the spin of each e_g electron at a Mn site aligns completely in parallel with the direction of the t_{2g} spin $S = 3/2$ at the same site. Thus, the spin degrees of freedom are effectively lost for the e_g electrons, and the spin index will be dropped hereafter. Since it is known experimentally that the t_{2g} spins are antiparallel along the c-axis, we can assume that the e_g electrons can move only in the (a, b) plane due to the DE mechanism.

The above situation is well described by the Hamiltonian [206],

$$H = -\sum_{ij\alpha\beta} t_{ij}^{\alpha\beta} c_{i\alpha}^\dagger c_{j\beta} + J_{AF} \sum_{\langle ij \rangle} \vec{S}_i \cdot \vec{S}_j$$
$$+ E_{\mathrm{JT}} \sum_i \left[2 \sum_{\alpha\beta} c_{i\alpha}^\dagger (Q_{2i}\sigma_i^x + Q_{3i}\sigma_i^z)_{\alpha\beta} c_{i\beta} + (Q_{2i}^2 + Q_{3i}^2) \right], \quad (263)$$

with the same notation as used in Eq. (15). The second term $\propto J_{AF}$ represents the AF coupling between nearest-neighbor classical t_{2g} spins which are normalized for convenience to $|\vec{S}_i| = 1$. The third term is controlled by the JT energy E_{JT} and describes the coupling of an e_g electron with the (x^2-y^2)- and $(3z^2-r^2)$-type (dimensionless) JT modes, given by Q_{2i} and Q_{3i} (Fig. 3), respectively.

Intuitively, it can be understood that the competition between kinetic and magnetic energies can produce an S-AF state. The ground state of the system described by Eq. (263) with $J_{AF} = 0$ is a 2D FM metal and optimizes the kinetic energy of e_g electrons, while it becomes a 2D AF insulator at $J_{AF} \geq t$ to exploit the magnetic energy of the t_{2g} spins. For smaller but nonzero values of J_{AF}, a competition between these states occurs which results in a mixture of FM and AF states. In this S-AF state, a 1D conducting path can be defined by connecting nearest-neighbor sites with parallel t_{2g} spins. A path with a large stabilization energy is needed to construct a stable 2D structure. Thus, Hotta et al. [206] concentrated themselves on specifying the shapes of (quasi-)stable 1D paths in the S-AF manifold.

Let us start with the case of $E_{\mathrm{JT}}=0$ and no electron correlation, which allows us to illustrate the importance of topology in the present problem. In a 2D FM metal, the kinetic-energy gain is reduced by $2J_{AF}$, due to the loss of magnetic energy per site. In contrast, in a 2D AF insulator, the magnetic energy gain per site is $2J_{AF}$. In S-AF states, the optimized periodicity M for a 1D path along the a-axis direction is numerically found to be given by $M=2/n$ [206], in agreement with Ref. [207], where $n(=1-x)$ is the e_g-electron number per site. The S-AF structure with zigzag path is nothing but the well-known CE phase, and the analysis of Ref. [206] predicts that this state is very stable at $x=1/2$. However, the same analysis for $x \geq 2/3$ does not lead to a zigzag path but instead to a straight line as the optimized structure, which disagrees with experiment. Thus, it is necessary to find a quantity other than the energy to discuss the possible preferred paths that may arise from a full calculation, including nonzero E_{JT} and Coulomb interactions. Reconsidering the results at $x=1/2$ led Hotta et al. [206] to the idea that the number of vertices along the path, N_{V}, may provide the key difference among paths. A confirmation of this idea is provided by the calculation of energies for the 2^6 and 2^8 paths at $x = 2/3$ (M=6) and 3/4 (M=8), respectively.

In the next step one may include the JT distortions, $E_{\rm JT} \neq 0$ in Eq. (263). By writing the JT modes in polar coordinates as $Q_{2i} = Q_i \sin\theta_i$ and $Q_{3i} = Q_i \cos\theta_i$, "phase-dressed" fermion operators, $\tilde{c}_{ix}$ and $\tilde{c}_{iz}$, are introduced as:

$$\tilde{c}_{ix} = e^{i\theta_i/2}[c_{ix}\cos(\theta_i/2) + c_{iz}\sin(\theta_i/2)], \tag{264}$$

$$\tilde{c}_{iz} = e^{i\theta_i/2}[-c_{ix}\sin(\theta_i/2) + c_{iz}\cos(\theta_i/2)], \tag{265}$$

with $e^{i\theta_i/2}$ representing the molecular Aharonov-Bohm effect. The amplitude Q_i is determined by a MF approximation [206], while the phases θ_i's are interrelated through the Berry-phase connection to provide the winding number w along the 1D path as $w = \oint_c dr \cdot \nabla\theta/(2\pi)$, where c forms a closed loop for the periodic lattice boundary conditions. Mathematically, the winding number w is proven to be an integer [208]. In this system, it may be decomposed into two terms as $w = w_{\rm g} + w_{\rm t}$. The former, $w_{\rm g}$, is the geometric term which becomes $w_{\rm g} = 0$ or 1, corresponding to the periodic or antiperiodic boundary condition in the e_g electron wave function. It may be shown that the kinetic energy is lower with $w_{\rm g}$=0 than that with $w_{\rm g}$=1 for $x \geq 1/2$. Thus, $w_{\rm g}$ is taken as zero hereafter. To show that only the number of vertices $N_{\rm v}$ along a given path determines the topological term $w_{\rm t}$, let us consider the transfer of a single e_g electron along the path shown in Fig. 62(a). On the straight line part in either a- or b-direction, the phase is fixed at $\theta_a = 2\pi/3$ ($\theta_b = 4\pi/3$), because the e_g-electron orbital is polarized along the direction of the bridge bonds. This effect may be called an *orbital double-exchange* in the sense that the orbitals align along the actual chain direction to maximize the kinetic energy, similarly as the FM alignment of t_{2g} spins in induced by the usual DE mechanism (Sec. II.B). Thus, $w_{\rm t}$ does not change when one of the bridge positions is passed. In contrast, when the electron passes by one of the vertex positions, the phase changes from θ_a to θ_b (from θ_b to θ_a), indicating that the electron picks up a phase change of $2\pi/3$ ($4\pi/3$). Since these two vertices appear in pairs, $w_{\rm t}(=w)$ is evaluated as $w_{\rm t} = (N_{\rm v}/2)(2\pi/3 + 4\pi/3)/(2\pi) = N_{\rm v}/2$. The phases at the corner positions are assigned as an average of the phases sandwiching those vertices, $\theta_\alpha = \pi$ and $\theta_\beta = 0$, to keep $w_{\rm g}$ invariant. Then, the phases are determined at all the sites once θ_x, θ_y, θ_α, and θ_β are known.

Now we include the cooperative JT effect, important ingredient to determine COO patterns in the actual manganites. Although its microscopic treatment is involved, we can treat it phenomenologically as a constraint for macroscopic distortions [206], energetically penalizing $w = 0$ and $M/2$ paths. In fact, it is numerically found that $w = 1, 2, \cdots, M/2 - 1$ paths constitute the lowest-energy band and they can be regarded as degenerate, since its bandwidth is about $0.01t$, much smaller than the interband energy difference ($\sim 0.1t$). Summarizing, the cooperative JT effect gives us two rules for the localization of e_g electrons [206]: (i) electrons never localize at vertices; (ii) electrons localize pairwise – any electron that localizes on one of the segments in the a-direction has a partner that localizes on one of the segments in the b-direction.

Applying these rules, we obtain a general structure for the lowest-energy path. Important features are the *renormalized vertices*, $\tilde{\alpha}$ and $\tilde{\beta}$, abbreviated notations to represent the set of straight-line parts that are not occupied by e_g electrons. The winding number assigned to $\tilde{\alpha}$ ($\tilde{\beta}$) is $1/3+w_\alpha$ ($2/3+w_\beta$), where the number of vertices included in $\tilde{\alpha}$ ($\tilde{\beta}$) is $1+2w_\alpha$ ($1+2w_\beta$). Thus, the lowest-energy path is labeled by the nonnegative integers w_α and w_β, leading to a total winding number $w=1+w_\alpha+w_\beta$. Although the topological argument does not determine the precise position at which an e_g electron localizes in space, it is enough to regard a charged straight-line part as a *quasi-charge*. Since the quasi-charges align at equal distance in the WC structure, the corresponding path is labeled by $w_\alpha = w_\beta = m$, with m a nonnegative integer. By increasing w_β keeping w_α fixed, we can produce any non-WC-structure paths with $w_\alpha = m$ and $w_\beta = m+1, m+2, \cdots$ [Fig. 62(c)]. In this way, the WC structure with $w = 2m+1$ can be considered the *mother state* for all non-WC-structure paths with $w = 2m+2, 2m+3, \cdots$, referred to as the *daughter states*. The states belonging to different m's are labeled by the same w, but a large energy barrier exists for the conversion among them, since an e_g electron must be

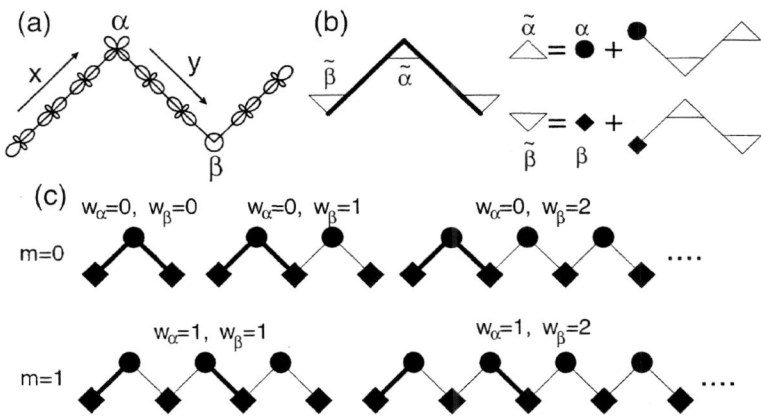

FIGURE 62. (a) A typical building block for a 1D path for an e_g electron with JT distortions. (b) General structure of the lowest-energy-state path and the renormalization scheme for the vertices α and β. The thick (thin) line denotes the straight-line part with (without) an e_g electron localized on it. The solid circle and diamond denote, respectively, the bare vertices, α and β, while open up-and down-triangles indicate the renormalized vertices, $\tilde{\alpha}$ and $\tilde{\beta}$. Note that the periodicity of the 1D path is given by $M = 2/n = 2/(1-x)$. (c) Groups of 1D paths derived from mother states with $m = 0$ and 1. Paths in the first column corresponding to the mother WC structures with $w=2m+1$, which produce daughter states with $w = 2m+2, 2m+3, \cdots$ (after Ref. [206]).

moved through a vertex in such a process. Thus, the state characterized by w in the group with m, once formed, it cannot decay, even if it is not the lowest-energy state.

Note that the topological argument works irrespective of the details of the Hamiltonian H, since w is a conserved quantity. However, it cannot single out the true ground state, since the quantitative discussion on the ground state energy depends on the choice of H and on the approximations employed. In fact, either the BS or WC structure can be stable, but in view of the small energy difference, their relative energy will likely change whenever a new ingredient is added to H. It may be expected that these phases are rather sensitive to the Coulomb interactions which will play an important role in stabilizing one of these two structures.

Now we analyze following Hotta et al. [206] the charge and orbital arrangement in $La_{1-x}Ca_xMnO_3$ ($x > 0.5$), in which the experimental appearance of the BS structure provides key information to specify the 1D path. Since the quasi-charges exist in a contiguous way in the BS structure, its path is produced from the mother state with $m = 0$ [see Fig. 62(c)]. In particular, the COO pattern in the shortest 1D path is uniquely determined as shown in Figs. 63. At $x = 1/2$, the path is characterized by $w = 1$ which is the basic mother state with $m = 0$. The COO pattern shown in Fig. 63(a) leads to the CE-type AF state [204]. The other paths with $w = 2$ and 3 are nothing but the BS structures experimentally observed at $x = 2/3$ and $3/4$ [202].

It may be assumed that the long-range Coulomb interaction V destabilizes the BS structure and transforms it to the WC structure, but this is not the case; for the BS $\to$ WC conversion with the help of V, an e_g electron must be on the vertex in the path with $w=2$ or 3 [see Figs. 63(b) and 63(c)]. This is against rule (i) and thus, the BS structure, once formed, is stable due to the topological condition, even including a weak repulsion V. In the group of $m = 0$, the WC structure appears only in the path with $w = 1$. Thus, the WC-structure paths with $w = 1$ at $x = 2/3$ and $3/4$ are obtained by simple extension of the straight-line part in the path at $x = 1/2$ [see Figs. 63(d) and 63(e)]. The detailed charge distribution inside the quasi-charge segment is determined by a self-consistent calculation with the JT effect, leading to the WC structure. Even if the non-WC structure occurs for $w = 1$, it is unstable in the sense that it is easily converted to the WC structure, because no energy barrier exists for an e_g electron shift along the straight-line part. The above topological analysis shows that: (i) the WC structure is made of $w_{WC}=1$ zigzag paths, and (ii) the BS structure contains a shorter-period zigzag path with $w_{BS} = M/2 - 1 = x/(1-x)$. Note that on the BS path, the less-distorted Mn^{4+} sites occupy all the vertices (N_v equals the number of Mn^{4+} ions), while the heavily distorted Mn^{3+} sites appear in pairs (the number of Mn^{3+} ions equal to 2). Thus, w_{BS} is rewritten as

$$w_{BS} = \frac{N_v}{2} = \frac{\text{Number of } Mn^{4+} \text{ ions}}{\text{Number of } Mn^{3+} \text{ ions}} = \frac{x}{1-x}. \tag{266}$$

Since w_{BS} is an integer, we can predict that at specific values of $x[=w_{BS}/(1+w_{BS})]$,

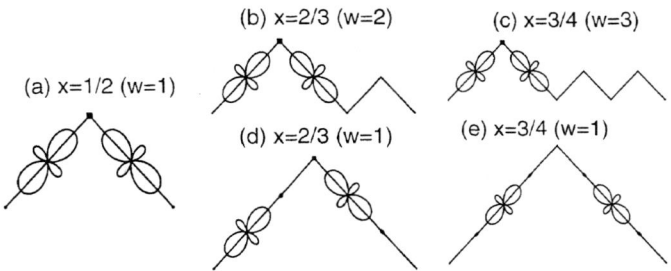

FIGURE 63. (a) Path with $w = 1$ at $x = 1/2$ for $E_{JT} = 2t$. At each site, the orbital shape is shown with its size in proportion to the orbital density. (b) The BS-structure path with $w = 2$ at $x = 2/3$. (c) The BS-structure path with $w = 3$ at $x = 3/4$. (d) The WC-structure path with $w = 1$ at $x = 2/3$. (e) The WC-structure path with $w = 1$ at $x = 3/4$ (after Ref. [206]).

such as 1/2, 2/3, 3/4, *etcetera*, nontrivial charge and orbital arrangement will be stabilized in agreement with the experimental observation [202].

C Orbital ordering and phase separation

Finally, we turn to the competition between different magnetic phases in doped manganites. There is no controversy on the fact that ferromagnetism at large doping is promoted by the DE mechanism – the kinetic energy of e_g electrons is maximized when the t_{2g} spins and their own spins are aligned. Within this scenario, one gets a natural explanation of the FM metallic phase. When an e_g electron moves in some region, it does not pay an energy J_H if all the t_{2g} spins in its neighborhood are parallel. The hole-spin scattering is reduced in this way, and one gains the kinetic energy. As the carrier concentration increases, the FM polarization clouds around the holes start to overlap and the ground state becomes metallic, with e_g electrons moving in the correlated degenerate band, inducing the saturated FM ordering. The understanding of the role played by the orbital degrees of freedom in the FM metallic phase is not complete, however. As suggested by the studies of DE both in the electron and hole doping regime [65,152,169], one possibility is that the orbital ordering of some type is stabilized in particular doping regimes. The transitions between different types of orbital ordering are promoted by the kinetic energy which is optimized by either planar $|x\rangle$-type or directional $|z\rangle$-type orbitals, while the kinetic energy in the phases with orbital ordering and occupied orbitals that are linear combinations of $|x\rangle$ and $|z\rangle$ states as in Eqs. (72) is lower [182,177]. This explains the mechanism of an orbital liquid discovered by Nagaosa *et al.* [164], and confirmed by numerical studies [179].

It has been found that due to a competition between the superexchange which favors orbital ordering and DE which favors a uniform phase, the magnetic phase

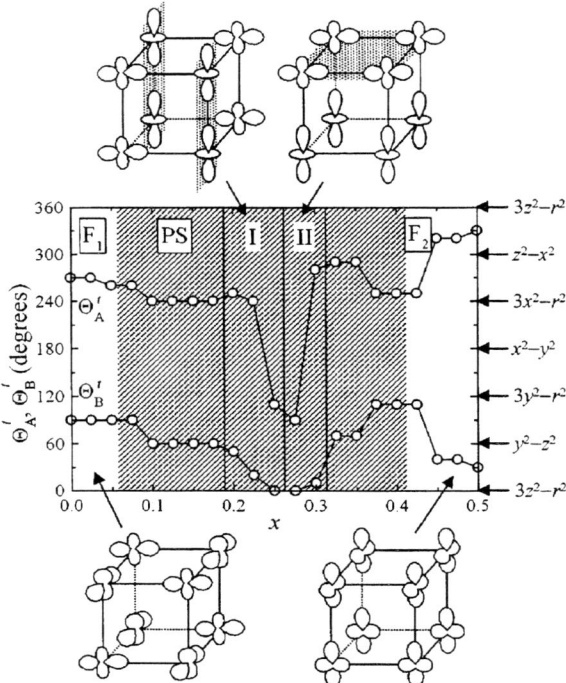

FIGURE 64. A sequential change of orbital states as a function of hole concentration x (after Ref. [152]). $\Theta^t_{A(B)}$ is the angle in the orbital space in the $A(B)$ orbital sublattice. Note that these angles are related to the angle used in Eq. (9) by $\theta_{A(B)} = 2\Theta^t_{A(B)}$. The schematic orbital states are shown in phase-I and phase-II; the dotted areas show the regions where the hole concentration is rich.

diagrams obtained from the models which include the degeneracy of e_g orbitals have orbital ordering also in the FM phase [152,169]. An example is shown in Fig. 64, where all FM states have some kind of orbital ordering, modified under increasing doping. Such states are in qualitative agreement with the experimental observations at low doping, where the orbital ordering was observed in $La_{0.88}Sr_{0.12}MnO_3$ [209], and may be concluded from the anisotropic exchange interactions found in $La_{0.85}Ca_{0.15}MnO_3$ [210]. The charge is uniformly distributed in this doping regime, while the orbital ordering occurs only in low temperatures in $La_{0.88}Sr_{0.12}MnO_3$, with a FM phase with disordered orbitals in the intermediate temperatures [209]. At higher doping, the transition to the orbital liquid state is indeed realized in double-layered manganites, where it is consistent with the observed lattice deformation along c axis and allows to explain the observed the observed spin ordering

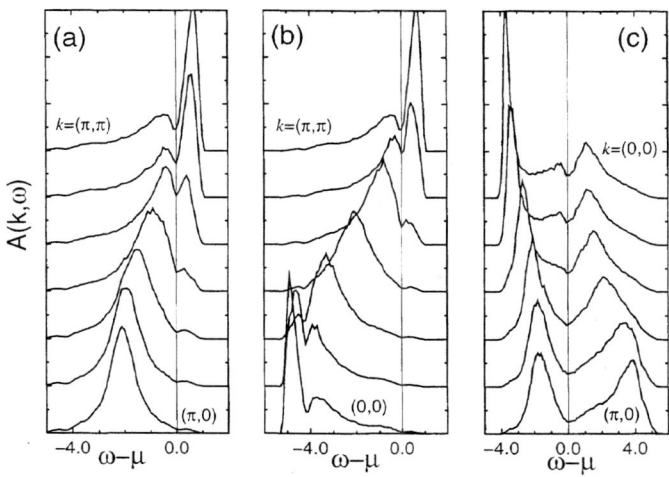

FIGURE 65. Spectral functions $A(\vec{k},\omega)$ at $J_H = \infty$ for a 2D one-orbital case with $T = 1/30$, $\langle n \rangle \sim 0.92$, on a 12×12 cluster for: (a) along $(\pi,0)$ to (π,π); (b) along $(0,0)$ to (π,π). Part (c) shows the results for the 2D two-orbital model with $T = 1/10$, $\lambda = 1.5$, $\langle n \rangle \sim 0.70$, on a 10×10 cluster along $(0,0)$ to $(\pi,0)$ (after Ref. [212]).

and its anisotropy [211]

In contrast, two phases (I and II) found for $0.18 < x < 0.32$ doping (Fig. 64) are characterized by the doped holes concentrated either in the regions of directional (I), or planar (II) orbitals, separated by the 1D or 2D structures with few holes and the orbitals being closer to the undoped situation. Whether or not such a competition between the phases with differently ordered orbitals really happens close to $x \sim 0.25$ is still controversial at the moment – it might be that the phase diagram of Okamoto, Ishihara, and Maekawa [152] is closer to the situation found in the insulating rather than metallic FM manganites. In fact, this competition suggests that there are also other ways of gaining the kinetic energy in the metallic phase, as for instance realized in uniform orbital phases with complex coefficients of e_g orbitals [177], and such states might be better candidates in the metallic phase. The experimental phase diagrams of manganites (Fig. 4) are typically richer than those obtained from model calculations. For instance, the phase diagram of Fig. 64 shows several different types of orbital ordering, but the magnetic state is FM in the whole range of doping $0 < x < 0.5$. Instead, the A-AF order is observed at small doping [163], and CE-phase (Sec. VIII.A) at doping $x \simeq 0.5$. It is rather difficult to reproduce this complex behavior in the theoretical models. However, the interplay of charge, spin, lattice and orbital degrees of freedom belongs to the generic features of this class of compounds, and the challenge in the theory is still to describe these various energy contributions with comparable and sufficient accuracy.

The competition between the fully polarized FM state and other types of order which occurs as temperature and electron concentration are varied is a central problem in the physics of manganite perovskites. This problem has attracted a lot of interest recently, especially due to the unusual experimental results obtained in the manganese oxides [22,213]. For instance, above the Curie temperature T_C and for a wide range of densities, several manganites show an insulating behavior of not completely clear origin that contributes to the large magnetoresistance effects. The low temperature phases have complex structures, not fully understood within the DE scheme, that includes different phases with AF and CO ordering, orbital ordering, FM insulating phase, and tendencies towards the formation of charge inhomogeneities even in the FM phase. Moreover, many experimental results show the occurrence of charge inhomogeneities in macroscopic form or in small clusters of one phase embedded into another. It turns out the metallic FM phase has regions where FM clusters coexist with another phase in a range of temperature and concentration [214,215], either in Sr- and Ca-doped manganese oxides.

In this context, De Gennes suggested rather early that the competition between the AF superexchange and the DE results in the canting of the AF state [216], that is the angle θ between the spins from different sublattices becomes smaller than π. The canting angle grows with the concentration of charge carriers, which might explain the increase of magnetization upon doping observed in $La_{1-x}Ca_xMnO_3$. Already rather long ago arguments were given against the stability of canted ground state [217,218]. In the De Gennes approach the local spins were treated classically. It was found that quantum corrections stabilize the AF state and the canted state appears only above a certain concentration of charge carriers [218]. By this mechanism the canted state might also disappear – in fact it was not observed in $La_{1-x}Ca_xMnO_3$ at increasing doping $0 < x < 0.15$ [163].

A more fundamental problem however is that at partial filling of the conduction band, the homogeneous ground state might be unstable against phase separation (PS). Experimentally the PS was recently observed by Babushkina et al. [219], and it is also obtained in theory. For instance, the phase diagram of Fig. 64 contains several regions with PS, either between different orbital states, or between differently doped phases. Another possibility is the PS between the regions with different magnetic ordering which accompanies differently doped regimes of the sample. This implies that many experimental data on doped manganites should be reinterpreted taking into account the inhomogeneity of the ground state. In particular, the charge transport and the metal-insulator transition should be more appropriately described in terms of percolation rather than by the properties of the pure states.

The problem of the nature of the PS has been studied extensively by means of QMC techniques within the FM Kondo model with one and two orbitals, including the JT effect for taking into account the occurrence of orbital order, though without fully considering the consequences of large Coulomb interactions [50,220]. Several unexpected results have been found in that study. In particular, either for the one- or two-orbital model, when calculating the density of e_g electrons as a function of varying chemical potential, one finds that some densities are unstable [50,220]; that is, the density is changing discontinuously at particular values of the chemical potential. Other calculations done in the canonical ensemble where the density is kept fixed, showed that the ground state is not homogeneous, while being constituted by separate regions with values of densities corresponding to the unstable regime. This phenomenon is usually referred to as *phase separation*, such as the familiar liquid-vapor coexistence in the phase diagram of water where it is known that the compressibility becomes negative negative. It is also similar to the stripe instability in the cuprates [221], but the present phenomenon is more difficult to study as it happens in a larger space which involves the orbitals and lattice.

In the regime of large Hund's coupling J_H and intermediate values of the JT coupling, PS occurs both for small and large hole densities. In particular at small e_g densities, the PS appears between an electron-undoped AF state and a metallic uniform orbital-ordered FM state. At small hole concentrations, the latter phase coexists with an insulating staggered orbital ordered FM phase. In the QMC results [50,220], the PS manifests itself as the occurrence of a macroscopic separation of two phases with different charge distribution. Actually, this possibility should be prevented by long-range Coulomb interactions, which is usually neglected. In fact, even taking into account polarization effects and screening, a complete separation would lead to a huge loss of energy. These considerations suggest that two large regions should split into many smaller pieces in order to distribute the charge more uniformly and reduce the energy loss in the Coulomb channel.

If the PS involves just single carriers (doped holes) with their local environment that has been distorted by the presence of a hole, such states are referred to as polarons. The distortion can be either in the magnetic channel (spin polaron), or in the phononic channel (lattice polaron), or in the orbital channel (orbital polaron), or may be even a combination of these effects. Furthermore, the case where such regions of reduced electron density involve more than one carrier are referred to as *clusters* or *droplets*. The competition between the long-range Coulomb repulsion and the magnetic interaction would determine the size and the shape of the resulting clusters. The stable state, obtained when the extended Coulomb interactions are included, is considered as a charge inhomogeneous state which has to be distinguished from the metastable state that appears in a standard first order transition. A large-scale PS is expected if the competing phases have approximately the same density, as it happens experimentally at $x = 0.5$ which is a singular point in the phase diagrams (Sec. VII.B). It is of fundamental importance to investigate how the properties of the ordered FM state are influenced by the vicinity of a PS regime, especially in connection to the CMR phenomenon. One of the consequences of this proximity to PS state is observed as an effective increase of the compressibility in the FM phase which implies the occurrence of strong charge fluctuations. This result implies that even within the FM phase, which is uniform when time averaged, there is a dynamical tendency toward cluster formation. This effect is expected to influence the transport property. Calculations done on finite size clusters show that the resistivity behaves as observed experimentally, in other words it is insulating at small x and rapidly decreases when x increases, and a metallic state is approached.

A very interesting observation was made recently by Moreo, Yunoki and Dagotto that the PS is responsible for a pseudogap formation in the spectral function $A(\vec{k}, \omega)$ calculated using Monte-Carlo techniques for the model of manganites which includes the coupling to JT phonons [212]. The pseudogaps was found to be a robust feature at all momenta along the Fermi surface and occurs due to the existence of FM clusters coexisting with the AF ordering. A particularly pronounced pseudogap was found in the model with doubly degenerate e_g orbitals which demonstrates again that the orbital degrees of freedom play an important role in manganites. The results have striking similarity with the angle resolved photoemission studies of $La_{1.2}Sr_{1.8}Mn_2O_7$ [223] which suggests that microscopic inhomogeneities exist in this system, both above and below T_C.

The PS scenario is likely to play an important role in the low and intermediate doping regime, where several mechanisms have been proposed to explain the CMR effects in manganese oxides. It adds to the simple DE idea the possibility of charge inhomogeneities as the main effect competing with ferromagnetism. This picture is different from that proposed by Millis et al. [59], where the interplay between spin disorder and the formation of lattice polarons via JT-coupling is fundamental for explaining the metal-insulator transition. In this respect, though the JT-coupling turns out to be important in both scenarios, in the PS scheme a state given by independent local lattice polarons is a special case of a more general situation where

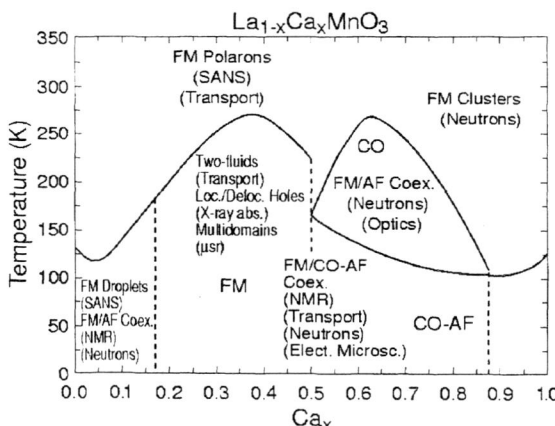

FIGURE 66. A schematic version of the phase diagram for $La_{1-x}Ca_xMnO_3$ [214]. The abbreviations Coex., Loc., Deloc., abs., μSR, and Elect. Microsc., stand for coexistence, localized, delocalized, absorption, muon spin relaxation, and electron microscopy, respectively (after Ref. [222]). Copyright 1999 American Association for the Advancement of Science.

droplets of various shape and size might form. Still, the dynamical fluctuation of charge inhomogeneities increase as T_C decreases, thus explaining the CMR effect at the boundary of the FM phase. Other theories emphasize the electronic localization effect induced by off-diagonal disorder in the spin correlations [224–226], or diagonal disorder due to chemical substitution.

Summarizing, the generic phase diagram of manganites shown in Fig. 66 represents still a very challenging and interesting problem in the field of correlated electrons in degenerate e_g bands. There are many open questions related to the understanding of the insulating state of manganese oxides, especially in the framework of tendencies towards charge inhomogeneous state. Analytical techniques beyond the local MF approximations are required to reproduce the essential physics of the nonuniform charge phase and to investigate the role of orbital ordering in doped manganites. More detailed analysis is needed to understand the shape and the size of droplets in such inhomogeneous states, and the crossover from the PS regime to polaronic regime which has been observed experimentally [60,213,227].

IX SUMMARY AND CONCLUSIONS

We have reported a systematic analysis of the consequences of *orbital degrees of freedom* in a class of correlated MHI: copper and manganese oxides. These systems are characterized by several different interactions which involve three different types of degrees of freedom in the undoped compounds: spins, orbitals, and lattice. If such systems are doped, one has to include in addition charge dynamics. The main question we have been dealing with through this report was: What is the role of the orbital degrees of freedom in presence of strong electron correlations and/or lattice effects, and what kind of *magnetic and orbital ordering* is promoted by them?

The essence of the problem posed by the superexchange interactions at orbital degeneracy is captured by the spin-orbital model (57)–(59) for the cuprates, defined for the d^9 ions as in $KCuF_3$. This model is of particular interest as it combines only two aspects of a more complex problem: magnetic (spin) and orbital degrees of freedom, which are discussed at integral filling, i.e., in the absence of doping, and without taking seriously into account the JT effect. The model Hamiltonian has already been proposed long ago [18], but its full consequences have been appreciated only recently [63,64].

The Hamiltonian derived for the cuprates from the spectroscopic information about the spectra of excited states has been first considered on the classical level, where we have shown in Sec. III that it gives a particular frustration of spin and orbital interactions. This is best represented by a singular point at the origin of the phase diagram for the d^9 spin-orbital model (Fig. 12) which is highly degenerate at the classical level, so that many axes separating different classical phases should emerge from it. In this respect the classical phase diagram is incomplete: the quantum mechanics is likely to take over in this regime and decide about the actual ordered or disordered state. As a main result of this part, we have shown that enhanced quantum fluctuations close to the transition lines between different classical phases destroy classical states (Sec. IV). Unlike in the frustrated spin models, the Gaussian fluctuations around classical states involve the orbital sector, probably yielding novel spin-orbital liquids in the form of generalized (R)VB states [63]. Such states turn out to be very good variational wave functions, and turn out to be exceptionally stable in the d^9 model with respect to the classical phases.

Whether the phase diagram presented in Fig. 31 is the qualitatively correct picture of the quantum disorder realized next to the orbital degeneracy, is still an open question. Another possibility would be an ordered state which results by the order-out-of disorder mechanism. Unfortunately, even the physics at the above multicritical point of the phase diagram of the d^9 model is not captured by the SU(4) model which, instead, turns out to be an idealized generalization of the Kugel-Khomskii model, putting the spin and orbital variables on the same footing. However, this highly symmetric problem is not easier to investigate except for the 1D case, and thus the final answer to the problem of the ground state of the spin-orbital d^9 model in the vicinity of the multicritical point is still an open issue.

The physical properties of a system described by the spin-orbital model (57)–(59) are mainly determined by the nature of the collective excitations. In addition to the usual magnon dispersion known from the spin systems, we have to consider the pure orbital (or excitonic) excitations, and the mixed modes which involve simultaneously both spin and orbital excitation (spin-and-orbital modes). These modes couple to the magnons and may thus have measurable consequences, predicted as an anomalous spin response detected in neutron scattering experiments. It is thus challenging for the experimentalists to investigate carefully any peculiarities which are expected to arise in the spin spectra of the MHI with orbital degeneracy.

The above ideas on the spin liquid near orbital degeneracy and on the role played by the orbital fluctuations should be verified by future experiments. It is not easy to find compounds whose microscopic parameters live in the region close to this peculiar point in the phase diagram, and a progress in *material science* in needed. Furthermore, though close to this region in the sector of the electronic interactions, the cooperative and local JT effects convey to remove the orbital degeneracy stabilizing a particular spin pattern, as for example, it is realized in $NaNiO_2$. In this respect, an interesting situation seems to happen in the case of $LiNiO_2$, where these competing lattice-induced phenomena that promote ordered states are absent, and a quantum critical state characterized by power-law behavior of the spin correlation functions appears instead.

The spin-orbital model (159) derived for the d^4 ions as in $LaMnO_3$ [23] has many common features with the simpler model in the d^9 case (Sec. V). The spin interactions are different as they concern large spins $S = 2$ of Mn^{3+} ions, and more excited states of Mn^{2+} ions are involved, even when the e_g part of superexchange (157) is considered alone. As a consequence of larger spins, the model is more classical and the disordered states are not likely taking the realistic situation in manganites, with the strong JT effect which drives the system away from the multicritical point of frustrated interactions. The classical phases in the d^4 case are similar to those found for the d^9 model, but quantitatively the regions of their stability are different. Therefore, the A-AF phase obtained by a particular interrelation between the magnetic and orbital ordering is obtained in the d^4 case in a natural way as a consequence of the e_g part of superexchange interactions – this phase is robust and agrees with experimentally observed ordering in manganites. An important finding in this respect is that the electronic mechanism *alone* is responsible of the A-AF ordering, and that the JT interaction is only changing this state quantitatively and tuning a somewhat different orbital ordering. Furthermore, when a FM (a, b) plane of the A-AF phase is considered, the spins are integrated out and one finds the same orbital part of the superexchange Hamiltonian discussed in Sec. VII.

The analysis of the d^9 and d^4 models may be treated as the first step on the way towards the understanding of one of the most challenging problems in the modern solid state physics: How the delicate balance of magnetic and orbital interactions is affected by doping in the manganese oxides, and which physical mechanisms are responsible for various types of ordering observed in the CMR materials? We have analyzed this issue in an extensive way in Secs. V-VIII. Firstly, we have discussed the low doping regime of the manganese oxides which is simpler because the lattice helps to restrict the doped holes to particular sites, and the model deals in principle still only with the same degrees of freedom as in undoped LaMnO$_3$: spin, orbital, and lattice, but they are modified by the added charges. This part of the phase diagram is dominated by many competing effects both in the magnetic and orbital channel. The central issue is the presence of an *orbital polaron* regime which mediates the crossover from the A-AF to the FM insulating phase. The DE mechanism stabilizes the polarons and their localization by new FM effective interactions that give rise to the transition to the FM insulating phase as the hole concentration is increased (Sec. V.B). The binding energy of these orbital-hole bound state depends on the scale of fluctuations that involve orbitals, and therefore on the concentration of doped holes. Different configurations of the polarons may be stabilized by the lattice effects; the polarons are separated from each other [228] and may induce interesting orbital ordering in their neighborhood [229], but finally start to overlap. Thus, when the hole concentration increases, the orbital fluctuations also do, breaking eventually the localization process and favoring instead a different kind of orbital (ordered or disordered) state in a FM metallic phase. This problem is at heart of the CMR phenomenon.

Naively, the transition to the FM metallic phase occurs due to the DE mechanism. This picture is oversimplified – it ignores the orbital degrees of freedom which are of importance in the doped regime and lead to several types of magnetic ordering when the DE model is generalized to the case of degenerate e_g orbitals (Sec. VI). We would like to emphasize that the DE via degenerate orbitals has very interesting special features that are very different from the conventional nondegenerate situation. The orbital degeneracy leads in general to the formation of anisotropic magnetic structures that depend on the actual doping concentration, and include the layered magnetic A-type structures, predominantly the $x^2 - y^2$ orbitals, and chain-like structures of the C-type. The stability of such states will be influenced by lattice distortions – in this case a compression of the MnO$_6$ octahedra along one of the cubic directions. One can show, however, that due to the anharmonicity the JT distortion always leads to a local elongation. If strong enough along the c-axis, this tendency would favor a structure in which the $3z^2 - r^2$ orbitals are occupied, in this case lowering the energy of C-type structures. These considerations demonstrate the importance of the JT effect. Cooperative JT coupling between the individual MnO$_6$ centers in the crystal leads to the simultaneous ordering of the octahedral distortions and the electronic orbitals. Part of the electronic energy is thereby lost and therefore an accurate description of the orbital state in manganites is possible only when the lattice effects are *explicitly* included in the model.

Unlike the DE Hamiltonian for nondegenerate orbitals, the realistic DE model which includes the degeneracy of e_g electrons cannot be considered without the strong Coulomb correlations. Electronic correlations are very important for the understanding CMR manganites and cannot be neglected. Very different results are obtained when local Coulomb correlations are not included – for instance: the tendency towards the FM metallic phase is strongest at the filling of one e_g electron per site [180], and the band gap practically vanishes without JT distortions [230], while in reality the gap is primarily due to large on-site Coulomb interactions [225] (Sec. VII). The main result of the DE model without Coulomb interactions is that it can explain FM metallic phase only for doping $x > 0.5$, where, however, the localized G-AF phases are typically found (Fig. 4). At lower values of x, the suppression of double occupations by the local Coulomb repulsion becomes more and more important, and leads to a crossover from DE to superexchange. First it reduces somewhat the effective FM interactions which follow from the DE mechanism, leading to the reduction of the magnon bandwidth [175] and of the Curie temperature T_C [231] with decreasing x, and eventually the superexchange between the Mn^{3+} ions becomes more important and stabilizes the A-AF ordering coexisting with the orbital ordering.

Recently it became clear that the nature of FM states may be studied by analyzing the spectrum of magnetic excitations. In recent neutron scattering experiments [172,173,181] the spin wave dispersion throughout the BZ was measured in various manganites in the FM phase. It has been found that the spin waves are clearly resolved at low temperature, and the stiffness constant shows a universal behavior. This universal behavior and also the higher-energy magnons in the metallic phase are well described by the DE model with degenerate e_g orbitals, supplemented by smaller SE terms (Sec. VI.B). At higher temperature, however, different charge dynamics in metallic and localized compounds manifest themselves in the heavy damping of high-frequency spin waves even below T_c. Therefore, the simple DE model does not describe well the observed spectra in such situations when the doped holes localize, high-frequency magnons soften, and the value of T_C is reduced. We have discussed that orbital and charge fluctuations are responsible for a strong modulation of the exchange bonds, leading to a softening of the magnon excitation spectrum close to the BZ boundary (Sec. VI.C). The presence of JT phonons further enhances this effect. This peculiar interplay between DE physics and orbital-lattice dynamics becomes dominant close to the instability towards an orbital-lattice ordered state. The unusual magnon dispersion experimentally observed in low-T_C FM manganites can hence be considered as a precursor effect of orbital-lattice ordering. A complete understanding of these complex phenomena is still an open problem [232] and may be achieved only by taking into account all relevant degrees of freedom, including the JT effect which is still present in the metallic phase [233].

The proper understanding of pure orbital excitations is also very important, both in the undoped regime, and for the FM phase where such excitations are responsible for the orbital dynamics. Starting from the uniform FM phase, where the spin operators can be integrated out, we considered the interaction which occurs only between the pairs of ions with singly occupied staggered e_g orbitals at nearest neighbor sites (Sec. VII). A gapless orbital-wave excitation is found for the 3D system at orbital degeneracy, a peculiarity related to the invariance of the classical ground state energy with respect to the orbital rotations by angle θ at $E_z = 0$. This rotational invariance is broken in the 2D systems by the lack of interactions along the c-axis, and the 2D orbital model is more classical, with a gap in the excitation spectrum and suppressed quantum fluctuations, unlike in spin systems which are more quantum when one comes to a lower dimension. At increasing orbital field E_z, however, the system resembles again the behavior of a 3D system, with finite quantum fluctuations even when the orbitals are aligned which follows from the nonconservation of the orbital quantum number in the subspace of e_g orbitals.

Another open problem is the mechanism of stability of the CE phase in half-doped manganites with $x = 0.5$. The insulating CO state has been observed in almost all such compounds at half-doping, accompanied by a peculiar magnetic structure: 1D FM zigzag chains coupled antiferromagnetically within the (a,b) planes, and along the c-axis. In addition, these systems show $d_{3x^2-r^2}/d_{3y^2-r^2}$ orbital ordering along the FM chains. The is no doubt that the degeneracy of e_g orbitals plays here once again a crucial role. It follows both from a different behavior of the effective DE models with nondegenerate and degenerate orbitals, and from the specific mechanism of an insulating behavior discussed in Sec. VIII.A. The stability of different magnetic states is changed by orbital degeneracy, and for instance the C-type spin ordering is never achieved in the two-orbital case due to its instability against the effective *dimerization* and formation of the zigzag FM order. The alternating $d_{3x^2-r^2}/d_{3y^2-r^2}$ orbital ordering along the FM zigzag chain follows naturally the topology of the hopping in the CE phase. However, the role of the JT effect played in this phase and the magnetic and orbital excitation spectra are not yet understood.

The competition between different phases represents a particularly challenging problem. It results in a PS between AF and FM domains which together with the tendency to form inhomogeneous structures, seems to be a generic property of systems with strongly correlated electrons. Depending on the specific situation tuned by the strength of the coupling to the lattice and the doping level, the instability of homogeneous state and the tendency to PS may result in a formation of different structures: either random, percolation-like networks or regular structures, e.g. stripes, Wigner crystals, and the like (Sec. VIII). Many properties of such phase-separated states differ markedly from those of homogeneous states, and the possibility of PS has to serve as a starting point for the future explanation of certain anomalous phenomena observed experimentally in manganites. The PS scenario should lead to certain extensions of the DE model for degenerate e_g orbitals by the ideas of charge inhomogeneities as the main effect competing with ferromagnetism, and by considering the insulating properties above the Curie temperature as a consequence of the formation of dynamical clusters. How the PS scenario occurs is still an open problem and it is related to the understanding of the insulating states in manganites and other materials with strongly correlated electron bands. Analytical techniques beyond the local MF approximations are required to get the essential physics of the non-uniform CO phase under control. In experiment, the photoemission spectroscopy is expected to be of particular importance, as such features as the pseudogap [212], and the changes of the spectra between the compounds with different number of adjacent magnetic layers, lattice distortions, and doping levels, would help to characterize the orbital and magnetic ordering in the underlying phases.

Summarizing, the realistic model for manganites has to include the superexchange of spin-orbital type, the coupling to the lattice due to the cooperative and local JT effect, and some form of the DE in the correlated e_g orbitals, either localized or itinerant. Several theoretical models have already been proposed, but in most cases they neglect at least one of the important aspects of the problem (as, for instance, either e_g orbital degeneracy, or coupling to the lattice, or electron correlations), and then use some parameters to fit such properties as the type of magnetic ordering, the stiffness constant, or the types of occupied orbitals to the experiment — this is usually successful as the parameter space is large enough. While the model proposed by Feiner and Oleś [23] can certainly still be improved, it not only includes all the essential aspects mentioned above, but also takes the parameters from spectroscopy, except for the orbital interaction which follows from the JT effect and was fixed by experiment. In this way, the parameters which determine the magnetic transitions are all fixed and the predictions of this model may by confronted with experiment. The quality of this model may be appreciated by looking at Table 3. The Néel temperatures of the Mott-Hubbard insulators: $LaMnO_3$ and $CaMnO_3$, of the weakly doped polaronic $La_{0.92}Ca_{0.08}MnO_3$ compound, as well as the Curie temperature of the metallic FM $La_{0.7}Pb_{0.3}MnO_3$, are all obtained within 30 % to the experimental value. Of course, there is still a lot of room for improvement and the understanding is far from complete, in particular at finite temperature. Future progress may be expected by including the JT effect in a better way, and by a more careful analysis of the orbital ordered and disordered phases in various doping regimes.

We believe that the present report makes the role played by the correlated e_g orbitals in the MHI more transparent, and demonstrates that they have to be included to obtain qualitatively correct answers in still open problems, such as:

TABLE 3. Magnetic transition temperatures for the A-AF and G-AF states in undoped compounds: $LaMnO_3$, $CaMnO_3$, and in polaronic A-AF phase in $La_{0.92}Ca_{0.08}MnO_3$ (T_N), and in the metallic FM $La_{0.7}Pb_{0.3}MnO_3$ (T_C), as obtained from the spin-orbital model in doped manganites presented in Refs. [23,175], including a reduction factor due to quantum fluctuations (theory), compared with the experimental values (exp). The physical mechanisms which contribute to the stability of these different magnetic phases are indicated by '+'; DE effect was included either by polaronic (P) or by itinerant (I) mechanism. Orbital ordering stabilized by a superposition of the e_g-superexchange and the JT effect above T_N plays a crucial role in driving the magnetic ordering in $LaMnO_3$ and in $La_{0.92}Ca_{0.08}MnO_3$ into the A-AF phase.

| compound | phase | superexchange | | DE | JT | T_N (T_C) | | Ref. |
		e_g part	t_{2g} part			theory	exp	
$LaMnO_3$	A-AF	+	+	−	+	106 K	139 K	[163]
$La_{0.92}Ca_{0.08}MnO_3$	A-AF	+	+	P	+	95 K	122 K	[163]
$La_{0.7}Pb_{0.3}MnO_3$	FM	+	+	I	−	390 K	355 K	[181]
$CaMnO_3$	G-AF	−	+	--	−	124 K	110 K	[145]

the stability of different phases including the CE phase and stripe phases, the mechanism of PS, the metal-insulator transition at finite temperature, and the mechanism of the CMR itself.

ACKNOWLEDGMENTS

A.M.O. acknowledges warmly numerous stimulating conversations and a very friendly collaboration with Louis Felix Feiner, Peter Horsch and Jan Zaanen, who contributed significantly to his present understanding of this fascinating subject. It is a pleasure to thank G. Aeppli, J. van den Brink, B. Keimer, A. Fujimori, G. Khaliullin, D. I. Khomskii, R. Micnas, F. Moussa, G. A. Sawatzky, T. M. Rice, H. Shiba, J. Spałek, Y. Tokura, and W. Weber for valuable discussions. We are grateful to Ferdinando Mancini for creating this opportunity to discuss the spin-orbital models and to the stuff of the International Institute for Advanced Studies "E.R. Caianiello" in Vietri sul Mare for their assistance and financial support. The financial support by the Committee of Scientific Research (KBN) of Poland, Project No. 2 P03B 175 14, is acknowledged.

N.B.P. acknowledges INTAS N97-963, INTAS N-9711066 and INFM of Salerno for the financial support and the warm hospitality.

REFERENCES

1. M. Imada, A. Fujimori, and Y. Tokura, Rev. Mod. Phys. **70**, 1039 (1998).
2. P. Fulde, *Electron Correlation in Molecules and Solids*, Springer, Berlin, 1995.
3. P. Fazekas, *Lecture Notes on Electron Correlations and Magnetism*, World Scientific, Singapore, 1999.
4. K. A. Chao, J. Spałek, and A. M. Oleś, J. Phys. C **10**, L271 (1977).
5. K. A. Chao, J. Spałek, and A. M. Oleś, Phys. Rev. B **18**, 3453 (1978).
6. J. Zaanen, G. A. Sawatzky, and J. W. Allen, Phys. Rev. Lett. **55**, 418 (1999).
7. J. Zaanen and G. A. Sawatzky, J. Sol. State Chem. **88**, 8 (1990).
8. M. Meinders, H. Eskes, and G. A. Sawatzky, Phys. Rev. B **48**, 3916 (1993).
9. F. C. Zhang and T. M. Rice, Phys. Rev. B **37**, 3759 (1988).
10. L. F. Feiner, J. H. Jefferson, and R. Raimondi, Phys. Rev. B **53**, 8751 (1996).
11. J. H. Jefferson, H. Eskes, and L. F. Feiner, Phys. Rev. B **45**, 7959 (1992).
12. R. Raimondi, L. F. Feiner, and J. H. Jefferson, Phys. Rev. B **53**, 8774 (1996).
13. J. Zaanen and A. M. Oleś, Phys. Rev. B **37**, 9423 (1988).
14. G. A. Gehring and K. A. Gehring, Rep. Prog. Phys. **38**, 1 (1975).
15. K. I. Kugel and D. I. Khomskii, Usp. Fiz. Nauk (Sov. Phys. JETP) **136 (25)**, 621 (1982).
16. J. Zaanen and A. M. Oleś, Phys. Rev. B **48**, 7197 (1993).
17. D. I. Khomskii and G. A. Sawatzky, Sol. State Commun. **102**, 87 (1997).
18. K. I. Kugel and D. I. Khomskii, Zh. Eksp. Teor. Fiz. (Sov. Phys. JETP) **64 (37)**, 1429 (1973).
19. J. L. García-Munoz, J. Rodriguez-Carvajal, and P. Lacorre, Europhys. Lett. **20**, 241 (1992).
20. M. L. Medarde, J. Phys.: Condens. Matter **9**, 1679 (1997).
21. J. Rodriguez-Carvajal, S. Rosenkranz, M. Medarde, P. Lacorre, M. T. Fernandez-Daz, F. Fauth, and V. Trounov, Phys. Rev. B **57**, 456 (1998).
22. A. P. Ramirez, J. Phys. Cond. Matter. **9**, 8171 (1997).
23. L. F. Feiner and A. M. Oleś, Phys. Rev. B **59**, 3295 (1999).
24. C. Castellani, C. R. Natoli, and J. Ranninger, Phys. Rev. B **18**, 4945 and 4967, and 5001 (1978).
25. W. Bao, C. Broholm, G. Aeppli, P. Dai, J. M. Honig, and P. Metcalf, Phys. Rev. Lett. **78**, 507 (1997).
26. W. Bao, C. Broholm, G. Aeppli, S. A. Carter, P. Dai, T. F. Rosenbaum, J. M. Honig, P. Metcalf, , and S. F. Trevino, Phys. Rev. B **58**, 12 727 (1998).
27. H. F. Pen, J. van den Brink, D. I. Khomskii, and G. A. Sawatzky, Phys. Rev. Lett. **78**, 1323 (1998).
28. S. Y. Ezhov, V. I. Anisimov, D. I. Khomskii, and G. A. Sawatzky, Phys. Rev. Lett. **83**, 4136 (1999).
29. B. Keimer, D. Casa, A. Ivanov, J. W. Lynn, M. v. Zimmermann, J. P. Hill, D. Gibbs, Y. Taguchi, and Y. Tokura, (unpublished).
30. A. E. Bocquet, T. Mizokawa, K. Morikawa, A. Fujimori, S. M. Barman, K. B. Maiti, D. D. Sarma, Y. Tokura, and M. Onoda, Phys. Rev. B **53**, 1161 (1996).
31. J. S. Griffith, *The Theory of Transition Metal Ions*, Cambridge University Press,

Cambridge, 1971.
32. A. M. Oleś, Phys. Rev. B **28**, 327 (1983).
33. A. M. Oleś and G. Stollhoff, Phys. Rev. B **29**, 314 (1984).
34. G. Stollhoff, A. M. Oleś, and V. Heine, Phys. Rev. B **41**, 7028 (1990).
35. V. I. Anisimov, J. Zaanen, and O. K. Andersen, Phys. Rev. B **44**, 943 (1991).
36. A. I. Liechtenstein, V. I. Anisimov, and J. Zaanen, Phys. Rev. B **52** (1995).
37. V. I. Anisimov, I. S. Elfimov, M. A. Korotin, and K. Terakura, Phys. Rev. B **55**, 15 494 (1997).
38. I. V. Solovyev, N. Hamada, and K. Terakura, Phys. Rev. Lett. **76**, 4825 (1996).
39. K. Hirota, N. Kaneko, A. Nishizawa, and Y. Endoh, J. Phys. Soc. Jpn. **65**, 3736 (1996).
40. F. Moussa, H. Henion, J. Rodriguez-Carvajal, H. Moudden, L. Pinsard, and A. Revcolevschi, Phys. Rev. B **54**, 15 149 (1996).
41. J. Kanamori, J. Appl. Phys. **31**, 14S (1960).
42. A. J. Millis, Phys. Rev. B **53**, 8434 (1996).
43. H. Röder, J. Zhang, and A. R. Bishop, Phys. Rev. Lett. **76**, 1356 (1996).
44. R. von Helmolt, J. Wecker, B. Holzapfel, L. Schultz, and K. Samwer, Phys. Rev. Lett. **71**, 2331 (1993).
45. C. Zener, Phys. Rev. **82**, 403 (1951).
46. P. W. Anderson and H. Hasegawa, Phys. Rev. **100**, 675 (1955).
47. T. Hotta, S. Yunoki, M. Mayr, and E. Dagotto, Phys. Rev. B **60**, R15 009 (1999).
48. J. C. Slater and G. F. Koster, Phys. Rev. **94**, 1498 (1954).
49. P. A. Allen and V. Perebeinos, Phys. Rev. B **60**, 10 747 (1999).
50. S. Yunoki, A. Moreo, and E. Dagotto, Phys. Rev. Lett. **81**, 5612 (1998).
51. J. C. Slater, Phys. Rev. **49**, 537 (1936).
52. P. W. Anderson, Phys. Rev. **115**, 2 (1959).
53. J. Nagaoka, Phys. Rev. **147**, 392 (1966).
54. J. H. Van Vleck, Rev. Mod. Phys. **25**, 220 (1953).
55. M. Cyrot and C. Lyon-Caen, J. Phys. (Paris) **36**, 253 (1975).
56. S. Inagaki, J. Phys. Soc. Jpn. **39**, 596 (1975).
57. K. A. Chao, J. Spałek, and A. M. Oleś, Phys. Stat. Solidi (b) **84**, 747 (1977).
58. J. Spałek and K. A. Chao, J. Phys. C **13**, 5241 (1980).
59. A. J. Millis, P. B. Littlewood, and B. I. Shraiman, Phys. Rev. Lett. **74**, 5144 (1995).
60. A. Urushibara, Y. Moritomo, T. Arima, A. Asamitsu, G. Kido, and Y. Tokura, Phys. Rev. B **51**, 14 103 (1995).
61. Y. Okimoto, T. Katsufuji, T. Ishikawa, T. Arima, and Y. Tokura, Phys. Rev. B **55**, 4206 (1997).
62. Y. Okimoto, T. Katsufuji, T. Ishikawa, A. Urushibara, T. Arima, and Y. Tokura, Phys. Rev. Lett. **75**, 109 (1995).
63. L. F. Feiner, A. M. Oleś, and J. Zaanen, Phys. Rev. Lett. **78**, 2799 (1997).
64. A. M. Oleś, L. F. Feiner, and J. Zaanen, Phys. Rev. B **61**, 6257 (2000).
65. J. van den Brink and D. I. Khomskii, Phys. Rev. Lett. **82**, 1016 (1999).
66. J. R. Schrieffer and P. A. Wolff, Phys. Rev. **149**, 491 (1966).
67. G. Baym, *Lectures on Quantum Mechanics*, Benjiamin, NY, 1974.
68. E. Dagotto, Rev. Mod. Phys. **66**, 763 (1994).

69. A. Georges, G. Kotliar, W. Krauth, and M. J. Rozenberg, Rev. Mod. Phys. **68**, 13 (1996).
70. N. Furukawa, in *Physics of Manganites*, edited by T. A. Kaplan and S. D. Mahanti, Kluwer/Plenum, New York, 1999.
71. N. Furukawa, J. Phys. Soc. Jpn. **65**, 1174 (1996).
72. K. Kubo and N. Ohata, J. Phys. Soc. Jpn. **33**, 21 (1972).
73. J. B. Grant and A. K. McMahan, Phys. Rev. B **46**, 8440 (1992).
74. A. I. Liechtenstein and D. I. Khomskii, Sov. Phys. JETP **52**, 501 (1980).
75. D. H. Tennant, T. G. Perring, R. A. Cowley, and S. E. Nagler, Phys. Rev. Lett. **70**, 4003 (1993).
76. H. Tennant, R. A. Cowley, S. E. Nagler, and A. M. Tsvelik, Phys. Rev. B **52**, 13 368 (1995).
77. D. H. Tennant, S. E. Nagler, D. Welz, G. Shirane, and K. Yamada, Phys. Rev. B **52**, 13 381 (1995).
78. L. F. Feiner, A. M. Oleś, and J. Zaanen, J. Phys.: Condens. Matter **10**, L555 (1998).
79. J. van den Brink, W. Stekelenburg, D. I. Khomskii, G. A. Sawatzky, and K. I. Kugel, Phys. Rev. B **58**, 10 276 (1998).
80. B. Sutherland, Phys. Rev. B **12**, 3795 (1975).
81. B. Frischmuth, F. Mila, and M. Troyer, Phys. Rev. Lett. **82**, 835 (1999).
82. A. Auerbach, *Interacting Electrons and Quantum Magnetism*, Springer, New York, 1994.
83. M. Takahashi, Phys. Rev. B **40**, 2494 (1989).
84. D. N. Zubarev, Sov. Usp. Phys. **3**, 320 (1960).
85. S. B. Haley and P. Erdös, Phys. Rev. B **5**, 1106 (1972).
86. G. Khaliullin and V. Oudovenko, Phys. Rev. B **56**, R14 243 (1997).
87. Y. Yamashita, N. Shibata, and K. Ueda, Phys. Rev. B **58**, 9114 (1998).
88. Y. Q. Li, M. Ma, D. N. Shi, and F. C. Zhang, Phys. Rev. Lett. **81**, 3527 (1998).
89. Y. Q. Li, M. Ma, D. N. Shi, and F. C. Zhang, Phys. Rev. B **60**, 12 781 (1999).
90. P. Azaria, A. O. Gogolin, P. Lecheminant, and A. A. Nersesyan, Phys. Rev. B **83**, 624 (1999).
91. A. Joshi, M. Ma, F. Mila, D. N. Shi, and F. C. Zhang, Phys. Rev. B **60**, 6584 (1999).
92. F. Mila, B. Frischmuth, A. Deppeler, and M. Troyer, Phys. Rev. Lett. **82**, 3697 (1999).
93. I. Affleck, Nucl. Phys. **B265**, 409 (1986).
94. G. Santoro, S. Sorella, L. Guidoni, A. Parola, and E. Tosatti, Phys. Rev. Lett. **83**, 3065 (1999).
95. H. Bethe, Z. Phys. B **71**, 205 (1930).
96. H. Evertz, G. Lana, and M. Marcu, Phys. Rev. Lett. **70**, 875 (1993).
97. B. Beard and U. Wiese, Phys. Rev. Lett. **77**, 5130 (1996).
98. M. van den Bossche, F. C. Zhang, and F. Mila, cond-mat/0001051.
99. S. K. Pati, R. Singh, and D. I. Khomskii, Phys. Rev. Lett. **81**, 5406 (1998).
100. S. K. Pati and R. Singh, Phys. Rev. B **61**, 5868 (2000).
101. P. Chandra and B. Doucot, Phys. Rev. B **38**, 9335 (1988).

102. A. Moreo, E. Dagotto, T. Jolicoeur, and J. Riera, Phys. Rev. B **42**, 6283 (1990).
103. N. Read and S. Sachdev, Phys. Rev. Lett. **66**, 1773 (1991).
104. A. V. Dotsenko and O. P. Sushkov, Phys. Rev. B **50**, 13 821 (1994).
105. N. B. Ivanov, S. E. Krüger, and J. Richter, Phys. Rev. B **53**, 2633 (1996).
106. A. Chubukov, Phys. Rev. B **44**, 392 (1991).
107. A. J. Millis and H. Monien, Phys. Rev. B **50**, 16606 (1994).
108. A. W. Sandvik and D. J. Scalapino, Phys. Rev. Lett. **72**, 2777 (1994).
109. A. V. Chubukov and D. K. Morr, Phys. Rev. B **52**, 3521 (1995).
110. P. Anderson, Science **235**, 1196 (1987).
111. C. K. Majumdar and D. K. Ghosh, J. Math. Phys. **10**, 1388 (1969).
112. I. Affleck, T. Kennedy, E. Lieb, and H.Tasaki, Phys. Rev. Lett. **59**, 799 (1987).
113. S. Sachdev, *Quantum Phase Transition*, Cambridge University Press, Cambridge, 1999.
114. J. Igarashi, J. Phys. Soc. Jpn. **62**, 4449 (1993).
115. H. Schulz, T. Ziman, and D. Poilblanc, *Magnetic systems with competing interactions*, World Scientific, Singapore, 1994.
116. H. Schulz, T. Ziman, and D. Poilblanc, J. Phys. I (Paris) **6**, 675 (1996).
117. S. Sachdev and R. Bhatt, Phys. Rev. B **41**, 9323 (1990).
118. M. Gelfand, R. Singh, and D. Huse, Phys. Rev. B **40**, 10 801 (1989).
119. V. Kotov, O. Sushkov, and R. Eder, Phys. Rev. B **59**, 6266 (1999).
120. E. Lieb, T. Schultz, and D. Mattis, Ann. Phys. (N.Y.) **16**, 407 (1961).
121. I. Affleck, Phys. Rev. B **37**, 5186 (1988).
122. S. Sachdev and M. Vojta, cond-mat/9908008.
123. N. Katoh and M. Imada, J. Phys. Soc. Jpn. **62**, 3728 (1993).
124. K. Ueda, H. Kontani, M. Sigrist, and P. A. Lee, Phys. Rev. Lett. **76**, 1932 (1996).
125. O. A. Starykh, M. E. Zhitomirsky, D. I. Khomskii, R. Singh, and K. Ueda, Phys. Rev. Lett. **77**, 2558 (1996).
126. M. Troyer, H. Kontani, and K. Ueda, Phys. Rev. Lett. **76**, 3822 (1996).
127. S. R. White, Phys. Rev. Lett. **77**, 3633 (1996).
128. E. Fradkin, *Field theories of condensed matter systems*, Addison-Wesley, NY, 1991.
129. F. Haldane, Phys. Rev. Lett. **61**, 1029 (1988).
130. A. Sandvik, A. Chubukov, and S. Sachdev, Phys. Rev. B **51**, 16483 (1995).
131. S. R. White, R. M. Noack, and D. J. Scalapino, Phys. Rev. Lett. **73**, 886 (1994).
132. E. Dagotto and T. M. Rice, Science **271**, 618 (1996).
133. M. Troyer, H. Tsunetsugu, and T. M. Rice, Phys. Rev. B **53**, 251 (1996).
134. B. Normand and T. M. Rice, Phys. Rev. B **54**, 7180 (1996).
135. K. Hirakawa, H. Kadowaki, and K. Ubokoshi, J. Phys. Soc. Jpn. **54**, 3526 (1985).
136. K. Yamaura, M. Takano, A. Hirano, and R. Kanno, J. Solid State Chem. **127**, 109 (1997).
137. Y. Y. Hsieh and M. Blume, Phys. Rev. B **6**, 2684 (1972).
138. T. Mizokawa and A. Fujimori, Phys. Rev. B **54**, 5368 (1996).
139. D. L. Cox, Phys. Rev. Lett. **59**, 1240 (1987).
140. J. Zaanen, L. F. Feiner, and A. M. Oleś, Mat. Sci. Eng. B **69**, 140 (1999).
141. S. A. Kivelson, D. S. Rokshar, and J. P. Sethna, Phys. Rev. B **35**, 8865 (1987).
142. L. F. Feiner, A. M. Oleś, and J. Zaanen, (unpublished).

143. P. L. Iske and W. J. Caspers, Physica A **146**, 151 (1987).
144. B. Sutherland, Phys. Rev. B **37**, 3786 (1988).
145. E. O. Wollan and W. C. Koehler, Phys. Rev. **100**, 545 (1955).
146. J. B. Goodenough, Phys. Rev. **100**, 564 (1955).
147. O. N. Mryasov, R. N. Sabiryanov, A. J. Freeman, and S. S. Jaswal, Phys. Rev. B **56**, 7255 (1997).
148. D. Feinberg, P. Germain, M. Grilli, and G. Seibold, Phys. Rev. B **57**, R5583 (1998).
149. P. Benedetti and R. Zeyher, Phys. Rev. B **59**, 9923 (1999).
150. A. J. Millis, R. Mueller, and B. I. Shraiman, Phys. Rev. B **54**, 5389 and 5405 (1996).
151. S. Ishihara, J. Inoue, and S. Maekawa, Phys. Rev. B **55**, 8280 (1997).
152. S. Okamoto, S. Ishihara, and S. Maekawa, Phys. Rev. B **61**, 451 (2000).
153. A. E. Bocquet, T. Mizokawa, H. N. T. Saitoh, and A. Fujimori, Phys. Rev. B **46**, 46 (1992).
154. T. Mizokawa and A. Fujimori, Phys. Rev. B **51**, 12 880 (1995).
155. R. Shiina, T. Nishitani, and H. Shiba, J. Phys. Soc. Jpn. **66**, 3159 (1997).
156. R. V. Ditzian, J. R. Banavar, G. S. Grest, and L. P. Kadanoff, Phys. Rev. B **22**, 2542 (1980).
157. H. Kawano, R. Kajimoto, M. Kubota, and H. Yoshizawa, Phys. Rev. B **53**, R14709 (1996).
158. Y. S. Su, T. A. Kaplan, S. D. Mahanti, and J. F. Harrison, Phys. Rev. B **61**, 1324 (2000).
159. L. F. Feiner and A. M. Oleś, Physica B **259-261**, 796 (1999).
160. B. Shraiman and E. Siggia, Phys. Rev. Lett. **61**, 467 (1998).
161. R. Kilian and G. Khaliullin, Phys. Rev. B **60**, 13 458 (1999).
162. J. van den Brink, P. Horsch, and A. M. Oleś, (unpublished).
163. F. Moussa, M. Hennion, G. Biotteau, J. Rodriguez-Carvajal, L. Pinsard, and A. Revcolevschi, Phys. Rev. B **60**, 12 299 (1999).
164. S. Ishihara, M. Yamanaka, and N. Nagaosa, Phys. Rev. B **56**, 686 (1997).
165. A. Maignan, C. Martin, F. Damay, and B. Raveau, Phys. Rev. B **58**, 2758 (1998).
166. T. Akimoto, Y. Maruyama, Y. Moritomo, A. Nakamura, K. Hirota, K. Ohoyama, and M. Ohashi, Phys. Rev. B **57**, R5594 (1998).
167. H. Kawano, R. Kajimoto, H. Yoshizawa, Y. Tomioka, H. Kuwarhara, and Y. Tokura, Phys. Rev. Lett. **78**, 4253 (1999).
168. A. Takahashi and H. Shiba, Eur. Phys. J. B **5**, 413 (1998).
169. R. Maezono, S. Ishihara, and N. Nagaosa, Phys. Rev. B **58**, 11 583 (1998).
170. J. van den Brink, G. Khaliullin, and D. I. Khomskii, Phys. Rev. Lett. **83**, 5118 (1999).
171. A. M. Oleś and L. F. Feiner, J. Supercond. **12**, 299 (1999).
172. T. G. Perring, G. Aeppli, S. M. Hayden, S. A. Carter, J. P. Remeika, and S. W. Cheong, Phys. Rev. Lett. **77**, 711 (1996).
173. Y. Endoh and K. Hirota, J. Phys. Soc. Jpn. **66**, 2264 (1997).
174. J. A. Fernandez-Baca, P. Dai, H. Hwang, C. Kloc, and S.-W. Cheong, Phys. Rev. Lett. **80**, 4012 (1998).
175. A. M. Oleś and L. F. Feiner, Acta Phys. Polon. A **97**, 193 (2000).

176. G. Kotliar and A. E. Ruckenstein, Phys. Rev. Lett. **57**, 1362 (1986).
177. L. F. Feiner and A. M. Oleś, (unpublished).
178. M. van Veenendaal and A. J. Fedro, Phys. Rev. B **59**, 1285 (1999).
179. F. Mack and P. Horsch, Phys. Rev. Lett. **82**, 3160 (1999).
180. I. V. Solovyev and K. Terakura, Phys. Rev. Lett. **82**, 2959 (1999).
181. H. Y. Hwang, P. Dai, S.-W. Cheong, G. Aeppli, D. A. Tennant, and H. A. Mook, Phys. Rev. Lett. **80**, 1368 (1998).
182. G. Khaliullin and R. Kilian, Phys. Rev. B **61**, 3494 (2000).
183. J. van den Brink, P. Horsch, F. Mack, and A. M. Oleś, Phys. Rev. B **59**, 6795 (1999).
184. J. Jaklič and P. Prelovšek, Phys. Rev. B **49**, 5065 (1994).
185. J. Jaklič and P. Prelovšek, Adv. Phys. **49**, 1 (2000).
186. P. Horsch, J. Jaklič, and F. Mack, Phys. Rev. B **59**, 6217 (1999).
187. G. Martínez and P. Horsch, Phys. Rev. B **44**, 317 (1991).
188. C. L. Kane, P. A. Lee, and N. Read, Phys. Rev. B **39**, 6880 (1989).
189. S. Schmitt-Rink, C. M. Varma, and A. E. Ruckenstein, Phys. Rev. Lett. **60**, 2793 (1989).
190. C. L. Kane, P. A. Lee, and N. Read, Phys. Rev. B **39**, 6880 (1989).
191. C. H. Chen and S.-W. Cheong, Phys. Rev. Lett. **76**, 4042 (1996).
192. Y. Tomioka, A. Asamitsu, Y. Moritomo, H. Kuwahara, and Y. Tokura, Phys. Rev. Lett. **74**, 5108 (1995).
193. H. Kuwahara, Y. Tomioka, A. Asamitsu, Y. Moritomo, and Y. Tokura, **270**, 961 (1995).
194. M. v. Zimmermann, J. P. Hill, D. Gibbs, M. Blume, D. Casa, B. Keimer, Y. Murakami, Y. Tomioka, and Y. Tokura, Phys. Rev. Lett. **83**, 4872 (1999).
195. M. Kubota, H. Yoshizawa, Y. Moritomo, H. Fujioka, K. Hirota, and Y. Endoh, cond-mat/9811192.
196. I. V. Solovyev and K. Terakura, Phys. Rev. Lett. **83**, 2825 (1999).
197. M. Cuoco, A. M. Oleś, and C. Noce, (unpublished).
198. G. Jackeli, N. B. Perkins, and N. M. Plakida, Phys. Rev. B **62** (2000).
199. T. Mutou and H. Kontani, Phys. Rev. Lett. **83**, 3685 (1999).
200. T. Mizokawa and A. Fujimori, Phys. Rev. B **56**, R493 (1997).
201. P. G. Radaelli, D. E. Cox, M. Marezio, and S.-W. Cheong, Phys. Rev. B **55**, 3015 (1997).
202. S. Mori and C. H. Chen and S.-W. Cheong, Phys. Rev. Lett. **81**, 3972 (1998).
203. C. H. Chen, S. Mori, and S.-W. Cheong, Phys. Rev. Lett. **83**, 4792 (1997).
204. Y. Murakami, J. P. Hill, D. Gibbs, M. Blume, I. Koyama, M. Tanaka, H. Kawata, T. Arima, Y. Tokura, K. Hirota, and Y. Endoh, Phys. Rev. Lett. **81**, 582 (1998).
205. P. G. Radaelli, D. E. Cox, L. Capogna, S.-W. Cheong, and M. Marezio, Phys. Rev. B **59**, 14 440 (1999).
206. T. Hotta, Y. Takada, H. Koizumi, and E. Dagotto, Phys. Rev. Lett. **84**, 2477 (2000).
207. H. Koizumi, T. Hotta, and Y. Takada, Phys. Rev. Lett. **80**, 4518 (1998).
208. T. Hotta, Y. Takada, and H. Koizumi, Int. J. Mod. Phys. B **12**, 3437 (1998).
209. Y. Endoh, K. Hirota, S. Ishikawa, S. Okamoto, Y. Murakami, A. Nishizawa,

T. Fukuda, H. Kimura, H. Nojiri, K. Kaneko, and S. Maekawa, Phys. Rev. Lett. **82**, 4328 (1999).
210. F. Moussa, private communication (2000).
211. R. Maezono and N. Nagaosa, Phys. Rev. B **61**, 1825 (2000).
212. A. Moreo, S. Yunoki, and E. Dagotto, Phys. Rev. Lett. **83**, 2773 (1999).
213. S.-W. Cheong and H. Y. Hwang, *Colossal Magnetoresistance Oxides*, Y. Tokura, Gordon & Breach, Monographs in Cond. Matt. Science.
214. P. Schiffer, A. P. Ramirez, W. Bao, and S.-W. Cheong, Phys. Rev. Lett. **75**, 3336 (1995).
215. A. P. Ramirez, P. Schiffer, S.-W. Cheong, C. H. Chen, W. Bao, T. T. M. Palstra, P. L. Gammel, D. J. Bishop, and B. Zegarski, Phys. Rev. Lett. **76**, 3188 (1996).
216. P.-G. de Gennes, Phys. Rev. **118**, 141 (1960).
217. E. Nagaev, *Physics of Magnetic Semiconductors*, Mir Publ., Moscow, 1979.
218. E. Nagaev, Sov. Phys. Uspekhi **166**, 833 (1996).
219. N. A. Babushkina, L. M. Belova, D. I. Khomskii, K. I. Kugel, O. Y. Gorbenko, and A. R. Kaul, Phys. Rev. B **59**, 6994 (1999).
220. S. Yunoki, J. Hu, A. L. Malvezzi, A. Moreo, N. Furukawa, and E. Dagotto, Phys. Rev. Lett. **80**, 845 (1998).
221. J. Zaanen, J. Phys. Chem. Solids **59**, 1769 (1998).
222. A. Moreo, S. Yunoki, and E. Dagotto, Science **283**, 2034 (1999).
223. D. S. Dessau, T. Saitoh, C.-H. Park, Z.-X. Shen, P. Villella, N. Hamada, Y. Moritomo, and Y. Tokura, Phys. Rev. Lett. **81**, 192 (1998).
224. E. Müller-Hartmann and E. Dagotto, Phys. Rev. B **54**, R6819 (1996).
225. C. M. Varma, Phys. Rev. B **54**, 7328 (1996).
226. M. Calderon and L. Brey, Phys. Rev. B **58**, 3286 (1998).
227. H. Fujishiro, M. Ikebe, and Y. Konno, J. Phys. Soc. Jpn. **67**, 1799 (1998).
228. K. H. Ahn and A. J. Millis, Phys. Rev. B **58**, 3697 (1998).
229. T. Mizokawa, D. I. Khomskii, and G. A. Sawatzky, Phys. Rev. B **61**, R3776 (2000).
230. Z. Popovic and S. Sapathy, Phys. Rev. Lett. **84**, 1603 (2000).
231. K. Held and D. Vollhardt, cond-mat/9909311.
232. R. Maezono and N. Nagaosa, Phys. Rev. B **61**, 1189 (2000).
233. D. Louca, T. Egami, E. L. Brosha, H. Röder, and A. R. Bishop, Phys. Rev. B **56**, R8475 (1997).

Author Index

B

Barabanov, A. F., 1

C

Cuoco, M., 226

M

Maksimov, L. A., 1
Mikheyenkov, A. V., 1

N

Nolting, W., 118

O

Oleś, A. M., 226

P

Perkins, N. B., 226